Linear Algebra, Matrices, and Their Applications

CONTEMPORARY MATHEMATICS

833

Linear Algebra, Matrices, and Their Applications

International Conference
on Linear Algebra and its Applications (ICLAA)
December 18–21, 2023
Center for Advanced Research
in Applied Mathematics and Statistics (CARAMS)
MAHE, Manipal, India

Shaun Michael Fallat
Surender K. Jain
Manjunatha Prasad Karantha
Stephen James Kirkland
T. S. S. R. K. Rao
Editors

AMS AMERICAN MATHEMATICAL SOCIETY

2020 *Mathematics Subject Classification.* Primary 15A09, 15B48, 15B33, 15A69, 05C50, 16S50, 46B28, 62J05, 62K05, 03B52.

Library of Congress Cataloging-in-Publication Data

Cataloging-in-Publication Data has been applied for by the AMS.
See http://www.loc.gov/cip/.

Contemporary Mathematics ISSN: 0271-4132 (print); ISSN: 1098-3627 (online)
DOI: https://doi.org/10.1090/conm/833

Visit the AMS home page at `https://www.ams.org/`

10 9 8 7 6 5 4 3 2 1 31 30 29 28 27 26

Contents

Preface

Linear Algebra and Statistics, among others, have many applications in almost every branch of science, including medical sciences, engineering, and social sciences. These areas are the focus of the activities in the Center for Advanced Research in Applied Mathematics and Statistics (CARAMS), MAHE, Manipal, India. The International Conference on Linear Algebra and its Applications (ICLAA 2023), to which this volume is dedicated, is the fifth in its series, organized at CARAMS. ICLAA 2023 was organized at CARAMS on December 18–21, 2023, and endorsed by the International Linear Algebra Society (ILAS). The theme of the conference focused on, but was not limited to, the topics including Classical Matrix Theory and Linear Algebra, Linear Operators on Banach Spaces, Nonnegative Matrices, Matrix Methods in Statistics as well as in Graph Theory, Combinatorial Matrix Theory, Matrix Methods in Statistics, Linear Statistical Inference, Matrices in Error Analysis and its applications.

Special Sessions in "Special Matrices", "Matrix Methods in Statistics", and "Matrices, Graphs, and Applications" were also organized. The invited speakers included Rafikul Alam (IIT Guwahati, India), Ravindra B. Bapat (ISI Delhi Center, India), Sudeep R. Bapat (IIT Bombay, India), Abraham Berman (Technion-Israel Institute of Technology, Israel), N. Eagambaram (Indian Statistical Service, Government of India), Shaun M. Fallat (University of Regina, Canada), Stephen J. Haslett (Massey University, New Zealand), S. K. Jain (Ohio University, USA), Debajit Kalita (Tezpur University, India), Rajesh Kannan (IIT Hyderabad, India), Apoorva Khare (IISc Bengaluru, India), Stephen J. Kirkland (University of Manitoba, Canada), Mika Mattila (Tampere University, Finland), David Raj Micheal (VIT Chennai, India), Bojan Mohar (Simon Fraser University, Canada), Samir K. Neogy (ISI Delhi Center, India), Sukanta Pati (IIT Guwahati, India), Simo Puntanen (Tampere University, Finland), T. E. S. Raghavan (University of Illinois at Chicago, USA), Balaji Ramamurthy (IIT Madras, India), T. S. S. R. K. Rao (Shiv Nadar Institute of Eminence, India), Dietrich von Rosen (Swedish University of Agricultural Sciences, Sweden), Arumugam S (Ramco Institute of Technology, India), Peter Šemrl (Institute of Mathematics, Physics and Mechanics, Slovenia), Ajit Iqbal Singh (The Indian National Science Academy, India), K. C. Sivakumar (IIT Madras, India), Sivaramakrishnan Sivasubramanian (IIT Bombay, India), and Murali K. Srinivasan (IIT Bombay, India).

ICLAA 2023 was inaugurated by Pro Vice Chancellor Dr. Narayana Sabhahit, MAHE, Manipal, along with Professors Peter Šemrl and Stephen J. Haslett as the chief guests. The ICLAA 2023 was partially supported by National Board for

Higher Mathematics (NBHM), Science and Engineering Research Board (SERB), and Indian National Science Academy (INSA).

The present CONM volume, dedicated to ICLAA 2023, consists of fifteen articles and focuses on the allied topics of linear algebra, such as recent developments in matrix theory, linear algebra and its applications, linear operators on Banach algebras, game theory, and optimal designs. Besides the research articles, the volume also contains expository articles, describing new avenues of research and open problems. It is believed that this would also serve as reference material for young researchers working in linear algebra and allied fields. The articles in the book volume are from diverse fields associated with Linear Algebra. The article "Vertical Block P- and P_0-matrices" by Aniekan A. Ebiefung explores the theoretical foundation and properties of vertical block P- and P_0-matrices, which arise in the study of complementary problems in mathematical optimization.

The article "The Block Matrices and Vertical Linear Complementarity Problem" by Punit Kumar Yadav et al., contains the results related to vertical block matrices and provides several characterizations of different classes of vertical block matrices. Another article entitled "Totally Positive Matrices" by T. E. S. Raghavan, discusses the properties of totally nonnegative (TN) and totally positive (TP) matrices, providing a comprehensive study of their characteristics, applications, and theoretical foundations. The article "Some Remarks on the Reparameterizations of Linear Models" (Matrix Methods in Statistics) by Simo Puntanen et al. explores the reparameterizations of linear models. Shubham Niphadkar et al., in their article "Studying Optimal Designs for Multivariate Crossover Trials" focuses on deriving A-, D-, and E-optimality results for multivariate crossover designs, where multiple responses are measured per period for each subject. S. K. Jain et al., in their article "Products of Idempotents of Linear Operators in Banach Algebras" explore the conditions as to when a linear operator on Banach algebras is a product of idempotents. Another article entitled "On Subspaces of Banach Spaces" which admit unique norm preserving extensions under various embeddings by Sainik Karak et al., investigates the concept of weak Hahn-Banach smoothness in Banach spaces. The theme of others includes topics from generalized inverses, edge-Sudoku number of a graph, optimal designs for multivariate crossover trials, etc. It is expected that the volume will be a good resource for persons working in these fields.

The editors acknowledge the incredible contribution of the authors, which, we hope, will make the present CONM volume in their series valuable. Also, the editors acknowledge the association of and the support from Vinay Madhusudanan (Manipal Institute of Technology, MAHE, Manipal, India), David Raj Micheal (Vellore Institute of Technology, Chennai, India), and Harikrishnan P. K. (Manipal Institute of Technology, MAHE, Manipal, India). The selfless work done on time by the reviewers cannot be underestimated. The editors thank them profusely. The editors acknowledge their sincere help without naming them.

Last but not least, we thank Umashankara Kelathaya and Savitha Varkady, who helped in the editorial work and putting together the articles and preface.

Our sincere thanks to Christine M. Thivierge and Bridget Manzella of American Mathematical Society for their invaluable support and guidance throughout the publication process.

Manipal, March 2025

Editors:
Shaun Michael Fallat
Surender K. Jain
Manjunatha Prasad Karantha
Stephen James Kirkland
Srinivasa Siva Rama Krishna Rao Taduri

Contemporary Mathematics
Volume **833**, 2026
https://doi.org/10.1090/conm/833/16659

Canonical forms of holomorphic and meromorphic matrices

Rafikul Alam and Jibrail Ali

ABSTRACT. Canonical forms are powerful tools for spectral analysis of matrices. Smith form of a matrix polynomial and Smith-McMillan form of a rational matrix are well known canonical forms which are used extensively in theory and applications. Although, local canonical forms such as local Smith forms of holomorphic and meromorphic matrices are well known in the literature, there has been relatively less research on global canonical forms of holomorphic and meromorphic matrices. Our aim is to fill this gap. We discuss, among other things, the global Smith form (resp., global Smith-McMillan form) of a holomorphic (resp., meromorphic) matrix which is akin to the Smith form (resp., Smith-McMillan form) of a matrix polynomial (resp., rational matrix).

1. Introduction

Canonical forms are powerful tools for spectral analysis of matrices whose entries are scalars, polynomials or rational functions. Let $A \in \mathbb{C}^{n\times n}$. Then the standard eigenvalue problem

$$Av = \lambda v$$

where $\lambda \in \mathbb{C}$ and $v \in \mathbb{C}^n$ is nonzero, can be analyzed by considering the Jordan canonical form (JCF) of A. The JCF of A states that there exists a nonsingular matrix $S \in \mathbb{C}^{n\times n}$ such that

$$S^{-1}(\lambda I - A)S = \lambda I - J$$

where $J \in \mathbb{C}^{n\times n}$ is a Jordan matrix, that is, J is a block diagonal matrix whose diagonal blocks are Jordan blocks containing eigenvalues of A [**18**]. On the other hand, for $A \in \mathbb{C}^{m\times n}$ and $B \in \mathbb{C}^{m\times n}$, the generalized eigenvalue problem

$$(A - \lambda B)v = 0$$

can be analyzed by considering the Kronecker canonical form(KCF) of the $m \times n$ matrix pencil $A - \lambda B$. The KCF states that [**10, 13, 19**] there exist nonsingular matrices $P \in \mathbb{C}^{m\times m}$ and $Q \in \mathbb{C}^{n\times n}$ such that

$$P(A - \lambda B)Q = \mathrm{diag}(L_{\varepsilon_1}, L_{\varepsilon_2}, \ldots, L_{\varepsilon_p}, L_{\eta_1}^T, L_{\eta_2}^T, \ldots, L_{\eta_s}^T, \lambda N - I_p, J - \lambda I_\ell),$$

2020 *Mathematics Subject Classification.* Primary 15A54, 15A21; Secondary 15B33, 15A18.
Key words and phrases. Holomorphic matrices, meromorphic matrices, eigenvalues, eigenvectors, zeros, poles, Smith form, Smith-McMillan form.
The second author was supported by IIT Guwahati through an Institute Fellowship.

where N and J are Jordan matrices with N being nilpotent, and L_k is the $k\times(k+1)$ bidiagonal pencil given by

$$L_k = \begin{bmatrix} -\lambda & 1 & & & \\ & -\lambda & 1 & & \\ & & \ddots & \ddots & \\ & & & -\lambda & 1 \end{bmatrix}.$$

The elementary divisors of $J-\lambda I$ and $\lambda N - I$, respectively, are the finite and infinite elementary divisors of the pencil $A-\lambda B$ and constitute the regular part of $A-\lambda B$. The L_k and $L_k^\top$ blocks constitute the singular part of $A-\lambda B$. The index sets $\{\varepsilon_1,\dots,\varepsilon_p\}$ and $\{\eta_1,\dots,\eta_s\}$ are the right and left minimal indices (Kronecker indices) of $A-\lambda B$, respectively; see [**10**, **13**, **19**]. Generalized eigenvalue problems arise, for instance, in the study of differential algebraic equations (DAE) [**21**] and in control theory [**6**, **19**].

An $m\times n$ polynomial eigenvalue problem (PEP) of degree ℓ of the form

$$P(\lambda)v := (A_0 + \lambda A_1 + \cdots + \lambda^\ell A_\ell)v = 0$$

arises in many applications; see [**13**, **19**, **28**, **29**]. However, $P(\lambda)$ does not have a KCF like canonical form when $\ell \geq 2$. Nevertheless, the Smith canonical form [**13**, **19**, **28**, **29**] of $P(\lambda)$ provides a complete spectral data of $P(\lambda)$. An $m\times m$ matrix polynomial $U(\lambda)$ is called unimodular if $\det(U(\lambda))$ is a nonzero constant for all $\lambda\in\mathbb{C}$. The Smith form states that there exist $m\times m$ unimodular matrix polynomial $U(\lambda)$ and $n\times n$ unimodular matrix polynomial $V(\lambda)$ such that

$$U(\lambda)P(\lambda)V(\lambda) = \left[\begin{array}{c|c} \mathrm{diag}(\phi_1(\lambda),\dots,\phi_r(\lambda)) & 0 \\ \hline 0 & 0 \end{array}\right] =: S_P(\lambda), \tag{1.1}$$

where r is the normal rank of $P(\lambda)$ and $\phi_1(\lambda),\dots,\phi_r(\lambda)$ are monic polynomials in $\mathbb{C}[\lambda]$ such that $\phi_i(\lambda)$ divides $\phi_{i+1}(\lambda)$ for $i = 1 : r-1$. The diagonal matrix $S_P(\lambda)$ is called the Smith form of $P(\lambda)$. The eigenvalues of $P(\lambda)$ are the roots of the polynomial $\Phi_P(\lambda) := \prod_{j=1}^r \phi_j(\lambda)$; see [**1**, **5**, **10**, **13**, **19**, **28**, **29**].

Eigenvalue problems involving rational matrices, that is, matrices whose entries are rational functions, also arise in applications. For example, the rational eigenvalue problem (REP)

$$G(\lambda)v := -Av + \lambda Bv + \lambda^2 \sum_{j=1}^{k} \frac{1}{w_j - \lambda} C_j v = 0$$

arises when a generalized linear eigenvalue problem condensed exactly [**24**]. The states-space analysis of linear time-invariant systems also leads to rational eigenvalue problems in which the rational matrices arise as the transfer functions, see [**6**, **19**, **29**]. The Smith-McMillan form provides a complete spectral data of a rational matrix. If $G(\lambda)$ is an $m\times n$ rational matrix then the Smith-McMillan form states that there exist $m\times m$ unimodular matrix polynomial $U(\lambda)$ and $n\times n$ unimodular matrix polynomial $V(\lambda)$ such that

$$U(\lambda)G(\lambda)V(\lambda) = \left[\begin{array}{c|c} \mathrm{diag}(\phi_1(\lambda)/\psi_1(\lambda),\dots,\phi_r(\lambda)/\psi_r(\lambda)) & 0 \\ \hline 0 & 0 \end{array}\right] =: \Sigma_G(\lambda), \tag{1.2}$$

where r is the normal rank of $G(\lambda)$, $\phi_1(\lambda),\dots,\phi_r(\lambda)$ and $\psi_1(\lambda),\dots,\psi_r(\lambda)$ are monic polynomials in $\mathbb{C}[\lambda]$ and, $\phi_i(\lambda)$ and $\psi_i(\lambda)$ are coprime for $i=1:r$. Further, $\phi_i(\lambda)$ divides $\phi_{i+1}(\lambda)$ and $\psi_{i+1}(\lambda)$ divides $\psi_i(\lambda)$ for $i = 1 : r-1$. The diagonal matrix

$\Sigma_G(\lambda)$ is called the Smith-McMillan form of $G(\lambda)$. The zeros of $G(\lambda)$ are the roots of the zero polynomial $\Phi_G(\lambda) := \prod_{j=1}^{r} \phi_j(\lambda)$ and the poles of $G(\lambda)$ are the roots of the pole polynomial $\Psi_G(\lambda) := \prod_{j=1}^{r} \psi_j(\lambda)$; see [**1, 2, 5, 13, 19, 28, 29**].

Nonlinear eigenvalue problems involving holomorphic matrices (i.e., matrices whose entries are holomorphic functions) arise in many applications and their numerical treatment is an active area of research; see [**3, 4, 17, 23–26**] and the references therein. For example, consider the delay-differential equation (DDE) [**26**]

$$\frac{d}{dt}x(t) = A_0 x(t) + \sum_{i=1}^{m} A_i x(t - \tau_i),$$

where $x(t) \in \mathbb{C}^n$ is a vector of state variable at time $t \in \mathbb{R}$, $A_1, \ldots, A_m$ are matrices in $\mathbb{C}^{n\times n}$ and $0 < \tau_1 < \tau_2 < \cdots < \tau_m$ are the time-delays. Seeking a solution of the form $x(t) = ve^{\lambda t}$ with $v \in \mathbb{C}^n$ and $\lambda \in \mathbb{C}$ yields the nonlinear eigenvalue problem (NEP)

$$\mathbf{T}(\lambda)v := (\lambda I - A_0 - \sum_{i=1}^{m} A_i e^{-\lambda \tau_i})v = 0.$$

More generally, nonlinear eigenvalue problems involving meromorphic matrices (i.e., matrices whose entries are meromorphic functions) also arise in applications such as in the study of controllability of delay-differential systems (DDS) [**7, 8, 30**]

$$\frac{\mathrm{d}\mathbf{x}(t)}{\mathrm{d}t} = A_0\mathbf{x}(t) + \sum_{j=1}^{N_1} A_j \mathbf{x}(t-\tau_j) + \sum_{j=1}^{N_2} B_j \mathbf{u}(t - t_j),$$

$$\mathbf{y}(t) = \sum_{j=1}^{N_3} C_j \mathbf{x}(t - s_j) + \sum_{j=1}^{N_4} D_j \mathbf{u}(t - h_j).$$

Here $\mathbf{x}(t) \in \mathbb{C}^r$ is a vector of state variables and $\mathbf{u}(t) \in \mathbb{C}^n$ is a vector of control variables. Further, $0 < \tau_1 < \cdots < \tau_{N_1}$, $0 \le t_1 < \cdots < t_{N_2}$, $0 \le s_1 < \cdots < s_{N_3}$ and $0 \le h_1 < \cdots < h_{N_4}$ are delay parameters. Furthermore, $A_0, \ldots, A_{N_1}$ are $r \times r$ matrices, $B_1, \ldots, B_{N_2}$ are $r \times n$ matrices, $C_1, \ldots, C_{N_3}$ are $m \times r$ matrices and $D_1, \ldots, D_{N_4}$ are $m \times n$ matrices.

For the ansatz $\begin{bmatrix}\mathbf{x}(t)\\ \mathbf{u}(t)\end{bmatrix} := \begin{bmatrix}\mathbf{x}\\ \mathbf{u}\end{bmatrix} e^{\lambda t}$ with $\mathbf{x} \in \mathbb{C}^{r\times r}$ and $\mathbf{u} \in \mathbb{C}^{n\times n}$, the DDS yields

$$\begin{bmatrix}-A(\lambda) & B(\lambda)\\ C(\lambda) & D(\lambda)\end{bmatrix}\begin{bmatrix}\mathbf{x}\\ \mathbf{u}\end{bmatrix} e^{\lambda t} = \begin{bmatrix}0\\ \mathbf{y}(t)\end{bmatrix} \tag{1.3}$$

where $A(\lambda) := \lambda I_r - A_0 - \sum_{j=1}^{N_1} A_j e^{-\lambda\tau_j}$, $B(\lambda) := \sum_{j=1}^{N_2} B_j e^{-\lambda t_j}$, $C(\lambda) := \sum_{j=1}^{N_3} C_j e^{-\lambda s_j}$, and $D(\lambda) := \sum_{j=1}^{N_4} D_j e^{-\lambda h_j}$. The holomorphic matrix (i.e., matrix whose entries are holomorphic functions)

$$\mathbf{S}(\lambda) := \begin{bmatrix}-A(\lambda) & B(\lambda)\\ C(\lambda) & D(\lambda)\end{bmatrix}$$

is called the system matrix of the DDS and the meromorphic matrix (i.e., matrix whose entries are meromorphic functions)

$$\mathbf{M}(\lambda) := D(\lambda) + C(\lambda)A(\lambda)^{-1}B(\lambda)$$

is called the transfer function of the DDS. Eliminating $\mathbf{x}e^{\lambda t}$ in (1.3), we have

$$\mathbf{y}(t) = [D(\lambda) + C(\lambda)A(\lambda)^{-1}B(\lambda)]\mathbf{u}e^{\lambda t} = \mathbf{M}(\lambda)\mathbf{u}e^{\lambda t}.$$

Observe that $\mathbf{y}(t) = 0$ whenever $\mathbf{M}(\lambda)\mathbf{u} = 0$, that is, when $(\lambda, \mathbf{u})$ is an eigenpair of $\mathbf{M}(\lambda)$. This leads to meromorphic eigenvalue problem (MEP)

$$\mathbf{M}(\lambda)v = 0.$$

Alternatively, it follows from (1.3) that $\mathbf{y}(t) = 0$ whenever $\mathbf{S}(\lambda)\begin{bmatrix}\mathbf{x}\\ \mathbf{u}\end{bmatrix} = 0$. This shows that zero output of the DDS can be analyzed via eigenvalues and eigenvectors of the holomorphic system matrix $\mathbf{S}(\lambda)$ as well as via eigenvalues and eigenvectors of the transfer function $\mathbf{M}(\lambda)$. Naturally, it is expected that canonical forms of holomorphic and meromorphic matrices will have an important role to play in analyzing holomorphic and meromorphic eigenvalue problems. However, canonical forms of holomorphic and meromorphic matrices which are akin to the canonical forms in (1.1) and (1.2), respectively, have not received much attention in the literature.

Local spectral theory for a regular holomorphic matrix has been developed in [**12**]. Also, local spectral analysis of a regular holomorphic operator-valued function has been investigated, for example, in [**11, 15, 16, 25**]. A local Smith form of a holomorphic matrix $\mathbf{T}(\lambda)$ has been studied in the literature when $\mathbf{T}(\lambda)$ is regular; see [**2, 12, 20, 25**]. Likewise, a local Smith form of a meromorphic matrix $\mathbf{M}(\lambda)$ has been studied in the literature when $\mathbf{M}(\lambda)$ is regular; see [**2, 14, 20**]. Suppose that $\mathbf{M}(\lambda)$ is meromorphic on a domain $\Omega \subset \mathbb{C}$. Then the existence of a global canonical form of $\mathbf{M}(\lambda)$ has been presented in [**22**] where it is shown that $\mathbf{M}(\lambda)$ is "analytically equivalent" to a diagonal matrix of the form $\text{diag}(f_1(\lambda), \ldots, f_r(\lambda), 0, \ldots, 0)$. The functions $f_1(\lambda), \ldots, f_r(\lambda)$ are meromorphic on Ω and have the property that $f_i(\lambda)$ divides $f_{i+1}(\lambda)$ (i.e., f_{i+1}/f_i is holomorphic on Ω) for $i = 1 : r-1$.

Canonical forms such as Smith forms of matrix polynomials and Smith-McMillan forms of rational matrices are important tools for spectral analysis of matrix polynomials and rational matrices and have been studied extensively in the literature. Local canonical forms such as local Smith forms of holomorphic and meromorphic matrices are also well known in the literature and have been studied extensively over the years. By contrast, global canonical forms of holomorphic and meromorphic matrices have not received much attention of researchers. In fact, barring [**22**], there has been little research on global canonical forms of holomorphic and meromorphic matrices. Our aim in this paper is to fill this gap. We present a rigorous treatment of global Smith forms and global Smith-McMillan forms of holomorphic and meromorphic matrices.

The rest of the paper is organized as follows. In Section 2, we undertake spectral analysis of holomorphic matrices. We discuss Weierstrass product, factorization of holomorphic functions and prove, among other things, the global Smith form of a holomorphic matrix. In Section 3, we undertake spectral analysis of meromorphic matrices and prove, among other things, the global Smith-McMillan form of a meromorphic matrix.

NOTATION. Let $\mathbb{C}^n := \left\{\begin{bmatrix}v_1 & \cdots & v_n\end{bmatrix}^\top : v_1, \ldots, v_n \in \mathbb{C}\right\}$ denote the vector space of all column vectors of length n. We denote the vector space of all $m \times n$ matrices with entries in $\mathbb{C}$ by $\mathbb{C}^{m\times n}$. Let $A \in \mathbb{C}^{m\times n}$. We denote the transpose of A by $A^\top$ and the conjugate transpose of A by A^*. The null space $N(A)$ and the range space $R(A)$ of A are given by

$$N(A) := \{v \in \mathbb{C}^n : Av = 0\} \text{ and } R(A) := \{Av : v \in \mathbb{C}^n\}.$$

We denote the $n \times n$ identity matrix by I_n and also by I when the size is clear from the context. We denote the j-th column of I_n by e_j for $j = 1 : n$.

2. Holomorphic matrices

Let $\mathcal{O} \subset \mathbb{C}$ be a **domain**, that is, $\mathcal{O}$ is a nonempty connected open set. Let X be a complex Banach space. A function $f : \mathcal{O} \longrightarrow X$ is said to be holomorphic (or analytic) in $\mathcal{O}$ if f is differentiable on $\mathcal{O}$. We denote the set of all X-valued holomorphic functions on $\mathcal{O}$ by $\mathbb{H}(\mathcal{O}, X)$, that is,

$$\mathbb{H}(\mathcal{O}, X) := \{f : \mathcal{O} \longrightarrow X \mid f \text{ is holomorphic in } \mathcal{O}\}.$$

Let $\Omega \subset \mathbb{C}$ be a **region**, that is, the interior of Ω is a domain and Ω possibly contains a part or all of its boundary $\partial\Omega$. A function $f : \Omega \longrightarrow X$ is said to be holomorphic in Ω if $f \in \mathbb{H}(\mathcal{O})$ for some open set $\mathcal{O} \subset \mathbb{C}$ such that $\Omega \subset \mathcal{O}$. We denote the set of all X-valued holomorphic functions on Ω by $\mathbb{H}(\Omega, X)$.

For simplicity of notation, when $X = \mathbb{C}, X = \mathbb{C}^n$ and $X = \mathbb{C}^{m\times n}$, we set

$$\mathbb{H}(\Omega) := \mathbb{H}(\Omega, \mathbb{C}),\ \mathbb{H}(\Omega)^n := \mathbb{H}(\Omega, \mathbb{C}^n) \text{ and } \mathbb{H}(\Omega)^{m\times n} := \mathbb{H}(\Omega, \mathbb{C}^{m\times n}).$$

We refer to the elements of $\mathbb{H}(\Omega)^n$ as **holomorphic vectors** and the elements of $\mathbb{H}(\Omega)^{m\times n}$ as **holomorphic matrices**. Note that $\mathbb{H}(\Omega)$ is a commutative ring of holomorphic functions on Ω and $\mathbb{H}(\Omega)^{n\times n}$ is a non-commutative ring of $n \times n$ holomorphic matrix-valued functions on Ω.

DEFINITION 2.1. Let $A \in \mathbb{H}(\Omega)^{n\times n}$. Then A is said to be **invertible** (or a unit element of $\mathbb{H}(\Omega)^{n\times n}$) if there exists $B \in \mathbb{H}(\Omega)^{n\times n}$ such that $AB = BA = I$, that is,

$$A(z)B(z) = B(z)A(z) = I_n \text{ for all } z \in \Omega.$$

In such a case, B is called an inverse of A.

REMARK 2.2. If $A \in \mathbb{H}(\Omega)^{n\times n}$ is invertible, then A has a unique inverse, which is denoted by A^{-1}. If $A(z)$ is invertible for all $z \in \Omega$ then defining $A^{-1}(z) := A(z)^{-1}$ for $z \in \Omega$, it follows that $A^{-1} \in \mathbb{H}(\Omega)^{n\times n}$ and $AA^{-1} = A^{-1}A = I$, that is, A^{-1} is the unique inverse of A.

Let $\mathrm{GL}_n(\mathbb{H}(\Omega))$ denote the group of invertible elements in $\mathbb{H}(\Omega)^{n\times n}$, that is,

$$\mathrm{GL}_n(\mathbb{H}(\Omega)) := \{A \in \mathbb{H}(\Omega)^{n\times n} \mid A \text{ is invertible }\}.$$

The elements in $\mathrm{GL}_n(\mathbb{H}(\Omega)) \subset \mathbb{H}(\Omega)^{n\times n}$ are called units or unit elements of $\mathbb{H}(\Omega)^{n\times n}$.

DEFINITION 2.3. Let $A \in \mathbb{H}(\Omega)^{m\times n}$. Then A is said to be **regular** if $m = n$ and the matrix $A(z)$ is invertible for some $z \in \Omega$. If A is not regular then A is said to be **singular.**

For a regular matrix-valued function, the spectrum is defined as follows.

DEFINITION 2.4. Let $A \in \mathbb{H}(\Omega)^{n\times n}$ be regular. The spectrum $\sigma_\Omega(A)$ and the resolvent set $\rho_\Omega(A)$ of A are given by

$$\sigma_\Omega(A) := \{z \in \Omega : \det(A(z)) = 0\} \text{ and } \rho_\Omega(A) = \Omega \setminus \sigma_\Omega(A),$$

where $\det(M)$ is the determinant of a matrix $M \in \mathbb{C}^{n\times n}$.

Note that $A \in \mathrm{GL}_n(\mathbb{H}(\Omega)) \iff \sigma_\Omega(A) = \emptyset$. For a nonzero scalar function $f \in \mathbb{H}(\Omega)$ the spectrum $\sigma_\Omega(f)$ is the set of zeros of f, that is, $\sigma_\Omega(f) = \{z \in \Omega : f(z) = 0\}$. Thus $\sigma_\Omega(f) = \emptyset \iff f$ is a unit element of $\mathbb{H}(\Omega) \iff 1/f \in \mathbb{H}(\Omega)$. It is well known that $\sigma_\Omega(f)$ is at most a countable set with no accumulation points in Ω. Consequently, $\sigma_\Omega(A)$ is at most a countable set with no accumulation points in Ω when $A \in \mathbb{H}(\Omega)^{n\times n}$ is regular.

For spectral analysis, we view a holomorphic matrix as a homomorphism between modules. Therefore, we briefly discuss some basic facts about $\mathbb{H}(\Omega)$-modules. Note that $\mathbb{H}(\Omega)^n$ is a $\mathbb{H}(\Omega)$-module. The span of a subset $\{f_1, \ldots, f_\ell\} \subset \mathbb{H}(\Omega)^n$ is given by

$$\mathrm{span}_{\mathbb{H}(\Omega)}(f_1, \ldots, f_\ell) := \{g_1 f_1 + \cdots + g_\ell f_\ell : \; g_1, \ldots, g_\ell \in \mathbb{H}(\Omega)\}.$$

Obviously, $\mathrm{span}_{\mathbb{H}(\Omega)}(f_1, \ldots, f_\ell)$ is a submodule of $\mathbb{H}(\Omega)^n$. A subset $\{f_1, \ldots, f_\ell\} \subset \mathbb{H}(\Omega)^n$ is said to be **linearly dependent** if there exist $g_1, \ldots, g_\ell$ in $\mathbb{H}(\Omega)$ not all are zero such that $g_1 f_1 + \cdots + g_\ell f_\ell = 0$. Thus $\{f_1, \ldots, f_\ell\}$ is said to be **linearly independent** if it is not linearly dependent, that is, if $g_1 f_1 + \cdots + g_\ell f_\ell = 0$ then $g_1 = \cdots = g_\ell = 0$.

DEFINITION 2.5. Let $\mathcal{V} \subset \mathbb{H}(\Omega)^n$ be a submodule. Then $\mathcal{B} \subset \mathbb{H}(\Omega)^n$ is said to be a **basis** of $\mathcal{V}$ if $\mathcal{B}$ is linearly independent and $\mathrm{span}_{\mathbb{H}(\Omega)}(\mathcal{B}) = \mathcal{V}$. The dimension of $\mathcal{V}$ is the number of elements in $\mathcal{B}$ and is denoted by $\dim_{\mathbb{H}(\Omega)}(\mathcal{V})$.

Obviously $\{e_1, \ldots, e_n\}$ is a basis of the module $\mathbb{H}(\Omega)^n$ and $\dim_{\mathbb{H}(\Omega)}(\mathbb{H}(\Omega)^n) = n$. In fact, $\mathbb{H}(\Omega)^n$ is a free module. It is easily seen that if $\mathcal{V} \subset \mathbb{H}(\Omega)^n$ is a submodule and $\mathcal{V} \neq \{0\}$ then $\mathcal{V}$ has a basis and $\dim_{\mathbb{H}(\Omega)}(\mathcal{V}) \leq n$.

Let $A \in \mathbb{H}(\Omega)^{m\times n}$. Then for each fixed $\omega \in \Omega$, we can talk about the rank of the matrix $A(\omega) \in \mathbb{C}^{m\times n}$ over the field $\mathbb{C}$, which we denote by $\mathrm{rank}(A(\omega))$ or $\mathrm{rank}_{\mathbb{C}}(A(\omega))$. Now consider A as a module homomorphism

$$A : \mathbb{H}(\Omega)^n \longrightarrow \mathbb{H}(\Omega)^m, f \longmapsto Af.$$

The rank of A as a module homomorphism is called the ***normal rank*** of A. Indeed, consider

$$N(A) := \{f \in \mathbb{H}(\Omega)^n : Af = 0\} \subset \mathbb{H}(\Omega)^n \text{ and } R(A) := \{Af : f \in \mathbb{H}(\Omega)^n\} \subset \mathbb{H}(\Omega)^m.$$

Then $N(A)$ is a submodule of $\mathbb{H}(\Omega)^n$ and $R(A)$ is a submodule of $\mathbb{H}(\Omega)^m$.

DEFINITION 2.6 (normal rank). Let $A \in \mathbb{H}(\Omega)^{m\times n}$. Then the rank of A is given by

$$\mathrm{rank}_{\mathbb{H}(\Omega)}(A) := \dim_{\mathbb{H}(\Omega)}(R(A)).$$

The rank $\mathrm{rank}_{\mathbb{H}(\Omega)}(A)$ is also called the normal rank of A and is denoted by $\mathrm{nrank}(A)$. Thus $\mathrm{nrank}(A) = \mathrm{rank}_{\mathbb{H}(\Omega)}(A) = \dim_{\mathbb{H}(\Omega)}(R(A))$.

Equivalently, $\mathrm{nrank}(A)$ is the number of linearly independent columns of A. Similarly, the dimension of the null space $N(A)$ is called the nullity of A, that is,

$$\mathrm{nullity}_{\mathbb{H}(\Omega)}(A) := \dim_{\mathbb{H}(\Omega)}(N(A)).$$

If $A \in \mathbb{H}(\Omega)^{m\times n}$ then it is easily seen that A is regular $\iff \mathrm{nrank}(A) = m = n$. Hence A is singular when $m \neq n$ or when $\mathrm{nrank}(A) < \min(m, n)$.

REMARK 2.7. Let $A \in \mathbb{H}(\Omega)^{m\times n}$. Then there are two ranks of A, namely, the normal rank $\mathrm{nrank}(A)$ and the rank of the matrix $A(\omega)$ for each fixed $\omega \in \Omega$. It can be easily shown that $\mathrm{nrank}(A) = \max\{\mathrm{rank}(A(\omega)) : \omega \in \Omega\}$.

REMARK 2.8. Let $B := \begin{bmatrix} f_1 & \cdots & f_\ell \end{bmatrix} \in \mathbb{H}(\Omega)^{n\times \ell}$. Then $\{f_1, \ldots, f_\ell\}$ is linearly independent $\iff \mathrm{nrank}(B) = \ell$. Hence B is an ordered basis of $R(B) \iff \mathrm{nrank}(B) = \ell$.

We now consider analytic equivalence between holomorphic matrices.

DEFINITION 2.9.

(a) Let $U \subset \Omega$ be open and $A, B \in \mathbb{H}(\Omega)^{m\times n}$. Then A and B are said to be equivalent on U if there exist $F \in \mathrm{GL}_m(\mathbb{H}(U))$ and $E \in \mathrm{GL}_n(\mathbb{H}(U))$ such that

$$A = FBE, \quad \text{that is,} \quad A(z) = F(z)B(z)E(z) \text{ for all } z \in U.$$

We write $A \sim_U B$ or $A(z) \sim_U B(z)$ when A and B are equivalent on U.

(b) Let $\lambda \in \Omega$ and $A, B \in \mathbb{H}(\Omega)^{m\times n}$. Then A and B are said to be equivalent at λ if there exists an open neighbourhood U of λ such that $A \sim_U B$. We write $A \sim_\lambda B$ or $A(z) \sim_\lambda B(z)$ when A and B are equivalent at λ.

REMARK 2.10. Let $A \in \mathbb{H}(\Omega)^{m\times n}$. Define the equivalent orbit $\mathrm{Orb}_\Omega(A)$ of A by

$$\mathrm{Orb}_\Omega(A) := \{UAV : U \in \mathrm{GL}_m(\mathbb{H}(\Omega)) \text{ and } V \in \mathrm{GL}_n(\mathbb{H}(\Omega))\}.$$

Note that $A \sim_\Omega B \iff B \in \mathrm{Orb}_\Omega(A) \iff A \in \mathrm{Orb}_\Omega(B) \iff \mathrm{Orb}_\Omega(A) = \mathrm{Orb}_\Omega(B)$. Similarly, $A \sim_\lambda B \iff \mathrm{Orb}_U(A) = \mathrm{Orb}_U(B)$ for some open neighbourhood U of λ.

Observe that if $A \sim_\Omega B$ then $\mathrm{nrank}(B) = \mathrm{nrank}(A)$. Thus analytic equivalence preserves the normal rank of A. Hence $B \in \mathrm{Orb}_\Omega(A) \Longrightarrow \mathrm{nrank}(B) = \mathrm{nrank}(A)$.

Obviously, $A \sim_\Omega B \Longrightarrow A \sim_\lambda B$ for each $\lambda \in \Omega$. The following result, which is a special case of [**22**, Theorem 5.2], shows that the converse also holds.

THEOREM 2.11. *Let $A, B \in \mathbb{H}(\Omega)^{m\times n}$. Then $A \sim_\Omega B \iff A \sim_\lambda B$ for each $\lambda \in \Omega$.*

We have defined the spectrum of $A \in \mathbb{H}(\Omega)^{n\times n}$ when A is regular. We now define the spectrum of an $m \times n$ holomorphic matrix which subsumes the notion of spectrum of a regular holomorphic matrix.

DEFINITION 2.12 (spectrum). Let $A \in \mathbb{H}(\Omega)^{m\times n}$. Then $\lambda \in \Omega$ is said to be an eigenvalue of A if $\mathrm{rank}(A(\lambda)) < \mathrm{nrank}(A)$. The spectrum $\sigma_\Omega(A)$ of A is given by

$$\sigma_\Omega(A) := \{\lambda \in \Omega : \mathrm{rank}(A(\lambda)) < \mathrm{nrank}(A).\}$$

Thus the spectrum $\sigma_\Omega(A)$ consists of eigenvalues of A irrespective of whether A is regular or singular. Observe that $\sigma_\Omega(A)$ is invariant under analytic equivalence transformation of A on Ω. Hence

$$A \sim_\Omega B \Longrightarrow \sigma_\Omega(A) = \sigma_\Omega(B).$$

We now briefly review Weierstrass product and Weierstrass theorem for holomorphic functions, which we utilize for proving canonical forms of holomorphic matrices.

2.1. Weierstrass factorization of holomorphic functions. Let $f \in \mathbb{H}(\Omega)$ and $\lambda \in \sigma_\Omega(f)$. A positive integer ℓ is said to be the order of λ as a zero of f if $f^{(\ell)}(\lambda) \neq 0$ and $f^{(j)}(\lambda) = 0$ for $j = 1 : \ell - 1$. Also, ℓ is called the multiplicity of λ as a zero of f. On the other hand, if $\mu \in \Omega$ and $f(\mu) \neq 0$ then μ is said to be a zero of f of order 0. Thus μ is a zero of f of order $0 \iff \mu \in \rho_\Omega(f)$.

We denote the set of integers by $\mathbb{Z}$ and the set of non-negative integers by $\mathbb{Z}_+$.

DEFINITION 2.13. Define $o : \Omega \times (\mathbb{H}(\Omega) \setminus \{0\}) \longrightarrow \mathbb{Z}_+, (z, f) \longmapsto o_z(f)$, by $o_z(f) := p$, where p is the order of z as a zero of f.

Let $\phi : \Omega \longrightarrow \mathbb{C}$. The support of ϕ denoted by $\mathrm{supp}(\phi)$ is given by

$$\mathrm{supp}(\phi) := \{z \in \Omega : \phi(z) \neq 0\}.$$

The support of ϕ is said to be **locally finite** if $\mathrm{supp}(\phi) \cap \mathrm{K}$ is a finite set for every compact set $\mathrm{K} \subset \Omega$.

DEFINITION 2.14 (divisor). A function $\partial : \Omega \longrightarrow \mathbb{Z}_+$ is said to be a divisor on Ω if $\mathrm{supp}(\partial)$ is locally finite. If ∂_1 and ∂_2 are divisors on Ω then we write $\partial_1 \leq \partial_2$ when $\partial_1(z) \leq \partial_2(z)$ for all $z \in \Omega$. We denote the set of all divisors on Ω by $\mathrm{Div}(\Omega)$.

Observe that if $\partial_1, \partial_2 \in \mathrm{Div}(\Omega)$ then $\partial_1 + \partial_2 \in \mathrm{Div}(\Omega)$. Also $\partial_{\max} := \max\{\partial_1, \partial_2\}$ and $\partial_{\min} := \min\{\partial_1, \partial_2\}$ are divisors on Ω, where

$$\partial_{\max}(z) := \max\{\partial_1(z), \partial_2(z)\} \text{ and } \partial_{\min}(z) := \min\{\partial_1(z), \partial_2(z)\} \text{ for } z \in \Omega.$$

Let $f \in \mathbb{H}(\Omega)$ be nonzero. Then $\sigma_\Omega(f)$ is a locally finite subset of Ω. Define $\partial : \Omega \longrightarrow \mathbb{Z}_+$ by $\partial(z) := o_z(f)$. Then $\mathrm{supp}(\partial) = \sigma_\Omega(f)$ is locally finite. Hence $\partial \in \mathrm{Div}(\Omega)$. Since $\partial(z) = o_z(f)$ is the order of z as a zero of f, it follows that $(z-\lambda)^{\partial(\lambda)}$ is a divisor of $f(z)$ for $\lambda \in \sigma_\Omega(f)$. Thus, multiplicities of zeros of f give a divisor on Ω.

DEFINITION 2.15. Let $\partial \in \mathrm{Div}(\Omega)$ and $f \in \mathbb{H}(\Omega)$ be nonzero. Then ∂ is said to be the principal divisor of f on Ω if $\partial(z) = o_z(f)$ for all $z \in \Omega$.

REMARK 2.16. Let $f \in \mathbb{H}(\Omega)$ be nonzero and ∂ be the principal divisor of f on Ω. Then $\sigma_\Omega(f) = \mathrm{supp}(\partial)$. Hence it follows that $\partial = 0 \iff f$ is a unit element of $\mathbb{H}(\Omega)$. Note that ∂ is the principal divisor of uf on Ω for any unit element u of $\mathbb{H}(\Omega)$. As we shall see, the converse is also true. If f and g in $\mathbb{H}(\Omega)$ have the same principal divisor ∂ on Ω then $g = uf$ for some unit element u of $\mathbb{H}(\Omega)$.

If ∂ is the principal divisor of f on Ω with finite support and $\mathrm{supp}(\partial) = \{\lambda_1, \ldots, \lambda_\ell\}$ then $f(z) = \prod_{j=1}^{\ell}(z-\lambda_j)^{\partial(\lambda_j)} g(z)$ for some unit element $g \in \mathbb{H}(\Omega)$. In contrast, if $\mathrm{supp}(\partial) = \{\lambda_n : n \in \mathbb{N}\}$ then an infinite product representation of f of the form $f(z) = \prod_{n=1}^{\infty}(z-\lambda_n)^{\partial(n)} g(z)$ for some unit element $g \in \mathbb{H}(\Omega)$ may not exist. However, an infinite product representation of f can be obtained by replacing the factors $(z-\lambda_n)^{\partial(n)}, n \in \mathbb{N}$, with Weierstrass factors.

DEFINITION 2.17 ([**27**]). An infinite product $\prod_{n=1}^{\infty} f_n$ with $f_n = 1 + g_n \in \mathbb{H}(\Omega)$ is said to be **normally convergent** in Ω if the series $\sum_{n=1}^{\infty} g_n$ converges absolutely on every compact set $\mathrm{K} \subset \Omega$.

Observe that if $f_1, \ldots, f_\ell \in \mathbb{H}(\Omega)$ are nonzero then

$$\sigma_\Omega\Big(\prod_{j=1}^{\ell} f_j\Big) = \cup_{j=1}^{\ell} \sigma_\Omega(f_j) \text{ and } o_\lambda\Big(\prod_{j=1}^{\ell} f_j\Big) = o_\lambda(f_1) + \cdots + o_\lambda(f_\ell) \text{ for } \lambda \in \Omega.$$

Now, let $f := \prod_{n=1}^{\infty} f_n, f_n \in \mathbb{H}(\Omega)$ and $f_n \neq 0$ for $n \in \mathbb{N}$, be a normally convergent product in Ω. Then it is known [**27**] that $f \in \mathbb{H}(\Omega)$ and $f \neq 0$. Further, for each $\lambda \in \Omega$ there exists N such that $f_n(\lambda) \neq 0$ for all $n \geq N$. Hence, for each $\lambda \in \Omega$ there exists $N \in \mathbb{N}$ such that

$$f = f_1 f_2 \cdots f_{N-1} F_N \text{ and } F_N(\lambda) \neq 0, \text{ where } F_N := \prod_{n=N}^{\infty} f_n \in \mathbb{H}(\Omega). \tag{2.1}$$

Consequently, we have $\sigma_\Omega(f) = \cup_{n=1}^{\infty} \sigma_\Omega(f_n)$ and $o_\lambda(f) = \sum_{n=1}^{\infty} o_\lambda(f_n)$ for $\lambda \in \Omega$.

DEFINITION 2.18. Let $\partial \in \mathrm{Div}(\Omega)$ and $\mathrm{S} \subset \mathrm{supp}(\partial)$. A sequence (λ_n) is said to be a ∂-sequence in S if $\lambda_n \in \mathrm{S}$ for all $n \in \mathbb{N}$ and each $\lambda \in \mathrm{S}$ appears exactly $\partial(\lambda)$ times in the sequence.

We now define Weierstrass product corresponding to a divisor on Ω.

DEFINITION 2.19 ([**27**], Weierstrass product). Let $\partial \in \mathrm{Div}(\Omega)$ and (λ_n) be a ∂-sequence in $\mathrm{supp}(\partial) \setminus \{0\}$. Then an infinite product

$$W(z) := z^{\partial(0)} \prod_{n=1}^{\infty} f_n(z),\ f_n \in \mathbb{H}(\Omega), \tag{2.2}$$

is called a Weierstrass product in Ω for the divisor ∂ if the following conditions hold:

(a) The Weierstrass factor f_n has no zeros in $\Omega \setminus \{\lambda_n\}$ and $o_{\lambda_n}(f_n) = 1$ for all $n \in \mathbb{N}$.
(b) The product $\prod_{n=1}^{\infty} f_n$ converges normally in Ω.

The Weierstrass product W in Ω for a divisor $\partial \in \mathrm{Div}(\Omega)$ is unique up to a unit element of $\mathbb{H}(\Omega)$. Nevertheless, we refer to W as the Weierstrass product in Ω for the divisor ∂. Note that if W is the Weierstrass product in Ω for $\partial \in \mathrm{Div}(\Omega)$ then ∂ is the principal divisor of W on Ω, that is, $\sigma_\Omega(W) = \mathrm{supp}(\partial)$ and $\partial(\lambda) = o_\lambda(W)$ for $\lambda \in \Omega$.

THEOREM 2.20 ([**27**],Weierstrass product theorem).

(a) *Let $\partial \in \mathrm{Div}(\mathbb{C})$ and (λ_n) be a ∂-sequence in $\mathrm{supp}(\partial) \setminus \{0\}$. If $\mathrm{supp}(\partial)$ has no accumulation points in $\mathbb{C}$ then there is a Weierstrass product in $\mathbb{C}$ for ∂ given by*

$$W(z) := z^{\partial(0)} \prod_{n=1}^{\infty} f_n(z),\quad f_n \in \mathbb{H}(\mathbb{C}),$$

where $f_n(z) := \left(1 - \dfrac{z}{\lambda_n}\right) \exp\left(\dfrac{z}{\lambda_n} + \dfrac{1}{2}\left(\dfrac{z}{\lambda_n}\right)^2 + \cdots + \dfrac{1}{n-1}\left(\dfrac{z}{\lambda_n}\right)^{n-1}\right).$

(b) *Let $\partial \in \mathrm{Div}(\Omega)$ and (λ_n) be a ∂-sequence in $\mathrm{supp}(\partial)$. If $\mathrm{supp}(\partial)$ has no accumulation points in Ω then there is a Weierstrass product in Ω for ∂ given by*

$$W(z) := \prod_{n=1}^{\infty} f_n(z),\quad f_n \in \mathbb{H}(\Omega)),$$

where $f_n, n \in \mathbb{N}$, are Weierstrass factors.

The following version of Theorem 2.20(b) is also known as Weierstrass theorem for holomorphic functions with prescribed zeros including their orders and is deduced from Mittag-Leffler theorem; see [**9**, p.358].

WEIERSTRASS THEOREM IN $\mathbb{H}(\Omega)$. *Let $(\lambda_n) \subset \Omega$ be a sequence of distinct complex numbers with no accumulation points in Ω. Let $(\ell_n) \subset \mathbb{N}$ be a sequence of natural numbers. Then there exists $f \in \mathbb{H}(\Omega)$ such that $\sigma_\Omega(f) = \{\lambda_n : n \in \mathbb{N}\}$ and $o_{\lambda_n}(f) = \ell_n$ for all $n \in \mathbb{N}$.*

PROOF. Define $\partial : \Omega \longrightarrow \mathbb{Z}_+$ by $\partial(\lambda_n) := \ell_n$ for all $n \in \mathbb{N}$ and $\partial(z) := 0$ for $z \neq \lambda_n$. Then we have $\partial \in \mathrm{Div}(\Omega)$. Hence by Theorem 2.20(b) there exists a Weierstrass product $f \in \mathbb{H}(\Omega)$ for ∂ such that $\sigma_\Omega(f) = \{\lambda_n : n \in \mathbb{N}\}$ and $o_{\lambda_n}(f) = \ell_n$ for all $\in \mathbb{N}$. $\square$

A consequence of the Weierstrass product (Theorem2.20) is that a holomorphic function $f \in \mathbb{H}(\Omega)$ can be factorized as $f = uW$, where u is a unit element of $\mathbb{H}(\Omega)$ and W is the Weierstrass product for the principal divisor ∂ of f.

THEOREM 2.21 (factorization, [**27**]).

(a) *Let f be a nonzero entire function and ∂ be the principal divisor of f on $\mathbb{C}$. Let (λ_n) be a ∂-sequence in* $\operatorname{supp}(\partial) \setminus \{0\}$. *Then f can be written in the form*

$$f(z) = e^{g(z)} z^{\partial(0)} \prod_{n=1}^{\infty} \left[\left(1 - \frac{z}{\lambda_n}\right) \exp\left(\frac{z}{\lambda_n} + \frac{1}{2}\left(\frac{z}{\lambda_n}\right)^2 + \cdots + \frac{1}{n-1}\left(\frac{z}{\lambda_n}\right)^{n-1} \right) \right],$$

where $g \in \mathbb{H}(\mathbb{C})$ and $z^{\partial(0)} \prod_{n=1}^{\infty}(\cdots)$ is a (possibly empty) Weierstrass product in $\mathbb{C}$ for the divisor ∂.

(b) *Let $f \in \mathbb{H}(\Omega)$ be nonzero and ∂ be the principal divisor of f on Ω. Let (λ_n) be a ∂-sequence in* $\operatorname{supp}(\partial)$. *Then f can be written in the form*

$$f = u \prod_{n=1}^{\infty} f_n, \; f_n \in \mathbb{H}(\Omega),$$

where $u \in \mathbb{H}(\Omega)$ is a unit element and $\prod_{n=1}^{\infty} f_n$ is a (possibly empty) Weierstrass product in Ω for the divisor ∂.

Let $f, g \in \mathbb{H}(\Omega)$. Then f is called a divisor of g if $g/f \in \mathbb{H}(\Omega)$, that is, there exists $h \in \mathbb{H}(\Omega)$ such that $g = fh$. A nonzero non-unit $u \in \mathbb{H}(\Omega)$ is called a prime of $\mathbb{H}(\Omega)$ if u divides fg for some $f, g \in \mathbb{H}(\Omega)$ then u divides f or u divides g. The functions $\phi_\lambda : z \mapsto (z - \lambda)$ for $\lambda \in \Omega$, up to unit factors, are the primes of $\mathbb{H}(\Omega)$. We have the following result which is a kind of prime factorization of f.

THEOREM 2.22 (prime factorization). *Let $f \in \mathbb{H}(\Omega)$ be nonzero and ∂ be the principal divisor of f on Ω. Let $\sigma_\Omega(f) = \{z_n : n \in \mathbb{N}\}$. Then f can be written as*

$$f(z) = u(z) \prod_{n=1}^{\infty} (z - z_n)^{\partial(z_n)} u_n(z) \;\; \textit{for all } z \in \Omega,$$

where $u, u_n \in \mathbb{H}(\Omega)$ are unit elements, $n \in \mathbb{N}$ and $\prod_{n=1}^{\infty}(z - z_n)^{\partial(z_n)} u_n(z)$ is a (possibly empty) product in Ω for the divisor ∂.

PROOF. Let (λ_n) be a ∂-sequence in $\operatorname{supp}(\partial)$. Then by Theorem 2.21(b), we have

$$f = u \prod_{n=1}^{\infty} f_n, \; f_n \in \mathbb{H}(\Omega),$$

where $u \in \mathbb{H}(\Omega)$ is a unit element. Note that the Weierstrass factor f_n has no zeros in $\Omega \setminus \{\lambda_n\}$ and that $o_{\lambda_n}(f_n) = 1$ for all $n \in \mathbb{N}$. Hence $f_n(z)/(z - \lambda_n)$ is a unit element of $\mathbb{H}(\Omega)$ as it has a removable singularity at λ_n for all $n \in \mathbb{N}$. Thus $f_n(z) = (z - \lambda_n)u_n(z)$ for some unit element $u_n \in \mathbb{H}(\Omega)$ for all $n \in \mathbb{N}$. Next, observe that each $\lambda \in \sigma_\Omega(f)$ is a simple zero of exactly $\partial(\lambda)$ Weierstrass factors $f_n, n \in \mathbb{N}$. Further, as $\prod_{n=1}^{\infty} f_n$ converges normally, for every bijection $\tau : \mathbb{N} \longrightarrow \mathbb{N}$ the rearrangement $\prod_{n=1}^{\infty} f_{\tau(n)}$ converges normally to the same limit [**27**, p.8]. Consequently, for each $z_n \in \sigma_\Omega(f)$, rearranging the Weierstrass product if necessary, we can collect together $\partial(z_n)$ Weierstrass factors with zero at z_n yielding the product $(z - z_n)^{\partial(z_n)} u_n(z)$ for some unit element $u_n \in \mathbb{H}(\Omega)$. Since $\sigma_\Omega(f) = \{z_n : n \in \mathbb{N}\}$, it follows that there exist unit elements $u_n \in \mathbb{H}(\Omega)$ for all $n \in \mathbb{N}$ such that

$$\prod_{n=1}^{\infty} f_n(z) = \prod_{n=1}^{\infty} (z - z_n)^{\partial(z_n)} u_n(z).$$

This completes the proof. □

2.2. Smith form of a holomorphic matrix. Let $\lambda \in \mathbb{C}$ and let $\mathbb{H}(\lambda)$ denote the set of all complex functions holomorphic at λ. Thus $f \in \mathbb{H}(\lambda) \Longleftrightarrow$ there exists an open neighbourhood U of λ such that $f \in \mathbb{H}(U)$. Let $f \in \mathbb{H}(\Omega)$ be nonzero and $\lambda \in \mathbb{C}$. Set $\nu(\lambda) := o_\lambda(f)$. Then there exists $g \in \mathbb{H}(\lambda)$ such that

$$f(z) = (z - \lambda)^{\nu(\lambda)} g(z) \quad \text{and} \quad g(\lambda) \neq 0. \tag{2.3}$$

Since $g(\lambda) \neq 0$, there exists an open neighbourhood U of λ such that g is a unit element of $\mathbb{H}(U)$. Hence the local factorization in (2.3) can be restated as

$$f(z) \sim_\lambda (z - \lambda)^{\nu(\lambda)}, \text{ where } \nu(\lambda) := o_\lambda(f).$$

In other words, $f(z)$ and $(z - \lambda)^{\nu(\lambda)}$ are locally equivalent at λ. A generalization of this result to the case of a matrix-valued holomorphic function is called the local Smith form.

The local Smith form is considered in [**22**] and [**2**, p.69] for a meromorphic matrix-valued function and in [**20**, p.414] for a holomorphic matrix valued-function. For completeness, we state and prove the local Smith form which will be useful later in proving the global Smith form of a holomorphic matrix-valued function.

THEOREM 2.23. *Let $A \in \mathbb{H}(\Omega)^{m\times n}$. Suppose that $nrank(A) = r$. Let $\lambda \in \Omega$. Then there exist non-negative integers $\nu_1(\lambda) \leq \cdots \leq \nu_r(\lambda)$ such that*

$$A(z) \sim_\lambda \left[\begin{array}{ccc|c} (z-\lambda)^{\nu_1(\lambda)} & & & \\ & \ddots & & \\ & & (z-\lambda)^{\nu_r(\lambda)} & \\ \hline & & & 0_{m-r\times n-r} \end{array}\right] =: S_\lambda(z).$$

PROOF. Let A be given by $A(z) = [a_{ij}(z)]$ for $z \in \Omega$. Define $\nu_1(\lambda) = \min\{o_\lambda(a_{ij})\}$ where the minimum is taken over all nonzero functions a_{ij}. Then $\nu_1(\lambda) \geq 0$. By permuting rows and columns, if necessary, assume that $\nu_1(\lambda) = o_\lambda(a_{11})$. Then by local factorization of a_{11} at λ we have $a_{11}(z) = (z - \lambda)^{\nu_1(\lambda)} f_{11}(z)$ with $f_{11}(\lambda) \neq 0$ and $f_{11} \in \mathbb{H}(\lambda)$. Thus, $a_{11}(z) \sim_\lambda (z - \lambda)^{\nu_1(\lambda)}$. Also, we have $a_{ij}(z) = (z - \lambda)^{\nu_1(\lambda)} f_{ij}(z)$ for some $f_{ij} \in \mathbb{H}(\lambda)$. Multiplying the first row of A by f_{i1}/f_{11} and subtracting from the i-th row for $i = 2 : m$ and then multiplying the first column of A by f_{1j}/f_{11} and subtracting from the j-th column for $j = 2 : n$, we have

$$A(z) \sim_\lambda \left[\begin{array}{c|ccc} (z-\lambda)^{\nu_1(\lambda)} & & & \\ \hline & g_{22}(z) & \cdots & g_{2n}(z) \\ & \vdots & \ddots & \vdots \\ & g_{m2}(z) & \cdots & g_{mn}(z) \end{array}\right] \quad \text{for some } g_{ij} \in \mathbb{H}(\lambda).$$

Again, let $\nu_2(\lambda) := \min\{o_\lambda(g_{ij})\}$ where the minimum is taken over all nonzero functions g_{ij}. Assume that $\nu_2(\lambda) = o_\lambda(g_{22})$. Then $g_{22}(z) = (z - \lambda)^{\nu_2(\lambda)} h_{11}(z)$ with

$h_{11}(\lambda) \neq 0$ and $g_{ij}(z) = (z-\lambda)^{\nu_2(\lambda)} h_{ij}(z)$ for some $h_{ij} \in \mathbb{H}(\lambda)$. By construction, we have $\nu_1(\lambda) \leq \nu_2(\lambda)$. Again, by row and column elimination we obtain

$$A(z) \sim_\lambda \left[\begin{array}{cc|ccc} (z-\lambda)^{\nu_1(\lambda)} & & & & \\ & (z-\lambda)^{\nu_2(\lambda)} & & & \\ \hline & & b_{33}(z) & \cdots & b_{3n}(z) \\ & & \vdots & \cdots & \vdots \\ & & b_{m3}(z) & \cdots & b_{mn}(z) \end{array}\right] \text{ for some } b_{ij} \in \mathbb{H}(\lambda).$$

Repeating the procedure, we obtain the desired result after r steps. □

We write a block diagonal matrix $\mathrm{diag}(A_1, \cdots, A_m)$ by $A_1 \oplus \cdots \oplus A_m$. Then

$$A(z) \sim_\lambda (z-\lambda)^{\nu_1(\lambda)} \oplus \cdots \oplus (z-\lambda)^{\nu_r(\lambda)} \oplus 0_{m-r \times n-r} = S_\lambda(z).$$

The diagonal matrix $S_\lambda(z)$ is called the **local Smith form** of $A(z)$ at $\lambda \in \Omega$. Observe that

$$\lambda \in \sigma_\Omega(A) \Longleftrightarrow \mathrm{rank}(S_\lambda(\lambda)) < r \Longleftrightarrow \nu_r(\lambda) \neq 0.$$

In particular, if $\nu_j(\lambda) \neq 0$ then $\lambda \in \sigma_\Omega(A)$. Suppose that $\lambda \in \sigma_\Omega(A)$. Then there exists $0 \leq \ell < r$ such that $\nu_j(\lambda) \neq 0$ for $j = \ell+1, \ldots, r$. The positive integers $\nu_{\ell+1}(\lambda), \ldots, \nu_r(\lambda)$ are called the **partial multiplicities** of the eigenvalue λ and

$$\nu(\lambda) := \nu_{\ell+1}(\lambda) + \cdots + \nu_r(\lambda)$$

is called the **algebraic multiplicity** of the eigenvalue λ.

DEFINITION 2.24 (index). Let $A \in \mathbb{H}(\Omega)^{m \times n}$. Suppose that $\mathrm{nrank}(A) = r$. Let $\lambda \in \Omega$. Then a tuple $\mathrm{ind}(\lambda, A) := (\nu_1(\lambda), \nu_2(\lambda), \ldots, \nu_r(\lambda)) \in \mathbb{Z}_+^r$ with $\nu_1(\lambda) \leq \cdots \leq \nu_r(\lambda)$ is said to be the index of $A(z)$ at λ if

$$A(z) \sim_\lambda \left[(z-\lambda)^{\nu_1(\lambda)} \oplus \cdots \oplus (z-\lambda)^{\nu_r(\lambda)} \oplus 0_{m-r \times n-r}\right].$$

We denote the zero tuple in $\mathbb{Z}_+^r$ by $\mathbf{0}$. Then it follows that

$$\lambda \in \sigma_\Omega(A) \Longleftrightarrow \mathrm{ind}(\lambda, A) \neq \mathbf{0}.$$

Let $f \in \mathbb{H}(\Omega)$ be nonzero with the principal divisor $\partial \in \mathrm{Div}(\Omega)$. Suppose that $\sigma_\Omega(f) = \{z_n : n \in \mathbb{N}\}$. Then by the factorization Theorem 2.22 we have

$$f(z) \sim_\Omega \prod_{n=1}^{\infty} (z - z_n)^{\partial(z_n)} u_n(z),$$

where $u_n \in \mathbb{H}(\Omega)$ is a unit element and $n \in \mathbb{N}$. We show that the global Smith form is a diagonal representation of a holomorphic matrix-valued function in which the diagonal elements are Weierstrass products.

We now state and prove the global Smith form of a holomorphic matrix-valued function using divisors on Ω which is akin to the Smith form of a matrix polynomial [**13, 19, 28, 29**].

THEOREM 2.25 (Smith form). *Let $A \in \mathbb{H}(\Omega)^{m \times n}$. Suppose that $nrank(A) = r$ and $\sigma_\Omega(A) = \{z_\ell : \ell \in \mathbb{N}\}$. Then there exist holomorphic functions $\phi_1, \ldots, \phi_r$ in*

$\mathbb{H}(\Omega)$ with principal divisors $\partial_1, \ldots, \partial_r$ on Ω such that $\partial_1 \leq \cdots \leq \partial_r$ and

$$A(z) \sim_\Omega \left[\begin{array}{ccc|c} \phi_1(z) & & & \\ & \ddots & & \\ & & \phi_r(z) & \\ \hline & & & 0_{m-r\times n-r} \end{array}\right] =: S_A(z).$$

The functions $\phi_1, \ldots, \phi_r$ are unique up to unit elements of $\mathbb{H}(\Omega)$ and are given by

$$\phi_j(z) = u_j(z) \prod_{\ell=1}^{\infty} (z - z_\ell)^{\partial_j(z_\ell)} u_{j\ell}(z) \quad \textit{for all } z \in \Omega \textit{ and } j = 1 : r,$$

where $u_j, u_{j\ell} \in \mathbb{H}(\Omega)$ are unit elements, $\ell \in \mathbb{N}$ and $\prod_{\ell=1}^{\infty}(z - z_\ell)^{\partial_j(z_\ell)} u_{j\ell}(z)$ is a (possibly empty) product in Ω for the divisor ∂_j. Further, ϕ_j divides ϕ_{j+1} for $j = 1 : r - 1$.

PROOF. Let $\lambda \in \Omega$ and $\mathrm{ind}(\lambda, A) = (\nu_1(\lambda), \ldots, \nu_r(\lambda))$ be the index of $A(z)$ at λ. Then we have the local Smith form

$$A(z) \sim_\lambda (z - \lambda)^{\nu_1(\lambda)} \oplus \cdots \oplus (z - \lambda)^{\nu_r(\lambda)} \oplus 0_{m-r\times n-r} = S_\lambda(z).$$

For $j = 1 : r$, define $\partial_j : \Omega \longrightarrow \mathbb{Z}$ by $\partial_j(\lambda) := \nu_j(\lambda)$ for all $\lambda \in \Omega$. Then we have $\partial_1 \leq \cdots \leq \partial_r$. Note that $\partial_j(\lambda) = \nu_j(\lambda) \neq 0 \Longrightarrow \lambda \in \sigma_\Omega(A) \Longrightarrow \mathrm{supp}(\partial_j) \subset \sigma_\Omega(A)$. This shows that $\mathrm{supp}(\partial_j)$ is locally finite. Hence $\partial_1, \ldots, \partial_r$ are divisors on Ω.

By Theorem 2.20(b), let $\phi_1, \ldots, \phi_r$ be Weierstrass products in Ω for $\partial_1, \ldots, \partial_r$, respectively. Then $\partial_1, \ldots, \partial_r$ are the principal divisors of $\phi_1, \ldots, \phi_r$ on Ω, that is, $\partial_j(\lambda) = o_\lambda(\phi_j)$ for $j = 1 : r$. Hence by local factorization $\phi_j(z) \sim_\lambda (z - \lambda)^{\partial_j(\lambda)}$ for $j = 1 : r$. This shows that for each $\lambda \in \Omega$ we have

$$A(z) \sim_\lambda (z-\lambda)^{\partial_1(\lambda)} \oplus \cdots \oplus (z-\lambda)^{\partial_r(\lambda)} \oplus 0_{m-r\times n-r} \sim_\lambda \phi_1(z) \oplus \cdots \oplus \phi_r(z) \oplus 0_{m-r\times n-r}.$$

Hence by Theorem 2.11, we have $A(z) \sim_\Omega \phi_1(z) \oplus \cdots \oplus \phi_r(z) \oplus 0_{m-r\times n-r} = S_A(z)$.

Since $\lambda \in \sigma_\Omega(A) \iff \mathrm{ind}(\lambda, A) \neq 0$, it follows that $\sigma_\Omega(A) = \cup_{j=1}^{r} \mathrm{supp}(\partial_j)$. Hence the factorization of ϕ_j follows from Theorem 2.22 for $j = 1 : r$. Since $\partial_1 \leq \cdots \leq \partial_r$, it follows that ϕ_j divides ϕ_{j+1}, that is, $\phi_{j+1}/\phi_j \in \mathbb{H}(\Omega)$ for $j = 1 : r - 1$. Indeed, $o_z(\phi_{i+1}/\phi_i) = o_z(\phi_{i+1}) - o_z(\phi_i) = \partial_{i+1}(z) - \partial_i(z) \geq 0$ for $z \in \Omega$ shows that $\phi_{i+1}/\phi_i \in \mathbb{H}(\Omega)$. This completes the proof. □

REMARK 2.26. The diagonal matrix $S_A(z)$ is called the global Smith canonical form of $A(z)$ on Ω. The functions $\phi_1, \ldots, \phi_r$ are called the **invariant functions (or factors)** of $A(z)$ on Ω. If ϕ_j is not a unit element then it is called a non-unit invariant function or a non-unit invariant factor of $A(z)$ on Ω. Define $\phi_A(z) := \prod_{j=1}^{r} \phi_j(z)$ for $z \in \Omega$. Then ϕ_A is unique up to a unit element of $\mathbb{H}(\Omega)$ and is called the **zero function** of $A(z)$ on Ω. It follows that $\lambda \in \Omega$ is an eigenvalue value of $A(z) \iff \phi_A(\lambda) = 0$. Hence $\sigma_\Omega(A) = \{\lambda \in \Omega : \phi_A(\lambda) = 0\}$.

In view of Theorem 2.25, we have the following result.

PROPOSITION 2.27. *Let $A \in \mathbb{H}(\Omega)^{m\times n}$. Let the Smith form of $A(z)$ be given by*

$$S_A(z) := \phi_1(z) \oplus \cdots \oplus \phi_r(z) \oplus 0_{m-r\times n-r}.$$

Then $\mathrm{ind}(\lambda, A) := (\nu_1(\lambda), \ldots, \nu_r(\lambda)) \subset \mathbb{Z}_+^r$ with $\nu_1(\lambda) \leq \cdots \leq \nu_r(\lambda)$ is the index of $A(z)$ at $\lambda \iff \nu_j(\lambda) = o_\lambda(\phi_j)$ for $j = 1 : r$.

PROOF. Suppose that $\nu_j(\lambda) = o_\lambda(\phi_j)$ for $j = 1 : r$. Then $\nu_1(\lambda) \leq \cdots \leq \nu_r(\lambda)$ and $\phi_j(\lambda) \sim_\lambda (z-\lambda)^{\nu_j(\lambda)}$ for $j = 1 : r$. Consequently, we have

$$A(z) \sim_\Omega S_A(z) \sim_\lambda (z-\lambda)^{\nu_1(\lambda)} \oplus \cdots \oplus (z-\lambda)^{\nu_r(\lambda)} \oplus 0_{m-r \times n-r}$$

which shows that $\operatorname{ind}(\lambda, A) = (\nu_1(\lambda), \ldots, \nu_r(\lambda))$ is the index of $A(z)$ at λ.

Conversely, if $\operatorname{ind}(\lambda, A) = (\nu_1(\lambda), \ldots, \nu_r(\lambda))$ is the index of $A(z)$ at λ then, as shown in the proof of Theorem 2.25, we have $\phi_j(z) \sim_\lambda (z-\lambda)^{\nu_j(\lambda)}$ which in turn implies that $\nu_j(\lambda) = o_\lambda(\phi_j)$ for $j = 1 : r$. This completes the proof. $\square$

Let $A \in \mathbb{H}(\Omega)^{m\times n}$. Then $A(z) \sim_\Omega S_A(z)$, where $S_A(z)$ is the Smith canonical form of $A(z)$. Hence there exist $F \in \mathrm{GL}_m(\mathbb{H}(\Omega))$ and $E \in \mathrm{GL}_n(\mathbb{H}(\Omega))$ such that

$$F(z)A(z)E(z) = S_A(z) \text{ for all } z \in \Omega.$$

If $\operatorname{nrank}(A) = r$ then the last $n-r$ columns of $E(z)$ form a basis of $N(A)$ and the first r columns of $F(z)^{-1}$ form a basis of $R(A)$. Indeed, let $E(z) = \begin{bmatrix} v_1(z) & \cdots & v_n(z) \end{bmatrix}$ and $F(z)^{-1} = \begin{bmatrix} u_1(z) & \cdots & u_m(z) \end{bmatrix}$ be column partitions of $E(z)$ and $F(z)^{-1}$. Since $S_A(z) = \operatorname{diag}(\phi_1(z), \cdots, \phi_r(z), 0_{m-r\times n-r})$, we have $Av_j = 0$ for $j = r+1 : n$. This shows that $N(A) = \operatorname{span}_{\mathbb{H}(\Omega)}(v_{r+1}, \ldots, v_n)$ and $\{v_{r+1}, \ldots, v_n\}$ is a basis of $N(A)$. Also, $Av_j = \phi_j u_j$ for $j = 1 : r$ which shows that $R(A) = \operatorname{span}_{\mathbb{H}(\Omega)}(u_1, \ldots, u_r)$ and $\{u_1, \ldots, u_r\}$ is a basis of $R(A)$.

3. Meromorphic matrices

Let X be a complex Banach space. A function $f : \Omega \longrightarrow X$ is said to be meromorphic in Ω if f is holomorphic in Ω except for poles. We denote the set of all X-valued meromorphic functions on Ω by $\mathbb{M}(\Omega, X)$, that is,

$$\mathbb{M}(\Omega, X) := \{f : \Omega \longrightarrow X \mid f \text{ is meromorphic in } \Omega\}.$$

Again, for simplicity of notation, when $X = \mathbb{C}$, $X = \mathbb{C}^n$ and $X = \mathbb{C}^{m\times n}$, we set

$$\mathbb{M}(\Omega) := \mathbb{M}(\Omega, \mathbb{C}),\ \mathbb{M}(\Omega)^n := \mathbb{M}(\Omega, \mathbb{C}^n) \text{ and } \mathbb{M}(\Omega)^{m\times n} := \mathbb{M}(\Omega, \mathbb{C}^{m\times n}).$$

Thus $\mathbb{M}(\Omega)$ is the set of meromorphic functions and $\mathbb{M}(\Omega)^{m\times n}$ is the set of $m \times n$ matrices whose entries are meromorphic functions. We refer to elements of $\mathbb{M}(\Omega)^n$ as meromorphic vectors and elements of $\mathbb{M}(\Omega)^{m\times n}$ as meromorphic matrices.

Let $M \in \mathbb{M}(\Omega)^{m\times n}$. We denote the set of poles of M by $\wp_\Omega(M)$, that is,

$$\wp_\Omega(M) := \{z \in \Omega : z \text{ is a pole of } M(z)\}.$$

If $f \in \mathbb{M}(\Omega)$ then λ is a pole of $f \iff \lambda$ is a zero of $1/f$. Hence $\wp_\Omega(f)$ is a locally finite subset of Ω. The multiplicity of $\lambda \in \wp(f)$ as a zero of $1/f$ is called the order of λ as a pole of f. We now define the order function $o_\lambda(f)$ for a meromorphic function f. This subsumes the order function defined in Definition 2.13.

DEFINITION 3.1. Define $o : \Omega \times (\mathbb{M}(\Omega) \setminus \{0\}) \longrightarrow \mathbb{Z}, (z, f) \longmapsto o_z(f)$, by

$$o_z(f) := \begin{cases} p & \text{if } p \text{ is the order of } z \text{ as a zero of } f, \\ -p & \text{if } p \text{ is the order of } z \text{ as a pole of } f. \end{cases}$$

We now define the divisors on Ω for analyzing zeros and poles of meromorphic functions. This subsumes the divisors defined in Definition 2.14

DEFINITION 3.2 (divisor). A function $\partial : \Omega \longrightarrow \mathbb{Z}$ is said to be a divisor on Ω if $\mathrm{supp}(\partial)$ is locally finite. If ∂_1 and ∂_2 are divisors on Ω then we write $\partial_1 \leq \partial_2$ when $\partial_1(z) \leq \partial_2(z)$ for all $z \in \Omega$. A divisor ∂ on Ω is said to be non-negative if $\partial \geq 0$, that is, $\partial(z) \geq 0$ for all $z \in \Omega$. We denote the set of all divisors on Ω by $\mathrm{Div}(\Omega)$.

Let $f \in \mathbb{M}(\Omega)$ be nonzero and $\partial \in \mathrm{Div}(\Omega)$. Then ∂ is said to the principal divisor of f on Ω if $\partial(z) := o_z(f)$ for $z \in \Omega$.

Let $\partial \in \mathrm{Div}(\Omega)$. Then $\partial = \partial^+ - \partial^-$ and $\mathrm{supp}(\partial^+) \cap \mathrm{supp}(\partial^-) = \emptyset$, where ∂^+ and ∂^- are non-negative divisors on Ω given by

$$\partial^+(z) := \max(0, \partial(z)) \text{ and } \partial^-(z) := \max(0, -\partial(z)) \text{ for } z \in \Omega. \tag{3.1}$$

The set of divisors $\mathrm{Div}(\Omega)$ on Ω is an additive abelian group.

REMARK 3.3. By convention 0 is a divisor with empty support, that is, $\mathrm{supp}(0) = \emptyset$. Let $f \in \mathbb{M}(\Omega)$ be nonzero and $\partial \in \mathrm{Div}(\Omega)$ be the principal divisor of f on Ω. Then it follows that $f \in \mathbb{H}(\Omega) \Longleftrightarrow \partial \geq 0$.

Let $f_1, f_2 \in \mathbb{H}(\Omega)$ be nonzero. Then $f_1/f_2 \in \mathbb{M}(\Omega)$. Let ∂_1 and ∂_2, respectively, be the principal divisors of f_1 and f_2 on Ω. Since $o_z(f_1/f_2) = o_z(f_1) - o_z(f_2)$ for $z \in \Omega$, it follows that $\partial := \partial_1 - \partial_2$ is the principal divisor of f_1/f_2 on Ω. Thus

$$f_2 \text{ divides } f_1 \Longleftrightarrow f_1/f_2 \in \mathbb{H}(\Omega) \Longleftrightarrow \partial \geq 0 \Longleftrightarrow \partial_2 \leq \partial_1.$$

On the other hand, let $g \in \mathbb{M}(\Omega)$ be nonzero and let $\partial \in \mathrm{Div}(\Omega)$ be the principal divisor of g. Then $\partial = \partial^+ - \partial^-$, where ∂^+ and ∂^- are the non-negative divisors on Ω given by (3.1). Now, by Weierstrass Theorem 2.20, there exist $f_1, f_2 \in \mathbb{H}(\Omega)$ such that ∂^+ and ∂^-, respectively, are the principal divisors of f_1 and f_2. Since $\mathrm{supp}(\partial^+) \cap \mathrm{supp}(\partial^-) = \emptyset$, the functions f_1 and f_2 do not have common zeros. Further, the principal divisor of f_1/f_2 on Ω is given by $\partial^+ - \partial^- = \partial$. Thus g and f_1/f_2 have the same principal divisor ∂ on Ω. Hence $g = f_1/f_2$ up to a unit element of $\mathbb{H}(\Omega)$. Indeed, the principal divisor of $g/(f_1/f_2)$ is $\partial - \partial = 0$ which implies that $g/(f_1/f_2)$ has no zeros in Ω and hence a unit element of $\mathbb{H}(\Omega)$. Thus $g = uf_1/f_2$ for some unit element $u \in \mathbb{H}(\Omega)$. This shows that if $g \in \mathbb{M}(\Omega)$ is nonzero then there exist $f_1, f_2 \in \mathbb{H}(\Omega)$ such that f_1 and f_2 do not have common zeros and $g = f_1/f_2$. In other words, $\mathbb{M}(\Omega)$ is the quotient field of the integral domain $\mathbb{H}(\Omega)$.

COROLLARY 3.4. *The set* $\mathbb{M}(\Omega)$ *is the quotient field of the integral domain* $\mathbb{H}(\Omega)$. *The set of meromorphic vectors* $\mathbb{M}(\Omega)^n$ *and the set of meromorphic matrices* $\mathbb{M}(\Omega)^{m \times n}$ *are vector spaces over the field* $\mathbb{M}(\Omega)$.

Let $\mathrm{S} \subset \mathbb{H}(\Omega) \setminus \{0\}$ be nonempty. Then $f \in \mathbb{H}(\Omega)$ is called a common divisor of S if f divides every element of S. A common divisor f of S is called a greatest common divisor (gcd) of S if every common divisor of S is a divisor of f. Greatest common divisors, when exist, are uniquely determined up to unit elements of $\mathbb{H}(\Omega)$. If f is a gcd of S then we write $f = \gcd(\mathrm{S})$. If $1 = \gcd(\mathrm{S})$ then S is called relatively prime. Note that S is relatively prime $\Longleftrightarrow \cap_{f \in \mathrm{S}} \sigma_\Omega(f) = \emptyset$. Also, if $\mathrm{U} \subset \mathbb{H}(\Omega)$ is nonempty and $g := \gcd(\mathrm{U})$ then $\gcd(\mathrm{S} \cup \mathrm{U}) = \gcd(f, g)$. If $f, g \in \mathbb{H}(\Omega)$ and $\gcd(f, g) = 1$ then by Wedderburn lemma [**27**, p.136] there exist $a, b \in \mathbb{H}(\Omega)$ such that $af + bg = 1$ and a has no zeros in Ω. Also, if $f \in \mathbb{H}(\Omega)$ is a gcd of the functions $f_1, \ldots, f_n$ in $\mathbb{H}(\Omega)$ then there exist [**27**, p.138] functions $a_1, \ldots, a_n$ in $\mathbb{H}(\Omega)$ such that $f = a_1 f_1 + \cdots + a_n f_n$. Consequently, the ideal $\langle f_1, \ldots, f_n \rangle$ in $\mathbb{H}(\Omega)$ generated by $f_1, \ldots, f_n$ is given by the principal ideal $\langle f \rangle$, where $f = \gcd(f_1, \ldots, f_n)$. In other words, every finitely generated ideal in $\mathbb{H}(\Omega)$ is a principal ideal. Similarly, lcm of S also exists and is unique up to a unit element of $\mathbb{H}(\Omega)$.

Let $\partial_1, \ldots, \partial_n$ be non-negative divisors in $\mathrm{Div}(\Omega)$. Define $\partial_{\min}, \partial_{\max} : \Omega \longrightarrow \mathbb{Z}$ by

$$\partial_{\min}(z) := \min\{\partial_j(z) : j = 1 : n\} \text{ and } \partial_{\max}(z) := \max\{\partial_j(z) : j = 1 : n\}.$$

Then $\partial_{\min}, \partial_{\max} \in \mathrm{Div}(\Omega)$. If $\partial_1, \ldots, \partial_n$ are the principal divisors of the functions $f_1, \ldots, f_n$ in $\mathbb{H}(\Omega)$ then by Theorem 2.22 every function $f \in \mathbb{H}(\Omega)$ with $\partial_{\min}$ as the principal divisor on Ω is a gcd of $\{f_1, \ldots, f_n\}$, that is, $f = \gcd(f_1, \ldots, f_n)$. In fact, every nonempty subset of $\mathbb{H}(\Omega) \setminus \{0\}$ has a gcd. In particular, gcd of any finite set $\{f_1, \ldots, f_n\} \subset \mathbb{H}(\Omega)$ exists and is a linear combination of $f_1, \ldots, f_n$. In other words, $\mathbb{H}(\Omega)$ is a Bezout domain [**27**, p.138]. Similarly, by Theorem 2.22 every function $f \in \mathbb{H}(\Omega)$ with $\partial_{\max}$ as the principal divisor on Ω is a lcm of $\{f_1, \ldots, f_n\}$, that is, $f = \mathrm{lcm}(f_1, \ldots, f_n)$.

The discussion above proves the following well known result; see also [**27**].

THEOREM 3.5. *Let $h \in \mathbb{M}(\Omega)$ be nonzero. Then there exist f and g in $\mathbb{H}(\Omega)$ such that $h = f/g$ and* $\gcd(f, g) = 1$.

The discussion above also yields Weierstrass theorem for meromorphic functions which can be deduced from Mittag-Leffler theorem; see [**9**].

THEOREM 3.6 (Weierstrass). *Let $\partial \in$* $\mathrm{Div}(\Omega)$ *be such that* $\mathrm{supp}(\partial)$ *has no accumulation points in Ω. Then there exists $f \in \mathbb{M}(\Omega)$ such that ∂ is the principal divisor of f on Ω. Further, $\sigma_\Omega(f) =$* $\mathrm{supp}(\partial)$ *and $o_\lambda(f) = \partial(\lambda)$ for $\lambda \in \Omega$.*

PROOF. By (3.1), we have $\partial = \partial^+ - \partial^-$ and $\mathrm{supp}(\partial^+) \cap \mathrm{supp}(\partial^-) = \emptyset$, where $\partial^+ \geq 0$ and $\partial^- \geq 0$ are divisors on Ω. Hence by Weierstrass Theorem 2.20, there exist $f_1, f_2 \in \mathbb{H}(\Omega)$ such that ∂^+ and ∂^-, respectively, are the principal divisors of f_1 and f_2. Since $\mathrm{supp}(\partial^+) \cap \mathrm{supp}(\partial^-) = \emptyset$, the functions f_1 and f_2 do not have common zeros, that is, $\gcd(f_1, f_2) = 1$. Define $f := f_1/f_2$. Then $f \in \mathbb{M}(\Omega)$ and ∂ is the principal divisor of f. Hence the results follow. □

The following version of Theorem 3.6 is known as Weierstrass theorem for meromorphic functions with prescribed zeros (including multiplicities) and poles (including orders) and is usually deduced from Mittag-Leffler theorem in [**9**, p.358]. Here is an alternative proof.

WEIERSTRASS THEOREM IN $\mathbb{M}(\Omega)$. *Let $(\lambda_n) \subset \Omega$ be a sequence of distinct complex numbers with no accumulation points in Ω. Let $(\ell_n) \subset \mathbb{Z}$ be a sequence of integers. Then there exists $f \in \mathbb{M}(\Omega)$ such that $\sigma_\Omega(f) = \{\lambda_n : n \in \mathbb{N}\}$ and $o_{\lambda_n}(f) = \ell_n$ for all $n \in \mathbb{N}$.*

PROOF. Define $\partial : \Omega \longrightarrow \mathbb{Z}$ by $\partial(\lambda_n) := \ell_n$ and $\partial(z) := 0$ when $z \neq \lambda_n$ for all $n \in \mathbb{N}$. Then we have $\partial \in \mathrm{Div}(\Omega)$. Hence by Theorem 3.6 we have the desired function $f \in \mathbb{M}(\Omega)$. □

3.1. Smith-McMillan form of a meromorphic matrix. Recall that $\mathbb{M}(\Omega)$ is a field and $\mathbb{M}(\Omega)^n$ is a vector space over the field $\mathbb{M}(\Omega)$. The span of a subset $\{f_1, \ldots, f_\ell\} \subset \mathbb{M}(\Omega)^n$ is given by

$$\mathrm{span}_{\mathbb{M}(\Omega)}(f_1, \ldots, f_\ell) := \{r_1 f_1 + \cdots + r_\ell f_\ell : \ r_1, \ldots, r_\ell \in \mathbb{M}(\Omega)\}.$$

A subset $\{f_1, \ldots, f_\ell\} \subset \mathbb{M}(\Omega)^n$ is said to be **linearly dependent** if there exist $r_1, \ldots, r_\ell$ in $\mathbb{M}(\Omega)$ not all are zero such that $r_1 f_1 + \cdots + r_\ell f_\ell = 0$. Thus $\{f_1, \ldots, f_\ell\}$ is said to be **linearly independent** if it is not linearly dependent. Let $\mathcal{V} \subset \mathbb{M}(\Omega)^n$ be a subspace. Then $\mathcal{B} \subset \mathbb{M}(\Omega)^n$ is said to be a **basis** of $\mathcal{V}$ if $\mathcal{B}$ is linearly independent and $\text{span}_{\mathbb{M}(\Omega)}(\mathcal{B}) = \mathcal{V}$. The dimension $\dim_{\mathbb{M}(\Omega)}(\mathcal{V})$ of $\mathcal{V}$ is the number of elements of a basis $\mathcal{B}$ of $\mathcal{V}$.

Let $S := \{f_1, \ldots, f_\ell\} \subset \mathbb{H}(\Omega)^n \subset \mathbb{M}(\Omega)^n$. Then $\text{span}_{\mathbb{H}(\Omega)}(S)$ is a submodule of $\mathbb{H}(\Omega)^n$ and $\text{span}_{\mathbb{M}(\Omega)}(S)$ is a subspace of $\mathbb{M}(\Omega)^n$. It is easy to see that S is linearly independent in the module $\mathbb{H}(\Omega)^n \iff S$ is linearly independent in the vector space $\mathbb{M}(\Omega)^n$.

PROPOSITION 3.7. *Let $S := \{f_1, \ldots, f_\ell\} \subset \mathbb{H}(\Omega)^n \subset \mathbb{M}(\Omega)^n$. Then S is a basis of $\text{span}_{\mathbb{H}(\Omega)}(S) \iff S$ is a basis of $\text{span}_{\mathbb{M}(\Omega)}(S)$. Hence*

$$\dim_{\mathbb{H}(\Omega)}\left(\text{span}_{\mathbb{H}(\Omega)}(S)\right) = \dim_{\mathbb{M}(\Omega)}\left(\text{span}_{\mathbb{M}(\Omega)}(S)\right).$$

PROOF. If S is linearly independent in $\mathbb{M}(\Omega)^n$ then it is immediate that S is linearly independent in $\mathbb{H}(\Omega)^n$. Conversely, suppose that S is linearly independent in $\mathbb{H}(\Omega)^n$. Now consider $r_1 f_1 + \cdots + r_\ell f_\ell = 0$ for $r_1, \ldots, r_\ell$ in $\mathbb{M}(\Omega)$. By Theorem 3.5, we have $r_j = g_j/h_j$ and $\gcd(g_j, h_j) = 1$ where $g_j, h_j \in \mathbb{H}(\Omega)$ for $j = 1 : \ell$. Let $h := \text{lcm}(h_1, \ldots, h_\ell)$. Then $hr_j \in \mathbb{H}(\Omega)$ for $j = 1 : \ell$. Hence $r_1 f_1 + \cdots + r_\ell f_\ell = 0 \implies hr_1 f_1 + \cdots + hr_\ell f_\ell = 0 \implies hr_1 = \cdots = hr_\ell = 0 \implies r_1 = \cdots = r_\ell = 0$. This shows that S is linearly independent in $\mathbb{M}(\Omega)^n$. Hence the results follow. □

Proposition 3.7 shows that if $S \subset \mathbb{H}(\Omega)^n$ then we can talk about linear independence of S without specifying $\mathbb{H}(\Omega)^n$ or $\mathbb{M}(\Omega)^n$.

DEFINITION 3.8 (analytic basis). Let $\mathcal{V}$ be subspace of $\mathbb{M}(\Omega)^n$ and $\mathcal{B} \subset \mathbb{H}(\Omega)^n$. Then $\mathcal{B}$ said to be an analytic basis of $\mathcal{V}$ if $\mathcal{B}$ is linearly independent and $\text{span}_{\mathbb{M}(\Omega)}(\mathcal{B}) = \mathcal{V}$.

Let $\mathcal{V} \subset \mathbb{M}(\Omega)^n$ be a subspace. If $\mathcal{V} \neq \{0\}$ then it has an analytic basis. Indeed, let $\{f_1, \ldots, f_\ell\} \subset \mathbb{M}(\Omega)^n$ be a basis of $\mathcal{V}$. Then by Theorem 3.5 entries of each f_j can be written as fractions of relatively coprime functions in $\mathbb{H}(\Omega)$. Now, taking lcm of the denominators of the entries of f_j, each f_j can be written as $f_j = g_j/h_j$ where $g_j \in \mathbb{H}(\Omega)^n$ and $h_j \in \mathbb{H}(\Omega)$ for $j = 1 : n$. Obviously, we have

$$\text{span}_{\mathbb{M}(\Omega)}(f_1, \ldots, f_\ell) = \text{span}_{\mathbb{M}(\Omega)}(g_1, \ldots, g_\ell).$$

Now $r_1 g_1 + \cdots + r_\ell g_\ell = 0 \implies r_1 h_1 f_1 + \cdots + r_\ell h_\ell f_\ell = 0 \implies r_1 h_1 = \cdots = r_\ell h_\ell = 0 \implies r_1 = \cdots = r_\ell = 0$ showing that $\{g_1, \ldots, g_\ell\}$ is linearly independent in $\mathbb{M}(\Omega)^n$. This shows that $\{g_1, \ldots, g_\ell\} \subset \mathbb{H}(\Omega)^n$ is a basis of $\mathcal{V}$ and $\dim_{\mathbb{M}(\Omega)}(\mathcal{V}) = \ell$. This proves the following result.

THEOREM 3.9. *Let $\mathcal{V} \subset \mathbb{M}(\Omega)^n$ be a subspace. If $\mathcal{V} \neq \{0\}$ then $\mathcal{V}$ has an analytic basis.*

Let $A \in \mathbb{M}(\Omega)^{m \times n}$. Then for each fixed $\lambda \notin \wp_\Omega(A)$, we can talk about the rank of the matrix $A(\lambda) \in \mathbb{C}^{m \times n}$ over the field $\mathbb{C}$, which we denote by $\text{rank}(A(\lambda))$. We can also talk about the rank of the matrix A when viewed as a linear transformation

$$A : \mathbb{M}(\Omega)^n \longrightarrow \mathbb{M}(\Omega)^m, f \longmapsto Af,$$

on vector spaces over the field $\mathbb{M}(\Omega)$. Consider the null and range spaces

$$N(A) := \{f \in \mathbb{M}(\Omega)^n : Af = 0\} \subset \mathbb{M}(\Omega)^n \text{ and } R(A) := \{Af : f \in \mathbb{M}(\Omega)^n\} \subset \mathbb{M}(\Omega)^m.$$

The dimensions of the null space $N(A)$ and the range space $R(A)$ over the field $\mathbb{M}(\Omega)$ are called nullity and normal rank of A, respectively.

DEFINITION 3.10 (normal rank). Let $A \in \mathbb{M}(\Omega)^{m\times n}$. Then the rank of A is given by

$$\text{rank}_{\mathbb{M}(\Omega)}(A) := \dim_{\mathbb{M}(\Omega)}(R(A)).$$

The rank $\text{rank}_{\mathbb{M}(\Omega)}(A)$ is also called the normal rank of A and is denoted by $\text{nrank}(A)$. Thus $\text{nrank}(A) = \text{rank}_{\mathbb{M}(\Omega)}(A) = \dim_{\mathbb{M}(\Omega)}(R(A))$.

DEFINITION 3.11 (regular). Let $A \in \mathbb{M}(\Omega)^{m\times n}$. Then A is said to be regular if $m = n$ and $A(z)$ is invertible for some $z \in \Omega \setminus \wp_\Omega(A)$.

If $A \in \mathbb{M}(\Omega)^{n\times n}$ then it is easily seen that A is regular $\Longleftrightarrow$ $\text{nrank}(A) = n$. Thus $A \in \mathbb{M}(\Omega)^{m\times n}$ is singular when $m \neq n$ or when $\text{nrank}(A) < \min(m, n)$.

REMARK 3.12. Let $A \in \mathbb{M}(\Omega)^{m\times n}$. Then there are two ranks of A, namely, the normal rank $\text{nrank}(A)$ and the rank of the matrix $A(\lambda)$ for each fixed $\lambda \notin \wp_\Omega(A)$. It can be shown that $\text{nrank}(A) = \max\{\text{rank}(A(\lambda)) : \lambda \notin \wp_\Omega(A)\}$.

We now define equivalence between two meromorphic matrices.

DEFINITION 3.13.

(a) Let $U \subset \Omega$ be open and $A, B \in \mathbb{M}(\Omega)^{m\times n}$. Then A and B are said to be equivalent on U if there exist $F \in \text{GL}_m(\mathbb{H}(U))$ and $E \in \text{GL}_n(\mathbb{H}(U))$ such that

$$A = FBE, \quad \text{that is,} \quad A(z) = F(z)B(z)E(z) \text{ for all } z \in U.$$

We write $A \sim_U B$ or $A(z) \sim_U B(z)$ when A and B are equivalent on U.

(b) Let $\lambda \in \Omega$ and $A, B \in \mathbb{M}(\Omega)^{m\times n}$. Then A and B are said to be equivalent at λ if there exists an open neighbourhood U of λ such that $A \sim_U B$. We write $A \sim_\lambda B$ or $A(z) \sim_\lambda B(z)$ when A and B are equivalent at λ.

Observe that if $A \sim_\Omega B$ then $\text{nrank}(B) = \text{nrank}(A)$. Thus analytic equivalence preserves the normal rank of A. Obviously, $A \sim_\Omega B \Longrightarrow A \sim_\lambda B$ for each $\lambda \in \Omega$. The following result, which is a special case of [**22**, Theorem 5.2], shows that the converse also holds.

THEOREM 3.14. *Let* $A, B \in \mathbb{M}(\Omega)^{m\times n}$. *Then* $A \sim_\Omega B \Longleftrightarrow A \sim_\lambda B$ *for each* $\lambda \in \Omega$.

We have defined eigenvalues and spectrum of an $m \times n$ holomorphic matrix. As in the case of holomorphic matrices, the eigenvalues $A \in \mathbb{M}(\Omega)^{m\times n}$ can be defined in terms of the normal rank of A. However, the notion of spectrum of A requires appropriate modification. See [**1**] for the spectrum of a rational matrix.

DEFINITION 3.15. Let $A \in \mathbb{M}(\Omega)^{m\times n}$. Then $\lambda \in \Omega$ is said to be an eigenvalue of $A(z)$ if $\lambda \notin \wp_\Omega(A)$ and $\mathrm{rank}(A(\lambda)) < \mathrm{nrank}(A)$. The eigenspectrum $\mathrm{eig}_\Omega(A)$ of A is given by

$$\mathrm{eig}_\Omega(A) := \{\lambda \in \Omega : \ \lambda \notin \wp_\Omega(A) \text{ and } \mathrm{rank}(A(\lambda)) < \mathrm{nrank}(A)\}.$$

Let $\mu \in \wp_\Omega(A)$. Then μ said to be an eigenpole of A if there exists $f \in \mathbb{H}(\Omega)^n$ such that $f(\mu) \neq 0, f \notin N(A)$ and $\lim_{z\to\mu} A(z)f(z) = 0$. We denote the set of eigenpoles of A by $\mathrm{eip}_\Omega(A)$. Then the spectrum $\sigma_\Omega(A)$ of $A(z)$ in Ω is given by

$$\sigma_\Omega(A) := \mathrm{eig}_\Omega(A) \cup \mathrm{eip}_\Omega(A).$$

If $\lambda \in \sigma_\Omega(A)$ then λ is said to be a zero of $A(z)$ in Ω.

Note that if $A \in \mathbb{H}(\Omega)^{m\times n}$ then $\sigma_\Omega(A) = \mathrm{eig}_\Omega(A)$. However, if $A \in \mathbb{M}(\Omega)^{m\times n}$ then $\mathrm{eig}_\Omega(A) \subset \sigma_\Omega(A)$ and the inclusion may be proper.

EXAMPLE 3.16. Consider $\Omega := \{z \in \mathbb{C} : |z| < 1\}$ and $A \in \mathbb{M}(\Omega)^{2\times 2}$ given by

$$A(z) := \begin{bmatrix} 1 & \frac{1}{\sin z} \\ 0 & 1 \end{bmatrix}, \quad z \in \Omega.$$

Then $\mathrm{eig}_\Omega(A) = \emptyset$ and $\wp_\Omega(A) = \{0\}$. Consider $f(z) := \begin{bmatrix} 1 & -\sin z \end{bmatrix}^\top$. Then $f(0) \neq 0$ and $f \notin N(A)$. Further, we have $\lim_{z\to 0} A(z)f(z) = 0$ which shows that 0 is an eigenpole of A. Hence $\sigma_\Omega(A) = \mathrm{eip}_\Omega(A) = \{0\}$. ■

REMARK 3.17. Let $M \in \mathbb{M}(\Omega)^{m\times n}$ be given by $M(z) := [a_{ij}(z)]$, $z \in \Omega$. Then by Theorem 3.5 there exist $f_{ij}, g_{ij} \in \mathbb{H}(\Omega)$ such that $\gcd(f_{ij}, g_{ij}) = 1$ and $a_{ij} = f_{ij}/g_{ij}$ for all $i = 1 : m$ and $j = 1 : n$. Let $g := \mathrm{lcm}\{g_{ij}, i = 1 : m, j = 1 : n\}$. Note that g is unique up to a unit element of $\mathbb{H}(\Omega)$. Then we have $M = [f_{ij}/g_{ij}] = [h_{ij}]/g$, where $h_{ij} \in \mathbb{H}(\Omega)$ for $i = 1 : m$ and $j = 1 : n$. It follows that λ is a pole of $M(z) \Longleftrightarrow g(\lambda) = 0$. Hence we have $\wp_\Omega(M) = \sigma_\Omega(g)$. The function g is called the pole function of M.

A canonical form of a meromorphic matrix-valued function $A \in \mathbb{M}(\Omega)^{m\times n}$ is derived in [**22**]. It is shown that $A(z) \sim_\Omega f_1(z) \oplus \cdots \oplus f_r(z) \oplus 0_{m-r\times n-r}$, where $r = \mathrm{nrank}(A)$ and $f_1, \ldots, f_r$ are in $\mathbb{M}(\Omega)$ such that f_j divides f_{j+1} for $j = 1 : r$.

The Smith-McMillan form of a rational matrix is a well-known canonical form which has wide applications [**19, 29**]. We now prove the Smith-McMillan form of a meromorphic matrix which is akin to the Smith-McMillan form of a rational matrix.

Consider $h \in \mathbb{M}(\Omega) \setminus \{0\}$. Then there exist $f, g \in \mathbb{H}(\Omega)$ such that $\gcd(f, g) = 1$ and $h = f/g$. Let ∂_1 and ∂_2 be the principal divisors of f and g on Ω, respectively. Then f and g can be written as Weierstrass products in Ω for the divisors ∂_1 and ∂_2. The Smith-McMillan form generalizes this fact to the case of meromorphic matrices.

THEOREM 3.18 (Smith-McMillan form). *Let $A \in \mathbb{M}(\Omega)^{m\times n}$. Suppose that*

$$\sigma_\Omega(A) = \{\lambda_n : n \in \mathbb{N}\} \text{ and } \wp_\Omega(A) = \{\mu_n : n \in \mathbb{N}\}.$$

Also suppose that nrank$(A) = r$. Then the following hold:

(a) *There exist holomorphic functions $\phi_1, \ldots, \phi_r$ in $\mathbb{H}(\Omega)$ with principal divisors $\partial_1, \ldots, \partial_r$ on Ω such that $\partial_1 \leq \cdots \leq \partial_r$.*

(b) *There exist holomorphic functions $\psi_1, \ldots, \psi_r$ in $\mathbb{H}(\Omega)$ with principal divisors $\delta_1, \ldots, \delta_r$ on Ω such that $\delta_1 \geq \cdots \geq \delta_r$.*

(c) *ϕ_j and ψ_j are relatively prime, that is, $\gcd(\phi_j, \psi_j) = 1$ for $j = 1 : r$ and*

$$A(z) \sim_\Omega \left[\begin{array}{ccc|c} \phi_1(z)/\psi_1(z) & & & \\ & \ddots & & \\ & & \phi_r(z)/\psi_r(z) & \\ \hline & & & 0_{m-r\times n-r} \end{array}\right] =: \Sigma_A(z),$$

where $\phi_1, \ldots, \phi_r$ and $\psi_1, \ldots, \psi_r$ are unique up to unit elements of $\mathbb{H}(\Omega)$. Further, ϕ_j divides ϕ_{j+1} and ψ_{j+1} divides ψ_j for $j = 1 : r-1$.

(d) *Furthermore, $\phi_1, \cdots, \phi_r$ and $\psi_1, \ldots, \psi_r$ are given by*

$$\phi_j(z) = u_j(z)\prod_{\ell=1}^{\infty}(z-\lambda_\ell)^{\partial_j(\lambda_\ell)}u_{j\ell}(z) \text{ and } \psi_j(z) = v_j(z)\prod_{\ell=1}^{\infty}(z-\mu_\ell)^{\delta_j(\mu_\ell)}v_{j\ell}(z)$$

for $j = 1 : r$, where $u_j, v_j, u_{j\ell}, v_{j\ell} \in \mathbb{H}(\Omega)$ are unit elements and $\ell \in \mathbb{N}$. Here $\prod_{\ell=1}^{\infty}(z-\lambda_\ell)^{\partial_j(\lambda_\ell)}u_{j\ell}(z)$ and $\prod_{\ell=1}^{\infty}(z-\mu_\ell)^{\delta_j(\mu_\ell)}v_{j\ell}(z)$ are (possibly empty) products in Ω for the divisors ∂_j and δ_j, respectively, for $j = 1 : r$.

PROOF. Let A be given by $A(z) := [a_{ij}(z)]$, $z \in \Omega$. Then there exist $f_{ij}, g_{ij} \in \mathbb{H}(\Omega)$ such that $\gcd(f_{ij}, g_{ij}) = 1$ and $a_{ij} = f_{ij}/g_{ij}$ for all $i = 1 : m$ and $j = 1 : n$. Let $g := \mathrm{lcm}(g_{ij}, i = 1 : m, j = 1 : n)$. Note that g is unique up to a unit element of $\mathbb{H}(\Omega)$. Then $A = [f_{ij}/g_{ij}] = [h_{ij}]/g = H/g$, where $H := [h_{ij}]$ and $h_{ij} \in \mathbb{H}(\Omega)$ for $i = 1 : m$ and $j = 1 : n$. Note that $\lambda \in \wp_\Omega(A) \iff g(\lambda) = 0$. Hence we have $\wp_\Omega(A) = \sigma_\Omega(g)$. By Theorem 2.25, the Smith form $S_H(z) = \widehat{\phi}_1(z) \oplus \cdots \oplus \widehat{\phi}_r \oplus 0_{m-r\times n-r}$ of H yields

$$A(z) = H(z)/g(z) \sim_\Omega \left[\begin{array}{ccc|c} \widehat{\phi}_1(z)/g(z) & & & \\ & \ddots & & \\ & & \widehat{\phi}_r(z)/g(z) & \\ \hline & & & 0_{m-r\times n-r} \end{array}\right],$$

where $\widehat{\phi}_1, \ldots, \widehat{\phi}_r$ are invariant factors of $H(z)$. Let $\phi_j, \psi_j \in \mathbb{H}(\Omega)$ be such that

$$\gcd(\phi_j, \psi_j) = 1 \text{ and } \frac{\phi_j}{\psi_j} = \frac{\widehat{\phi}_j}{g} \text{ for } j = 1 : r.$$

Since $\widehat{\phi}_j(z)$ divides $\widehat{\phi}_{j+1}(z)$, that is, $\widehat{\phi}_{j+1}/\widehat{\phi}_j \in \mathbb{H}(\Omega)$ there exists $f_j \in \mathbb{H}(\Omega)$ such that $\widehat{\phi}_{j+1} = f_j\widehat{\phi}_j$ for $j = 1 : r$. Hence for $j = 1 : r-1$, we have

$$\frac{\widehat{\phi}_{j+1}}{g} = f_j\frac{\widehat{\phi}_j}{g} \Longrightarrow \frac{\phi_{j+1}}{\psi_{j+1}} = f_j\frac{\phi_j}{\psi_j} \Longrightarrow \frac{\phi_{j+1}}{\psi_{j+1}}\frac{\psi_j}{\phi_j} = f_j.$$

Since ϕ_j and ψ_j are relatively prime and ϕ_j divides ϕ_{j+1}, it follows that ψ_{j+1} divides ψ_j for $j = 1, \ldots, r-1$.

Let $\partial_1, \ldots, \partial_r$ be the principal divisors of $\phi_1, \ldots, \phi_r$ on Ω. Also, let $\delta_1, \ldots, \delta_r$ be the principal divisors of $\psi_1, \ldots, \psi_r$ on Ω. As ϕ_j divides ϕ_{j+1} and ψ_{j+1} divides ψ_j for $j = 1 : r-1$, we have $\partial_1 \leq \cdots \leq \partial_r$ and $\delta_1 \geq \cdots \geq \delta_r$.

Next, observe that $\sigma_\Omega(A) = \sigma_\Omega(\Sigma_A)$ and $\wp_\Omega(A) = \wp_\Omega(\Sigma_A)$. Consequently, we have $\delta_j(\lambda) \neq 0 \Longrightarrow \psi_j(\lambda) = 0 \Longrightarrow \lambda \in \wp_\Omega(A)$. This shows that $\text{supp}(\delta_j) \subset \wp_\Omega(A)$ for $j = 1 : r$. Hence the factorization of ψ_j follows from Theorem 2.22. On the other hand, $\partial_j(\lambda) \neq 0 \Longrightarrow \phi_j(\lambda) = 0$. Now either $\psi_i(\lambda) \neq 0$ for all $i = 1 : r$ or $\psi_i(\lambda) = 0$ for some $i \neq j$. In the first case, we have $\lambda \in \text{eig}_\Omega(\Sigma_A) = \text{eig}_\Omega(A)$. In the second case, $\lambda \in \wp_\Omega(\Sigma_A)$. Now, considering $v(z) := e_j$ we have $v(\lambda) \neq 0, v \notin N(\Sigma_A)$ and $\lim_{z\to\lambda} \Sigma_A(z)v(z) = \lim_{z\to\lambda} \phi_j(z)/\psi_j(z)e_j = 0$ as $\psi_j(\lambda) \neq 0$ which shows that $\lambda \in \text{eip}_\Omega(\Sigma_A) = \text{eip}_\Omega(A) \subset \sigma_\Omega(A)$. Thus, in either case, $\partial_j(\lambda) \neq 0 \Longrightarrow \lambda \in \sigma_\Omega(A)$. This shows that $\text{supp}(\partial_j) \subset \sigma_\Omega(A)$ for $j = 1 : r$. Hence the factorization of ϕ_j follows from Theorem 2.22. This completes the proof. □

REMARK 3.19. The diagonal matrix $\Sigma_A(z)$ is called the global Smith-McMillan form of $A(z)$ on Ω. The functions $\phi_1, \ldots, \phi_r$ are called the invariant zero functions (or factors) of $A(z)$ and the functions $\psi_1, \ldots, \psi_r$ are called the invariant pole functions (or factors) of $A(z)$ on Ω. Define

$$\phi_A(z) := \prod_{j=1}^{r} \phi_j(z) \text{ and } \psi_A(z) := \prod_{j=1}^{r} \psi_j(z).$$

The function $\phi_A(z)$ is called the zero function of $A(z)$ and the function $\psi_A(z)$ is called the pole function of $A(z)$ on Ω. It follows that

$$\wp_\Omega(A) = \{\mu \in \Omega : \psi_A(\mu) = 0\} = \sigma_\Omega(\psi_A).$$

It also follows from the proof of Theorem 3.18 that

$$\sigma_\Omega(\phi_A) = \{\lambda \in \Omega : \phi_A(\lambda) = 0\} \subset \sigma_\Omega(A).$$

In fact, the equality holds, that is, $\sigma_\Omega(A) = \sigma_\Omega(\phi_A)$.

THEOREM 3.20. *Let $A \in \mathbb{M}(\Omega)^{m\times n}$. Let ϕ_A and ψ_A be the zero and pole functions of A on Ω. Then $\sigma_\Omega(A) = \sigma_\Omega(\phi_A) = \text{eig}_\Omega(A) \cup \text{eip}_\Omega(A)$ and $\wp_\Omega(A) = \sigma_\Omega(\psi_A)$. Further,*

$$\text{eig}_\Omega(A) = \{\lambda \in \Omega \; : \; \phi_A(\lambda) = 0 \text{ and } \psi_A(\lambda) \neq 0\},$$
$$\text{eip}_\Omega(A) = \{\lambda \in \Omega : \; \phi_A(\lambda) = 0 \text{ and } \psi_A(\lambda) = 0\}.$$

PROOF. Let the Smith-McMillan form $\Sigma_A(z)$ of $A(z)$ be as in Theorem 3.18. Since $A(z) \sim_\Omega \Sigma_A(z)$, it follows that $\sigma_\Omega(A) = \sigma_\Omega(\Sigma_A)$ and $\wp_\Omega(A) = \wp_\Omega(\Sigma_A)$. Next, observe that $\lambda \in \wp_\Omega(\Sigma_A) \Longleftrightarrow \psi_i(\lambda) = 0$ for some $i \Longleftrightarrow \psi_A(\lambda) = 0$. This shows that $\wp_\Omega(A) = \sigma_\Omega(\psi_A)$.

It follows that $\lambda \in \text{eig}_\Omega(\Sigma_A) \Longleftrightarrow \phi_j(\lambda) = 0$ for some j and $\psi_i(\lambda) \neq 0$ for $i = 1 : r \Longleftrightarrow \phi_A(\lambda) = 0$ and $\psi_A(\lambda) \neq 0$. This shows that

$$\text{eig}_\Omega(A) = \text{eig}_\Omega(\Sigma_A) = \{\lambda \in \Omega : \phi_A(\lambda) = 0 \text{ and } \psi_A(\lambda) \neq 0\}.$$

Now suppose that $\phi_A(\lambda) = 0$ and $\psi_A(\lambda) = 0$. Then $\phi_j(\lambda) = 0$ and $\psi_j(\lambda) \neq 0$ for some j. Note that $\lambda \in \wp_\Omega(\Sigma_A)$. Consider $v(z) := e_j$ for $z \in \Omega$, where e_j is the j-th column of I_n. Then $v \in \mathbb{H}(\Omega)^n$ and we have $v(\lambda) \neq 0, v \notin N(\Sigma_A)$ and

$$\lim_{z\to\lambda} \Sigma_A(z)v(z) = \lim_{z\to\lambda} \frac{\phi_j(z)}{\psi_j(z)} e_j = 0$$

which shows that $\lambda \in \text{eip}_\Omega(\Sigma_A)$.

Conversely, let $\mu \in \mathrm{eip}(\Sigma_A)$. Then there exists $v \in \mathbb{H}(\Omega)^n$ such that $v(\mu) \neq 0$, $v \notin N(\Sigma_A)$ and $\lim_{z\to\mu} \Sigma_A(z)v(z) = 0$. Let $v(z)$ be given by $v(z) := [v_1(z) \cdots v_n(z)]^\top$ for $z \in \Omega$. Then $\lim_{z\to\mu} \Sigma_A(z)v(z) = 0 \Longrightarrow \lim_{z\to\mu} \phi_j(z)v_j(z)/\psi_j(z) = 0$ for $j = 1 : r$. As $v(\mu) \neq 0$ and $v \notin N(\Sigma_A)$, we have $v_i(\mu) \neq 0$ for some $i \leq r$. This shows that $\lim_{z\to\mu} \phi_i(z)/\psi_i(z) = \lim_{z\to\mu} \frac{1}{v_i(\mu)} \lim_{z\to\mu} \phi_i(z)v_i(z)/\psi_i(z) = 0$. Since ϕ_i and ψ_i are coprime, we have $\psi_i(\mu) \neq 0$ and $\phi_i(\mu) = 0$. Hence we have $\phi_A(\mu) = 0$ and $\psi_A(\mu) = 0$. This proves that $\mathrm{eip}_\Omega(\Sigma_A) = \{\lambda \in \Omega : \phi_A(\lambda) = 0 \text{ and } \psi_A(\lambda) = 0\}$.

Finally, $\sigma_\Omega(A) = \sigma_\Omega(\Sigma_A) = \sigma_\Omega(\phi_A)$ follows from the fact that if $\phi_A(\lambda) = 0$ then either $\psi_A(\lambda) \neq 0$ or $\psi_A(\lambda) = 0$. □

Notice that the spectrum $\sigma_\Omega(A)$ of a meromorphic matrix $A \in \mathbb{M}(\Omega)^{m\times n}$ consists of eigenvalues and eigenpoles. If $\lambda \in \sigma_\Omega(A)$ then λ is called a ***zero*** of $A(z)$. Thus, a zero of $A(z)$ is either an eigenvalue or an eigenpole of $A(z)$. We now define partial multiplicities of zeros and poles of $A(z)$. Let

$$\Sigma_A(z) = \phi_1/\psi_1 \oplus \cdots \oplus \phi_r/\psi_r \oplus 0_{m-r\times n-r}$$

be the Smith-McMillan form of $A(z)$. Recall that ϕ_i divides ϕ_{i+1} and ψ_{i+1} divides ψ_i for $i = 1 : r-1$. Let $\phi_A(z)$ and $\psi_A(z)$, respectively, be the zero function and pole function of $A(z)$ on Ω. Let $\lambda \in \Omega$. Define $\nu_j(\lambda) := o_\lambda(\phi_j)$ and $\kappa_j(\lambda) := o_\lambda(\psi_j)$ for $j = 1 : r$. Then $\nu_1(\lambda) \leq \cdots \leq \nu_r(\lambda)$ and $\kappa_1(\mu) \geq \cdots \geq \kappa_r(\mu)$ are non-negative integers such that

$$\phi_j(z) \sim_\lambda (z-\lambda)^{\nu_j(\lambda)} \text{ and } \psi_j(z) \sim_\lambda (z-\mu)^{\kappa_j(\mu)} \text{ for } j = 1 : r.$$

Define $\tau_j(\lambda) := o_\lambda(\phi_j/\psi_j)$ for $j = 1 : r$. Then it follows that $\tau_j(\lambda) = \nu_j(\lambda) - \kappa_j(\lambda) \in \mathbb{Z}$ for $j = 1 : r$. Since ϕ_j and ψ_j are coprime, either $\tau_j(\lambda) = \nu_j(\lambda)$ or $\tau_j(\lambda) = -\kappa_j(\lambda)$ for $j = 1 : r$. Further, we have $\tau_1(\lambda) \leq \cdots \leq \tau_r(\lambda)$ and

$$\phi_j(z)/\psi_j(z) \sim_\lambda (z-\lambda)^{\tau_j(\lambda)} \text{ for } j = 1 : r.$$

Hence it follows that the following statements are equivalent.

(a) $\lambda \in \mathrm{eig}_\Omega(A)$.
(b) $\phi_A(\lambda) = 0$ and $\psi_A(\lambda) \neq 0$.
(c) $\nu_r(\lambda) \neq 0$ and $\kappa_1(\lambda) = 0$.
(d) $\tau_j(\lambda) = \nu_j(\lambda)$ for $j = 1 : r$.

Let $\lambda \in \mathrm{eig}_\Omega(A)$. Then $\phi_A(\lambda) = 0$ and hence there exists $0 \leq \ell < r$ such that

$$0 = \nu_1(\lambda) = \cdots = \nu_\ell(\lambda) < \nu_{\ell+1}(\lambda) \leq \cdots \leq \nu_r(\lambda). \tag{3.2}$$

The positive integers $\nu_{\ell+1}(\lambda), \ldots, \nu_r(\lambda)$ are called the ***partial multiplicities*** of λ. Further, $\nu(\lambda) := \nu_1(\lambda) + \cdots + \nu_r(\lambda) = \nu_{\ell+1}(\lambda) + \cdots + \nu_r(\lambda)$ is called the ***algebraic multiplicity*** of λ and $\nu_r(\lambda)$ is called the ***ascent*** of λ. Note that the algebraic multiplicity $\nu(\lambda)$ of an eigenvalue λ of A is the multiplicity of λ as a zero of $\phi_A(z)$.

Next, it follows that the following statements are equivalent for poles of $A(z)$ which are not zeros of $A(z)$.

(a) $\lambda \in \wp_\Omega(A) \setminus \mathrm{eip}_\Omega(A)$.
(b) $\phi_A(\lambda) \neq 0$ and $\psi_A(\lambda) = 0$.
(c) $\nu_r(\lambda) = 0$ and $\kappa_1(\lambda) \neq 0$.
(d) $\tau_j(\lambda) = -\kappa_j(\lambda)$ for $j = 1 : r$.

Let $\lambda \in \wp_\Omega(A)$. Then $\psi_A(\lambda) = 0$ and hence there exists $1 \leq \ell \leq r$ such that

$$\kappa_1(\lambda) \geq \cdots \geq \kappa_\ell(\lambda) > \kappa_{\ell+1}(\lambda) = \cdots = \kappa_r(\lambda) = 0. \tag{3.3}$$

The positive integers $\kappa_1(\lambda), \ldots, \kappa_\ell(\lambda)$ are called the ***partial multiplicities*** of the pole λ of $A(z)$ and $\kappa_1(\lambda)$ is the order of the pole λ. Further,

$$\kappa(\lambda) := \kappa_1(\lambda) + \cdots + \kappa_\ell(\lambda)$$

is the total multiplicity of λ as a zero of the pole function $\psi_A(z)$.
Finally, it follows that the following statements are equivalent.

(a) $\lambda \in \mathrm{eip}_\Omega(A)$.
(b) $\phi_A(\lambda) = 0$ and $\psi_A(\lambda) = 0$.
(c) $\nu_r(\lambda) \neq 0$ and $\kappa_1(\lambda) \neq 0$.
(d) $\tau_1(\lambda) = -\kappa_1(\lambda) < 0$ and $\tau_r(\lambda) = \nu_r(\lambda) > 0$.

Let $\lambda \in \mathrm{eip}_\Omega(A)$. Then $\phi_A(\lambda) = 0$ and $\psi_A(\lambda) = 0$. Hence there exist positive integers ℓ and p such that $1 \leq \ell < p \leq r$ such that

$$\tau_1(\lambda) \leq \cdots \leq \tau_\ell(\lambda) < 0 < \tau_p(\lambda) \leq \cdots \leq \tau_r(\lambda). \tag{3.4}$$

The nonzero integers $\tau_1(\lambda), \ldots, \tau_\ell(\lambda), \tau_p(\lambda), \ldots, \tau_r(\lambda)$ are called the ***partial multiplicities*** of the eigenpole λ. Further,

$$\kappa(\lambda) := -(\tau_1(\lambda) + \cdots + \tau_\ell(\lambda)) = \kappa_1(\lambda) + \cdots + \kappa_\ell(\lambda)$$

is the total multiplicity of λ as a zero of $\psi_A(z)$ and

$$\nu(\lambda) := \tau_p(\lambda) + \cdots + \tau_r(\lambda) = \nu_p(\lambda) + \cdots + \nu_r(\lambda)$$

is the total multiplicity of λ as a zero of $\phi_A(z)$. Obviously, $-\tau_1(\lambda)$ is the order of λ as a pole of $A(z)$ and $\tau_r(\lambda)$ is the ascent of λ as a zero of $A(z)$.

In view of the observations above, we now define three indices of $A(z)$ at $\lambda \in \Omega$.

DEFINITION 3.21. Let $A \in \mathbb{M}(\Omega)^{m\times n}$. Let the Smith-McMillan form $\Sigma_A(z)$ of $A(z)$ be given by $\Sigma_A(z) = \phi_1/\psi_1 \oplus \cdots \oplus \phi_r/\psi_r \oplus 0_{m-r\times n-r}$.

(a) **Zero index:** Let $\lambda \in \mathrm{eig}_\Omega(A)$. A tuple $\mathrm{ind}_e(\lambda, A) := (\nu_1(\lambda), \ldots, \nu_r(\lambda)) \in \mathbb{Z}_+^r$ with $\nu_1(\lambda) \leq \cdots \leq \nu_r(\lambda)$ is called the zero index of $A(z)$ at λ if $\nu_j(\lambda) = o_\lambda(\phi_j)$ for $j = 1 : r$.
(b) **Pole index:** Let $\mu \in \wp_\Omega(A)$. A tuple $\mathrm{ind}_p(\mu, A) := (\kappa_1(\mu), \ldots, \kappa_r(\mu)) \in \mathbb{Z}_+^r$ with $\kappa_1(\mu) \geq \cdots \geq \kappa_r(\mu)$ is called the pole index of $A(z)$ at μ if $\kappa_j(\mu) = o_\mu(\psi_j)$ for $j = 1 : r$.
(c) **Pole-zero index:** Let $\mu \in \sigma_\Omega(A) \cup \wp_\Omega(A)$. Then a tuple

$$\mathrm{ind}(\mu, A) := (\tau_1(\mu), \ldots, \tau_r(\mu)) \in \mathbb{Z}^r$$

with $\tau_1(\mu) \leq \cdots \leq \tau_r(\mu)$ is said to be the pole-zero index of $A(z)$ at μ if $\tau_j(\mu) = o_\mu(\phi_j/\psi_j)$ for $j = 1 : r$.

REMARK 3.22.

(a) Observe that the pole-zero index $\mathrm{ind}(\lambda, A)$ of $A(z)$ at λ is determined by the zero index $\mathrm{ind}_e(\lambda, A)$ and the pole index $\mathrm{ind}_p(\lambda, A)$ of $A(z)$ at λ. Indeed, $\mathrm{ind}(\lambda, A) = \mathrm{ind}_e(\lambda, A) - \mathrm{ind}_p(\lambda, A) = (\nu_1(\lambda) - \kappa_1(\lambda), \ldots, \nu_r(\lambda) - \kappa_r(\lambda))$.

(b) Note that if $\lambda \in \mathrm{eip}_\Omega(A)$ then λ is a zero as well as a pole of $A(z)$. Hence $A(z)$ has two indices at λ, namely, the zero index $\mathrm{ind}_e(\lambda, A)$ and the pole index $\mathrm{ind}_p(\lambda, A)$ which separately provide information about zero structure and pole structure of $A(z)$ at λ. On the other hand, the pole-zero index $\mathrm{ind}(\lambda, A)$ provides information about pole-zero structure of $A(z)$ at the eigenpole λ.

(c) Let $\mathrm{ind}(\lambda, A) := (\tau_1(\lambda), \ldots, \tau_r(\lambda)) \in \mathbb{Z}^r$ be the pole-zero index of $A(z)$ at $\lambda \in \Omega$. Let $\Sigma_A(z) := \phi_1(z)/\psi_1(z) \oplus \cdots \oplus \phi_r(z)/\psi_r(z) \oplus 0_{m-r,n-r}$ be the Smith-McMillan form of $A(z)$. Then $\phi_j(z)/\psi_j(z) \sim_\lambda (z-\lambda)^{\tau_j(\lambda)}$ for $j = 1 : r$. Consequently, we have

$$A(z) \sim_\Omega \Sigma_A(z) \sim_\lambda (z-\lambda)^{\tau_1(\lambda)} \oplus \cdots \oplus (z-\lambda)^{\tau_r(\lambda)} \oplus 0_{m-r\times n-r} =: \Sigma_\lambda(z).$$

The diagonal matrix $\Sigma_\lambda(z)$ is called the local Smith form of $A(z)$ at λ.

4. Conclusion

Canonical forms are powerful tools for spectral analysis of matrices. The Smith form of a matrix polynomial and the Smith-McMillan form of a rational matrix are well known canonical forms which are used extensively in theory and applications. Unlike a matrix polynomial (resp., rational matrix), a holomorphic (resp., meromorphic) matrix can have an infinite number of eigenvalues (resp., zeros and poles). Hence it is natural to approach canonical forms of holomorphic and meromorphic matrices via tools from analysis so as to deal with infinite number of eigenvalues and poles. We have described such a framework by utilizing Weierstrass product and Weierstrass factorization of holomorphic functions and proved, among other things, the global Smith form (resp., Smith-McMillan form) of a holomorphic (resp., meromorphic) matrix which is akin to the Smith form (resp., Smith-McMillan form) of a matrix polynomial (resp., rational matrix). We have also presented multiplicity indices of eigenvalues (resp., zeros and poles) of holomorphic (resp., meromorphic) matrices which provide structural indices of eigenvalues (resp., zeros and poles) and local structures of holomorphic (resp., meromorphic) matrices.

References

[1] R. Alam and N. Behera, *Linearizations for rational matrix functions and Rosenbrock system polynomials*, SIAM J. Matrix Anal. Appl. **37** (2016), no. 1, 354–380, DOI 10.1137/15M1008622. MR3477318

[2] J. A. Ball, I. Gohberg, and L. Rodman, *Interpolation of rational matrix functions*, Operator Theory: Advances and Applications, vol. 45, Birkhäuser Verlag, Basel, 1990, DOI 10.1007/978-3-0348-7709-1. MR1083145

[3] T. Betcke, N. J. Higham, V. Mehrmann, C. Schröder, and F. Tisseur, *NLEVP: a collection of nonlinear eigenvalue problems*, ACM Trans. Math. Software **39** (2013), no. 2, Art. 7, 28, DOI 10.1145/2427023.2427024. MR3031626

[4] W.-J. Beyn, *An integral method for solving nonlinear eigenvalue problems*, Linear Algebra Appl. **436** (2012), no. 10, 3839–3863, DOI 10.1016/j.laa.2011.03.030. MR2914550

[5] R. K. Das and R. Alam, *Affine spaces of strong linearizations for rational matrices and the recovery of eigenvectors and minimal bases*, Linear Algebra Appl. **569** (2019), 335–368, DOI 10.1016/j.laa.2019.02.001. MR3908736

[6] B. N. Datta, *Numerical methods for linear control systems: Design and analysis*, Elsevier Academic Press, San Diego, CA, 2004. With 1 CD-ROM (Windows, Macintosh and UNIX). MR2260346

[7] M. G. Frost and C. Storey, *A note on the controllability of linear constant delay-differential systems*, Internat. J. Control **28** (1978), no. 5, 673–679, DOI 10.1080/00207177808922488. MR517166

[8] M. G. Frost and C. Storey, *Further remarks on the controllability of linear constant delay-differential systems*, Internat. J. Control **30** (1979), no. 5, 863–870, DOI 10.1080/00207177908922818. MR552311

[9] T. W. Gamelin, *Complex analysis*, Undergraduate Texts in Mathematics, Springer-Verlag, New York, 2001, DOI 10.1007/978-0-387-21607-2. MR1830078

[10] F. R. Gantmacher, *The Theory of Matrices*. Vols. 1 & 2, Chelsea Publishing Co., New York, 1959.

[11] I. Gohberg, S. Goldberg, and M. A. Kaashoek, *Classes of linear operators. Vol. I*, Operator Theory: Advances and Applications, vol. 49, Birkhäuser Verlag, Basel, 1990, DOI 10.1007/978-3-0348-7509-7. MR1130394

[12] I. Gohberg, M. A. Kaashoek, and F. van Schagen, *On the local theory of regular analytic matrix functions*, Linear Algebra Appl. **182** (1993), 9–25, DOI 10.1016/0024-3795(93)90488-A. MR1207071

[13] I. Gohberg, P. Lancaster, and L. Rodman, *Matrix polynomials*, Computer Science and Applied Mathematics, Academic Press, Inc. [Harcourt Brace Jovanovich, Publishers], New York-London, 1982. MR662418

[14] I. Gohberg and L. Rodman, *Analytic matrix functions with prescribed local data*, J. Analyse Math. **40** (1981), 90–128 (1982), DOI 10.1007/BF02790157. MR659786

[15] I. C. Gohberg, M. A. Kaashoek, and D. C. Lay, *Equivalence, linearization, and decomposition of holomorphic operator functions*, J. Functional Analysis **28** (1978), no. 1, 102–144, DOI 10.1016/0022-1236(78)90081-2. MR482317

[16] I. C. Gohberg and E. I. Sigal, *An operator generalization of the logarithmic residue theorem and Rouché's theorem* (Russian), Mat. Sb. (N.S.) **84(126)** (1971), 607–629. MR313856

[17] S. Güttel and F. Tisseur, *The nonlinear eigenvalue problem*, Acta Numer. **26** (2017), 1–94, DOI 10.1017/S0962492917000034. MR3653850

[18] R. A. Horn and C. R. Johnson, *Matrix analysis*, 2nd ed., Cambridge University Press, Cambridge, 2013. MR2978290

[19] T. Kailath, *Linear systems*, Prentice Hall Information and System Sciences Series, Prentice-Hall, Inc., Englewood Cliffs, NJ, 1980. MR569473

[20] V. Kozlov and V. Maz'ya, *Differential equations with operator coefficients*, Springer, 1999.

[21] P. Kunkel and V. Mehrmann, *Differential-algebraic equations: Analysis and numerical solution*, EMS Textbooks in Mathematics, European Mathematical Society (EMS), Zürich, 2006, DOI 10.4171/017. MR2225970

[22] J. Leiterer, *Local and global equivalence of meromorphic operator functions. II*, Math. Nachr. **84** (1978), 145–170, DOI 10.1002/mana.19780840107. MR518118

[23] V. Mehrmann and C. Schröder, *Nonlinear eigenvalue and frequency response problems in industrial practice*, J. Math. Ind. **1** (2011), Art. 7, 18, DOI 10.1186/2190-5983-1-7. MR2853230

[24] V. Mehrmann and H. Voss, *Nonlinear eigenvalue problems: a challenge for modern eigenvalue methods*, GAMM Mitt. Ges. Angew. Math. Mech. **27** (2004), no. 2, 121–152 (2005), DOI 10.1002/gamm.201490007. MR2124762

[25] R. Mennicken and M. Möller, *Non-self-adjoint boundary eigenvalue problems*, North-Holland Mathematics Studies, vol. 192, North-Holland Publishing Co., Amsterdam, 2003. MR1995773

[26] W. Michiels and S.-I. Niculescu, *Stability and stabilization of time-delay systems: An eigenvalue-based approach*, Advances in Design and Control, vol. 12, Society for Industrial and Applied Mathematics (SIAM), Philadelphia, PA, 2007, DOI 10.1137/1.9780898718645. MR2384531

[27] R. Remmert, *Classical topics in complex function theory*, Graduate Texts in Mathematics, vol. 172, Springer-Verlag, New York, 1998. Translated from the German by Leslie Kay, DOI 10.1007/978-1-4757-2956-6. MR1483074

[28] H. H. Rosenbrock, *State-space and multivariable theory*, John Wiley & Sons, Inc. [Wiley Interscience Division], New York, 1970. MR325201

[29] A. I. G. Vardulakis, *Linear multivariable control: Algebraic analysis and synthesis methods*, John Wiley & Sons, Ltd., Chichester, 1991. MR1104222

[30] L. Weiss, *On the controllability of delay-differential systems*, SIAM J. Control **5** (1967), 575–587. MR237918

Department of Mathematics, IIT Guwahati, Guwahati - 781039, India
Email address: rafik@iitg.ac.in, rafikul68@gmail.com

Department of Mathematics, IIT Guwahati, Guwahati - 781039, India
Email address: jibrail@iitg.ac.in, zibrailali@gmail.com

Contemporary Mathematics
Volume **833**, 2026
https://doi.org/10.1090/conm/833/16660

Stress centrality measure of a vertex

S. Arumugam, K. Arathi Bhat, Manjunatha Prasad Karantha, and Raksha Poojary

ABSTRACT. Centrality measures are the tools used in most of the network analysis to rank the vertices of the network. Depending on different perspectives, many important centrality measures are defined and used for network analysis. Stress is one such important centrality measure of graphs applicable to the study of social and biological networks. In this chapter, we explore the theory describing various properties of the stress centrality measure. Since the adjacency matrix plays an important role in finding the number of shortest paths between a given pair of vertices, an alternate expression to find the stress of a vertex using the adjacency matrix of the graph is presented in this chapter.

1. Introduction

A network is a structure consisting of a collection of points called vertices, and a set of connections between these vertices, called edges. Analysis of network helps us in decision making, as it helps us in understanding different characteristics of the network. A visual representation of data gives a better insight into the network. If the network size is very small, then visually one can analyze the network data and obtain vital information, which is otherwise invisible from the raw data. However, if the network is very large, visualizing its properties becomes quite challenging. In such cases, some graph parameters describing the properties become useful. Relevant parameters can be derived from specific algorithms, which has significant role in network analysis. In an analysis of the network, consider the following problems.

(i) How a particular vertex/edge is better or not in comparison with another vertex/edge? This information helps in ranking the vertices/edges in the given network.
(ii) Find a value that describes certain characteristics of the network which remain invariant under graph isomorphism. This information helps in comparing the graphs with reference to the characteristics under discussion.

For example, the vertices with the maximum degree could be considered as central vertices and $\deg_G$ is a real valued function on the vertex set, which helps in ranking

2020 *Mathematics Subject Classification.* Primary 05C12, 05C50, 05C90.
Key words and phrases. Centrality measure, ranking of vertices, adjacency matrix, cut vertex.

the vertices in a graph G. Further, $\max_{v \in V} \deg_G(v)$ assigns a real value to the graph G which describes the maximum of number of adjacency for any vertex in G.

A vertex centrality measure assigns a real number to each vertex of a graph and it quantifies the importance or criticality of a vertex from a particular perspective. Different centrality measures describe the importance of a vertex from different perspectives. A few examples of vertex centrality measures are betweenness, closeness, degree, eigenvector centrality and stress.

Centrality measures in a graph are used to rank the vertices, and the vertices with higher rank are considered to be more important than the others. For further results on centrality measures, readers are referred to [**4**, **5**, **7**] and [**8**]. Applications of centrality measures include identification of the most influential person in a social network, proteins which play a significant role in a biological process, and key infrastructure vertices in an urban network or internet etc. For a brief survey of centrality measures with emphasis on applications in the study of biological networks, we refer to [**3**].

Let $G = (V(G), E(G))$ be a finite simple graph with vertex set $V(G) = \{v_1, v_2, \ldots, v_n\}$ and edge set $E(G) = \{e_1, e_2, \ldots, e_m\}$. If two vertices v_i and v_j are adjacent, we write $v_i \sim v_j$, and the edge between them is denoted by e_{ij}. The degree of a vertex v in the graph G is denoted by $\deg_G(v)$ or simply by $\deg(v)$ when the graph under discussion is clear.

Let $P = (u = v_0, v_1, \ldots, v_k = v)$ be a u-v path of length k in G with *origin* $u = v_0$ and *terminus* $v = v_k$. The vertices v_i, $1 \leq i \leq k-1$ are called the *internal* vertices of the path P. The length of a shortest u-v path, denoted by $d(u, v)$ is called the *distance* between u and v. The *diameter* of G, denoted by $\operatorname{diam}(G)$, is given by $\max\{d(u,v) : u, v \in V\}$. For $v \in V$, $e(v)$ denotes the eccentricity of the vertex v and is given by $\max\{\mathrm{d}(u,v) : u \in V\}$. The vertices with minimum eccentricity are called '*central vertices*' (with reference to eccentricity) of the graph and the eccentricity of any vertex at center is called the '*radius of the graph*'. A graph G is called a '*block graph*' if each block (a maximal nonseparable subgraph) of G forms a complete graph.

In order to compute '*stress measure*' in a graph G different classes of shortest paths are used. For convenience, the notations given in the following list are used to represent those classes of distinct shortest paths:

- $\mathbb{P}_G$ (or simply $\mathbb{P}$) - the collection of all distinct shortest paths,
- $\mathbb{P}_G^k$ (or simply $\mathbb{P}^k$) - the set of all distinct shortest paths of length k,
- $\mathbb{P}_G(u, w)$ (or simply $\mathbb{P}(u, w)$) - the set of all shortest paths between the vertices u and w,
- $\mathbb{P}_G^k(u, w)$ (or simply $\mathbb{P}^k(u, w)$) - the set of all shortest paths of length k between the vertices u and w,
- $\mathbb{P}_v^0(u, w)$ - the set of all distinct shortest paths between u and w with v among their internal vertices,
- $\mathcal{P}_G^0(v)$ (or $\mathcal{P}^0(v)$) - the set of all distinct shortest paths with v among their internal vertices,
- $\mathcal{P}_G^k(v)$ or $\mathcal{P}^k(v)$ $(k \geq 2)$ - the set of all distinct paths of length k having v among their internal vertices,
- $\mathcal{S}_G(v)$ (or simply $\mathcal{S}(v)$) - the set of all distinct shortest paths which originates (or terminates) at v.

For graph theoretic terminology we refer to Chartrand and Lesniak [**9**].

In the present chapter, we focus on the study of stress, which is a vertex centrality measure studied to some extent in [**1**, **2**] and [**6**].

DEFINITION 1.1 (Stress). Given a graph G, the *stress measure* st_G is a real valued function on the set of vertices $V(G)$, where $\mathrm{st}_G(v)$ represents the number of distinct shortest paths in the graph G with v among the internal vertices of those paths. In other words,

$$\mathrm{st}_G(v) = \#\, \mathcal{P}^0(v) \quad \text{for } v \in V(G) \tag{1.1}$$

$$= \frac{1}{2} \sum_{(u,w)\in V(G)\times V(G)} \#\, \mathbb{P}^0_v(u,w). \tag{1.2}$$

Further, *stress of a graph* G, denoted by $\mathrm{st}(G)$, is given by

$$\mathrm{st}(G) = \sum_{v\in V(G)} \mathrm{st}_G(v). \tag{1.3}$$

The 'stress sequence' is a notion which describes the distribution of stress measure in a graph, which helps in comparing the vertices by their stress measure like in the case of 'degree' with reference to 'degree sequence'. The definition of stress sequence of vertices is given in the following.

DEFINITION 1.2. Given the ordering $(v_{i1}, v_{i2}, \cdots, v_{in})$ of $V(G) = \{v_1, v_2, \ldots, v_n\}$ such that $\mathrm{st}(v_{i1}) \geq \mathrm{st}(v_{i2}) \geq \cdots \geq \mathrm{st}(v_{in})$, the *stress sequence* of vertices in the graph G is defined by

$$\left(\mathrm{st}(v_{i1}), \mathrm{st}(v_{i2}), \cdots, \mathrm{st}(v_{in})\right). \tag{1.4}$$

A graph is termed as a 'stress regular graph' if its sequence of stress measures for vertices forms a constant sequence. This definition is presented as follows.

DEFINITION 1.3. A graph G is said to be a k*-stress regular* if $\mathrm{st}_G(v) = k$ for all $v \in V(G)$. In other words, the stress sequence of vertices is with constant value equal to k.

The results appearing in this chapter are taken from the research article [**10**, **12**]. The stress measure of vertices in some modified wheel graphs, namely gear graph, helm graph, sunflower graph, friendship graph, flower graph, cocktail party graph, crown graph are discussed in [**11**].

When examining a vertex v within the subgraph H of a graph G, it is not guaranteed that $\mathrm{st}_H(v)$ is anyway comparable with $\mathrm{st}_G(v)$. The following two examples of graphs illustrates the same.

EXAMPLE 1.4. Consider a subgraph H of a graph G, formed by eliminating a vertex v_1 as illustrated in Figure 1.

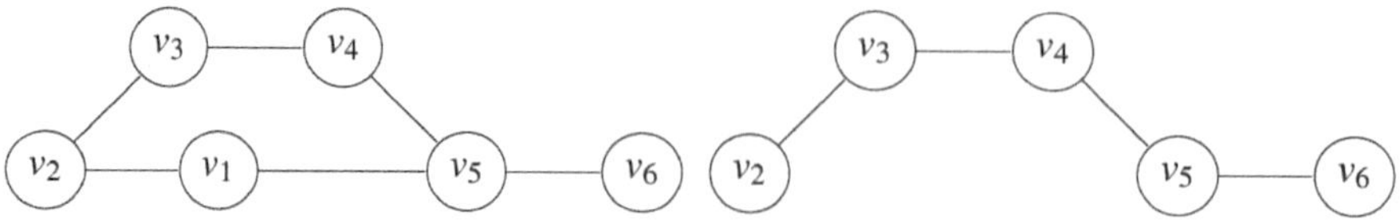

FIGURE 1. Graph G and $H = G - v_1$

In case of v_5, a common vertex from G and H, by observation, we have that

$$\mathcal{P}^0_G(v_5) = \{v_1\text{-}v_5\text{-}v_6,\ v_1\text{-}v_5\text{-}v_4,\ v_2\text{-}v_1\text{-}v_5\text{-}v_6,\ v_3\text{-}v_4\text{-}v_5\text{-}v_6,\ v_4\text{-}v_5\text{-}v_6\},$$

where as

$$\mathcal{P}^0_H(v_5) = \{v_2\text{-}v_3\text{-}v_4\text{-}v_5\text{-}v_6,\ v_3\text{-}v_4\text{-}v_5\text{-}v_6,\ v_4\text{-}v_5\text{-}v_6\}.$$

Now by Definition 1.1, $\mathrm{st}_G(v_5) = 5$ and $\mathrm{st}_H(v_5) = 3$. Hence $\mathrm{st}_H(v_5) < \mathrm{st}_G(v_5)$.

Now, for another common vertex v_3, observe that $\mathcal{P}^0_G(v_3) = \{v_2\text{-}v_3\text{-}v_4\}$ and $\mathcal{P}^0_H(v_3) = \{v_2\text{-}v_3\text{-}v_4, v_2\text{-}v_3\text{-}v_4\text{-}v_5,\ v_2\text{-}v_3\text{-}v_4\text{-}v_5\text{-}v_6\}$. Therefore, $\mathrm{st}_G(v_3) = 1$, $\mathrm{st}_H(v_3) = 3$, and $\mathrm{st}_H(v_3) > \mathrm{st}_G(v_3)$.

The next example, demonstrates that the stress of a given graph and the stress of an induced subgraph may not always be directly comparable.

EXAMPLE 1.5. Consider the graph G and its subgraph H induced by $\{v_2, v_3, v_4, v_5, v_6\}$ as shown in Figure 1 of Example 1.4. The stress sequence of vertices of G is given by

$$(\mathrm{st}_G(v_5), \mathrm{st}_G(v_1), \mathrm{st}_G(v_4), \mathrm{st}_G(v_2), \mathrm{st}_G(v_3), \mathrm{st}_G(v_6)) = (5, 2, 2, 1, 1, 0)$$

and the stress sequence of the vertices in H is given by

$$(\mathrm{st}_H(v_4), \mathrm{st}_H(v_3), \mathrm{st}_H(v_5), \mathrm{st}_H(v_2), \mathrm{st}_H(v_6)) = (4, 3, 3, 0, 0).$$

So, by Definition 1.3, $\mathrm{st}(G) = 11$ and $\mathrm{st}(H) = 10$ and hence $\mathrm{st}(G) > \mathrm{st}(H)$.

Now consider a graph $\hat{G}$ and the induced subgraph $\hat{H}$ of $\hat{G}$ obtained by removing an vertex v_1 from G, as shown in Figure 2.

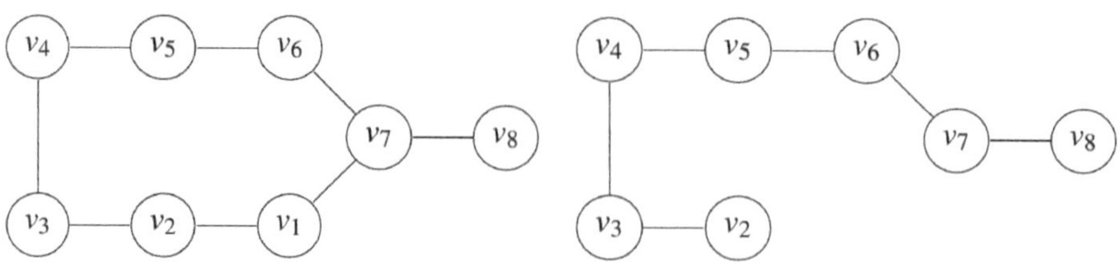

FIGURE 2. Graph $\hat{G}$ and $\hat{H} = \hat{G} - v_1$

The stress sequence of vertices of $\hat{G}$ is given by

$$(\mathrm{st}_{\hat{G}}(v_7), \mathrm{st}_{\hat{G}}(v_1), \mathrm{st}_{\hat{G}}(v_6), \mathrm{st}_{\hat{G}}(v_2), \mathrm{st}_{\hat{G}}(v_5), \mathrm{st}_{\hat{G}}(v_3), \mathrm{st}_{\hat{G}}(v_4), \mathrm{st}_{\hat{G}}(v_8)) \\ = (9, 5, 5, 4, 4, 3, 3, 0)$$

and the stress sequence of vertices in $\hat{H}$ is given by

$$(\mathrm{st}_{\hat{H}}(v_5), \mathrm{st}_{\hat{H}}(v_4), \mathrm{st}_{\hat{H}}(v_6), \mathrm{st}_{\hat{H}}(v_3), \mathrm{st}_{\hat{H}}(v_7), \mathrm{st}_{\hat{H}}(v_2), \mathrm{st}_{\hat{H}}(v_8)) = (9, 8, 8, 5, 5, 0, 0).$$

From the above stress sequences, $\mathrm{st}(\hat{G}) = 33$, $\mathrm{st}(\hat{H}) = 35$ and therefore $\mathrm{st}(\hat{G}) < \mathrm{st}(\hat{H})$.

However, if every shortest path in H is a shortest path in G too, then the stress measure of common vertices are comparable. In fact in such a case, $\mathrm{st}(H) \leq \mathrm{st}(G)$. The following lemma, describing the same, is immediate from Definition 1.1.

LEMMA 1.6. *If H is a subgraph of G such that every shortest path in H is also a shortest path in G, then*

$$\mathrm{st}_H(v) \leq \mathrm{st}_G(v), \quad v \in V(H).$$

Further, $\mathrm{st}(H) \leq \mathrm{st}(G)$. *In particular, for $H = P(u, w)$, a shortest path between $u, w \in V(G)$,*

$$\mathrm{st}_{P(u,w)}(v) \leq \mathrm{st}_G(v), \quad v \in V(P(u, w)).$$

Observe that 'stress' is a graph parameter. The following lemma can be readily derived from the property of graph isomorphism.

LEMMA 1.7. *If f is a graph isomorphism from a graph G onto a graph H, then*

$$\mathrm{st}_G(v) = \mathrm{st}_H(f(v)) \quad \textit{for all } v \in V(G).$$

PROOF. Consider $f(v) = w$, where $v \in V(G)$ and $w \in V(H)$. Now from adjacency preserving property of a graph isomorphism, note that any path P in the graph G is a shortest path with v as an internal vertex if and only if $f(P)$ is a shortest path with w as an internal vertex. Therefore, there is an one-one correspondence between $\mathcal{P}^0_G(v)$ and $\mathcal{P}^0_H(f(v))$ and hence $\mathrm{st}_G(v) = \mathrm{st}_H(w)$. □

The following corollaries directly follow from Lemma 1.7 and the definition of stress.

COROLLARY 1. *The stress of a graph is a graph parameter.*

COROLLARY 2. *If there exists an automorphism α of a graph G and $\alpha(v) = w$, for some v and w, then* $\mathrm{st}(v) = \mathrm{st}(w)$. *Consequently, if G is a vertex transitive graph, then the stress measure is same for all the vertices, in other words, the graph G is stress regular.*

Noting that the stress measure quantifies the responsibility of concerned vertex in the connectivity between the vertices therein the graph, the problems of characterizing vertices with zero stress measure and obtaining the lower bounds for the stress measure are included in Subsection 1.1. Comparing the stress measure with degree and eccentricity is included in Subsection 1.2.

The expression for stress of standard graphs namely complete graph, tree, path, complete bipartite graph, cycle, block graph, etc., and analyze the distribution of the stress measure on the set of vertices of those graphs are discussed in Section 2. Among the family of all trees and all complete bipartite graphs with fixed order, identifying the graphs with the highest and lowest stress within each respective family is also discussed in the same section. The graphs with diameter three and lesser and analyze the distribution of the stress measure is discussed in Section 2.

Checking whether there exists any association between regularity and stress regularity of a graph along with the characterization of graphs that are k-stress regular for $k = 0, 1, 2$ are discussed in Section 3.

Observing that a cut-vertex in a graph, if present, consistently possesses a nonzero stress measure and therefore, the expression for the stress measure of such a cut-vertex and the vertices therein with reference to their relation with subgraphs associated with the cut-vertex are given in Section 4.

A method to compute the stress measure of a vertex using the adjacency matrix of given graph is discussed in Section. Rather than counting the number of shortest paths passing through the vertex, how the adjacency matrix of a given graph is an effective tool to compute stress is given in Section 5, which is drawn from the article [**12**].

1.1. Bounds for Stress Measure. A natural question one comes across is that what could be the bounds for the stress measure of any vertex in a graph. Though one tempts to suggest a positive correlation between the degree measure and stress measure of a given vertex, a result in this section describes that every vertex in a complete graph, where each vertex possesses maximum degree, has zero stress. This part of section, investigates if the number of adjacency among vertices in the neighborhood has any influence on the stress measure of given vertex. The following lemma describes the lower bound and an expression in some special case for the stress measure of a given vertex.

LEMMA 1.8. *Given a vertex v in a graph G. Then,*

(i) *If* $\deg(v) = 1$, *then* $\mathrm{st}(v) = 0$.
(ii) $\mathrm{st}(v) = 0$ *if and only if* $\langle N(v)\rangle$, *the subgraph induced by* $N(v)$, *is complete.*
(iii) *If v is any arbitrary vertex, then*

$$\mathrm{st}(v) \geq \binom{\deg(v)}{2} - \mu(v), \tag{1.5}$$

where $\mu(v)$ denotes the number of edges in $\langle N(v)\rangle$.

(iv) *If v is any vertex with eccentricity one (i.e., $e(v) = 1$), then*

$$\mathrm{st}(v) = \binom{\deg(v)}{2} - \mu(v). \tag{1.6}$$

PROOF. Part (i) of the lemma is trivial from the definition of stress measure and the property of pendant vertex.

Suppose that $\langle N(v)\rangle$ is not complete and the v_1 and v_2 are two nonadjacent vertices from $N(v)$. Note that v_1-v-v_2 is a path of length two from v_1 to v_2 having v as an internal vertex. Therefore, $\mathcal{P}^0(v)$ is nonempty. By the definition of stress measure, $\mathrm{st}(v)$ is nonzero. Thus, $\mathrm{st}(v) = 0$ implies $\langle N(v)\rangle$ is complete, proving the 'only if' part of (ii).

To prove 'if part', consider any vertex v such that $\langle N(v)\rangle$ is complete. Now, any path $P(u\text{-}w)$ between the vertices u and w in G, with v among its internal vertex, is represented as $P(u\text{-}w) = u\text{-}\cdots\text{-}v_1\text{-}v\text{-}v_2\text{-}\cdots\text{-}w$, where $v_1, v_2 \in N(v)$. Since v_1 and v_2 are adjacent in G, the $P(u\text{-}w)$ cannot be a shortest path in G, proving that $\mathcal{P}^0(v) = \emptyset$ and $\mathrm{st}(v) = 0$. Thus, Part (ii) of the lemma is proved.

Part (iii) of lemma follows from the fact that for every nonadjacent vertices v_1 and v_2 from $N(v)$, there is a path v_1-v-v_2 of length two between v_1 and v_2 with v among its internal vertices.

Part (iv) is deduced from the fact that $e(v) = 1$, which suggests that every vertex in G apart from v is in $N(v)$, and for every pair $v_1 \nsim v_2$ in $N(v)$, v_1-v-v_2 is the only path of length two between v_1 and v_2 with v as an internal vertex. □

In view of Part (i) of above lemma, $\mathrm{st}(v) = 0$ whenever the vertex v is a pendant vertex in the graph G and the case is trivial. Henceforth, we always consider the case that $\deg(v) \geq 2$ in the context of discussing the stress measure, unless indicated otherwise.

The following corollary, which characterizes the graphs with zero stress, is immediate from Part (ii) of Lemma 1.8.

COROLLARY 3. *Given a graph G, $\mathrm{st}(G) = 0$ if and only if $G = K_n$, a complete graph on n vertices. In other words, a graph is 0-stress regular if and only if it is a complete graph.*

REMARK 1.9. The converse of Part (iv) of Lemma 1.8 need not be true. For example, every vertex v in the graph $G = \mathbf{C}_5$ is with $\mu(v) = 0$ and Eq (1.6) is satisfied. But, the eccentricity of every vertex v in $\mathbf{C}_5$ is two.

REMARK 1.10. Part (iv) of Lemma 1.8 describes that among the class of vertices with eccentricity one and same degree, the stress measure decreases as the number of adjacency among the neighbors increases. In general, among the class of vertices with eccentricity one, the stress measure exhibits a direct relationship with the discrepancy between the vertex degree and its adjacency with neighbors.

Given a shortest path $P : v_1$-v_2-v_3-v_4 of length three in a graph G, note that $\mathrm{st}(v_2)$ and $\mathrm{st}(v_3)$ are at least two. Therefore, the following result is obtained.

LEMMA 1.11. *If G is a graph such that $\mathrm{diam}(G)$ is three or more, then there exist at least two vertices with stress measure two or more.*

1.2. Comparison: Stress Measure with other centrality measures. Given a graph, comparison of degree and eccentricity with the stress measure and observing that their associations are not consistent are discussed in this section.

1.3. Stress and Degree. Degree being a centrality measure and a vertex with maximal degree is among central vertices with reference to degree measure, one may be tempted to conjecture that vertices with maximal stress are among the vertices with maximal degree and vice versa. However, it does not hold in general. The following example describes this fact.

EXAMPLE 1.12 (Stress and Degree). Consider a graph G as given in Figure 3.

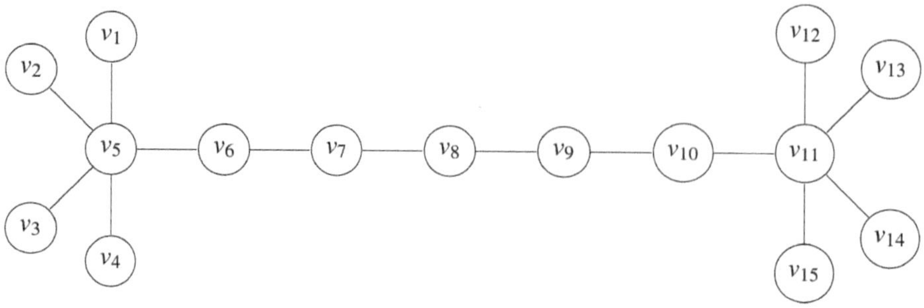

FIGURE 3. Graph G

The stress sequence of vertices in G is given by

(1.7)

$$(\mathrm{st}(v_8), \mathrm{st}(v_7), \mathrm{st}(v_9), \mathrm{st}(v_5), \mathrm{st}(v_{11}), \mathrm{st}(v_6), \mathrm{st}(v_{10}), \mathrm{st}(v_1), \cdots, \mathrm{st}(v_4), \mathrm{st}(v_{12}), \cdots, \mathrm{st}(v_{15}))$$
$$= (49, 48, 48, 46, 46, 45, 45, 0, \cdots, 0).$$

So, the vertex with maximal stress is v_8 with $\mathrm{st}(v_8) = 49$, where as the vertices with the maximal degree are v_5 and v_{11}.

1.4. Stress and Eccentricity. It may be noted that eccentricity being a real valued function on the set of vertices, it is a centrality measure. Further, the vertices with minimal eccentricity are among the central vertices. In general, a vertex with maximal stress measure need not be at the center of graph with the minimum eccentricity.

EXAMPLE 1.13 (Stress and eccentricity). Consider the broom graph G with diameter 6 as shown in Figure 4.

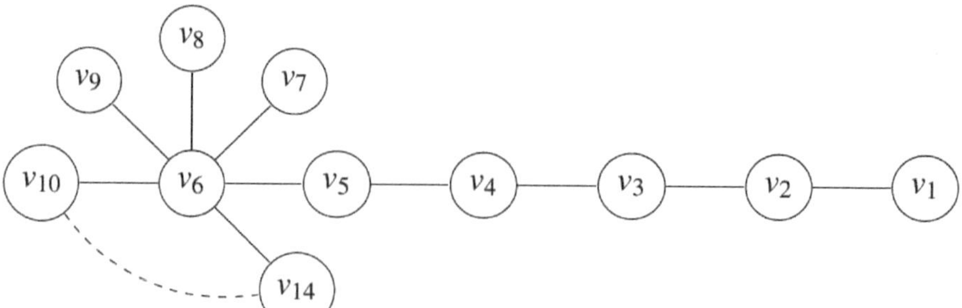

FIGURE 4. Graph G

The stress sequence of vertices in G is given by

$$(\mathrm{st}(v_6), \mathrm{st}(v_5), \mathrm{st}(v_4), \mathrm{st}(v_3), \mathrm{st}(v_2), \mathrm{st}(v_1), \mathrm{st}(v_7), \cdots, \mathrm{st}(v_{14}))$$
$$= (68, 36, 30, 22, 12, 0, 0, \cdots, 0).$$

Note that the central vertex with reference to eccentricity in G is v_4 with $\text{st}(v_4) = 30$, where as the vertex with the maximum stress is v_6 with $\text{st}(v_6) = 68$.

2. Stress Measure in some Families of Graphs

In this section, various families of graphs, including cycles, complete bipartite graphs, and tress, and examine the stress measure of vertices in these graphs are discussed. Also, characterizing the graphs in those families with the maximum stress and the minimum stress is discussed.

2.1. Family of Trees. Note that any graph belonging to the family of trees has several interesting properties such as (i) each vertex is a cut-vertex, (ii) there exists exactly one path between every pair of vertices, (iii) the components of the graph obtained by removing any vertex are also trees (ignoring the isolated vertices, if any), and (iv) a new graph, derived by adding a pendant vertex to the given tree, remains to be a tree, etc. So, these properties are valuable in determining the stress measure of any vertex in a tree graph.

LEMMA 2.1. *Let v be a vertex in a tree G with $\deg(v) = k$, and let $T_1, T_2, \cdots, T_k$ be the k components of $G - v$ with $\#V(T_i) = n_i$, $1 \leq i \leq k$. Then*

$$\text{st}(v) = \sum_{\substack{i,j=1 \\ i<j}}^{k} n_i n_j. \tag{2.1}$$

PROOF. Proof stems from the fact that only the paths connecting two vertices from different components of $G - v$ can have v among their internal vertices, and the path between such a pair of vertices is unique. □

Next, the path graph $\mathbf{P}_n$, a very important graph in the family of trees is considered.

THEOREM 2.2. *Given a path graph $\mathbf{P}_n = (v_1, v_2, \cdots, v_n)$ on n vertices, the stress measure of any vertex v_i is given by*

$$\text{st}(v_i) = (i-1)(n-i), \quad 1 \leq i \leq n. \tag{2.2}$$

Further, the stress sequence of the vertices in $\mathbf{P}_n$ is given by

$$(\text{st}(v_{k+1}), \text{st}(v_{k+2}), \text{st}(v_k), \text{st}(v_{k+3}), \text{st}(v_{k-1}), \ldots, \text{st}(v_{2k+1}), \text{st}(v_1)), \quad \textit{if } n = 2k+1 \tag{2.3}$$

and

$$(\text{st}(v_{k+1}), \text{st}(v_k), \text{st}(v_{k+2}), \text{st}(v_{k-1}), \cdots, \text{st}(v_{2k}), \text{st}(v_1)), \quad \textit{if } n = 2k. \tag{2.4}$$

PROOF. If $i = 1$ or $i = n$, then Eq (2.2) trivially holds, as v_1 and v_n are the pendant vertices of the graph $\mathbf{P}_n$. Otherwise, Eq (2.2) follows from Lemma 2.1 as the graph $\mathbf{P}_n - v$ consists of exactly two components, one with $(i-1)$ number of vertices and the other with $(n-i)$ number of vertices.

To obtain stress sequence of vertices in the path graph, even and odd cases of n are considered separately.

Case 1: If n is odd and $n = 2k+1$ for some integer k, then for $1 \leq i \leq k$,

$$\begin{aligned} \text{st}(v_{k+i}) - \text{st}(v_{k+i+1}) &= [(k+i)-1][n-(k+i)] - [(k+i+1)-1][n-(k+i+1)] \\ &= (2k-n) + 2i = 2i - 1 > 0. \end{aligned}$$

Therefore, $\mathrm{st}(v_{k+i}) > \mathrm{st}(v_{k+i+1})$. Now from Eq (2.2), it is clear that $\mathrm{st}(v_{k+1+i}) = \mathrm{st}(v_{k+1-i})$ and therefore the stress sequence of vertices is as given in Eq (2.3).

Case 2: If n is even and $n = 2k$ for some integer k.

In the similar lines of argument as in Case 1, observe that $\mathrm{st}(v_{k-i+1}) = \mathrm{st}(v_{k+i})$ and $\mathrm{st}(v_{k+i}) - \mathrm{st}(v_{k+i+1}) = 2i > 0$ for $1 \leq i \leq k$. Therefore, the stress sequence of vertices in the $\mathbf{P}_n$ is as provided in Eq (2.4).

Hence the theorem is proved. □

REMARK 2.3. Referring to Eq (2.1), in a path graph $G = \mathbf{P}_n$, the vertices with maximal stress measure are the central vertices with the minimum eccentricity. In other words, the central vertices with reference to both stress measure and eccentricity measure are the same. In fact, for those central vertices,

$$\max_{v \in V(G)} \mathrm{st}(v) = \left\lfloor \frac{n-1}{2} \right\rfloor \left\lceil \frac{n-1}{2} \right\rceil,$$

where as

$$\min_{v \in V(G)} \mathrm{st}(v) = \left\lfloor \frac{n-1}{2} \right\rfloor + 1.$$

Noting that $\sum_{i=1}^{n} (i-1)(n-i) = \binom{n}{3}$, the following theorem which provides the stress of a path graph in relation to the number of vertices, emerges as a direct corollary to Theorem 2.2.

THEOREM 2.4. *For a path graph* $\mathbf{P}_n$,

$$\mathrm{st}(\mathbf{P}_n) = \binom{n}{3}. \tag{2.5}$$

The star graph $K_{1,n-1}$ on n vertices is another simple example for a tree, and the stress measure of vertices in that graph is given in the following.

THEOREM 2.5. *If* $G = K_{1,n-1}$, *then*

(i) $\mathrm{st}(v) = \begin{cases} \binom{n-1}{2} & \text{if } v \text{ is the central vertex} \\ 0 & \text{if } v \text{ is any pendant vertex.} \end{cases}$

(ii) $\mathrm{st}(K_{1,n-1}) = \binom{n-1}{2}$.

PROOF. Since the central vertex v in $K_{1,n-1}$ is of degree $n-1$ and with $\mu(v) = 0$, referring (iv) of Lemma 1.8, $\mathrm{st}(v) = \binom{n-1}{2}$ and in the other case of pendant vertex refer to (i) of the same lemma.

Part (ii) of the theorem follows from Part (i) and Definition 1.1. □

Among all the trees on n vertices, the star graph $K_{1,n-1}$ is with the minimum Wiener index and the path graph $\mathbf{P}_n$ is with the maximum Wiener index. Interestingly, the stress measure attains the minimum and the maximum for the same pair of graphs in the family of trees.

Let $\mathcal{F}$ denote the family of trees on n vertices, where $n \geq 4$. Note that for $n = 3$ or lesser, a connected graph on n vertices is uniquely determined by a path.

THEOREM 2.6. *In a family* $\mathcal{F}$ *of tree graphs on* $n \geq 4$ *vertices,*

$$G_{\max}(\mathcal{F}) = \{\mathbf{P}_n\}.$$

PROOF. Consider a graph G in $G_{max}(\mathcal{F})$. Suppose that $G \neq \mathbf{P}_n$, in which case, there is at least one vertex v with degree three or more. In other words, the set $S = \{v \in V(T) : \deg(v) \geq 3\} \neq \emptyset$. To establish the result, we construct a tree having stress strictly greater than $\mathrm{st}(G)$.

Firstly, observe that there exist a vertex v with degree three or more such that one of the components of $G - v$ consists of all the remaining vertices of the graph with degree three or more. Alternatively, there exists at least two components of $G - v$ which are among the path graphs. For the purpose, choose a vertex among S such that $e(v) \geq e(x)$ for all $x \in S$. Without loss of generality, let a eccentric vertex w of v be in the first component C_1 among r components $C_1, C_2, \cdots, C_r$ of $G - v$, where r is the degree of v in G. Now, we claim that all the remaining vertices other than v, if any, from the set S are in C_1. Otherwise, for any $x \in S$ in C_k and $k \neq 1$, note that $d(x, w) = d(x, v) + d(v, w) > d(v, w) = e(v)$, which is a contradiction for our choice of v having maximal eccentricity among all the vertices from S. So, the components $C_2, \cdots, C_r$ are path graphs and C_1 is the only possible component of $G - v$ which may contain all the vertices with degree three or more from $V(G)$.

Let the components $C_1, C_2, \cdots, C_r$ of orders $n_1, n_2, \cdots, n_r$ respectively and let $V(C_k) = \{v_{k,1}, \cdots, v_{k,n_k}\}$ and $v_{k,1}$ be adjacent with v for all $1 \leq k \leq r$. Now, construct a graph T by removing the edge $vv_{2,1}$ from G and add $v_{2,1}v_{3,n_3}$. In other words $T = G - vv_{2,1} + v_{3,n_3}v_{2,1}$, in which the number of components of $T - x$ and the number of vertices in the corresponding components are same as those in $G - x$, for all x from $V(T) - \{v, v_{3,1}, \cdots v_{3,n_1}\}$. Therefore,

$$\mathrm{st}_G(x) = \mathrm{st}_T(x) \quad \text{for all } x \in V(T) - \{v, v_{3,1}, \cdots v_{3,n_1}\}.$$

From Lemma 2.1 and Eq (2.1),

$$\begin{aligned}\mathrm{st}_G(v) &= \sum_{\substack{i,j=1 \\ i<j}}^{r} n_i n_j \\ &= \sum_{\substack{i,j \in \{1,4,\ldots,r\} \\ i<j}} n_i n_j + \sum_{i \in \{1,4,\ldots,r\}} (n_2 + n_3) n_i + n_2 n_3\end{aligned}$$

and

$$\mathrm{st}_T(v) = \sum_{\substack{i,j \in \{1,4,\ldots,r\} \\ i<j}} n_i n_j + \sum_{i \in \{1,4,\ldots,r\}} (n_2 + n_3) n_i.$$

Further by taking the difference,

$$\mathrm{st}_T(v) - \mathrm{st}_G(v) = -n_2 n_3. \tag{2.6}$$

Now, for any vertex $v_{3,j}$, $1 \leq j \leq n_3$,

$$\mathrm{st}_G(v_{3,j}) = (n_3 - j)(n - n_3 + j - 1),$$

$$\mathrm{st}_T(v_{3,j}) = (n_2 + n_3 - j)(n - n_2 - n_3 + j - 1)$$

and therefore

$$\sum_{j=1}^{n_3} (\mathrm{st}_T(v_{3,j}) - \mathrm{st}_G(v_{3,j})) = \sum_{j=1}^{n_3} -n_2(n_3 - j) + n_2 (n - n_2 - n_3 + j - 1). \tag{2.7}$$

Now by adding Eq (2.6) and Eq (2.7),

$$\begin{aligned}\mathrm{st}(T)-\mathrm{st}(G) &= \mathrm{st}_{T^1}(v)-\mathrm{st}_T(v)+\sum_{j=1}^{n_3}\left[\mathrm{st}_{T^1}(v_{3,j})-\mathrm{st}_T(v_{3,j})\right]\\ &= -n_2n_3+\sum_{j=1}^{n_3}\left[-n_2(n_3-j)+n_2\left(n-n_2-n_3+j-1\right)\right]\\ &= n_2n_3\left[(n-1)-(n_3+n_2)\right]>0.\end{aligned}$$

This proves the theorem. □

The theorem below demonstrates that the minimum stress is observed for a star graph in the family of trees.

THEOREM 2.7. *In a family $\mathcal{F}$ of trees on $n \geq 4$ vertices,*

$$G_{\min}(\mathcal{F})=\{K_{1,n-1}\}.$$

PROOF. Suppose that $G \neq K_{1,n-1}$ is a tree with the minimum stress in the family of trees. In other words, $G \in G_{\min}(\mathcal{F})$ and there exist at least two vertices in the nonempty set $S=\{x \in V(G): 1<\deg(x)<n-1\}$, otherwise, for $v \in S$ the $\deg(v)$ is $n-1$ which is not possible. To establish the result, we construct a tree T and prove that it is with the stress strictly less than the $\mathrm{st}(G)$.

Consider a vertex v of degree r, $1<r<n-1$ from S and with the maximum eccentricity among all the vertices from S. If $C_1, C_2, \cdots, C_r$ are r components of $G-v$ and of order $n_1, n_2, \cdots, n_r$ respectively, then all the vertices from S are in the same component in which we find an eccentric vertex u of v. Otherwise, for u in C_1 and for any x from S in C_i, $i \neq 1$, note that $d(x,u)=d(x,v)+d(v,u)>e(v)$, contradicting the maximality of $e(v)$ among the elements of S. This proves that $n_1=n-r$ and $n_i=1$ for all $i=2,\cdots,r$.

Now, for the vertices w_i from C_i which are adjacent with v, derive a new graph by removing the adjacency between v and w_i for all $i=2,\cdots,r$ and adopting the adjacency between those w_i's and w_1. In other words,

$$T=G-\{vw_2,\cdots,vw_r\}+\{w_1w_2,\cdots,w_1w_r\}.$$

Now referring to Lemma 2.1 and Eq (2.1), stress measure in the new graph T remains same as that in G for all the vertices other than v and w_1. In fact,

$$\mathrm{st}_G(v)=\sum_{\substack{i,j=1\\ i<j}}^{r} n_in_j=\binom{r-1}{2}+(r-1)(n-r), \quad \mathrm{st}_T(v)=0.$$

Now in the case of vertex w_2,

$$\mathrm{st}_G(w_2)=r(n-r-1), \quad \mathrm{st}_T(w_2)=r(n-r-1)+\binom{r}{2}.$$

Further, substituting the above we get

$$\mathrm{st}(G)-\mathrm{st}(T)=\left[\mathrm{st}_G(v)+\mathrm{st}_G(w_2)\right]-\left[\mathrm{st}_T(v)+\mathrm{st}_T(w_2)\right]=(r-1)(n-r-1)>0.$$

This proves the theorem. □

2.2. Cycle Graph. In this subsection, the stress measure of any vertex in a cycle graph are discussed. Note that a cycle graph is a 2-regular graph and also vertex transitive. So, from Corollary 2, a cycle graph is stress regular and it is enough to find the stress measure of any arbitrary vertex to find the stress sequence of vertices, which is constant. Like we have that every vertex in a cycle is a central vertex with the minimum eccentricity, in case of stress measure we have that every vertex is with maximal stress.

THEOREM 2.8. *Given a cycle graph* $\mathbf{C}_n = (v_1, v_2, \cdots, v_n, v_1)$ *on* n *vertices,*

$$\mathrm{st}(v_i) = \frac{d(d-1)}{2}, \quad \textit{for all} \quad 1 \leq i \leq n, \tag{2.8}$$

where $d = \mathrm{diam}(C_n) = \lfloor \frac{n}{2} \rfloor$. *Further,*

$$\mathrm{st}(\mathbf{C}_n) = n\left(\frac{d(d-1)}{2}\right). \tag{2.9}$$

PROOF. Since the diameter of cycle $\mathbf{C}_n$ is $d = \lfloor \frac{n}{2} \rfloor$, any shortest path of length $k \leq d$, with v_1 among internal vertex, is of the form $(v_j, v_{j+1}, \cdots, v_n, v_1, v_2, \cdots, v_i)$, where $2 \leq i \leq k$ in which case $j = n - k + i$. So for each $2 \leq k \leq d$, there are $k-1$ choices of i. In other words, there are $k-1$ paths of length k with v_1 among their internal vertices. Therefore,

$$\mathrm{st}(v_1) = \sum_{k=2}^{d}(k-1) = \frac{d(d-1)}{2},$$

proving Eq (2.8). Now, Eq (2.9) follows from the definition of stress of a graph. □

2.3. Graphs with Diameter Two. From Corollary 3, we have that a graph is with 0-stress if and only if it is a complete graph K_n. In this subsection, we consider the graphs with diameter two and obtain the stress measure of vertices in those graphs. Such a study is useful as many standard graphs such as complete bipartite graphs, $G + K_1$, etc., are examples of graphs with diameter two.

The next theorem, is about the graphs with diameter two.

THEOREM 2.9. *For a graph* G *with diameter two, we have the following.*

$$\mathrm{st}(v) = \begin{cases} 0; & \textit{if } \deg(v) = 1 \\ \binom{\deg(v)}{2} - \mu(v); & \textit{otherwise}, \end{cases} \tag{2.10}$$

Further,

$$\mathrm{st}(G) = \sum_{v_i : \deg(v_i) \neq 1} \binom{\deg(v_i)}{2} - \mu(v_i) \tag{2.11}$$

and if G *is* k*-regular, then*

$$\mathrm{st}(G) = n\binom{k}{2} - \sum_{i=1}^{n} \mu(v_i). \tag{2.12}$$

PROOF. If the vertex v is such that $\deg(v) = 1$, then Eq (2.10) trivially holds. Since the diameter of the graph is two, length of any shortest path in the graph can not exceed two. Therefore the shortest paths of length two passing through a vertex v are exactly the paths between the nonadjacent vertices among $N(v)$ which pass through v. Therefore, $\mathrm{st}(v) = \binom{\deg(v)}{2} - \mu(v)$. Eq (2.11) and Eq (2.12) follow from the definition of stress of the graph and Eq (2.10). □

REMARK 2.10. As a consequence of Theorem 2.9, Theorem 2.5 follows as an immediate corollary. In fact, an expression for the stress measure of any vertex from a bipartite graph could be derived easily from Eq (2.10).

2.4. Complete Bipartite Graphs. Throughout this subsection, we consider the graphs from the family of complete bipartite graphs $K_{r,s}$, where $r+s=n$, and $U=\{u_1,\cdots,u_r\}$ and $W=\{w_1,\cdots,w_s\}$ are the partite sets of V. Observe that any complete bipartite graph is an interesting example for a graph with diameter two.

In the following theorem G is $K_{r,s}$ with $r,s\neq 1$.

THEOREM 2.11. *For $G=K_{r,s}$ with partitions U and W of V, we have the following.*

$$\mathrm{st}(v) = \begin{cases} \binom{s}{2}; & \text{if } v\in U \\ \binom{r}{2}; & \text{if } v\in W \end{cases} \tag{2.13}$$

and if $r\leq s$, then the stress sequence of vertices of $K_{r,s}$ is given by

$$(\mathrm{st}(u_1),\mathrm{st}(u_2),\cdots,\mathrm{st}(u_s),\mathrm{st}(v_1),\mathrm{st}(v_2),\cdots,\mathrm{st}(v_r)). \tag{2.14}$$

Further,

$$\mathrm{st}(G) = r\binom{s}{2} + s\binom{r}{2}. \tag{2.15}$$

PROOF. Since $\mathrm{diam}(G)=2$ and $\mu(v)=0$ for all $v\in V$, Eq (2.13) immediately follwos from Eq (2.10). The rest follows from the definitions of stress sequence of vertices and stress of a graph. □

Now in the following theorem we consider $\mathcal{F}$ to be the family of all complete bipartite graphs $K_{r,s}$, where $r+s=n$, and characterize $G_{\min}(\mathcal{F})$ and $G_{\max}(\mathcal{F})$. Since the graphs $K_{r,s}$ and $K_{s,r}$ are isomorphic to each other and have the same stress values, with no loss of generality, assume that $r\leq s$.

THEOREM 2.12. *In the family $\mathcal{F}$ of complete bipartite graphs $K_{r,s}$, we have*

(i) $G_{\max}(\mathcal{F}) = \{K_{r,s}\}$, *where $s-r=0$ or 1,*
(ii) $G_{\min}(\mathcal{F}) = \{K_{1,n-1}\}$.

In fact,

$$\max_{r,s}\ \mathrm{st}(K_{r,s}) = \left\lfloor\frac{n}{2}\right\rfloor\left\lceil\frac{n}{2}\right\rceil\frac{n-2}{2} \quad \text{and} \quad \min_{r,s}\ \mathrm{st}(K_{r,s}) = \binom{n-1}{2}.$$

PROOF. For $r=1$ and $s=n-1$, referring Theorem 2.5, we observe that $\mathrm{st}(K_{1,n-1}) = \binom{n-1}{2}$. Further, substituting $r=2$ in Eq (2.15), we get that $\mathrm{st}(K_{2,n-2}) = 2\frac{(n-2)(n-3)}{2}+(n-2) = (n-2)^2$, proving that $\mathrm{st}(K_{2,n-2}) > \mathrm{st}(K_{1,n-1})$.

Now let $r\geq 2$ and $r+1\leq s-1$. Referring to Eq (2.15), $\mathrm{st}(K_{r,s}) = \frac{rs(n-2)}{2}$ and $\mathrm{st}(K_{r+1,s-1}) = \frac{(r+1)(s-1)(n-2)}{2}$. Further, by taking the difference,

$$\mathrm{st}(K_{r+1,s-1}) - \mathrm{st}(K_{r,s}) = \frac{1}{2}(n-2r-1) > 0.$$

Therefore, for any r such that $r + s = n$ and $r \leq s$, we have that $\mathrm{st}(K_{r,s})$ is the minimum for $r = 1$ and $\mathrm{st}(K_{r,s})$ is the maximum for $n - 2r = s - r$, which is either zero or one, proving the initial part of the theorem.

The second part follows immediately follows from direct substitution in Eq (2.15) and Theorem 2.5. □

2.5. Graphs with Diameter three. In this subsection, we consider the graphs G with diameter three and investigate the stress measure of vertices in those graphs. Observe that if a vertex v is with $e(v) = 1$, then the stress measure of such v is discussed in Part (iv) of Lemma 1.8 and Eq (1.6). Since the diameter of a given graph G is three, any shortest path in the graph is of length at most three and hence, any possible shortest path of length three having v among its internal vertices is the shortest path connecting a vertex in the neighborhood of v and a vertex at distance two from it. In fact, such a path is of the form v_1-v-v_2-v_3, where v_1, v_2 are some nonadjacent vertices of v, and v_3 is at distance two from v. For convenience, for any $z \in N_2(v)$, a notation $\eta_v(z)$, where $N_2(v)$ is any vertex at distance two from v, to denote the number of path of length three between z and any vertex in the neighborhood of v is introduced. In fact $\eta_v(z)$ is given by

$$\eta_v(z) = l_2(v,z)k(v,z), \quad \text{for all } z \in N_2(v), \tag{2.16}$$

where $l_2(v,z)$ denotes the number of paths of length two between v and z and $k(v,z)$ denotes the number of vertices $w \in N(v)$ such that $d(z,w) = 3$. If $e(v) = 1$, then clearly $N_2(v) = \emptyset$.

In the subsequent theorem, the stress measure of any non-pendant vertex in a graph G of diameter three is discussed, from which we could write the stress sequence of vertices in G.

THEOREM 2.13. *If G is a graph with diameter three and v is a vertex such that $\deg(v) \neq 1$, then*

$$\mathrm{st}(v) = \binom{\deg(v)}{2} - \mu(v) + \sum_{z \in N_2(v)} \eta_v(z). \tag{2.17}$$

PROOF. Noting that for any non-pendant vertex v, the number of paths of length two with v among their internal vertices is given by $\binom{\deg(v)}{2} - \mu(v)$, and the number of paths of length three having v among their internal vertices is given by $\sum_{z \in N_2(v)} \eta_v(z)$.

Further, Eq (2.17) follows from the definition of the stress measure, as every shortest path in the graph is of length three or lesser. □

Let $B(s,r)$ be a bistar graph with central vertices u, w which are adjacent, $V(G) = \{u, u_1, \cdots, u_r, w, w_1, \cdots, w_s\}$, $N(u) = \{u_1, \cdots, u_r\}$, $N(w) = \{w_1, \cdots, w_s\}$ and $n = r + s + 2$. $B(s,r)$ is a nontrivial example of a graph with diameter three. Clearly, $N(u)$ and $N(w)$ are the pendant vertices of the graph with zero stress measure, u, w are the central vertices with $\mu(u) = \mu(w) = 0$, $l_2(u,y) = 1, k(u,y) = r = \eta_v(y)$ for all $y \in N(w) = N_2(u)$, and $l_2(w,z) = 1, k(w,z) = s = \eta_w(z)$ for all $z \in N(u) = N_2(w)$. Hence, the theorem below is an immediate corollary to Theorem 2.13.

THEOREM 2.14. *Given a bistar graph $B(s,r)$, the stress measure of a vertex v is given by*

$$\operatorname{st}(v) = \begin{cases} \binom{r+1}{2} + sr & \text{if } v = u \\ \binom{s+1}{2} + sr & \text{if } v = w \\ 0 & \text{otherwise} \end{cases} \tag{2.18}$$

and for $r \geq s$, the stress sequence of vertices of $B_{s,r}$ is given by

$$(\operatorname{st}(u), \operatorname{st}(w), \operatorname{st}(u_1), \cdots, \operatorname{st}(u_r), \operatorname{st}(w_1), \cdots, \operatorname{st}(w_s)). \tag{2.19}$$

Further,

$$\operatorname{st}(B_{s,r}) = \binom{r+1}{2} + \binom{s+1}{2} + 2sr. \tag{2.20}$$

REMARK 2.15. Note that in the case of a bistar graph, the vertex with maximal stress measure is among the central vertices having maximal degree.

3. Stress Regular Graphs

In Corollary 3, we have observed that a graph is 0-stress regular if and only if the graph is a complete graph. In the present section, characterization of 1-stress regular and 2-stress regular graphs are presented. We begin with the following subsection, where we make certain observations on stress regularity.

3.1. Is a Stress Regular Graph Essentially a Regular Graph? A natural problem one comes across is to probe if a stress regular graph is essentially a regular graph and vice-versa. Unfortunately, neither of the cases is true and we have the counter examples given in the following.

EXAMPLE 3.1. Consider the graph G depicted in Figure 5, which is a 3-regular graph and is of diameter three.

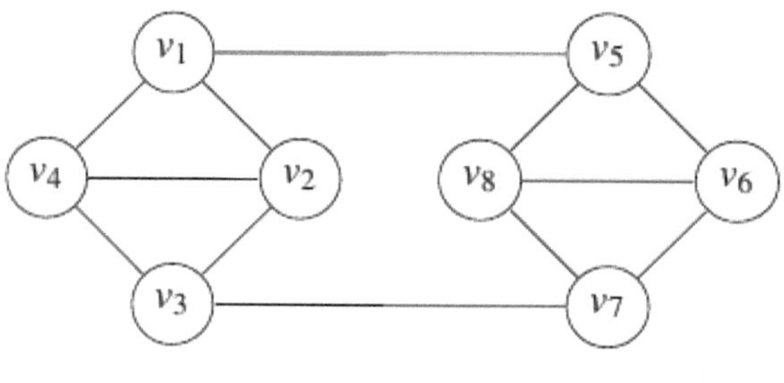

FIGURE 5. Graph G which is a 3-regular graph

Now, consider the vertex v_1 for which $\mu(v_1) = 1$, $N_2(v_1) = \{v_3, v_6, v_8\}$. Further, $l_2(v_1, v_3) = 2, l_2(v_1, v_6) = l_2(v_1, v_8) = 1$ and $k(v_1, v_3) = 1, k(v_1, v_6) = k(v_1, v_8) = 2$. Substituting these values in Eq (2.17), $\operatorname{st}(v_1) = 8$. For v_4, observe that $\mu(v_4) = 2$ and $N_2(v_4) = \{v_5, v_7\}$. Now, using Eq (2.17), $\operatorname{st}(v_4) = 3 \neq \operatorname{st}(v_1)$.

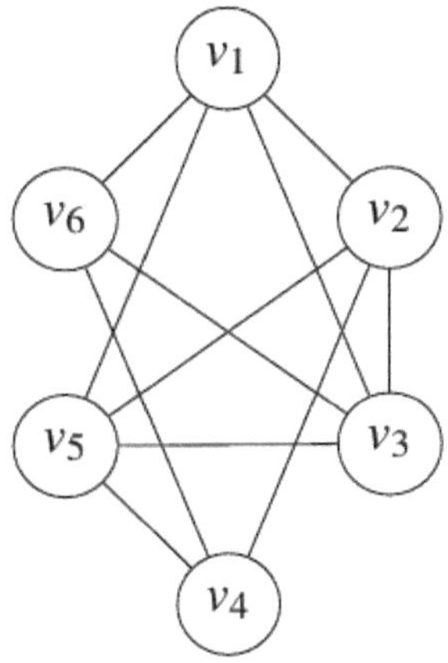

FIGURE 6. 2-stress regular graph H

Now, consider the graph H, given in Figure 6. Note that $\deg(v_1) = \deg(v_2) = \deg(v_3) = \deg(v_5) = 4$ and $\deg(v_4) = \deg(v_6) = 3$. Therefore H is not a regular graph. Observe that $\mu(v_1) = \mu(v_2) = \mu(v_3) = \mu(v_5) = 4$ and $\mu(v_4) = \mu(v_6) = 0$. So, by using Eq (2.10) of Theorem 2.9, $\mathrm{st}(v_i) = 2$ for all i and hence the graph H is a 2-stress regular which is not a regular graph.

In the upcoming theorem, few observations regarding stress regularity is given.

THEOREM 3.2. *If G is a k-stress regular graph with $\mathrm{diam}(G) = d$, then $k \geq \lfloor \frac{d^2}{4} \rfloor$.*

PROOF. Let u and w be any two vertices in G such that there exists a shortest path $P(u,w)$ of length $\mathrm{d}(u,w) = d$. If v is any central vertex of the path graph $P(u,w)$, then referring to Eq (2.2) of Theorem 2.2,

$$\mathrm{st}_{P(u,w)}(v) = \begin{cases} \frac{d^2}{4} & \text{if } d \text{ is even} \\ \frac{(d-1)(d+1)}{4} & \text{if } d \text{ is odd.} \end{cases}$$

Combining both the cases, $\mathrm{st}_{P(u,w)}(v) = \lfloor \frac{d^2}{4} \rfloor$. Again, referring to Lemma 1.8, we get that $\mathrm{st}_{P(u,w)}(v) \leq \mathrm{st}_G(v)$ and hence theorem follows due to stress regularity. □

With reference to Eq (2.8) in Theorem 2.8, in a cycle graph $G = \mathbf{C}_n$ of diameter d, we get that the graph is a k-stress regular graph, where $k = \lfloor \frac{d^2}{4} \rfloor$.

3.2. Characterization of Stress Regular Graphs. In this subsection, characterization of k-stress regular graphs, for $k = 1, 2$ is presented. It is quite interesting to observe that there are only two graphs which are 1-stress regular. In fact, those graphs are 2-regular graphs and the cycles of length 4 and 5, respectively.

LEMMA 3.3. *If G is a k-stress regular graph and $k \neq 0$, then we have the following.*

(i) *No vertex is a pendant vertex, in other words, $\deg(v) \geq 2$.*

(ii) *For every vertex $v \in V(G)$, the induced graph $\langle N(v) \rangle$ is not a complete graph.*
(iii) *There could be at most k pairs of nonadjacent vertices in $N(v)$.*
(iv) *If $k = 1$ or 2, then $d = \mathrm{diam}(G) = 2$.*

PROOF. Parts (i) and (ii) of the theorem follow from the characterization of vertices with zero stress measure. Part (iii) follows from Eq (1.5) of Lemma 1.8. Part (iv) derives from Theorem 3.2. □

The following theorem, characterizes 1-stress regular graphs.

THEOREM 3.4. *A connected graph G is 1-stress regular if and only if $G = \mathbf{C}_4$ or $\mathbf{C}_5$.*

PROOF. If $G = \mathbf{C}_4$ or $\mathbf{C}_5$, then it is easily verified that the graph G is 1-stress regular graph.

Now, consider any 1-stress regular graph G. Referring to Part (iv) of Lemma 3.3, the diameter of the graph G is exactly two.

First of all note that from Part (iii) of Lemma 3.3, G is 1-stress regular implies that there exists precisely one pair of nonadjacent vertices in the neighborhood of v, say v_1, v_2. In other words, all the remaining pairs of vertices, if any, in the $N(v)$ are adjacent to each other. The same arguments hold for the vertices v_1 and v_2, and there exists a unique path u-v_1-w of length two having v_1 as the internal vertex. Since all the vertices from $N[v] - \{v_1\}$ are adjacent to each other, at least one among the u and w is not in $N[v]$, in which case the other is v itself. Without loss of generality, let the vertex $w \notin N[v]$ such that w-v_1-v is the shortest path of length two and w, v are the unique pair of nonadjacent vertices in the neighborhood of v_1.

Now for any arbitrary vertex v in the graph G, we shall prove that $\deg(v) = 2$. On the contrary, let v_3 be any vertex from $N(v)$, different from v_1 and v_2. In case of $w \sim v_3$, we get that $(v_1, v_2), (v, w)$ are two pairs of nonadjacent vertices in the neighborhood of v_3, and in case of $w \not\sim v_3$, we get that $(w, v_3), (w, v)$ are two pairs of nonadjacent vertices in the neighborhood of v_1, getting contradiction to the fact that stress measure of vertices in G is one. Therefore $\deg(v) = 2$, proving that the graph G is of diameter two and G is 2-regular. This proves that G either $\mathbf{C}_4$ or $\mathbf{C}_5$ and hence the theorem. □

From the lines within the proof of Theorem 3.4, the following theorem is trivially true.

THEOREM 3.5. *For a graph G with diameter two, the following statements are equivalent.*

(i) *G is a 1-stress regular graph.*
(ii) *For any vertex v in the graph G, there exists exactly one pair of nonadjacent vertices in $N(v)$.*

Now, we will continue to characterize the graphs which are 2-stress regular. The following lemmas are useful for the purpose.

LEMMA 3.6. *If G is a 2-stress regular graph, then there exists exactly 2 pairs of nonadjacent vertices in $\langle N(v) \rangle$.*

PROOF. The proof immediately follows from Lemma 3.3 and Theorem 3.5. □

LEMMA 3.7. *If G is a 2-stress regular graph, then $\delta(G) \geq 3$ and $\Delta(G) \leq 4$, in other words,* $\deg(v) = 3$ *or* 4 *for all* $v \in V(G)$.

PROOF. From Lemma 3.6, it follows that there are exactly two pairs of nonadjacent vertices in the neighborhood and therefore $\delta(G) \geq 3$. Further if $\Delta(G) \geq 5$, consider a vertex v for which $\deg(v) \geq 5$ and $v_1, \cdots, v_5$ are in $N(v)$. Without loss of generality consider the two cases,

(i) $(v_1, v_2), (v_1, v_3)$ are the two pairs of nonadjacent vertices from $N(v)$, and
(ii) $(v_1, v_2), (v_3, v_4)$ are the two pairs of nonadjacent vertices from $N(v)$.

In either of the cases, there exists at least one vertex $w \notin N[v]$ such that $w \sim v_1$ as the stress measure of v_1 is two. If w is adjacent to at least one of v_4 and v_5 (say, adjacent to v_5), in case of (i), we get $(v_1, v_2), (v_1, v_3), (v, w)$ are three pairs of nonadjacent vertices in the neighborhood of v_5 and in case of (ii), $(v_1, v_2), (v_3, v_4), (v, w)$ are three pairs of nonadjacent vertices in the neighborhood of v_5. In both the cases we get a contradiction to the fact that every vertex has exactly two pairs of nonadjacent vertices in its neighborhood. If w is not adjacent with any of v_4 and v_5, then $(v, w), (v_4, w), (v_5, w)$ are three pairs of nonadjacent vertices in the $N(v_1)$, again a contradiction. Therefore $\deg(v) \leq 4$. Hence the lemma is proved. □

In the following theorem, we characterize 2-stress regular graphs.

THEOREM 3.8. *If G is a 2-stress regular graph, then G is isomorphic to one of the graphs G_1, G_2, or G_3 given in Figure 7.*

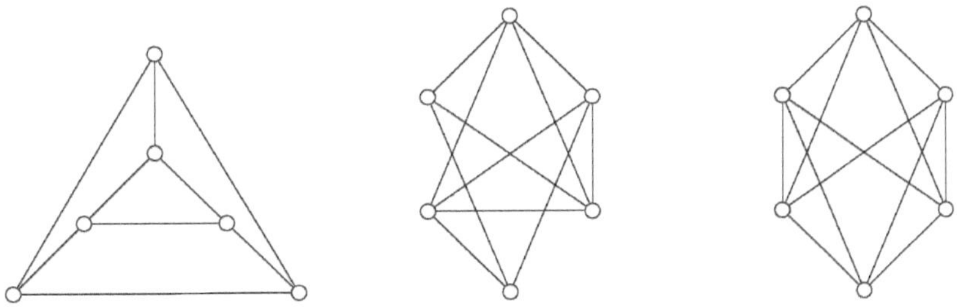

(a) G_1 (b) G_2 (c) G_3

FIGURE 7. 2-stress regular graphs

PROOF. We shall establish the theorem by constructing all the possible 2-stress regular graphs and see the resulting graphs could be only among G_1, G_2 and G_3.

If G is a 2-stress regular graph, then from Part (iv) of Lemma 3.3, G is of diameter two. Further from Lemma 3.7, we have that $3 \leq \deg(v) \leq 4$. Also, from the same Lemma 3.7, we have that given a vertex, there exist exactly two pairs of nonadjacent vertices in its neighborhood. In other words, we have that

$$\mu(v) = \begin{cases} 1 & \text{if } \deg(v) = 3 \\ 4 & \text{if } \deg(v) = 4. \end{cases} \tag{3.1}$$

We now examine the following cases.

Case 1: ($\delta(G) = \Delta(G) = 3$, in other words, G is a 3-regular graph). In this case, consider a vertex v_1 of G, where $\deg(v_1) = 3$ and $\mu(v_1) = 1$. Let $N(v_1) = \{v_2, v_3, v_4\}$ be such that v_2, v_3 and v_2, v_4 are the pairs of nonadjacent vertices. Since $\deg(v_2) = 3$ and v_1 is the only vertex adjacent with v_2 in $N[v_1]$, there exist two vertices $v_5, v_6 \notin N[v_1]$ (say) such that $v_5, v_6 \sim v_2$. Now, referring to Eq (3.1), $\mu(v_2) = 1$ implies that $v_5 \sim v_6$, as $v_1 \not\sim v_5, v_6$. Since $\mu(v_5) = 1$ and $v_2, v_5 \sim v_6$, $\mu(v_5) = 1$ implies that there is no vertex $z \neq v_2$ such that $z \sim v_5, v_6$. For the similar reason, there is no common adjacent vertex $z \neq v_1$ for v_3 and v_4.

Now, we shall claim that each one of v_3 and v_4 is adjacent with distinct vertices of the set $\{v_5, v_6\}$. On the contrary, let there be a vertex $y \neq v_1, v_4, v_5, v_6$ such that $y \sim v_3$ and $N(v_3) = \{v_1, v_4, y\}$. Since $y \not\sim v_2, v_4$ and y is adjacent with at most one of v_5, v_6, we have that $\deg(y) = 3$ implies that there exists at least one vertex from $N[y]$ which is at distance three from v_1. The resulting graph is as in Figure 8.

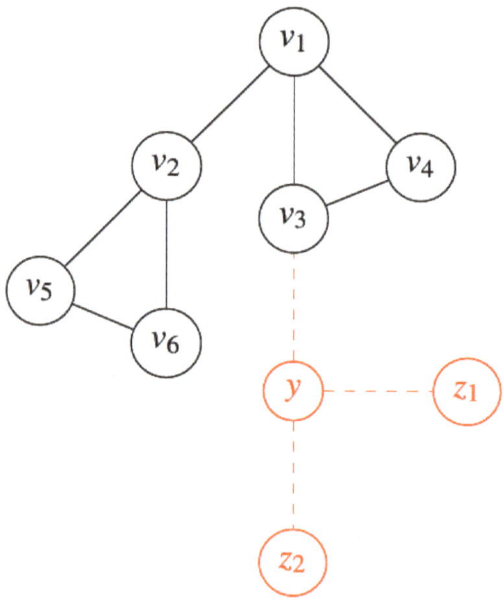

FIGURE 8

This contradicts the fact that $\operatorname{diam}(G) = 2$. Therefore, v_3 is adjacent with exactly one among v_5 and v_6, in which case, v_4 is adjacent with the other. Now, the resulting graph is isomorphic to G_1 given in Figure 7 and the presence of any vertex v_7 having adjacency with any of $v_1, \cdots, v_6$ contradicts 3-regularity of the graph.

Case 2: ($\Delta(G) = 4$).

Consider a vertex v_1 such that $\deg(v_1) = 4$ and $N(v_1) = \{v_2, v_3, v_4, v_5\}$. Since $\langle N(v_1) \rangle$ has exactly two pairs of nonadjacent vertices, we consider following two distinct sub-cases and characterize 2-stress regular graphs in each of those cases.

Case 2a: ((v_2, v_3) and (v_2, v_4) are the two pairs of nonadjacent vertices).

In this case, $v_5 \sim v_1, v_2, v_3, v_4$ and therefore $\Delta(G) = 4$ implies that $\deg(v_5) = 4$ and $N(v_5) = \{v_1, v_2, v_3, v_4\}$. Since v_2 is adjacent with only v_1 and v_5 in $N[v_1]$, $\delta(G) \geq 3$ implies that there exists a w such that $w \notin N[v_1]$, $w \sim v_2$ and $w \not\sim v_5$ (because $\deg(v_5)$ can not exceed 4). If there exist two such w's, (say w_1 and w_2), then $N(v_2) = \{v_1, v_5, w_1, w_2\}$ and $\deg(v_2) = 4$, but $\mu(v_2) < 4$, a contradiction. Therefore, there exists only one such w, say $w = v_6$.

Since the vertices v_1, v_4, v_5 in $N(v_3)$ induces a complete sub graph and therefore $\mu(v_3) \geq 3$, which implies that $\deg(v_3) = 4$. Same argument holds for the vertex v_4. Now, for any $z \notin N[v_1]$ such that $v_3 \sim w$, we have that $N(v_3) = \{v_1, v_4, v_5, w\}$ and therefore $\mu(v_3) = 4$ implies that $w \sim v_4$. Now, we prove that if w is any vertex such that $w \notin N[v_1]$, but adjacent with both v_3 and v_4 , then $w = v_6$.

Suppose that v_6 is adjacent with neither of v_3 and v_4, then a shortest path of length two connecting v_3 and v_6 is given by $y \notin N[v_1]$, such that $v_3, v_4 \sim y \sim v_6$. Now, the resulting graph is isomorphic as in Figure 9.

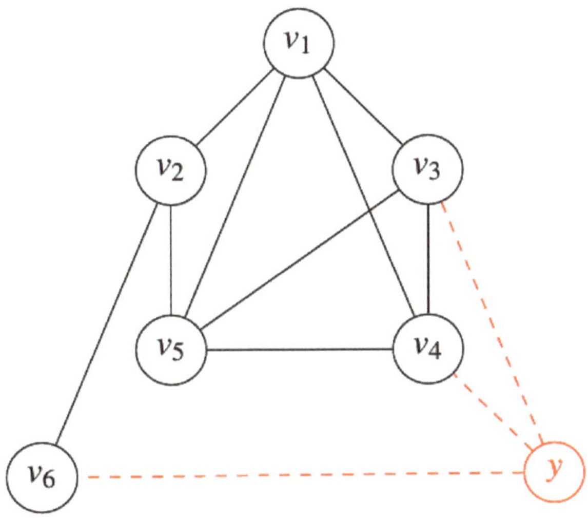

FIGURE 9

Since $\deg(v_6)$ is at least three, there exists a vertex $z \notin N[v_1]$ such that $v_6 \sim z$, proving that $d(v_1, z) = 3$ which contradicts to the fact that $\operatorname{diam}(G) = 2$. Now, the resulting graph is isomorphic to the graph G_2 given in Figure 7. Any additional vertex v_7 in the graph can not be adjacent with $v_1, \cdots, v_6$, the only 2-stress regular graph in this case is isomorphic to G_1.

Case 2b: $((v_2, v_3), (v_4, v_5)$ are the pairs of nonadjacent vertices).
Since $v_1 \sim v_5$ and $v_1 \sim v_3$ in the neighborhood of v_2, we have that $\mu(v_2) \geq 2$ and hence $\deg(v_2)$ must be four. So, there exists one vertex, say $v_6 \notin N[v_1]$, such that $v_2 \sim v_6$. Since $\deg(v_1) = \deg(v_5) = \deg(v_2) = 4$, $\mu(v_2) = 4$ implies that $v_6 \sim v_3, v_4$. Now the resulting graph is 4-regular and isomorphic to G_3. Any additional vertex v_7 can not be adjacent to the vertices $v_1, \cdots, v_6$ and therefore only connected 2-stress regular graph in this case could be isomorphic to G_3 given in Figure 7.

Hence the proof of the theorem. □

Characterizing the k-stress regular graphs for $k \geq 3$ is an interesting problem for the future research.

4. Stress Measure of a Cut-Vertex

Cut-vertices in a graph have a greater significance due to their importance in the connectivity of the graph. Therefore, it is of general interest to understand the stress measure of any cut-vertex in a graph. One may note that a vertex with the maximum stress measure need not be among the cut-vertices. As an example, consider the graph G given in Figure 10.

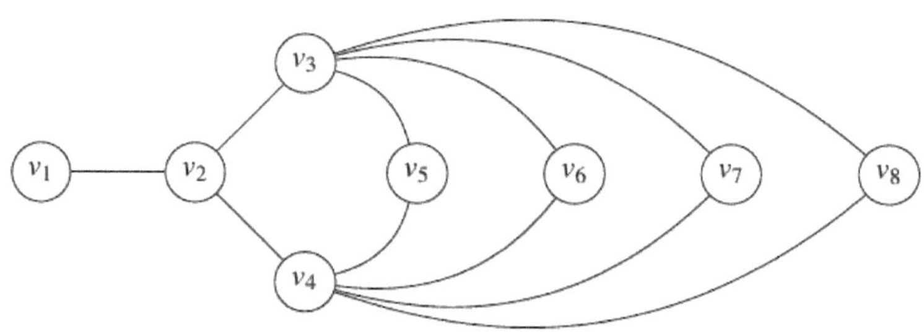

FIGURE 10. Graph G

The stress sequence of vertices in G is

$$(\text{st}(v_3), \text{st}(v_4), \text{st}(v_2), \text{st}(v_5), \text{st}(v_6), \text{st}(v_7), \text{st}(v_8), \text{st}(v_1)) = (14, 14, 11, 1, 1, 1, 1, 0),$$

from which we observe that v_3 and v_4 are with maximal stress of 14, whereas the cut-vertex is v_2 which is different from either of v_3 and v_4.

To start with, in this section, we make the simple observations and then proceed to find an expression for the stress measure of a cut-vertex. Given a cut-vertex v in the graph G, we have

Obs 1. $\text{st}_G(v) \neq 0$,

Obs 2. if there are k components in $G - v$, then $\mathrm{st}_G(v) \geq \binom{k}{2}$.

We now consider a graph with a cut-vertex and determine the stress measure of that cut-vertex.

If v is a cut-vertex of graph G such that C_i, $1 \leq i \leq k$ are the components of the graph $G - v$, then obtain B_i, for each i, the subgraph induced by the vertices of C_i and v.

THEOREM 4.1. *Given a graph G with a cut-vertex v, let B_i be the subgraph induced by the vertices of the component C_i and v, and let $\#\{\mathcal{S}_{B_i}(v)\} = t_i$ (the number of shortest paths in B_i which originate from v), $1 \leq i \leq k$. Then the stress measure of v is given by*

$$\mathrm{st}_G(v) = \sum_{i=1}^{k} \mathrm{st}_{B_i}(v) + \sum_{1 \leq i < j \leq k} t_i t_j. \tag{4.1}$$

Further, for any distinct vertices $u, w \in V(C_i)$, if λ_u denotes $\#\mathcal{P}^0_w(u, v)$, then the stress measure of the vertex w is given by

$$\mathrm{st}_G(w) = \mathrm{st}_{B_i}(w) + \sum_{1 \leq j \neq i \leq k} \; \sum_{u \in V(C_k);\, u \neq w} \lambda_u t_j. \tag{4.2}$$

PROOF. From the definition of the stress measure, $\mathrm{st}_G(v)$ is the number of distinct shortest paths in G with v among their internal vertices, i.e., $\mathrm{st}_G(v) = \#\mathcal{P}^0_G(v)$. Any u-w shortest path $P(u\text{-}w) \in \mathcal{P}^0_G(v)$ is either a path in $\mathcal{P}^0_{B_i}(v)$ for some i or else, $u \in B_i$ and $w \in B_j$ for some $1 \leq i < j \leq k$. Therefore Eq (4.1) follows from the definition.

Further, Eq (4.2) follows from the facts that for $x, y \notin V(C_i)$ any shortest path between the vertices x and y is not passing through w and for $x \in V(C_i)$ and $y \in V(C_j)$, $i \neq j$, every x, y path must pass through v. □

Eq (4.1) and Eq (4.2) could be simplified further to a great extent, using the following observations. If n_i is the order of C_i, $1 \leq i \leq k$, such that $\sum n_i + 1 = n$, we get that

$$t_i = \begin{cases} n_i - 1 & \text{if } B_i \text{ is complete} \\ n_i - 1 & \text{if } B_i \text{ is a tree} \\ 2d_i & \text{if } B_i \text{ is a cycle of diameter } d, \end{cases} \tag{4.3}$$

$$\mathrm{st}_{B_i}(v) = \begin{cases} 0 & \text{if } B_i \text{ is complete} \\ 0 & \text{if } B_i \text{ is a tree} \\ \binom{d}{2} & \text{if } B_i \text{ is a cycle of diameter } d. \end{cases} \tag{4.4}$$

We now consider two special cases in the following observations, in which we have two circular (wheel) blocks and one cut-vertex.

OBSERVATION 1. Consider a graph G with two circular blocks $B_1 = (v = v_1, v_2, \cdots, v_r, v)$ and $B_2 = (v = w_1, w_2, \cdots, w_s, v)$, and with unique cut-vertex v, a common vertex between B_1 and B_2. Clearly, the cycles B_1 and B_2 are the cycles of length r and s, respectively. Now, referring to Eq (4.3) and Eq (4.4), we get that $\mathrm{st}_{B_1}(v) = \binom{d_1}{2}$, $\mathrm{st}_{B_2}(v) = \binom{d_2}{2}$, $t_1 = 2d_1$, and $t_2 = 2d_2$, where $d_1 = \lfloor \frac{r}{2} \rfloor$ and $d_2 = \lfloor \frac{s}{2} \rfloor$. Now, substituting in Eq (4.1), we get that

$$\mathrm{st}_G(v) = \binom{d_1}{2} + \binom{d_2}{2} + 4d_1d_2.$$

OBSERVATION 2. Let G be a graph with a unique cut-vertex v and two blocks of wheel graphs, say B_1 and B_2, where $B_1 = W_{r+1} = (v_1, v_2, \cdots, v_r, v_1) + x$ and $B_2 = W_{s+1} = (w_1, w_2, \cdots, w_s, w_1) + y$ with central vertices x and y, respectively. Note that the unique cut-vertex could appear in the four different ways.

Case 1: $(v = x = y)$.

Referring to Eq (2.10), $\mathrm{st}_{B_1}(v) = \binom{r}{2} - r$ and $\mathrm{st}_{B_2}(v) = \binom{s}{2} - s$. If t_1 and t_2 denote the number of shortest paths originating from v in B_1 and B_2, respectively, then we observe that $t_1 = r$ and $t_2 = s$. Now, substituting these in Eq (4.1), we get that

$$\mathrm{st}_G(v) = \binom{r}{2} + \binom{s}{2} + (r-1)(s-1).$$

Case 2: $(v = x = w_1)$.

In this case, v is at the center of B_1 and at the rim of B_2. Again referring to Eq (2.10), we get that $\mathrm{st}_{B_1}(v) = \binom{r}{2} - r$, $\mathrm{st}_{B_2}(v) = 1$, $t_1 = r$ and $t_2 = s + 2$. So,

$$\mathrm{st}_G(v) = \binom{r}{2} + (r+2)(s-1) + 3.$$

Case 3: $(v = v_1 = y)$.

Similar to the Case 2, we get that

$$\mathrm{st}_G(v) = \binom{s}{2} + (s+2)(r-1) + 3.$$

Case 4: $(v = v_1 = w_1)$.

In this case, v is at rims of both B_1 and B_2. Referring to Eq (2.10), we get that $\mathrm{st}_{B_1}(v) = 1 = \mathrm{st}_{B_2}(v)$, $t_1 = r + 2$ and $t_2 = s_2$. So,

$$\mathrm{st}_G(v) = (r+2)(s+2) + 2.$$

Now, we shall consider a graph G with two cut-vertices to find the stress measure of those cut-vertices with reference to the other. For a graph G with two cut-vertices v and w, consider $B_1, \cdots, B_r$ the subgraphs induced by the vertex v and the r components of $G - \{v, w\}$ which are adjacent with v alone, $D_1, \cdots, D_s$ the subgraphs induced by the vertex w and the s components of $G - \{v, w\}$ which are adjacent with w alone, and H the subgraph induced by $\{v, w\}$ and all the components of $G - \{v, w\}$ which are adjacent with both u and w. Let

- l_i, $1 \le i \le r$, denote the cardinality of $\mathcal{S}_{B_i}(v)$
- k_j, $1 \le j \le s$, denote the cardinality of $\mathcal{S}_{D_j}(w)$
- l denote the cardinality of $\mathcal{S}_H(v)$
- k denote the cardinality of $\mathcal{S}_H(w)$
- t denote the cardinality of $\mathbb{P}_H(v, w)$.

THEOREM 4.2. *Given two cut-vertices v, w of a graph G,*

$$\mathrm{st}_G(v) = \mathrm{st}_H(v) + \sum_{i=1}^{r} \mathrm{st}_{B_i}(v) + \sum_{\substack{i,j=1 \\ i\neq j}}^{r} l_i l_j + l\sum_{i=1}^{r} l_i + t\sum_{i=1}^{r}\sum_{j=1}^{s} l_i k_j \tag{4.5}$$

and

$$\mathrm{st}_G(w) = \mathrm{st}_H(w) + \sum_{j=1}^{s} \mathrm{st}_{D_j}(w) + \sum_{\substack{i,j=1 \\ i\neq j}}^{s} k_i k_j + k\sum_{j=1}^{s} k_j + t\sum_{i=1}^{r}\sum_{j=1}^{s} l_i k_j. \tag{4.6}$$

PROOF. For any $P(x\text{-}y) \in \mathcal{P}_G^0(v)$, observe that $P(x\text{-}y)$ is as one of the following:

(i) x, y are in one of B_i, $1 \le i \le r$
(ii) $x \in B_i, y \in B_j$, where $1 \le i < j \le r$
(iii) $x \in B_i$, $1 \le i \le r$ and $y \in H$
(iv) $x \in B_i$, $1 \le i \le r$ and $y \in D_j$, $1 \le j \le s$
(v) $x, y \in H$
(vi) $x \in H$ and $y \in D_j$, $1 \le j \le s$
(vii) $x \in D_i$ and $y \in D_j$, $1 \le j \le s$

The number of $P(x\text{-}y)$ satisfying (vi) or (vii) is zero, as no such shortest paths pass through v. In fact, $\mathbb{P}_v^0(x\text{-}y) = \emptyset$, whenever $x \in H$ and $y \in D_j$, $1 \le j \le s$ or $x \in D_i$ and $y \in D_j$, $1 \le j \le s$. The number of $P(x\text{-}y) \in \mathcal{P}_G^0(v)$ satisfying (i) is exactly $\mathrm{st}_{B_i}(v)$, satisfying (ii) is $l_i l_j$, satisfying (iii) is $l_i l$, satisfying (iv) is $l_i l k_j$, and satisfying (v) is $\mathrm{st}_H(v)$. Now Eq (4.5) follows as we know that $\mathrm{st}_G(v)$ is given by the cardinality of $\mathcal{P}_G^0(v)$. Eq (4.6) is proved in the similar lines. □

The following theorem characterizes the graph having unique vertex with nonzero stress measure.

THEOREM 4.3. *In a graph G, there exists unique vertex with nonzero stress measure if and only if G is a block graph, in which every block is complete, with unique cut-vertex. In fact, in this case Eq* (4.1) *reduces to*

$$\mathrm{st}_G(v) = \sum_{1\le i<j\le k} t_i t_j, \tag{4.7}$$

where v is the cut-vertex of G and t_i, t_j are as defined in Theorem 4.1.

PROOF. Consider any graph G such that there exists a vertex v with $\mathrm{st}(v) > 0$ and $\mathrm{st}(w) = 0$ for all $w \in V - \{v\}$. Graph G is not complete as there exists a vertex $v \in V$ such that $\mathrm{st}(v) > 0$. Now, for any two vertices x, y in G such that $x \not\sim y$, there exists a path $P(x\text{-}y)$ of length two or more. Since v is the only vertex in G with $\mathrm{st}(v) > 0$, $P(x\text{-}y)$ is of length two, $\mathbb{P}(x\text{-}y) = \{P(x\text{-}y)\}$ and v is the only internal vertex in the $P^0(x\text{-}y)$. Therefore, every pair of vertices (x, y) such that $x \not\sim y$ are in different components of $G - v$, which implies that v is the unique cut-vertex of G and the blocks are complete.

Now, we consider a graph G with unique cut-vertex v and the blocks are complete, to prove the converse part. From 1, it is clear that $\mathrm{st}(v) \neq 0$. Now, for any vertex $x \neq v$ in the graph G, let x be in a block B_i of graph G and observe that $N_{B_i}(x) = N_G(x)$. Therefore B_i is complete implies that $N_G(x)$ is also complete. Now, referring to Part (ii) of Lemma 1.8, we have that $\mathrm{st}_G(x) = 0$ for all $x \neq v$. This proves the first part of the lemma.

The second part follows by substituting $\mathrm{st}_{B_i}(v) = 0$ in Eq (4.1). □

5. Adjacency matrix and the Stress Measure

This section provides a method to compute the stress measure of a vertex using the adjacency matrix of a given graph. Additionally, an algorithm for computing the stress measure is presented. Note that for $i \neq j$, the notation $\sigma(v_i, v_j)$ denotes the number of shortest paths between v_i and v_j, and $\sigma_v(v_i, v_j)$ denotes the number of shortest paths between v_i and v_j having v as an internal vertex i.e, $\sigma(v_i, v_j) = \#\mathbb{P}(v_i, v_j)$ and $\sigma_v(v_i, v_j) = \#\mathbb{P}_v(v_i, v_j)$. In the following definition, a shortest path matrix, denoted by $N(G)$ is introduced in which the non-diagonal entries describe the number of shortest paths between the vertices.

DEFINITION 5.1 (Shortest path matrix). Given a connected graph G, the shortest path matrix of G denoted by $N(G)$ is an $n \times n$ matrix in which the ijth entries are given by

$$\eta_{ij}^G = \begin{cases} \sigma(v_i, v_j) & \text{for } i \neq j \\ 1 & \text{otherwise.} \end{cases}$$

The notations $N(G)$ and η_{ij}^G used in the above definition are replaced by N and η_{ij}, respectively, when the graph under discussion is clear from the context. Writing $\eta_{ii} = 1$ in the above definition is for our computational convenience (representing the number of paths/walks of length zero from a vertex to itself!).

Note that by joining any two paths having one common terminal vertex need not result in a shortest path. In fact, a u-v-w path is a shortest path between u and w if and only if $\mathrm{d}(u, v) + \mathrm{d}(v, w) = \mathrm{d}(u, w)$. The following lemma relates the stress measure of any vertex of graph G with the entries of $N(G)$.

LEMMA 5.2. *If G is a connected graph with $N(G) = (\eta_{ij})$, then for any vertex $v_k \in V(G) = \{v_1, v_2, \cdots, v_n\}$,*

$$\mathrm{st}_G(v_k) = \sum_{\substack{1 \leq i,j \leq n \\ i<j,\ i,j \neq k \\ \mathrm{d}(v_i,v_k)+\mathrm{d}(v_k,v_j)=\mathrm{d}(v_i,v_j)}} \eta_{ik}\eta_{kj}. \tag{5.1}$$

PROOF. For any two vertices $v_i, v_j, i \neq j$, it may be noted that a shortest v_i-v_j path will pass through a vertex v_k if and only if

$$\mathrm{d}(v_i, v_k) + \mathrm{d}(v_k, v_j) = \mathrm{d}(v_i, v_j). \tag{5.2}$$

In the above case, any shortest v_i-v_j path passing through v_k is obtained by joining a shortest v_i-v_k path with a shortest v_k-v_j path, and vice versa. In other words, there is one-one correspondence between shortest v_i-v_j paths passing through v_k and pairs of shortest v_i-v_k and v_k-v_j paths. Therefore, if η_{ik} and η_{kj} denote the number of shortest v_i-v_k paths and v_k-v_j paths, respectively, then the number of shortest paths between v_i and v_j passing through v_k is given by the product $\eta_{ik}\eta_{kj}$ i.e, $\sigma_{v_k}(v_i, v_j) = \sigma(v_i, v_k)\sigma(v_k, v_j)$. Now by referring to Definition 1.1, the stress measure of v_k is given by the sum of the products $\eta_{ik}\eta_{kj}$, where the summation is taken over all pair of vertices $v_i, v_j, i < j$ satisfying Eq (5.2), proving Eq (5.1). Hence the lemma. □

It is well known that the entries of A^p provide the number of walks of length p between the distinct vertices. The following theorem follows immediately.

THEOREM 5.3. *If G is a connected graph, then $\mathrm{d}(v_i, v_j)$, the distance between v_i and v_j for $i \neq j$ is the least integer p for which the ijth entry of A^p is nonzero. In such a case, $\eta_{ij} = (A^p)_{ij}$.*

In view of the above theorem, the shortest path matrix $N(G)$ could be redefined as the matrix $N = (\eta_{ij})$ such that

$$\eta_{ij} = \begin{cases} 1 & \text{for } i = j \\ (A^p)_{ij} & \text{otherwise,} \end{cases}$$

where p is the least integer for which the ijth entry of A^p is nonzero. Now the following corollary is immediate.

COROLLARY 4. *If G is a graph with the adjacency matrix A, then for any vertex $v_k \in V(G) = \{v_1, v_2, \cdots, v_n\}$,*

$$\mathrm{st}_G(v_k) = \sum_{\substack{i,j=1 \\ i<j,\ i,j\neq k \\ \mathrm{d}(v_i,v_k)+\mathrm{d}(v_k,v_j)=\mathrm{d}(v_i,v_j)}}^{n} \eta_{ik}\eta_{kj}, \tag{5.3}$$

where $\eta_{ij} = \begin{cases} 1 & \text{for } i = j \\ (A^p)_{ij} & \text{otherwise} \end{cases}$ and p is the least integer for which the ijth entry of A^p is nonzero.

An algorithm to find the shortest path matrix N of a graph G is given below.

Algorithm 1

INPUT: Adjacency matrix A and order n of the graph G
OUTPUT: Shortest path matrix N of G

1: **function** SPMATRIX(A, n)
2: Declare N as the identity matrix of size n
3: Declare k as a variable with initial value 1
4: Declare L as a list of all ordered pairs (i, j), $1 \leq i \neq j \leq n$
5: **for** (i, j) in L **do**
6: **if** ijth entry of A^k is nonzero **then**
7: Assign ijth entry of A to ijth entry of N
8: Remove (i, j) from L
9: **end if**
10: Increase k by 1
11: Go to Step 5 if L is non empty
12: **end for**
13: **return** N
14: **end function**

In the above algorithm, Step 4 requires $\frac{n(n-1)}{2}$ number of iterations. Matrix multiplication, which is of $O(n^3)$ and the time required to execute the for loop in Step 4 is at most $\frac{n(n-1)}{2}$, which will be executed only if the list L is non empty. Considering the maximum possible iterations, i.e, considering the path graph, which requires $(n-1)$ iterations to make L an empty set, the total time required is less than $(n-1)[n^3+\binom{n}{2}+\binom{n-1}{2}+\binom{n-2}{2}+\cdots+1]$. So, it follows that the time complexity of Algorithm 1 is $O(n^4)+O(n^3)=O(n^4)$.

Using Algorithm 1, the algorithm to find the stress measure of a given vertex in a graph G is given below.

Algorithm 2

INPUT: Adjacency matrix A and a vertex v_k of G
OUTPUT: Stress measure s of v_k

```
1: function StrVertex(A,k)
2:     Declare variable s with initial value 0
3:     Find the shortest path matrix N of G
4:     for all i, j, 1 ≤ i < j ≤ n do
5:         if i, j ≠ k and D_ik + D_kj = D_ij then
6:             Update s by s + (N_ik * N_kj)
7:         end if
8:     end for
9:     return s
10: end function
```

Initially, to find the shortest path matrix in Step 3, the Algorithm 1 is used. The number of iterations required to execute the for loop in Step 4 of Algorithm 2 is $\left[\frac{n(n-1)}{2}\right]$. So the time complexity of Algorithm 2 will be $O(n^4)+O(n^2)=O(n^4)$.

6. Conclusion

The stress measure of a vertex in the subgraph of a given graph has no definite pattern of comparison with the stress of that vertex in the given graph. It has been observed that stress is a graph parameter. Observing that every pendant vertex is with stress measure zero, characterized that the vertices with zero stress are the ones with neighborhood vertices inducing a complete subgraph.

The stress sequence of vertices in some standard graphs such as, complete graph, path graph, cycle graph, etc. are presented. Though a vertex with maximal stress need not coincide either with a vertex with maximal degree or a vertex at the center or a vertex with maximal eigenvector centrality measure, in general, it is observed that in the case of a path graph, central vertices and the vertices with maximal stress coincide.

Graphs with the maximum/minimum stress in a given family of trees and block graphs of fixed order are characterized. The stress measure of vertices in graphs with diameter three or less is discussed and expression for the stress of a vertex in each of the cases is provided. The characterization of k-stress regular graphs, for $k=0,1,2$ is provided.

The stress measure of cut-vertices is discussed and also an expression for the stress measure of cut-vertices in a graph is derived. Further, the graphs with unique vertex having nonzero stress are characterized as the complete block graph with unique cut-vertex.

An alternate expression to compute the stress of a vertex in the graph using the adjacency matrix of the graph is discussed along with the algorithm to compute the stress measure of vertices in a graph using the adjacency matrix of the graph.

References

[1] C. Kang, C. Molinaro, S. Kraus, Y. Shavitt, and V.S. Subramaniam, *Diffusion Centrality in Social Networks*, 2012 IEEE/ACM International Conference on Advances in Social Networks Analysis and Mining, Istanbul, 558–564, 2012.

[2] D. Koschutzki and F. Schreiber, Comparison of Centralities for Biological Networks, *Proc. German Conf. Bioinformatics*, vol. 53, pp. 199–206, 2004.

[3] M. Indhumathy, S. Arumugam, Veeky Baths, and Tarkeshwar Singh, *Graph theoretic concepts in the study of biological networks*, Applied analysis in biological and physical sciences, Springer Proc. Math. Stat., vol. 186, Springer, New Delhi, 2016, pp. 187–200, DOI 10.1007/978-81-322-3640-5_1. MR3613737

[4] Ruth Lopez, Jacob Worrell, Henry Wickus, Rigoberto Flórez, and Darren A. Narayan, *Towards a characterization of graphs with distinct betweenness centralities*, Australas. J. Combin. **68** (2017), 285–303. MR3646041

[5] P. Shannon, A. Markiel, O. Ozier, N.S. Baliga, J.T. Wang, D. Ramage, N. Amin, B. Schwikowski, and T. Idekar, Cytoscape: a software environment for integrated model of biomolecular interaction networks, *Genome Res.*, vol. 13, no. 11, pp. 2498–2504, 2003.

[6] Alfonso Shimbel, *Structural parameters of communication networks*, Bull. Math. Biophys. **15** (1953), 501–507, DOI 10.1007/bf02476438. MR58954

[7] S.P. Borgatti and M.G. Everett, A Graph-theoretic perspective on centrality, *Social Networks*, vol. 28, pp. 466–484, 2006.

[8] Kumar Raghavan Unnithan Sunil, Balakrishnan Kannan, and Madambi Jathavedan, *Betweenness centrality in some classes of graphs*, Int. J. Comb., posted on 2014, Art. ID 241723, 12, DOI 10.1155/2014/241723. MR3294894

[9] G. Chartrand and L. Lesniak, *Graphs and digraphs*, 4th ed., Chapman & Hall/CRC, Boca Raton, FL, 2005. MR2107429

[10] Raksha Poojary, Arathi Bhat K., S. Arumugam, and Manjunatha Prasad Karantha, *The stress of a graph*, Commun. Comb. Optim. **8** (2023), no. 1, 53–65. MR4503650

[11] Raksha Poojary, K. Arathi Bhat, S. Arumugam, and Manjunatha Prasad Karantha, *Stress of wheel related graphs*, Nat. Acad. Sci. Lett. **45** (2022), no. 5, 427–431, DOI 10.1007/s40009-022-01149-z. MR4487640

[12] Raksha Poojary, Arathi Bhat K., Subramanian Arumugam, and Manjunatha Prasad Karantha, *Stress of a graph and its computation*, AKCE Int. J. Graphs Comb. **20** (2023), no. 2, 200–208, DOI 10.1080/09728600.2023.2243630. MR4637192

Ramco Institute of Technology, Tamil Nadu-626117, India
Email address: s.arumugam.klu@gmail.com

Department of Mathematics, Manipal Institute of Technology, Manipal Academy of Higher Education, Manipal, Karnataka-576104, India
Email address: arathi.bhat@manipal.edu

Department of Applied Statistics and Data Science, Prasanna School of Public Health, Manipal Academy of Higher Education, Manipal, Karnataka-576104, India
Email address: km.prasad@manipal.edu

Postgraduate Department of Mathematics, Poornaprajna College and Postgraduate Centre, Udupi, Karnataka-576101, India
Email address: rakshayogeesh@gmail.com

Contemporary Mathematics
Volume **833**, 2026
https://doi.org/10.1090/conm/833/16661

Stability of quaternion matrix polynomials

Pallavi Basavaraju, Shrinath Hadimani, and Sachindranath Jayaraman

Abstract. A right quaternion matrix polynomial is an expression of the form $P(\lambda) = \sum_{i=0}^{m} A_i\lambda^i$, where A_i's are $n \times n$ quaternion matrices with $A_m \neq 0$. The aim of this manuscript is to determine the location of right eigenvalues of $P(\lambda)$ relative to certain subsets of the set of quaternions. In particular, we extend the notion of (hyper)stability of complex matrix polynomials to quaternion matrix polynomials and obtain location of right eigenvalues of $P(\lambda)$ using the following methods: (1) we give a relation between (hyper)stability of a quaternion matrix polynomial and its complex adjoint matrix polynomial, (2) we prove that $P(\lambda)$ is stable with respect to an open (closed) ball in the set of quaternions, centered at a complex number if and only if it is stable with respect to its intersection with the set of complex numbers and (3) as a consequence of (1) and (2), we prove that right eigenvalues of $P(\lambda)$ lie between two concentric balls of specific radii in the set of quaternions centered at the origin. A generalization of the Eneström-Kakeya theorem to quaternion matrix polynomials is obtained as an application. We identify classes of quaternion matrix polynomials for which stability and hyperstability are equivalent. We finally deduce hyperstability of certain univariate quaternion matrix polynomials via stability of certain multivariate quaternion matrix polynomials.

1. Introduction

One of the prominent results in mathematics is the well-known fundamental theorem of algebra, which says any scalar polynomial with complex coefficients and degree $m \geq 1$ has a zero in the set of complex numbers and the number of zeros is equal to m. Although this theorem gives a precise number of zeros of a polynomial, when the degree of the polynomial exceeds 4, it is difficult to calculate the zeros algebraically. Therefore, finding regions in which the zeros lie plays a crucial role in certain iterative methods [**25**]. There is a vast literature on the location of zeros of polynomials. Some of these are [**11**], [**12**], [**18**], [**30**], [**32**] and references cited therein.

A natural generalization of scalar polynomials is to allow the coefficients to be complex matrices. Polynomials, whose coefficients are complex matrices, also known as matrix polynomials in the literature, arise in several problems in applied

---

2020 *Mathematics Subject Classification.* Primary 15A18, 15B33; Secondary 12E15, 15A66.
Key words and phrases. Right quaternion matrix polynomials, right eigenvalues of quaternion matrix polynomials, complex adjoint matrix polynomials, stability and hyperstability of quaternion matrix polynomials, multivariate quaternion matrix polynomials.

mathematics. An excellent monograph on this subject is the one by Gohberg et al. [**13**]. The eigenvalues of a matrix polynomial are the same as the zeros of the determinant (which is scalar complex polynomial) of the matrix polynomial. Therefore, one can apply methods given in the above cited references and scalar stability tests (see [**17**], [**21**]) to this complex polynomial to obtain the location of eigenvalues. However, when the size of coefficient matrices is large, computing the determinant is difficult. Therefore, the location of eigenvalues of complex matrix polynomials has been an interesting problem. The early attempts to find bounds on eigenvalues of complex matrix polynomials are due to Higham and Tisseur [**15**], by associating block matrices and scalar polynomials to the given matrix polynomial. Various other bounds obtained by associating scalar polynomials to the given matrix polynomial can be found in [**9**], [**20**], [**22**] and [**29**]. Bounds for eigenvalues of matrix polynomials, very similar to those of Higham and Tisseur were obtained recently in [**14**]. The readers may refer to [**8**] and [**23**] for various applications of matrix polynomials and related computational issues. In [**5**] and [**6**], the authors have extended some of these methods to rational matrices.

Recently, polynomials whose coefficients are from the noncommutative ring of quaternions has received interest of some mathematicians. Locating the zeros of such polynomials pose interesting questions due to the noncommutativity of quaternion multiplication. Readers may refer to the following recent articles [**3**] and [**24**] and the references cited therein.

For quaternion matrices as well as matrix polynomials whose coefficients have entries from quaternions, (henceforth known as quaternion matrix polynomials), the definition of the determinant is not same as that of the complex case. However, there are notions of determinant that have been used in the literature (see for instance, [**26**] and [**28**]). Bounds on the moduli of eigenvalues of quaternion matrices and quaternion matrix polynomials can be determined by localization and perturbation theorems such as the Geršgorin theorem, Bauer-Fike theorem and Ostrowski's theorem. Bounds are also determined by associating a block matrix and the norm of the coefficient matrices. These results can be found in [**2**], [**3**] and [**4**]. One can also introduce notions of stability of (complex) matrix polynomials, which does not involve the determinant. In [**31**], the authors give two such notions for complex matrix polynomials to locate eigenvalues.

This present work is an attempt to extend the notions of stability and hyperstability for quaternion matrix polynomials. Due to the noncommutativity of quaternions, these definitions are not the same as in the complex case, although some of the results for quaternion matrix polynomials are similar to the ones presented in [**31**] for complex matrix polynomials. A more elaborate description of the results obtained in this manuscript will be given later, after introducing some preliminaries.

2. Preliminaries

Preliminary ideas, definitions and basic facts needed in the main results are presented in this section. We divide this section into two subsections. We collect basic notions in the complex case in the first subsection. The second one concerns notions from the ring of quaternions.

2.1. The complex case.

$\mathbb{C}$ stands for the field of complex numbers. The space of all $n \times n$ matrices over $\mathbb{C}$ is denoted by $M_n(\mathbb{C})$. A complex matrix polynomial of degree m and size $n \times n$ is a function $P\colon \mathbb{C} \to M_n(\mathbb{C})$ defined as $P(\lambda) = A_m\lambda^m + A_{m-1}\lambda^{m-1} + \cdots + A_1\lambda + A_0$, where coefficients $A_0, A_1, \ldots, A_m \in M_n(\mathbb{C})$ and $A_m \neq 0$. If the map $\lambda \mapsto \det P(\lambda)$ is not identically the zero function, then the matrix polynomial $P(\lambda)$ is said to be regular. A complex number λ_0 is called an eigenvalue of a regular matrix polynomial $P(\lambda)$, if there is a vector $y \in \mathbb{C}^n \setminus \{0\}$ such that $P(\lambda_0)y$ is a zero vector in $\mathbb{C}^n$. The vector $y \in \mathbb{C}^n$ is called an eigenvector of $P(\lambda)$ corresponding to the eigenvalue λ_0. Equivalently, if $\det P(\lambda_0) = 0$, then $\lambda_0 \in \mathbb{C}$ is an eigenvalue for a given matrix polynomial $P(\lambda)$. We now introduce two important definitions that set the tone for this work. Let Ω be any nonempty subset of $\mathbb{C}$ and $P(\lambda)$ be an $n \times n$ regular complex matrix polynomial of degree m.

DEFINITION 2.1. If for each nonzero vector $y \in \mathbb{C}^n$ and each $\mu \in \Omega$, there exists a nonzero vector $z \in \mathbb{C}^n$ such that $z^*P(\mu)y \neq 0$, then the matrix polynomial $P(\lambda)$ is said to be stable with respect to Ω.

It is easy to verify that $P(\lambda)$ has no eigenvalue in Ω if and only if $P(\lambda)$ is stable with respect to Ω. A much stronger notion, namely, hyperstability of $P(\lambda)$ is defined as follows:

DEFINITION 2.2. If for each nonzero vector $y \in \mathbb{C}^n$, there exists a nonzero vector $z \in \mathbb{C}^n$ such that $z^*P(\mu)y \neq 0$ for all $\mu \in \Omega$, then $P(\lambda)$ is said to be hyperstable with respect to Ω.

Notice that if $P(\lambda)$ is hyperstable with respect to Ω, then it is stable with respect to Ω, although they are not equivalent in general (see for instance Example 3 from [**31**]). Some results on hyperstability along with certain inequalities for matrix polynomials can also be found in [**27**].

2.2. The quaternions setting.

The space of real quaternions is defined by $\mathbb{H} := \{a_0 + a_1 i + a_2 j + a_3 k : a_i \in \mathbb{R}\}$ with $i^2 = j^2 = k^2 = ijk = -1, ij = k = -ji, jk = i = -kj$ and $ki = j = -ik$. Quaternions, introduced by the Irish mathematician William Rowan Hamilton in 1843, are an extension of the complex numbers. They play crucial roles in several areas such as $3D$ computer graphics, signal processing, computer visions and robotics, quantum mechanics and so on (see [**1**] and [**19**] for more details). For $q = a_0 + a_1 i + a_2 j + a_3 k \in \mathbb{H}$, $|q| := \sqrt{a_0^2 + a_1^2 + a_2^2 + a_3^2}$ denotes the modulus of q and $\overline{q} := a_0 - a_1 i - a_2 j - a_3 k$ denotes the conjugate of q. Quaternions p and q are similar if $s^{-1}qs = p$ for some $s \in \mathbb{H} \setminus \{0\}$. If quaternions p and q are similar, then $|p| = |q|$. Let $M_n(\mathbb{H})$ denote the set of all $n \times n$ matrices over $\mathbb{H}$. For $A = (q_{ij}) \in M_n(\mathbb{H})$, the adjoint of A is defined as $A^* = (\overline{q}_{ij})$.

Let $\mathbb{H}^n$ denote the set of all column vectors with entries from the quaternions. For $v = \begin{bmatrix} v_1 \\ \vdots \\ v_n \end{bmatrix} \in \mathbb{H}^n$, $v_j \in \mathbb{H}$ and $\alpha \in \mathbb{H}$, let $v\alpha = \begin{bmatrix} v_1\alpha \\ \vdots \\ v_n\alpha \end{bmatrix}$. With respect to the standard addition and the scalar multiplication defined above, $\mathbb{H}^n$ is a right quaternion vector space over the division ring $\mathbb{H}$. Standard linear algebra concepts such as linear independence, spanning sets, basis, direct sums, and so on work in exactly the same way as for finite dimensional vector spaces over commutative division rings (see [**10**] and [**28**] for details). The subspace spanned by $v_1, v_2, \ldots, v_n \in \mathbb{H}^n$ is denoted by $\text{span}_{\mathbb{H}}\{v_1, v_2, \ldots, v_n\} = \{v_1q_1 + v_2q_2 + \cdots + v_nq_n : q_i \in \mathbb{H}\}$. The inner product on $\mathbb{H}^n$ is defined by $\langle u, v\rangle = v^*u$ for $u, v \in \mathbb{H}^n$.

The set of quaternions being a noncommutative division ring, there are notions of left and right eigenvalues for matrices in $M_n(\mathbb{H})$. If $A \in M_n(\mathbb{H})$, a quaternion λ_0 is a right (left) eigenvalue if there exists a vector $y \in \mathbb{H}^n \setminus \{0\}$ such that $Ay = y\lambda_0$ ($Ay = \lambda_0 y$). It is important to note that if λ_0 is a right eigenvalue of A, any quaternion similar to λ_0 is also a right eigenvalue of A. Thus there are infinitely many right eigenvalues for $A \in M_n(\mathbb{H})$. However, given $A \in M_n(\mathbb{H})$, there are exactly n complex numbers with nonnegative imaginary parts that are right eigenvalues of A (Theorem 5.4, [**33**]). These right eigenvalues are called the standard eigenvalues of A. Every quaternion that is a right eigenvalue of A falls within one of the equivalence classes of these standard eigenvalues. In contrast, determining the number of left eigenvalues and their equivalence classes remains an open question. Unlike right eigenvalues, a quaternion similar to a left eigenvalue need not be a left eigenvalue. In [**16**], the authors prove that for a 2×2 quaternion matrix the number of left eigenvalues is either at most 2 or infinite and for an $n \times n$ complex matrix the number of left quaternion eigenvalues is either at most n or infinite. However, it remains unknown if an $n \times n$ quaternion matrix with a finite number of left eigenvalues has at most n elements. Thus, left eigenvalues have received less attention in the literature than right eigenvalues.

We now introduce quaternion matrix polynomials. Due to noncommutativity of $\mathbb{H}$ there are notions of right and left quaternion matrix polynomials.

DEFINITION 2.3. An $n \times n$ right quaternion matrix polynomial of degree m is a function $P \colon \mathbb{H} \to M_n(\mathbb{H})$, where $P(\lambda) = \sum_{i=0}^{m} A_i\lambda^i$, with $A_i \in M_n(\mathbb{H})$ and $A_m \neq 0$.

Note that the indeterminate λ appears on the right side of the matrix coefficients. We now define eigenvalues of a right quaternion matrix polynomial.

DEFINITION 2.4. A quaternion $\lambda_0 \in \mathbb{H}$ is a right (left) eigenvalue of $P(\lambda)$ if there exists a nonzero vector $y \in \mathbb{H}^n$ such that $\sum_{i=0}^{m} A_i y \lambda_0^i = 0$ $\left(\sum_{i=0}^{m} A_i \lambda_0^i y = 0\right)$.

Similarly, one can define a left quaternion matrix polynomial and its left eigenvalues as follows.

DEFINITION 2.5. An $n \times n$ left quaternion matrix polynomial of degree m is a map $Q : \mathbb{H} \to M_n(\mathbb{H})$ given by $Q(\lambda) = \sum_{i=0}^{m} \lambda^i A_i$, with $A_i \in M_n(\mathbb{H})$ and $A_m \neq 0$ (the indeterminate λ is on the left of the matrix coefficients). A quaternion λ_0 is a left eigenvalue of $Q(\lambda)$ if there exists a nonzero vector $y \in \mathbb{H}^n$ such that $\sum_{i=0}^{m} \lambda_0^i A_i y = 0$.

However, defining right eigenvalues for left quaternion matrix polynomials is not possible in the same manner. The exact number of left eigenvalues for a quaternion matrix polynomial remains undetermined, making this an underexplored area in mathematical literature. As a result, our focus in this manuscript is entirely on right eigenvalues of Right quaternion matrix polynomials, which we henceforth refer to as quaternion matrix polynomials. We also refer to the right eigenvalues as eigenvalues, to avoid confusion. Additional notations and definitions will be introduced as and when needed later on.

In the following two examples, we illustrate the difficulty in finding the eigenvalues of quaternion matrices and quaternion matrix polynomials due to noncommutativity of quaternion multiplication. The first one is an example of a quaternion matrix.

EXAMPLE 2.6. Let $A = \begin{bmatrix} 0 & j \\ i & 0 \end{bmatrix}$. If $\lambda_0 \in \mathbb{H}$ is a right eigenvalue of A, then there is a nonzero vector $x = \begin{bmatrix} x_1 \\ x_2 \end{bmatrix} \in \mathbb{H}^2$ such that $Ax = x\lambda_0$. On expanding, we get $\begin{bmatrix} 0 & j \\ i & 0 \end{bmatrix} \begin{bmatrix} x_1 \\ x_2 \end{bmatrix} = \begin{bmatrix} x_1 \\ x_2 \end{bmatrix} \lambda_0$. This implies $jx_2 = x_1\lambda_0$, $ix_1 = x_2\lambda_0$ and both x_1, x_2 are nonzero quaternions. On simplification, we get $kx_2 = x_2\lambda_0^2$. Due to noncommutativity of quaternions, solving this equation is challenging. Similar challenges occur in finding left eigenvalues as well.

The following is an example of a quaternion matrix polynomial where computing the eigenvalues is not obvious. We skip the calculations.

EXAMPLE 2.7. Consider $P(\lambda) = \begin{bmatrix} 1 & 0 \\ 0 & 0 \end{bmatrix} \lambda^2 + \begin{bmatrix} 0 & 0 \\ i & 0 \end{bmatrix} \lambda + \begin{bmatrix} j & 0 \\ 0 & 1 \end{bmatrix}$.

3. A brief discussion of the results obtained

A brief description of the results obtained in this manuscript follows. We define the notions of stability and hyperstability for quaternion matrix polynomials. Due to noncommutativity of quaternion multiplication, these definitions are not the same as in the complex case. We begin by proving some results for quaternion matrix polynomials that are similar to those that appear in [**31**] for complex matrix polynomials. In particular, we prove that if a matrix polynomial is hyperstable with respect to a subset Ω of $\mathbb{H}$, then it is stable with respect to Ω. Since multiplication in quaternions is noncommutative, the proofs are not straightforward generalizations from the complex case, and hence we give detailed proofs (see Propositions 4.4 and 4.5). An important result (Theorem 4.8) of this manuscript gives a relation between (hyper)stability of a quaternion matrix polynomial and (hyper)stability of the corresponding complex adjoint matrix polynomial (see Definition 4.7). We use this result to prove that stability and hyperstability are generally not equivalent for quaternion matrix polynomials. We then proceed to prove that in order to verify stability of a quaternion matrix polynomial with respect to an open or closed ball in the set of quaternions, centered at any complex number, it suffices to verify its stability with respect to a smaller set in $\mathbb{C}$ (Theorem 4.12). As an application of Theorems 4.8 and 4.12, we prove stability of a quaternion matrix polynomial with respect to two concentric balls centered at the origin (Theorem 4.19). This can be viewed as an analogue of a similar result obtained by Higham and Tisseur for complex matrix polynomials (Lemma 3.1 of [**15**]). A generalization of the Eneström-Kakeya theorem to quaternion matrix polynomials (Theorem 4.23) is yet another interesting application of the results obtained in this section. Some classes of quaternion matrix polynomials for which stability and hyperstability are equivalent are also brought out. We end with a result on verifying hyperstability of a quaternion matrix polynomial by introducing these notions for certain specific multivariate quaternion matrix polynomials.

4. Main results

The main results are presented in this section.

4.1. Stability and hyperstability of quaternion matrix polynomials in one variable.

We deal with single variable quaternion matrix polynomials and prove results that are analogous to those from [**31**]. We begin with the definitions of stability, hyperstability and numerical range of quaternion matrix polynomials. Since multiplication in $\mathbb{H}$ is noncommutative, these definitions are not the same as in the complex case. Let Ω denote any nonempty subset of the set of quaternions $\mathbb{H}$ and

$$P(\lambda) = A_m\lambda^m + A_{m-1}\lambda^{m-1} + \cdots + A_1\lambda + A_0 \tag{4.1}$$

be an $n \times n$ matrix polynomial of degree m, where $A_0, A_1, \ldots, A_m \in M_n(\mathbb{H})$ and $A_m \neq 0$.

DEFINITION 4.1. $P(\lambda)$ is said to be stable with respect to Ω if for any nonzero vector $y \in \mathbb{H}^n$ and for any $\mu \in \Omega$ there exists a nonzero vector $z \in \mathbb{H}^n$ such that

$$z^*A_m y\mu^m + z^*A_{m-1}y\mu^{m-1} + \cdots + z^*A_1 y\mu + z^*A_0 y \neq 0. \tag{4.2}$$

DEFINITION 4.2. $P(\lambda)$ is said to be hyperstable with respect to Ω if for any nonzero vector $y \in \mathbb{H}^n$ there exists a nonzero vector $z \in \mathbb{H}^n$ such that

$$z^*A_m y\mu^m + z^*A_{m-1}y\mu^{m-1} + \cdots + z^*A_1 y\mu + z^*A_0 y \neq 0 \text{ for all } \mu \in \Omega. \tag{4.3}$$

DEFINITION 4.3. The numerical range of $P(\lambda)$ as given in (4.1) is the set

$$W(P) = \{\lambda \in \mathbb{H} : y^*A_m y\lambda^m + \cdots + y^*A_1 y\lambda + y^*A_0 y = 0 \text{ for some nonzero } y \in \mathbb{H}^n\}.$$

The following result gives a relation between the location of eigenvalues and the above notion of stability with respect to any subset of $\mathbb{H}$.

PROPOSITION 4.4. *Let $P(\lambda)$ be as in (4.1) and $\Omega \subseteq \mathbb{H}$ be nonempty. Then $P(\lambda)$ has no eigenvalues in Ω if and only if it is stable with respect to Ω.*

PROOF. Assume that $P(\lambda)$ has no eigenvalues in Ω. If $P(\lambda)$ is not stable with respect to Ω, then there exists a nonzero vector $y \in \mathbb{H}^n$ and a quaternion $\mu \in \Omega$ such that for any $z \in \mathbb{H}^n \setminus \{0\}$,

$$z^*A_m y\mu^m + z^*A_{m-1}y\mu^{m-1} + \cdots + z^*A_1 y\mu + z^*A_0 y = 0. \tag{4.4}$$

This implies

$$z^*\left(A_m y\mu^m + A_{m-1}y\mu^{m-1} + \cdots + A_1 y\mu + A_0 y\right) = 0. \tag{4.5}$$

Taking $z = e_i$, where e_i is the vector in $\mathbb{H}^n$ with 1 in the ith component and zeros elsewhere, we get

$$A_m y\mu^m + A_{m-1}y\mu^{m-1} + \cdots + A_1 y\mu + A_0 y = 0, \tag{4.6}$$

contradicting the assumption that $P(\lambda)$ has no eigenvalue in Ω.

Conversely, let $P(\lambda)$ be stable with respect to Ω. If $\mu \in \Omega$ is an eigenvalue of $P(\lambda)$, then there exists a nonzero vector $y \in \mathbb{H}^n$ such that

$$A_m y\mu^m + A_{m-1}y\mu^{m-1} + \cdots + A_1 y\mu + A_0 y = 0. \tag{4.7}$$

Then for any nonzero vector $z \in \mathbb{H}^n$,

$$z^*A_m y\mu^m + z^*A_{m-1}y\mu^{m-1} + \cdots + z^*A_1 y\mu + z^*A_0 y = 0, \tag{4.8}$$

which contradicts the assumption that $P(\lambda)$ is stable with respect to Ω. This completes the proof. □

We now prove that hyperstability of a matrix polynomial is stronger than stability with respect to Ω, even when its numerical range does not intersect Ω.

PROPOSITION 4.5. *Let $P(\lambda)$ be as in* (4.1) *and let $\Omega \subseteq \mathbb{H}$ be nonempty. Consider the following conditions:*

(a) $\Omega \cap W(P) = \emptyset$.
(b) *$P(\lambda)$ is hyperstable with respect to Ω.*
(c) *$P(\lambda)$ is stable with respect to Ω.*

Then (a) $\implies$ (b) $\implies$ (c).

PROOF. We first prove (a) $\implies$ (b). Let $y \in \mathbb{H}^n \setminus \{0\}$. Since $W(P) \cap \Omega = \emptyset$, for any $\lambda \in \Omega$, we have $\sum_{i=0}^{m} y^*A_i y\lambda^i \neq 0$. In particular, taking $z = y$, we obtain $\sum_{i=0}^{m} z^*A_i y\lambda^i \neq 0$ for all $\lambda \in \Omega$. This proves that $P(\lambda)$ is hyperstable with respect to Ω. The second implication follows from the definition of stability. □

As in the complex case, the reverse implications are not true in Proposition 4.5. Our first major result of this manuscript (Theorem 4.8), which gives a relation between (hyper)stability of a quaternion matrix polynomial and its complex adjoint matrix polynomial illustrates this. We begin with the definition of the complex adjoint matrix of a quaternion matrix.

DEFINITION 4.6. Any matrix $A \in M_n(\mathbb{H})$ can be expressed as $A = A_1 + A_2 j$, where $A_1, A_2 \in M_n(\mathbb{C})$. The complex adjoint matrix of A is a $2n \times 2n$ complex block matrix defined as $\chi_A := \begin{bmatrix} A_1 & A_2 \\ -\bar{A}_2 & \bar{A}_1. \end{bmatrix}$.

Readers may refer to [**28**] or [**33**] for more details about various properties of the complex adjoint matrix. The above definition suggests that given a quaternion matrix polynomial $P(\lambda)$, we can associate to $P(\lambda)$, a complex matrix polynomial $P_\chi(\lambda)$ as follows.

DEFINITION 4.7. Given an $n \times n$ quaternion matrix polynomial $P(\lambda) = \sum_{i=0}^{m} A_i\lambda^i$ we associate a $2n \times 2n$ complex matrix polynomial $P_\chi : \mathbb{C} \to M_{2n}(\mathbb{C})$ defined by $P_\chi(\lambda) = \sum_{i=0}^{m} \chi_{A_i}\lambda^i$. We call $P_\chi(\lambda)$ as the complex adjoint matrix polynomial of $P(\lambda)$.

THEOREM 4.8. *Let $P(\lambda)$ be as in (4.1) and $\Omega \subseteq \mathbb{H}$ be nonempty. Then (i) $P(\lambda)$ is stable with respect to $\Omega \cap \mathbb{C}$ if and only if its complex adjoint matrix polynomial $P_\chi(\lambda)$ is stable with respect to $\Omega\cap\mathbb{C}$. (ii) $P(\lambda)$ is hyperstable with respect to $\Omega\cap\mathbb{C}$ if and only if its complex adjoint matrix polynomial $P_\chi(\lambda)$ is hyperstable with respect to $\Omega \cap \mathbb{C}$.*

PROOF. We only prove the first statement as the proof of the second statement follows similarly. (i) Let $P(\lambda)$ be stable with respect to $\Omega\cap\mathbb{C}$. To prove that $P_\chi(\lambda)$ is stable with respect to $\Omega \cap \mathbb{C}$, consider a nonzero vector $Y = \begin{bmatrix} y_1 \\ y_2 \end{bmatrix} \in \mathbb{C}^{2n}$ and $\mu \in \Omega\cap\mathbb{C}$, where $y_1,\, y_2 \in \mathbb{C}^n$. Define $y := y_1 - \overline{y}_2 j$. Then $y \in \mathbb{H}^n \setminus \{0\}$. Since $P(\lambda)$ is stable with respect to $\Omega\cap\mathbb{C}$, there exists $z \in \mathbb{H}^n \setminus \{0\}$ such that $\sum_{i=0}^{m} z^* A_i y \mu^i \neq 0$. Writing $z = z_1 + z_2 j$, where $z_1,\, z_2 \in \mathbb{C}^n$ and $A_i = A_{i1} + A_{i2}$, where $A_{i1},\, A_{i2} \in M_n(\mathbb{C})$ for $1 \le i \le m$, we have $z^* = z_1^* - z_2^T j$ and

$$\sum_{i=0}^{m} z^* A_i y \mu^i = \sum_{i=0}^{m} (z_1^* - z_2^T j)(A_{i1} + A_{i2} j)(y_1 - \overline{y}_2 j)\mu^i \neq 0.$$

On expanding, we get

$$\sum_{i=0}^{m} \left(z_1^* A_{i1} y_1 + z_1^* A_{i2} y_2 + z_2^T \overline{A}_{i2} y_1 - z_2^T \overline{A}_{i1} y_2\right) \mu^i +$$
$$\left(\sum_{i=0}^{m} \left(z_1^* A_{i2} \overline{y}_1 - z_1^* A_{i1} \overline{y}_2 - z_2^T \overline{A}_{i1} \overline{y}_1 - z_2^T \overline{A}_{i2} \overline{y}_2\right) \overline{\mu}^i \right) j \neq 0.$$

We therefore have

$$\sum_{i=0}^{m} \left(z_1^* A_{i1} y_1 + z_1^* A_{i2} y_2 + z_2^T \overline{A}_{i2} y_1 - z_2^T \overline{A}_{i1} y_2\right) \mu^i \neq 0 \tag{4.9}$$

or

$$\sum_{i=0}^{m} \left(z_1^* A_{i2} \overline{y}_1 - z_1^* A_{i1} \overline{y}_2 - z_2^T \overline{A}_{i1} \overline{y}_1 - z_2^T \overline{A}_{i2} \overline{y}_2\right) \overline{\mu}^i \neq 0. \tag{4.10}$$

From Equations (4.9) and (4.10) we have

$$\sum_{i=0}^{m} \begin{bmatrix} z_1^* & -z_2^T \end{bmatrix} \begin{bmatrix} A_{i1} & A_{i2} \\ -\overline{A}_{i2} & \overline{A}_{i1} \end{bmatrix} \begin{bmatrix} y_1 \\ y_2 \end{bmatrix} \mu^i \neq 0, \tag{4.11}$$

or

$$\sum_{i=0}^{m} \begin{bmatrix} -z_2^* & -z_1^T \end{bmatrix} \begin{bmatrix} A_{i1} & A_{i2} \\ -\overline{A}_{i2} & \overline{A}_{i1} \end{bmatrix} \begin{bmatrix} y_1 \\ y_2 \end{bmatrix} \mu^i \neq 0. \tag{4.12}$$

This implies $\begin{bmatrix} z_1 & -\overline{z_2} \end{bmatrix}^* P_\chi(\mu) \begin{bmatrix} y_1 \\ y_2 \end{bmatrix} \neq 0$ or $\begin{bmatrix} -z_2 & -\overline{z_1} \end{bmatrix}^* P_\chi(\mu) \begin{bmatrix} y_1 \\ y_2 \end{bmatrix} \neq 0$. This proves that $P_\chi(\lambda)$ is stable with respect to $\Omega \cap \mathbb{C}$.

Conversely, assume that $P_\chi(\lambda)$ is stable with respect to $\Omega \cap \mathbb{C}$. To prove that $P(\lambda)$ is stable with respect to $\Omega \cap \mathbb{C}$, consider $y \in \mathbb{H}^n \setminus \{0\}$ and $\mu \in \Omega \cap \mathbb{C}$. Write $y = y_1 + y_2 j$, where $y_1,\, y_2 \in \mathbb{C}^n$. Define $Y := \begin{bmatrix} y_1 \\ -y_2 \end{bmatrix} \in \mathbb{C}^{2n} \setminus \{0\}$. Since $P_\chi(\lambda)$ is stable with respect to $\Omega \cap \mathbb{C}$, there exists $Z \in \mathbb{C}^{2n} \setminus \{0\}$ such that $Z^* P_\chi(\mu) Y \neq 0$. If $Z = \begin{bmatrix} z_1 \\ z_2 \end{bmatrix}$, where z_1, $z_2 \in \mathbb{C}^n$, then

$$Z^* P_\chi(\mu) Y = \sum_{i=0}^{m} \begin{bmatrix} z_1^* & z_2^* \end{bmatrix} \begin{bmatrix} A_{i1} & A_{i2} \\ -\overline{A}_{i2} & \overline{A}_{i1} \end{bmatrix} \begin{bmatrix} y_1 \\ -\overline{y}_2 \end{bmatrix} \mu^i \neq 0.$$

On multiplication we get

$$\sum_{i=0}^{m} \left(z_1^* A_{i1} y_1 - z_1^* A_{i2} \overline{y}_2 - z_2^* \overline{A}_{i2} y_1 - z_2^* \overline{A}_{i1} \overline{y}_2 \right) \mu^i \neq 0. \tag{4.13}$$

Define $z := z_1 - \overline{z}_2 j \in \mathbb{H}^n \setminus \{0\}$. Then,

$$\begin{aligned} \sum_{i=0}^{m} z^* A_i y \mu^i &= \sum_{i=0}^{m} (z_1^* + z_2^* j)(A_{i1} + A_{i2} j)(y_1 + y_2 j) \mu^i \\ &= \sum_{i=0}^{m} \left(z_1^* A_{i1} y_1 - z_1^* A_{i2} \overline{y}_2 - z_2^* \overline{A}_{i2} y_1 - z_2^* \overline{A}_{i1} \overline{y}_2 \right) \mu^i \\ &\quad + \left(\sum_{i=0}^{m} \left(z_1^* A_{i1} y_2 + z_1^* A_{i2} \overline{y}_1 - z_2^* \overline{A}_{i2} y_2 + z_2^* \overline{A}_{i1} \overline{y}_1 \right) \overline{\mu}^i \right) j. \end{aligned} \tag{4.14}$$

From Equations (4.13) and (4.14), we get $\sum_{i=0}^{m} z^* A_i y \mu^i \neq 0$. This proves that $P(\lambda)$ is stable with respect to $\Omega \cap \mathbb{C}$. □

The following examples illustrate that the reverse implication is not true in Proposition 4.5.

EXAMPLE 4.9. The numerical range of a matrix A and the numerical range of a linear matrix polynomial $P(\lambda) = I\lambda - A$ coincide. Therefore, condition (a) may not necessarily be equivalent to conditions (b) and (c). This is supported by the following example. Let $P(\lambda) = I\lambda - A$, where $A = \begin{bmatrix} 1 & 0 \\ 0 & 0 \end{bmatrix}$. It is easy to verify that the numerical range of $P(\lambda)$ is the interval $[0,1]$ on the real line. Note that from Theorem 5.4 of [**33**], the only right eigenvalues of $P(\lambda)$ are 0 and 1. Therefore $P(\lambda)$ is stable with respect to $\Omega = \mathbb{H} \setminus \{0,1\}$. We now show that $P(\lambda)$ is also hyperstable with respect to Ω. Let $y = \begin{bmatrix} y_1 \\ y_2 \end{bmatrix} \in \mathbb{H}^2 \setminus \{0\}$. For $y_2 \neq 0$, consider $z = \begin{bmatrix} 0 \\ 1 \end{bmatrix}$. Then for any $\mu \in \Omega$,

$$\begin{bmatrix} 0 & 1 \end{bmatrix} \begin{bmatrix} 1 & 0 \\ 0 & 1 \end{bmatrix} \begin{bmatrix} y_1 \\ y_2 \end{bmatrix} \mu - \begin{bmatrix} 0 & 1 \end{bmatrix} \begin{bmatrix} 1 & 0 \\ 0 & 0 \end{bmatrix} \begin{bmatrix} y_1 \\ y_2 \end{bmatrix} = y_2 \mu \neq 0. \tag{4.15}$$

If $y_2 = 0$, then $y_1 \neq 0$. In this case, take $z = \begin{bmatrix} 1 \\ 0 \end{bmatrix}$. Then for any $\mu \in \Omega$,

$$\begin{bmatrix} 1 & 0 \end{bmatrix} \begin{bmatrix} 1 & 0 \\ 0 & 1 \end{bmatrix} \begin{bmatrix} y_1 \\ y_2 \end{bmatrix} \mu - \begin{bmatrix} 1 & 0 \end{bmatrix} \begin{bmatrix} 1 & 0 \\ 0 & 0 \end{bmatrix} \begin{bmatrix} y_1 \\ y_2 \end{bmatrix} = y_1(\mu + 1)\mu \neq 0. \tag{4.16}$$

Thus $P(\lambda)$ is hyperstable with respect to Ω. Note that $W(P) \cap \Omega \neq \emptyset$.

EXAMPLE 4.10. This example shows that stability need not imply hyperstability. Let $P\colon \mathbb{H} \to M_2(\mathbb{H})$ be defined by $P(\lambda) = \begin{bmatrix} 1 & \lambda \\ \lambda & \lambda^2+1 \end{bmatrix}$. Let $y = \begin{bmatrix} y_1 \\ y_2 \end{bmatrix} \in \mathbb{H}^2$. If $\begin{bmatrix} 0 & 0 \\ 0 & 1 \end{bmatrix} \begin{bmatrix} y_1 \\ y_2 \end{bmatrix} \lambda^2 + \begin{bmatrix} 0 & 1 \\ 1 & 0 \end{bmatrix} \begin{bmatrix} y_1 \\ y_2 \end{bmatrix} \lambda + \begin{bmatrix} 1 & 0 \\ 0 & 1 \end{bmatrix} \begin{bmatrix} y_1 \\ y_2 \end{bmatrix} = \begin{bmatrix} y_2\lambda + y_1 \\ y_2\lambda^2 + y_1\lambda + y_2 \end{bmatrix} = \begin{bmatrix} 0 \\ 0 \end{bmatrix}$, then $y_2\lambda + y_1 = 0$ and $y_2\lambda^2 + y_1\lambda + y_2 = 0$. This gives $y_1 = 0 = y_2$ and so y is the zero vector. Therefore, no quaternion is an eigenvalue of $P(\lambda)$ and it is stable with respect to any subset $\Omega \subseteq \mathbb{H}$. Let $\Omega = \{q \in \mathbb{H} : |q| \leq 1\}$ and $\Omega' = \{c \in \mathbb{C} : |c| \leq 1\}$. Note that $\Omega \cap \mathbb{C} = \Omega'$. We show that the complex adjoint matrix polynomial $P_\chi(\lambda)$ is not hyperstable with respect to Ω'. Note that $P_\chi(\lambda) = \begin{bmatrix} 0 & 0 & 0 & 0 \\ 0 & 1 & 0 & 0 \\ 0 & 0 & 0 & 0 \\ 0 & 0 & 0 & 1 \end{bmatrix} \lambda^2 + \begin{bmatrix} 0 & 1 & 0 & 0 \\ 1 & 0 & 0 & 0 \\ 0 & 0 & 0 & 1 \\ 0 & 0 & 1 & 0 \end{bmatrix} \lambda + \begin{bmatrix} 1 & 0 & 0 & 0 \\ 0 & 1 & 0 & 0 \\ 0 & 0 & 1 & 0 \\ 0 & 0 & 0 & 1 \end{bmatrix}$. If $y = \begin{bmatrix} 0 & 0 & 0 & 1 \end{bmatrix}^T$, then for any nonzero vector $z = \begin{bmatrix} z_1 & z_2 & z_3 & z_4 \end{bmatrix}^T \in \mathbb{C}^4$, $z^* P_\chi(\lambda) y$ can be computed as $\overline{z}_3\lambda + \overline{z}_4(\lambda^2+1)$. We then have the following cases:

(i) If $z_3 = z_4 = 0$, then for any $\lambda_0 \in \Omega'$, $z^* P_\chi(\lambda_0) y = 0$.
(ii) If $z_3 \neq 0$, $z_4 = 0$, then for $\lambda_0 = 0 \in \Omega'$, $z^* P_\chi(\lambda_0) y = 0$.
(iii) If $z_3 = 0$, $z_4 \neq 0$, then for $\lambda_0 = i \in \Omega'$, $z^* P_\chi(\lambda_0) y = 0$.
(iv) If $z_3 \neq 0$, $z_4 \neq 0$, then $z^* P_\chi(\lambda) y = \overline{z}_3\lambda + \overline{z}_4(\lambda^2+1) = \overline{z}_4\lambda^2 + \overline{z}_3\lambda + \overline{z}_4$, a quadratic polynomial with complex coefficients. Since the product of its roots is equal to $\frac{\overline{z}_4}{\overline{z}_4} = 1$, one of the roots has to lie inside Ω'. Therefore, with respect to Ω', $P_\chi(\lambda)$ is not hyperstable. From Theorem 4.8, $P(\lambda)$ is not hyperstable with respect to Ω', which in turn implies that $P(\lambda)$ is not hyperstable with respect to Ω.

In the result that follows, we simplify stability analysis of a quaternion matrix polynomial by reducing verification of quaternion stability to a subset of complex numbers. In particular we prove that in order to check for stability of a quaternion matrix polynomial with respect to an open (or closed) ball in the set of quaternions, that is centered at any complex number, it is sufficient to check its stability with respect to a smaller set contained in the set of complex numbers. We begin with a few notations. Let $B(p;r) = \{q \in \mathbb{H} : |q-p| < r\}$ represent the open ball in $\mathbb{H}$ centered at $p \in \mathbb{H}$ and radius $r > 0$ and let $\overline{B(p;r)} = \{q \in \mathbb{H} : |q-p| \leq r\}$ be its closure. We prove this result for open balls and the proof follows verbatim for closed balls. We make use of the following lemma.

LEMMA 4.11. *Let $P(\lambda)$ be as in* (4.1) *and $B(a;r)$ be an open ball in $\mathbb{H}$ centered at $a \in \mathbb{C}$ of radius $r > 0$. Then $P(\lambda)$ is stable with respect to $B(a;r) \cap \mathbb{C}$ if and only if $P(\lambda)$ is stable with respect to $B(\overline{a};r) \cap \mathbb{C}$.*

PROOF. Suppose $P(\lambda)$ is stable with respect to $B(a;r) \cap \mathbb{C}$. Take a nonzero vector $y \in \mathbb{H}^n$ and let $\mu \in B(\overline{a};r) \cap \mathbb{C}$ so that $|\mu - \overline{a}| < r$. Since $|\overline{\mu} - a| = |\mu - \overline{a}|$ we have $\overline{\mu} \in B(a;r) \cap \mathbb{C}$. Therefore, for the nonzero vector $w := -yj \in \mathbb{H}^n$ and $\overline{\mu} \in B(a;r) \cap \mathbb{C}$, there exists a nonzero vector $z \in \mathbb{H}^n$ such that $z^*A_m w\overline{\mu}^m + \cdots + z^*A_1 w\overline{\mu} + z^*A_0 w = z^*A_m(-yj)\overline{\mu}^m + \cdots + z^*A_1(-yj)\overline{\mu} + z^*A_0(-yj) \neq 0$. Post-multiplying the above identity by j we have

$$z^*A_m(-yj)\overline{\mu}^m j + \cdots + z^*A_1(-yj)\overline{\mu} j + z^*A_0(-yj)j \neq 0.$$

Since $jq = \overline{q}j$ and $j^2 = -1$ we get

$$z^*A_m y\mu^m + \cdots + z^*A_1 y\mu + z^*A_0 y \neq 0.$$

This proves that $P(\lambda)$ is stable with respect to $B(\overline{a}; r) \cap \mathbb{C}$. The converse follows from a similar argument. □

We are now in a position to prove the aforesaid theorem.

THEOREM 4.12. *Let $P(\lambda)$ be as in (4.1) and $\Omega = B(a; r)$ be an open ball in $\mathbb{H}$ centered at $a \in \mathbb{C}$ of radius $r > 0$. Then $P(\lambda)$ is stable with respect to Ω if and only if $P(\lambda)$ is stable with respect to $\Omega \cap \mathbb{C}$.*

PROOF. Assume that $P(\lambda)$ is stable with respect to $\Omega \cap \mathbb{C}$. Suppose $P(\lambda)$ is not stable with respect to Ω. We first consider the case when a is a complex number with nonnegative imaginary part. Then $P(\lambda)$ has an eigenvalue $\mu \in \Omega$ which is not a complex number. Since every eigenvalue lies in one of the equivalence classes of standard eigenvalues, μ is similar to a standard eigenvalue $\mu_0 \in \mathbb{C}$ of $P(\lambda)$. Let q be a nonzero quaternion such that $\mu = q^{-1}\mu_0 q$. By Lemma 3.3 of [**4**], we have $|\mu_0 - a| \leq |q^{-1}\mu_0 q - a| = |\mu - a|$. Since $|\mu - a| < r$, we get $|\mu_0 - a| < r$, a contradiction to the fact that $P(\lambda)$ is stable with respect to $\Omega \cap \mathbb{C}$. This proves that $P(\lambda)$ is stable with respect to Ω and the first step is complete.

If a is a complex number whose imaginary part is negative, then $\overline{a}$ is a complex number with positive imaginary part. By Theorem 2.1(5) of [**33**], a and $\overline{a}$ are similar, and hence $p^{-1}\overline{a}p = a$ for some $p \in \mathbb{H} \setminus \{0\}$. Once again, by Lemma 3.3 of [**4**], we have

$$|\mu_0 - \overline{a}| \leq |q^{-1}\mu_0 q - p^{-1}\overline{a}p| = |\mu - a| < r. \tag{4.17}$$

We thus get $\mu_0 \in B(\overline{a}; r) \cap \mathbb{C}$. This shows that $P(\lambda)$ is not stable with respect to $B(\overline{a}; r) \cap \mathbb{C}$. From Lemma 4.11, we conclude that $P(\lambda)$ is not stable with respect to $\Omega \cap \mathbb{C}$ as well. Therefore for any $a \in \mathbb{C}$, $P(\lambda)$ is stable with respect to $B(a; r)$. For the converse, it is clear that if $P(\lambda)$ is stable with respect to Ω, then it is stable with respect to $\Omega \cap \mathbb{C}$, thereby completing the proof. □

Some remarks are in order.

REMARK 4.13. Theorem 4.12 is not true if $a \in \mathbb{H} \setminus \mathbb{C}$, as the following example illustrates. Let $P(\lambda) = \begin{bmatrix} 1 & 0 \\ 0 & 1 \end{bmatrix} \lambda + \begin{bmatrix} j & 0 \\ 0 & j \end{bmatrix}$. It is easy to verify that the set of eigenvalues of $P(\lambda)$ is $\sigma(P) = \{\mu \in \mathbb{H} : \mu = s^{-1}js \text{ for some nonzero } s \in \mathbb{H}\}$. Let $\Omega = B(j; 1)$. Notice that $P(\lambda)$ is not stable with respect to Ω as $j \in \Omega$. From Theorem 2.2 of [**33**], we have $\sigma(P) \cap \mathbb{C} = \{-i, i\}$. Therefore, $-i$ and i are the only complex eigenvalues of $P(\lambda)$. Since $-i, i \notin \Omega \cap \mathbb{C}$, $P(\lambda)$ is stable with respect to $\Omega \cap \mathbb{C}$. This example also suggests that Theorem 4.12 does not hold for an arbitrary convex set.

4.2. Stable sets for arbitrary quaternion matrix polynomials.

In this section, we make use of Theorems 4.8 and 4.12 to obtain particular sets with respect to which quaternion matrix polynomials are stable. The main results (Theorems 4.14, 4.18 and 4.19) are analogous to Lemma 3.1 of [**15**], due to Higham and Tisseur for complex matrix polynomials. We first prove that a quaternion matrix polynomial is stable with respect to a particular open ball centered at zero.

For $A \in M_n(\mathbb{H})$, the spectral norm is denoted by $\|A\|$ and is defined as $\|A\| = \sup_{x \neq 0} \left\{ \frac{\|Ax\|_2}{\|x\|_2} : x \in \mathbb{H}^n \right\}$. Note that $\|\chi_A\| = \|A\|$.

THEOREM 4.14. *Let $P(\lambda)$ be as in (4.1) with A_0 invertible. Let $l(z) = \|A_m\| z^m + \cdots + \|A_1\| z - \|A_0^{-1}\|^{-1}$ be a complex polynomial. Then $P(\lambda)$ is stable with respect to $\Omega = B(0; r) \subseteq \mathbb{H}$, where r is the unique positive real zero of $l(z)$.*

PROOF. Consider the complex adjoint matrix polynomial of $P(\lambda)$ given by $P_\chi(\lambda) = \chi_{A_m}\lambda^m + \cdots + \chi_{A_1}\lambda + \chi_{A_0}$. Since A_0 is invertible, it follows from Theorem 4.2(5) of [**33**] that χ_{A_0} is invertible and $\chi_{A_0}^{-1} = \chi_{A_0^{-1}}$. Therefore, by Lemma 3.1 of [**15**], $P_\chi(\lambda)$ is stable with respect to $D = \{z \in \mathbb{C} : |z| < r\}$, where r is the unique positive real root of the complex polynomial $\|\chi_{A_m}\| z^m + \cdots + \|\chi_{A_1}\| z - \|\chi_{A_0}^{-1}\|^{-1} = \|A_m\| z^m + \cdots + \|A_1\| z - \|A_0^{-1}\|^{-1} = l(z)$. This proves stability of $P_\chi(\lambda)$ with respect to $\Omega \cap \mathbb{C}$. From Theorem 4.8, we conclude that $P(\lambda)$ is stable with respect to $\Omega \cap \mathbb{C}$. Finally, Theorem 4.12, yields that $P(\lambda)$ is stable with respect to Ω. □

We now proceed to prove in Theorem 4.18, that an arbitrary quaternion matrix polynomial is stable with respect to the complement of a closed ball centered at zero in $\mathbb{H}$. The following set of lemmas set the tone for this. We prove them in the order of preference.

LEMMA 4.15. *Let $P(\lambda)$ be as in (4.1). Then zero is an eigenvalue of $P(\lambda)$ if and only if A_0 is non-invertible.*

PROOF. Suppose zero is an eigenvalue of $P(\lambda)$ and $y \in \mathbb{H}^n$ is a nonzero vector such that $A_m y 0^m + \cdots + A_1 y 0 + A_0 y = 0$. Thus $A_0 y = 0$, thereby proving that A_0 is not invertible. Conversely, if A_0 is not invertible, then χ_{A_0} is not invertible. Therefore, zero is an eigenvalue of χ_{A_0}, which in turn implies that zero is an eigenvalue of A_0. We thus have $A_0 y = 0$ for some nonzero vector $y \in \mathbb{H}^n$ so that $A_m y 0^m + \cdots + A_1 y 0 + A_0 y = 0$. This proves that zero is an eigenvalue of $P(\lambda)$. □

LEMMA 4.16. *Let $P(\lambda)$ be as in (4.1) with A_m and A_0 invertible. Then $\lambda_0 \in \mathbb{H}$ is an eigenvalue of $P(\lambda)$ if and only if $\frac{1}{\lambda_0}$ is an eigenvalue of $Q(\lambda) = P\left(\frac{1}{\lambda}\right)\lambda^m = A_0\lambda^m + A_1\lambda^{m-1} + \cdots + A_m$.*

PROOF. Let $\lambda_0 \in \mathbb{H}$ be an eigenvalue of $P(\lambda)$. By Lemma 4.15, $\lambda_0 \neq 0$. Then, $A_m y \lambda_0^m + \cdots + A_1 y \lambda_0 + A_0 y = 0$ for some $y \in \mathbb{H}^n \setminus \{0\}$. Post-multiplying by $\frac{1}{\lambda_0^m}$ on both the sides, we get $A_0 y \frac{1}{\lambda_0^m} + A_1 y \frac{1}{\lambda_0^{m-1}} + \cdots + A_m y = 0$. This implies $\frac{1}{\lambda_0}$ is an eigenvalue of $Q(\lambda)$. The converse follows in a similar argument. □

LEMMA 4.17. *Let $P(\lambda)$ be as in (4.1) with A_m, A_0 invertible and let $\Omega = \overline{B(0; r)} \subseteq \mathbb{H}$ with $r > 0$. If $P(\lambda)$ is stable with respect to $(\mathbb{H} \setminus \Omega) \cap \mathbb{C}$, then $P(\lambda)$ is stable with respect to $\mathbb{H} \setminus \Omega$.*

PROOF. Suppose $P(\lambda)$ is not stable with respect to $\mathbb{H} \setminus \Omega$. Then, there exists an eigenvalue μ of $P(\lambda)$ in $\mathbb{H} \setminus \Omega$. By Lemma 4.16, $\frac{1}{\mu}$ is an eigenvalue of $Q(\lambda) = P\left(\frac{1}{\lambda}\right)\lambda^m$. Then $Q(\lambda)$ is not stable with respect to $B\left(0; \frac{1}{r}\right)$. Therefore, by Theorem

4.12, $Q(\lambda)$ is not stable with respect to $B\left(0; \frac{1}{r}\right) \cap \mathbb{C}$. This in turn implies there exists an eigenvalue $\delta \in \mathbb{C}$ of $Q(\lambda)$, with $|\delta| < \dfrac{1}{r}$. Once again by Lemma 4.16, $\dfrac{1}{\delta}$ is an eigenvalue of $P(\lambda)$, thereby implying that $P(\lambda)$ is not stable with respect to $(\mathbb{H} \setminus \Omega) \cap \mathbb{C}$. This contradiction proves that $P(\lambda)$ is stable with respect to $\mathbb{H} \setminus \Omega$. $\square$

We now prove the main result stated at the beginning of this section.

THEOREM 4.18. *Let $P(\lambda)$ be as in (4.1) with A_m and A_0 invertible. Let $u(z) = \|A_m^{-1}\|^{-1} z^m - \|A_{m-1}\| z^{m-1} - \cdots - \|A_0\|$ be a complex polynomial. Then $P(\lambda)$ is stable with respect to $\left(\mathbb{H} \setminus \overline{B(0; R)}\right) \subseteq \mathbb{H}$, where R is the unique positive real zero of $u(z)$.*

PROOF. Consider the complex adjoint matrix polynomial $P_\chi(\lambda) = \chi_{A_m} \lambda^m + \cdots + \chi_{A_1} \lambda + \chi_{A_0}$ of $P(\lambda)$. Invertibility of A_m and A_0 implies χ_{A_m} and χ_{A_0} are invertible. Therefore, by Lemma 3.1 of [**15**], $P_\chi(\lambda)$ is stable with respect to $D = \{z \in \mathbb{C} : |z| > R\}$, where R is the unique positive real zero of the complex polynomial $\|\chi_{A_m}^{-1}\|^{-1} z^m - \|\chi_{A_{m-1}}\| z^{m-1} - \cdots - \|\chi_{A_0}\| = \|A_m^{-1}\|^{-1} z^m - \|A_{m-1}\| z^{m-1} - \cdots - \|A_0\| = u(z)$. That is, $P_\chi(\lambda)$ is stable with respect to $\left(\mathbb{H} \setminus \overline{B(0; R)}\right) \bigcap \mathbb{C}$. From Theorem 4.8, we conclude that $P(\lambda)$ is stable with respect to $\left(\mathbb{H} \setminus \overline{B(0; R)}\right) \bigcap \mathbb{C}$. Finally, the required conclusion follows from Lemma 4.17. $\square$

We thus have the following theorem as an application of Theorems 4.14 and 4.18. We state this below.

THEOREM 4.19. *Right eigenvalues of a quaternion matrix polynomial $P(\lambda)$ with invertible leading and constant coefficients lie in the set $\{q \in \mathbb{H} : r \leq |q| \leq R\}$, where r and R are as given in Theorems 4.14 and 4.18 respectively.*

REMARK 4.20. In [**7**], the authors prove that if $P(\lambda)$ is a quaternion matrix polynomial with unitary matrix coefficients, then the eigenvalues of $P(\lambda)$ lie in the set $\{q \in \mathbb{H} : \frac{1}{2} \leq |q| \leq 2\}$. This result was proved while studying the Hoffman-Wielandt inequality for quaternion matrix polynomials.

As the following example illustrates, both the lower and upper bounds r and R are attained.

EXAMPLE 4.21. In this example, both the lower and the upper bounds obtained from Theorems 4.14 and 4.18 coincide with the minimum and the maximum of the moduli of eigenvalues respectively. Let $P(\lambda) = A_2 \lambda^2 + A_1 \lambda + A_0$ where $A_2 = A_0 = \begin{bmatrix} 1 & 0 \\ 0 & 1 \end{bmatrix}$ and $A_1 = \begin{bmatrix} i & 0 \\ 0 & j \end{bmatrix}$. It is easy to verify that the minimum and the maximum of the moduli of eigenvalues of $P(\lambda)$ are $\dfrac{-1+\sqrt{5}}{2}$ and $\dfrac{1+\sqrt{5}}{2}$ respectively. From Theorem 4.14, the lower bound on the set of moduli of eigenvalues is the unique positive zero of the complex polynomial $l(z) = \|A_2\| z^2 + \|A_1\| z - \|A_0^{-1}\|^{-1} = z^2 + z - 1$ given by $r = \dfrac{-1+\sqrt{5}}{2}$. Similarly, from Theorem 4.18, the upper bound is the unique positive zero of the complex polynomial $u(z) = \|A_2^{-1}\|^{-1} z^2 - \|A_1\| z - \|A_0\| = z^2 - z - 1$ given by $R = \dfrac{1+\sqrt{5}}{2}$.

We now bring out yet another application of the results obtained in this section. Recall that $A \in M_n(\mathbb{H})$ is said to be positive semidefinite (positive definite) if for every $x \in \mathbb{H}^n$, $x^*Ax \geq 0$ ($x^*Ax > 0$ for all $0 \neq x \in \mathbb{H}^n$). It is customary to denote positive semidefiniteness and positive definiteness by the notations $A \succeq 0$ and $A \succ 0$ respectively. It turns out that a quaternion matrix A is positive (semi)definite if and only if its complex adjoint matrix is positive (semi)definite (see Proposition 3.4.1, [**28**]). It follows from the definition that any right eigenvalue of a positive (semi)definite matrix is nonnegative(positive). One of the classical results on the location of the roots of complex polynomials inside the unit disc or more generally an annulus is the Eneström-Kakeya theorem. There is a vast literature on this result as one can infer from MathSciNet. In Theorem 2.3 of [**20**], the authors prove a generalization of this for complex matrix polynomials. We state this below.

THEOREM 4.22. *Let $P(\lambda) = A_m\lambda^m + \cdots + A_1\lambda + A_0$ be a matrix polynomial whose coefficients $A_i \in M_n(\mathbb{C})$ satisfy $A_m \succeq A_{m-1} \succeq \cdots \succeq A_0 \succeq 0; A_m \succ 0$. Then each eigenvalue λ_0 of $P(\lambda)$ satisfies $\frac{\lambda_{\min}(A_0)}{2\lambda_{\max}(A_m)} \leq |\lambda_0| \leq 1$, where $\lambda_{\min}(A_0)$ denotes the smallest eigenvalue of A_0 and $\lambda_{\max}(A_m)$ denotes the largest eigenvalue of A_m.*

We now derive an analogue of the above theorem for quaternion matrix polynomials.

THEOREM 4.23. *Let $P(\lambda) = A_m\lambda^m + \cdots + A_1\lambda + A_0$ be an $n \times n$ quaternion matrix polynomial whose coefficients $A_i \in M_n(\mathbb{H})$ satisfy $A_m \succeq A_{m-1} \succeq \cdots \succeq A_0 \succ 0$. Then each right eigenvalue λ_0 of $P(\lambda)$ satisfies $\frac{\lambda_{\min}(A_0)}{2\lambda_{\max}(A_m)} \leq |\lambda_0| \leq 1$, where $\lambda_{\min}(A_0)$ denotes the smallest right eigenvalue of A_0 and $\lambda_{\max}(A_m)$ denotes the largest right eigenvalue of A_m.*

PROOF. Observe that $\frac{\lambda_{\min}(A_0)}{2\lambda_{\max}(A_m)} = \frac{\lambda_{\min}(\chi_{A_0})}{2\lambda_{\max}(\chi_{A_m})}$. Let Ω_1 and Ω_2 denote respectively the open and closed balls in $\mathbb{H}$ of radii $\frac{\lambda_{\min}(A_0)}{2\lambda_{\max}(A_m)}$ and 1. Positive (semi)definiteness of the matrix coefficients (in $\mathbb{H}$) imply positive (semi)definiteness of the respective complex adjoint matrices; moreover the latter are ordered in the same way as the A_i's. From Theorem 4.22, it follows that the complex adjoint matrix polynomial $P_\chi(\lambda)$ of $P(\lambda)$ is stable with respect to $\Omega_1 \cap \mathbb{C}$ and $\left(\mathbb{H} \setminus \Omega_2\right) \cap \mathbb{C}$. Resorting to Theorems 4.8 and 4.12, we deduce that the quaternion matrix polynomial $P(\lambda)$ is stable with respect to Ω_1. From Proposition 4.4, we infer that any right eigenvalue λ_0 of $P(\lambda)$ satisfies $\frac{\lambda_{\min}(A_0)}{2\lambda_{\max}(A_m)} \leq |\lambda_0|$. A similar argument yields that the quaternion matrix polynomial $P(\lambda)$ is stable with respect to $\mathbb{H} \setminus \Omega_2$ and therefore any right eigenvalue λ_0 of $P(\lambda)$ is also of absolute value at most 1. This completes the proof of the theorem. □

4.3. Class of matrix polynomials for which stability implies hyperstability.

We now proceed to identify classes of quaternion matrix polynomials for which stability implies hyperstability. Our first result in this direction concerns block upper triangular matrix polynomials, where each of the diagonal blocks are hyperstable with respect to a common set Ω.

THEOREM 4.24. *Let $P(\lambda)$ be a block upper triangular quaternion matrix polynomial and $\Omega \subseteq \mathbb{H}$ be nonempty. If each of the diagonal blocks of $P(\lambda)$ are hyperstable with respect to Ω, then $P(\lambda)$ is hyperstable with respect to Ω.*

PROOF. Let $P(\lambda)$ be an $n \times n$ quaternion matrix polynomial of degree m and of the form

$$P(\lambda) = \begin{bmatrix} P_{11}(\lambda) & P_{12}(\lambda) & \cdots & P_{1t}(\lambda) \\ 0 & P_{22}(\lambda) & \cdots & P_{2t}(\lambda) \\ \vdots & \vdots & \ddots & \vdots \\ 0 & 0 & \cdots & P_{tt}(\lambda) \end{bmatrix},$$

where $P_{ij}(\lambda) = \sum_{l=0}^{m} A_l^{(ij)} \lambda^l$ are matrix polynomials of size $k_i \times k_j$ for $1 \leq i, j \leq t$ and $\sum_{j=1}^{t} k_j = n$. Assume that $P_{ii}(\lambda)$'s are hyperstable with respect to Ω for $1 \leq i \leq t$. Consider a nonzero vector $y = \begin{bmatrix} y_1^T & y_2^T & \cdots & y_t^T \end{bmatrix}^T \in \mathbb{H}^n$, where $y_i \in \mathbb{H}^{k_i}$, for $1 \leq i \leq t$. Let r denote the index of the last nonzero y_i. Since $P_{rr}(\lambda)$ is hyperstable with respect to Ω, there exists $z_r \in \mathbb{H}^{k_r} \setminus \{0\}$ such that for all $\lambda \in \Omega$

$$z_r^* A_m^{(rr)} y_r \lambda^m + \cdots + z_r^* A_1^{(rr)} y_r \lambda + z_r^* A_0^{(rr)} y_r \neq 0.$$

Set $z := \begin{bmatrix} 0 & \cdots & z_r^T & \cdots & 0 \end{bmatrix}^T \in \mathbb{H}^n \setminus \{0\}$. On expanding $P(\lambda)$, we have

$$(4.18) \qquad P(\lambda) = \begin{bmatrix} \sum_{l=0}^{m} A_l^{(11)} \lambda^l & \sum_{l=0}^{m} A_l^{(12)} \lambda^l & \cdots & \sum_{l=0}^{m} A_l^{(1t)} \lambda^l \\ 0 & \sum_{l=0}^{m} A_l^{(22)} \lambda^l & \cdots & \sum_{l=0}^{m} A_l^{(2t)} \lambda^l \\ \vdots & \vdots & \ddots & \vdots \\ 0 & 0 & \cdots & \sum_{l=0}^{m} A_l^{(tt)} \lambda^l \end{bmatrix} = \sum_{l=0}^{m} A_l \lambda^l,$$

where $A_l = \begin{bmatrix} A_l^{(11)} & A_l^{(12)} & \cdots & A_l^{(1t)} \\ 0 & A_l^{(22)} & \cdots & A_l^{(2t)} \\ \vdots & \vdots & \ddots & \vdots \\ 0 & 0 & 0 & A_l^{(tt)} \end{bmatrix}$ for $1 \leq l \leq m$. For $1 \leq l \leq m$, consider

(4.19)

$$z^* A_l y = \begin{bmatrix} 0 & \cdots & z_r^* & \cdots & 0 \end{bmatrix} \begin{bmatrix} A_l^{(11)} & A_l^{(12)} & \cdots & A_l^{(1t)} \\ 0 & A_l^{(22)} & \cdots & A_l^{(2t)} \\ \vdots & \vdots & \ddots & \vdots \\ 0 & 0 & 0 & A_l^{(tt)} \end{bmatrix} \begin{bmatrix} y_1 \\ \vdots \\ y_r \\ \vdots \\ 0 \end{bmatrix} = z_r^* A_l^{(rr)} y_r.$$

We thus have $\sum_{l=0}^{m} z^* A_l y \lambda^i = \sum_{l=0}^{m} z_r^* A_l^{(rr)} y_r \lambda^i$. Since $P_{rr}(\lambda)$ is hyperstable with respect to Ω, for any $\lambda \in \Omega$, $\sum_{l=0}^{m} z^* A_l y \lambda^i = \sum_{l=0}^{m} z_r^* A_l^{(rr)} y_r \lambda^i \neq 0$. This proves that $P(\lambda)$ is hyperstable with respect to Ω. $\square$

We now discuss equivalence of stability and hyperstability for quaternion polynomials. This will be used in the theorem that follows.

THEOREM 4.25. *Let $p\colon \mathbb{H} \to \mathbb{H}$ be a quaternion scalar polynomial and Ω be any subset of $\mathbb{H}$. Then $p(\lambda)$ is hyperstable with respect to Ω if and only if $p(\lambda)$ is stable with respect to Ω.*

PROOF. Let $p(\lambda) = a_m\lambda^m + \cdots + a_1\lambda + a_0$, where $a_i \in \mathbb{H}$, for $i = 0, 1, \ldots, m$. If $p(\lambda)$ is hyperstable with respect to Ω, then certainly, it is stable with respect to Ω by definition. Suppose $p(\lambda)$ is stable with respect to Ω. Let $y \in \mathbb{H} \setminus \{0\}$ be arbitrary. Taking $z = 1$, we have $z^*a_my\mu^m + \cdots + z^*a_1y\mu + z^*a_0y \neq 0$ for all $\mu \in \Omega$. For otherwise, $a_my\mu_0^m + \cdots + a_1y\mu_0 + a_0y = 0$ for some $\mu_0 \in \Omega$. Since $y \neq 0$, μ_0 is an eigenvalue of $p(\lambda)$, a contradiction to Proposition 4.4. Therefore, $z^*a_my\mu^m + \cdots + z^*a_1y\mu + z^*a_0y \neq 0$ for all $\mu \in \Omega$. This proves hyperstability of $p(\lambda)$ with respect to Ω. □

REMARK 4.26. Note that, in the complex case, a scalar is an eigenvalue of a scalar polynomial if and only if it is a zero of that polynomial. In the quaternion case, this condition does not generally hold. For example, consider $p(\lambda) = \lambda - k$. It is easy to verify that $-k$ is an eigenvalue of $p(\lambda)$ with eigenvector $y = i + j$, but not a zero of $p(\lambda)$. However, we have the following result for quaternion scalar polynomials.

LEMMA 4.27. *Let $p(\lambda) = \lambda^m + a_{m-1}\lambda^{m-1} + \cdots + a_1\lambda + a_0$ be a quaternion monic scalar polynomial. Then λ_0 is an eigenvalue of $p(\lambda)$ if and only if λ_0 is a zero of the scalar polynomial $\sum_{k=1}^{2m} \sum_{i+j=k} a_i\overline{a_j}\lambda^k$.*

PROOF. Consider the companion matrix $A = \begin{bmatrix} 0 & 1 & \cdots & 0 \\ \vdots & \vdots & \ddots & \vdots \\ 0 & 0 & \cdots & 1 \\ -a_0 & -a_1 & \cdots & -a_{m-1} \end{bmatrix}$ of $p(\lambda)$. Theorem 5.1 of [**2**] tells us that λ_0 is an eigenvalue of $p(\lambda)$ if and only if λ_0 is an eigenvalue of A. The desired conclusion follows from Theorem 4.2 of [**26**]. □

We are now in a position to prove the main result of this section.

THEOREM 4.28. *Let $P(\lambda)$ be as in (4.1) with leading coefficient to be the identity matrix, and let $\Omega \subseteq \mathbb{H}$ be nonempty. If $P(\lambda)$ is an upper-triangular matrix polynomial, then $P(\lambda)$ is hyperstable with respect Ω if and only if it is stable with respect to Ω.*

PROOF. Let

$$\begin{aligned} P(\lambda) &= \begin{bmatrix} p_{11}(\lambda) & p_{12}(\lambda) & \cdots & p_{1n}(\lambda) \\ 0 & p_{22}(\lambda) & \cdots & p_{2n}(\lambda) \\ \vdots & \vdots & \ddots & \vdots \\ 0 & 0 & \cdots & p_{nn}(\lambda) \end{bmatrix} \\ &= I\lambda^m + A_{m-1}\lambda^{m-1} + \cdots + A_1\lambda + A_0, \end{aligned}$$

where $p_{ij}(\lambda) = \lambda^m + a_{m-1}^{(ij)}\lambda^{m-1} + \cdots + a_1^{(ij)}\lambda + a_0^{(ij)}$ are quaternion monic scalar polynomials and $A_i \in M_n(\mathbb{H})$ are upper-triangular matrices. Suppose $P(\lambda)$ is stable with respect to Ω. We then claim that $p_{ll}(\lambda)$ are stable with respect to Ω for all $l = 1, 2, \ldots, n$. If $p_{ll}(\lambda)$ is not stable with respect Ω for some $l = 1, 2, \ldots, n$, then by Proposition 4.4, there exists an eigenvalue $\lambda_0 \in \Omega$ of $p_{ll}(\lambda)$. From Lemma 4.27, it follows that λ_0 is a zero of the scalar polynomial $\sum_{k=1}^{2m} \sum_{i+j=k} a_i^{(ll)}\overline{a_j^{(ll)}}\lambda^k$. This implies that λ_0 is a zero of the scalar polynomial $\prod_{l=1}^{n}\left(\sum_{k=1}^{2m} \sum_{i+j=k} a_i^{(ll)}\overline{a_j^{(ll)}}\lambda^k\right)$. From Corollary 4.3 of [**26**], λ_0 is an eigenvalue of the block matrix

$$C = \begin{bmatrix} 0 & I & 0 & \cdots & 0 \\ 0 & 0 & I & \cdots & 0 \\ \vdots & \vdots & \vdots & \ddots & \vdots \\ 0 & 0 & 0 & \cdots & I \\ -A_0 & -A_1 & -A_2 & \cdots & -A_{m-1} \end{bmatrix}.$$

Thus, by Theorem 5.1 of [**2**], λ_0 is an eigenvalue of $P(\lambda)$, thereby proving that $P(\lambda)$ is not stable with respect to Ω. This contradiction proves that $p_{ll}(\lambda)$ are stable with respect to Ω for all $l = 1, 2, \ldots, n$. From Theorem 4.25, we then infer that $p_{ll}(\lambda)$ are hyperstable with respect to Ω for all $l = 1, 2, \ldots, n$. The desired conclusion then follows from Theorem 4.24. The converse follows from the definition. □

REMARK 4.29. In Theorem 3.7 of [**31**], the authors prove that stability and hyperstability are equivalent for complex matrix polynomials of the form $P(\lambda) = p(\lambda)A + q(\lambda)B$ with some scalar complex polynomials $p(\lambda)$, $q(\lambda)$ and $A, B \in M_n(\mathbb{C})$. The proof uses the well known generalized Schur form. However, due to noncommutativity of quaternions this is not amenable in the quaternion setting unless the variable itself is real and A and B are complex matrices. We wish to point out that Theorem 4.28 is one interesting result which avoids this classical technique and noncommutativity of quaternions comes into picture.

4.4. Stability and hyperstability of multivariate quaternion matrix polynomials.

We now define multivariate quaternion matrix polynomials and extend the notions of stability and hyperstability for the same. These are analogous to the definitions given in [**31**] for multivariate complex matrix polynomials. In particular, we establish hyperstability of univariate quaternion matrix polynomials from stability of multivariate quaternion matrix polynomials. Although some of the results are very similar to those from [**31**] in the complex case and are amenable for generalization to the quaternion setting, not everything goes through due to noncommutativity of the variables. We begin with the definitions and present some examples to illustrate the same.

DEFINITION 4.30. A multivariate right quaternion matrix polynomial in k variables of degree m and size n is a function $P(\lambda_1, \lambda_2, \ldots, \lambda_k)\colon \mathbb{H}^k \to M_n(\mathbb{H})$ defined by

$$P(\lambda_1, \lambda_2, \ldots, \lambda_k) = \sum_{w \in W} A_w w(\lambda_1, \lambda_2, \ldots, \lambda_k), \tag{4.20}$$

where W is the set of all possible words of finite length formed from the noncommuting letters $\{\lambda_1, \lambda_2, \ldots, \lambda_k\}$ such that the sum of the powers does not exceed m.

DEFINITION 4.31. Let Ω be a subset of $\mathbb{H}$. We say that the multivariate right quaternion matrix polynomial $P(\lambda_1, \lambda_2, \ldots, \lambda_k) = \sum_{w \in W} A_w w(\lambda_1, \lambda_2, \ldots, \lambda_k)$ is stable with respect to Ω^k, if for any $y \in \mathbb{H}^n \setminus \{0\}$ and $(\mu_1, \mu_2, \ldots, \mu_k) \in \Omega^k$, there exists $z \in \mathbb{H}^n \setminus \{0\}$ such that

$$\sum_{w \in W} z^* A_w y w(\mu_1, \mu_2, \ldots, \mu_k) \neq 0. \tag{4.21}$$

DEFINITION 4.32. Let Ω be a subset of $\mathbb{H}$. We say that the multivariate right quaternion matrix polynomial $P(\lambda_1, \lambda_2, \ldots, \lambda_k) = \sum_{w \in W} A_w w(\lambda_1, \lambda_2, \ldots, \lambda_k)$ is hyperstable with respect to Ω^k, if for any $y \in \mathbb{H}^n \setminus \{0\}$, there exists $z \in \mathbb{H}^n \setminus \{0\}$ such that

$$\sum_{w \in W} z^* A_w y w(\mu_1, \mu_2, \ldots, \mu_k) \neq 0 \quad \text{for all } (\mu_1, \mu_2, \ldots, \mu_k) \in \Omega^k. \tag{4.22}$$

In the following two examples, we illustrate these definitions for multivariate right quaternion matrix polynomials.

EXAMPLE 4.33. Let $P(\lambda_1, \lambda_2) = \begin{bmatrix} 1 & 0 \\ 0 & 1 \end{bmatrix} \lambda_1 \lambda_2 + \begin{bmatrix} 1 & 0 \\ 0 & 1 \end{bmatrix} \lambda_2$ and $\Omega \subseteq \mathbb{H}$ such that $0, -1 \notin \Omega$. Let us take an element $y = \begin{bmatrix} y_1 \\ y_2 \end{bmatrix} \in \mathbb{H}^2 \setminus \{0\}$ and let $(\mu_1, \mu_2) \in \Omega^2$ be arbitrary. Consider

$$\begin{bmatrix} 1 & 0 \\ 0 & 1 \end{bmatrix} \begin{bmatrix} y_1 \\ y_2 \end{bmatrix} \mu_1 \mu_2 + \begin{bmatrix} 1 & 0 \\ 0 & 1 \end{bmatrix} \begin{bmatrix} y_1 \\ y_2 \end{bmatrix} \mu_2 = \begin{bmatrix} y_1(\mu_1 + 1)\mu_2 \\ y_2(\mu_1 + 1)\mu_2 \end{bmatrix}.$$

If $y_1 \neq 0$, choose $z = \begin{bmatrix} 1 \\ 0 \end{bmatrix}$. Then, we have

$$\begin{aligned} &\begin{bmatrix} 1 & 0 \end{bmatrix} \begin{bmatrix} 1 & 0 \\ 0 & 1 \end{bmatrix} \begin{bmatrix} y_1 \\ y_2 \end{bmatrix} \mu_1 \mu_2 + \begin{bmatrix} 1 & 0 \end{bmatrix} \begin{bmatrix} 1 & 0 \\ 0 & 1 \end{bmatrix} \begin{bmatrix} y_1 \\ y_2 \end{bmatrix} \mu_2 \\ &= y_1(\mu_1 + 1)\mu_2 \neq 0 \text{ for any } (\mu_1, \mu_2) \in \Omega^2. \end{aligned}$$

If $y_2 \neq 0$, choose $z = \begin{bmatrix} 0 \\ 1 \end{bmatrix}$. Then, we have

$$\begin{aligned} &\begin{bmatrix} 0 & 1 \end{bmatrix} \begin{bmatrix} 1 & 0 \\ 0 & 1 \end{bmatrix} \begin{bmatrix} y_1 \\ y_2 \end{bmatrix} \mu_1 \mu_2 + \begin{bmatrix} 0 & 1 \end{bmatrix} \begin{bmatrix} 1 & 0 \\ 0 & 1 \end{bmatrix} \begin{bmatrix} y_1 \\ y_2 \end{bmatrix} \mu_2 \\ &= y_2(\mu_1 + 1)\mu_2 \neq 0 \text{ for any } (\mu_1, \mu_2) \in \Omega^2. \end{aligned}$$

Thus, in either of the cases, $P(\lambda_1, \lambda_2)$ is hyperstable with respect to Ω^2.

Here is an example of a matrix polynomial which is not stable with respect to a given set.

EXAMPLE 4.34. Let $P(\lambda_1,\lambda_2) = \begin{bmatrix}1 & 0\\0 & 1\end{bmatrix}\lambda_1\lambda_2 + \begin{bmatrix}1 & 0\\0 & 0\end{bmatrix}\lambda_2\lambda_1 + \begin{bmatrix}1 & 0\\0 & 1\end{bmatrix}\lambda_1 + \begin{bmatrix}1 & 0\\0 & 1\end{bmatrix}$ and $\Omega = B(0;1) \subseteq \mathbb{H}$. Consider $y = \begin{bmatrix}1\\0\end{bmatrix} \in \mathbb{H}^2$ and $(\mu_1,\mu_2) = (-\frac{1}{2},\frac{1}{2}) \in \Omega^2$. Then

$$\begin{aligned}&\begin{bmatrix}1 & 0\\0 & 1\end{bmatrix}\begin{bmatrix}1\\0\end{bmatrix}\mu_1\mu_2 + \begin{bmatrix}1 & 0\\0 & 0\end{bmatrix}\begin{bmatrix}1\\0\end{bmatrix}\mu_2\mu_1 + \begin{bmatrix}1 & 0\\0 & 1\end{bmatrix}\begin{bmatrix}1\\0\end{bmatrix}\mu_1 + \begin{bmatrix}1 & 0\\0 & 1\end{bmatrix}\begin{bmatrix}1\\0\end{bmatrix}\\ &= \begin{bmatrix}1\\0\end{bmatrix}\left(-\frac{1}{4}\right) + \begin{bmatrix}1\\0\end{bmatrix}\left(-\frac{1}{4}\right) + \begin{bmatrix}1\\0\end{bmatrix}\left(-\frac{1}{2}\right) + \begin{bmatrix}1\\0\end{bmatrix}\\ &= \begin{bmatrix}-\frac{1}{4}-\frac{1}{4}-\frac{1}{2}+1\\0\end{bmatrix} = \begin{bmatrix}0\\0\end{bmatrix}.\end{aligned}$$

Therefore, for the above chosen y and $(\mu_1,\mu_2) \in \Omega^2$, there exists no $z \in \mathbb{H}^2 \setminus \{0\}$ such that Equation (4.21) holds. Therefore, $P(\lambda_1,\lambda_2)$ is not stable with respect to Ω^2.

We present some sufficient conditions for hyperstability of quaternion matrix polynomials in one variable via stability of multivariate quaternion matrix polynomials.

THEOREM 4.35. *Let $P(\lambda) = A_2\lambda^2 + A_1\lambda + A_0$ be a quaternion matrix polynomial and $\Omega \subseteq \mathbb{H}$ be nonempty. Then $P(\lambda)$ is hyperstable with respect to Ω if any one of the following holds:*

(i) *$P(\lambda_1,\lambda_2) = A_2\lambda_1^2 + A_1\lambda_2 + A_0$ is stable with respect to Ω^2.*
(ii) *$P(\lambda_1,\lambda_2) = A_2\lambda_1\lambda_2 + A_1\lambda_2 + A_0$ is stable with respect to Ω^2 and $0 \notin \Omega$.*

PROOF. We present here only the proof of (*i*), as the proof of (*ii*) is similar. Suppose $P(\lambda_1,\lambda_2) = A_2\lambda_1^2 + A_1\lambda_2 + A_0$ is stable with respect to Ω. Note that, for $\lambda_1 = \lambda_2 = \lambda$, we have $P(\lambda_1,\lambda_2) = P(\lambda)$. Therefore, $P(\lambda)$ is stable with respect to Ω. Let $y \in \mathbb{H}^n \setminus \{0\}$ be arbitrary but fixed. If there is a $z \in \mathbb{H}^n \setminus \{0\}$ such that $\langle A_2y, z\rangle = 0, \langle A_1y, z\rangle = 0$ and $\langle A_0y, z\rangle \neq 0$, then

$$z^*A_2y\lambda^2 + z^*A_1y\lambda + z^*A_0y = z^*A_0y \neq 0 \text{ for any } \lambda \in \Omega.$$

Therefore $P(\lambda)$ is hyperstable with respect to Ω.

If there is no $z \in \mathbb{H}^n \setminus \{0\}$ such that $\langle A_2y, z\rangle = 0, \langle A_1y, z\rangle = 0$ and $\langle A_0y, z\rangle \neq 0$, then $A_0y \in \text{span}_{\mathbb{H}}\{A_2y, A_1y\}$, so that $A_0y = A_2yq_2 + A_1yq_1$ for some $q_1, q_2 \in \mathbb{H}$. Then, we have

$$A_2y\lambda^2 + A_1y\lambda + A_0y = A_2y(\lambda^2 + q_2) + A_1y(\lambda + q_1). \tag{4.23}$$

Suppose $\mu_1, \mu_2 \in \Omega$ are zeros of $\lambda^2 + q_2$ and $\lambda + q_1$ respectively. Then, $A_2y\mu_1^2 + A_1y\mu_2 + A_0y = A_2y\mu_1^2 + A_1y\mu_2 + A_2yq_2 + A_1yq_1 = A_2y(\mu_1^2 + q_2) + A_1y(\mu_2 + q_1) = 0$. This implies $P(\lambda_1,\lambda_2)$ is not stable with respect to Ω^2, which is a contradiction to our assumption. Therefore, atleast one of $\lambda^2 + q_2$ and $\lambda + q_1$ do not have zeros in Ω.

Case (1): Assume that $\lambda^2 + q_2$ does not have zeros in Ω.

If there is a vector $z \in \mathbb{H}^n \setminus \{0\}$ such that $\langle A_1 y, z\rangle = 0$ and $\langle A_2 y, z\rangle \neq 0$, then

$$\begin{aligned} z^*A_2y\lambda^2 + z^*A_1y\lambda + z^*A_0y &= z^*A_2y(\lambda^2 + q_1) + z^*A_1y(\lambda + q_1) \quad \text{by (4.23)} \\ &= z^*A_2y(\lambda^2 + q_2) \\ &\neq 0 \quad \text{for all } \lambda \in \Omega. \end{aligned}$$

This implies $P(\lambda)$ is hyperstable with respect to Ω.

Suppose there is no vector $z \in \mathbb{H}^n \setminus \{0\}$ such that $\langle A_1 y, z\rangle = 0$ and $\langle A_2 y, z\rangle \neq 0$. Then $A_2 y \in \mathrm{span}_{\mathbb{H}}\{A_1 y\}$ and $A_2 y = A_1 y q_3$ for some $q_3 \in \mathbb{H}$. Substituting in (4.23), we get

$$A_2y\lambda^2 + A_1y\lambda + A_0y = A_1y(q_3\lambda^2 + \lambda + q_3q_2 + q_1). \tag{4.24}$$

Since $P(\lambda)$ is stable with respect to Ω, we have $A_1 y \neq 0$ and $q_3\lambda^2 + \lambda + q_3 q_2 + q_1 \neq 0$ for any $\lambda \in \Omega$. Taking $z = A_1 y$, we have

$$\begin{aligned} z^*A_2y\lambda^2 + z^*A_1y\lambda + z^*A_0y &= z^*A_1y(q_3\lambda^2 + \lambda + q_3q_2 + q_1) \quad \text{by (4.24)} \\ &= (A_1y)^*(A_1y)(q_3\lambda^2 + \lambda + q_3q_2 + q_1) \\ &\neq 0 \quad \text{for all } \lambda \in \Omega. \end{aligned}$$

Therefore, $P(\lambda)$ is hyperstable with respect to Ω.

Case (2): Assume that $\lambda + q_1$ does not have zeros in Ω.

If there exists $z \in \mathbb{H}^n \setminus \{0\}$ such that $\langle A_2 y, z\rangle = 0$ and $\langle A_1 y, z\rangle \neq 0$, then

$$\begin{aligned} z^*A_2y\lambda^2 + z^*A_1y\lambda + z^*A_0y &= z^*A_2y(\lambda^2 + q_1) + z^*A_1y(\lambda + q_1) \quad \text{by (4.23)} \\ &= z^*A_1y(\lambda + q_1) \\ &\neq 0 \quad \text{for all } \lambda \in \Omega. \end{aligned}$$

This implies $P(\lambda)$ is hyperstable with respect to Ω. Therefore, we may assume that there is no $z \in \mathbb{H}^n \setminus \{0\}$ such that $\langle A_2 y, z\rangle = 0$ and $\langle A_1 y, z\rangle \neq 0$. This implies $A_1 y \in \mathrm{span}_{\mathbb{H}}\{A_2 y\}$, so that $A_1 y = A_2 y q_4$ for some $q_4 \in \mathbb{H}$. Substituting in (4.23), we get

$$A_2y\lambda^2 + A_1y\lambda + A_0y = A_2y(\lambda^2 + q_4\lambda + q_2 + q_4q_1). \tag{4.25}$$

Once again, since $P(\lambda)$ is stable with respect to Ω, we see that $A_2 y \neq 0$ and the scalar polynomial $\lambda^2 + q_4\lambda + q_2 + q_4 q_1$ does not have any zeros in Ω. Now by taking $z = A_2 y$ we have,

$$\begin{aligned} z^*A_2y\lambda^2 + z^*A_1y\lambda + z^*A_0y &= z^*A_2y(\lambda^2 + q_4\lambda + q_2 + q_4q_1) \quad \text{by (4.25)} \\ &= (A_2y)^*(A_2y)(\lambda^2 + q_4\lambda + q_2 + q_4q_1) \\ &\neq 0 \quad \text{for all } \lambda \in \Omega. \end{aligned}$$

Thus, $P(\lambda)$ is hyperstable with respect to Ω. $\square$

We end with the following theorem where a sufficient condition for hyperstability of a cubic matrix polynomial in one variable via stability of a multivariate quaternion matrix polynomial can be derived. We skip the proof as the proof technique is similar to that of the quadratic case presented in the previous theorem.

THEOREM 4.36. *Let* $P(\lambda) = A_0\lambda^3 + A_2\lambda^2 + A_1\lambda + A_0$ *be a cubic quaternion matrix polynomial. Let* $\Omega \subseteq \mathbb{H}$ *be nonempty and* $0 \notin \Omega$. *Then* $P(\lambda)$ *is hyperstable with respect to* Ω, *if the multivariate right matrix polynomial* $P(\lambda_1, \lambda_2) = A_0\lambda_2^3 + A_2\lambda_1\lambda_2 + A_1\lambda_1 + A_0$ *is stable with respect to* Ω^2.

We end the manuscript with the following remark.

REMARK 4.37. Note that in [**31**], the authors have considered other multivariate matrix polynomials as well to obtain hyperstability of a given univariate matrix polynomial. However, in the quaternion case, due to noncommutativity of variables one cannot conclude hyperstability of a given univariate matrix polynomial.

Acknowledgments

The authors thank the referees for a meticulous reading of the manuscript and for several suggestions that has improved the manuscript. Thanks are also due to the organizers of ICLAA-2023 for allowing us to publish this work in this monograph. Pallavi Basavaraju and Shrinath Hadimani acknowledge the Council of Scientific and Industrial Research (CSIR) and the University Grants Commission (UGC), Government of India, for financial support through research fellowships.

References

[1] Stephen L. Adler, *Quaternionic quantum mechanics and quantum fields*, International Series of Monographs on Physics, vol. 88, The Clarendon Press, Oxford University Press, New York, 1995. MR1333599

[2] Sk. Safique Ahmad and Istkhar Ali, *Bounds for eigenvalues of matrix polynomials over quaternion division algebra*, Adv. Appl. Clifford Algebr. **26** (2016), no. 4, 1095–1125, DOI 10.1007/s00006-016-0640-7. MR3564461

[3] Sk. Safique Ahmad and Istkhar Ali, *Localization theorems for matrices and bounds for the zeros of polynomials over quaternion division algebra*, Filomat **32** (2018), no. 2, 553–573, DOI 10.2298/fil1802553a. MR3812498

[4] Sk. Safique Ahmad, Istkhar Ali, and Ivan Slapničar, *Perturbation analysis of matrices over a quaternion division algebra*, Electron. Trans. Numer. Anal. **54** (2021), 128–149, DOI 10.1553/etna_vol54s128. MR4207060

[5] Pallavi Basavaraju, Shrinath Hadimani, and Sachindranath Jayaraman, *Bounds on the moduli of eigenvalues of rational matrices*, Results Math. **79** (2024), no. 5, Paper No. 206, 23, DOI 10.1007/s00025-024-02238-9. MR4775786

[6] P. Basavaraju, S. Hadimani, and S. Jayaraman, *Spectral bounds for certain special type of rational matrices*, arXiv:2302.02894, 2023.

[7] P. Basavaraju, S. Hadimani, and S. Jayaraman, *The Hoffman-Wielandt inequality for quaternion matrices and quaternion matrix polynomials*, arXiv:2302.07638, 2023.

[8] Timo Betcke, Nicholas J. Higham, Volker Mehrmann, Christian Schröder, and Françoise Tisseur, *NLEVP: a collection of nonlinear eigenvalue problems*, ACM Trans. Math. Software **39** (2013), no. 2, Art. 7, 28, DOI 10.1145/2427023.2427024. MR3031626

[9] Dario A. Bini, Vanni Noferini, and Meisam Sharify, *Locating the eigenvalues of matrix polynomials*, SIAM J. Matrix Anal. Appl. **34** (2013), no. 4, 1708–1727, DOI 10.1137/120886741. MR3144796

[10] Daniel Alpay, Fabrizio Colombo, and Irene Sabadini, *Inner product spaces and Krein spaces in the quaternionic setting*, Recent advances in inverse scattering, Schur analysis and stochastic processes, Oper. Theory Adv. Appl., vol. 244, Birkhäuser/Springer, Cham, 2015, pp. 33–65, DOI 10.1007/978-3-319-10335-8_4. MR3380262

[11] Matthias Dehmer, *On the location of zeros of complex polynomials*, JIPAM. J. Inequal. Pure Appl. Math. **7** (2006), no. 1, Article 26, 13. MR2217189

[12] Hunduma Legesse Geleta and Oluma Ararso Alemu, *Location of the zeros of certain complex-valued harmonic polynomials*, J. Math., posted on 2022, Art. ID 4886522, 5, DOI 10.1155/2022/4886522. MR4470446

[13] I. Gohberg, P. Lancaster, and L. Rodman, *Matrix polynomials*, Classics in Applied Mathematics, vol. 58, Society for Industrial and Applied Mathematics (SIAM), Philadelphia, PA, 2009. Reprint of the 1982 original [MR0662418], DOI 10.1137/1.9780898719024. MR3396732

[14] Sunil Hans and Samir Raouafi, *Annulus containing all the eigenvalues of a matrix polynomial*, Indian J. Pure Appl. Math. **53** (2022), no. 2, 405–412, DOI 10.1007/s13226-021-00211-8. MR4421150

[15] Nicholas J. Higham and Françoise Tisseur, *Bounds for eigenvalues of matrix polynomials*, Linear Algebra Appl. **358** (2003), 5–22, DOI 10.1016/S0024-3795(01)00316-0. Special issue on accurate solution of eigenvalue problems (Hagen, 2000). MR1942721

[16] Liping Huang and Wasin So, *On left eigenvalues of a quaternionic matrix*, Linear Algebra Appl. **323** (2001), no. 1-3, 105–116, DOI 10.1016/S0024-3795(00)00246-9. MR1809762

[17] A. Hurwitz, *Ueber die Bedingungen, unter welchen eine Gleichung nur Wurzeln mit negativen reellen Theilen besitzt* (German), Math. Ann. **46** (1895), no. 2, 273–284, DOI 10.1007/BF01446812. MR1510884

[18] A. Joyal, G. Labelle, and Q. I. Rahman, *On the location of zeros of polynomials*, Canad. Math. Bull. **10** (1967), 53–63, DOI 10.4153/CMB-1967-006-3. MR213513

[19] Jack B. Kuipers, *Quaternions and rotation sequences: A primer with applications to orbits, aerospace, and virtual reality*, Princeton University Press, Princeton, NJ, 1999. MR1670862

[20] Công-Trình Lê, Thi-Hoa-Binh Du, and Tran-Duc Nguyen, *On the location of eigenvalues of matrix polynomials*, Oper. Matrices **13** (2019), no. 4, 937–954, DOI 10.7153/oam-2019-13-66. MR4043867

[21] Morris Marden, *The Geometry of the Zeros of a Polynomial in a Complex Variable*, Mathematical Surveys, No. 3, American Mathematical Society, New York, 1949. MR31114

[22] C. Mehl, V. Mehrmann, and M. Wojtylak, *Matrix pencils with coefficients that have positive semidefinite Hermitian parts*, SIAM J. Matrix Anal. Appl. **43** (2022), no. 3, 1186–1212, DOI 10.1137/21M1439997. MR4451916

[23] Volker Mehrmann and Heinrich Voss, *Nonlinear eigenvalue problems: a challenge for modern eigenvalue methods*, GAMM Mitt. Ges. Angew. Math. Mech. **27** (2004), no. 2, 121–152 (2005), DOI 10.1002/gamm.201490007. MR2124762

[24] Abdullah Mir, *On the zeros of a quaternionic polynomial: an extension of the Eneström-Kakeya theorem*, Czechoslovak Math. J. **73(148)** (2023), no. 3, 649–662, DOI 10.21136/CMJ.2023.0097-22. MR4632850

[25] Victor Y. Pan, *Solving a polynomial equation: some history and recent progress*, SIAM Rev. **39** (1997), no. 2, 187–220, DOI 10.1137/S0036144595288554. MR1453318

[26] Ricardo Pereira and Paula Rocha, *On the determinant of quaternionic polynomial matrices and its application to system stability*, Math. Methods Appl. Sci. **31** (2008), no. 1, 99–122, DOI 10.1002/mma.901. MR2373924

[27] P. Pikul, O. J. Szymanski, and M. Wojtylak, *The Szász inequality for matrix polynomials and functional calculus*, `arXiv:2406.08965`, 2024.

[28] Leiba Rodman, *Topics in quaternion linear algebra*, Princeton Series in Applied Mathematics, Princeton University Press, Princeton, NJ, 2014, DOI 10.1515/9781400852741. MR3241695

[29] Nandita Roy and Shreemayee Bora, *Locating eigenvalues of quadratic matrix polynomials*, Linear Algebra Appl. **649** (2022), 452–490, DOI 10.1016/j.laa.2022.05.014. MR4432340

[30] Péter Gábor Szabó, *On the roots of the trinomial equation*, CEJOR Cent. Eur. J. Oper. Res. **18** (2010), no. 1, 97–104, DOI 10.1007/s10100-009-0130-2. MR2593126

[31] Oskar Jakub Szymański and Michał Wojtylak, *Stability of matrix polynomials in one and several variables*, Linear Algebra Appl. **670** (2023), 42–67, DOI 10.1016/j.laa.2023.04.003. MR4578670

[32] E. B. Van Vleck, *On the location of roots of polynomials and entire functions*, Bull. Amer. Math. Soc. **35** (1929), no. 5, 643–683, DOI 10.1090/S0002-9904-1929-04794-3. MR1561789

[33] Fuzhen Zhang, *Quaternions and matrices of quaternions*, Linear Algebra Appl. **251** (1997), 21–57, DOI 10.1016/0024-3795(95)00543-9. MR1421264

School of Mathematics, Indian Institute of Science Education and Research Thiruvananthapuram, Maruthamala P.O., Vithura, Thiruvananthapuram – 695 551, Kerala, India.

Email address: pallavipoorna20@iiservtvm.ac.in

School of Mathematics, Indian Institute of Science Education and Research Thiruvananthapuram, Maruthamala P.O., Vithura, Thiruvananthapuram – 695 551, Kerala, India.

Email address: srinathsh3320@iiservtvm.ac.in

School of Mathematics, Indian Institute of Science Education and Research Thiruvananthapuram, Maruthamala P.O., Vithura, Thiruvananthapuram – 695 551, Kerala, India.

Email address: sachindranathj@iisertvm.ac.in; sachindranathj@gmail.com

Contemporary Mathematics
Volume **833**, 2026
https://doi.org/10.1090/conm/833/16662

Vertical block P- and P_0-matrices

Aniekan A. Ebiefung

ABSTRACT. The vertical block P- and P_0-matrix classes arise in the context of solving the complementarity problem, an optimization problem that brings together disparate areas of Mathematics. Potential application areas include economics, mathematical programming, engineering, dynamical system theory, and ecology, since the square P-matrix has useful applications in these areas. The objective of this chapter is to bring together available results in these two matrix classes. We hope that making these results available to a wider audience will lead to the expansion of research and application areas of these generalized P- and P_o-matrix classes.

1. Introduction

The vertical generalized linear complementarity problem has been used as the mathematical model of many applications. Some of the applications include the formulation of the generalized Leontief input-output open model and its application to the choice of new technology [**1**], the generalization of the Nash Equilibrium Theorem [**2**], and the formulation of the global stability of a two-species linear Volterra ecosystem [**3**]. In many of these application instances, the associated matrices are the vertical block P-, P_o-, and Z-matrices. The objective of this chapter is to bring together properties of the vertical block P- and P_0-matrix classes, and to discuss their intersection with the related vertical block Z-matrix class. We start with some notation and definitions that will be used in the rest of the chapter.

DEFINITION 1.1. An $n \times n$ matrix $\mathbf{M}$ is a P-matrix if and only if all its principal minors are positive. If all the principal minors are nonnegative, then M is a P_o-matrix. Hence, the set of P-matrices is a subset of the set of P_0-matrices.

A vertical generalization of the square matrix is given below. We call this a vertical block matrix of type $(m_1, \ldots, m_n)$.

2020 *Mathematics Subject Classification.* Primary 90-02, 90C33.

Key words and phrases. P-matrix, P_0-matrix, vertical block P-matrix, vertical block P_0-matrix.

The author's research was partially supported by the University of Tennessee Chattanooga College of Arts and Sciences Research Grant, Summer 2024.

DEFINITION 1.2. A matrix N of dimensions $m \times n, m \geq n$, is called a vertical block matrix of type $(m_1, \ldots, m_n)$ if it can be partitioned row-wise into n blocks:

$$N = \begin{bmatrix} N^1 \\ \vdots \\ N^n \end{bmatrix},$$

where the jth block, N^j, is of dimension $m_j \times n$ and $m = \sum_{j=1}^{n} m_j$.

If $m_j = 1, j = 1, \ldots, n$, then N is an $n \times n$ square matrix. Thus, a vertical block matrix is a vertical generalization of a square matrix.

Associated with a vertical block matrix is the concept of representative submatrices. A representative submatrix plays important roles in proving theorems and finding algorithms for solving the vertical generalized linear complementarity problems. We start with the definitions below.

DEFINITION 1.3. Let N be a vertical block matrix of type $(m_1, \ldots, m_n)$. A square matrix M of size n is a representative submatrix of N if its jth row is from the jth block, N^j, of N.

A vertical block matrix N of type $(m_1, \ldots, m_n)$ has $\prod_{j=1}^{n} m_j$ representative submatrices.

DEFINITION 1.4. Let N be a vertical block matrix of type $(m_1, \ldots, m_n)$. A principal submatrix of N is a principal submatrix of a representative submatrix of N. The determinant of such a matrix is a principal minor of N.

DEFINITION 1.5. Let N be a vertical block matrix of type $(m_1, \ldots, m_n)$. N is called a Π-matrix of type $(m_1, \ldots, m_n)$ if and only if each representative submatrix is a square Π-matrix.

For example, N is a vertical block P-matrix of type $(m_1, \ldots, m_n)$ if each representative submatrix is a square P-matrix of size n. If the type of the matrix is clear from the context, the phrase type $(m_1, \ldots, m_n)$ may be omitted.

EXAMPLE 1.6. Let

$$N = \begin{bmatrix} 1 & 2 \\ 5 & 1 \\ -9 & 4 \\ 6 & -2 \\ 8 & 7 \end{bmatrix}, q = \begin{bmatrix} 3 \\ 2 \\ 4 \\ -1 \\ 5 \end{bmatrix}.$$

If N is a vertical block matrix of type $(2, 3)$, then N and q are partitioned into two blocks as given below.

$$N = \begin{bmatrix} N^1 \\ N^2 \end{bmatrix}, \quad q = \begin{bmatrix} q^1 \\ q^2 \end{bmatrix},$$

where

$$N^1 = \begin{bmatrix} 1 & 2 \\ 5 & 1 \end{bmatrix}, \quad N^2 = \begin{bmatrix} -9 & 4 \\ 6 & -2 \\ 8 & 7 \end{bmatrix}, \quad q^1 = \begin{bmatrix} 1 \\ 2 \end{bmatrix}, \quad q^2 = \begin{bmatrix} 4 \\ -1 \\ 5 \end{bmatrix},$$

The corresponding representative matrices are:

$$M^1 = \begin{bmatrix} 1 & 2 \\ -9 & 4 \end{bmatrix}, \quad M^2 = \begin{bmatrix} 1 & 2 \\ 6 & -2 \end{bmatrix}, \quad M^3 = \begin{bmatrix} 1 & 2 \\ 8 & 7 \end{bmatrix}.$$

$$M^4 = \begin{bmatrix} 5 & 1 \\ -9 & 4 \end{bmatrix}, \quad M^5 = \begin{bmatrix} 5 & 1 \\ 6 & -2 \end{bmatrix}, \quad M^6 = \begin{bmatrix} 5 & 1 \\ 8 & 7 \end{bmatrix}.$$

Observe that the number of representative matrices is six, that is, $m_1 \times m_2 = 2 \times 3 = 6$.

2. The Vertical Block P-matrix

Recall that the matrix N is a vertical block P-matrix of type $(m_1, \dots, m_n)$ if each representative matrix is a P-matrix of size n. The definition below is useful in obtaining properties of the vertical block P-matrix of type $(m_1, \dots, m_n)$.

DEFINITION 2.1. Let N be a vertical block matrix of type $(m_1, \dots, m_n)$. For each j, $(j = 1, \dots, n)$, let u^j be a nonzero positive row vector of dimension $1 \times m_j$. Define the matrix U by

$$U = \begin{bmatrix} u^1 & \dots & 0 \\ \vdots & \ddots & \vdots \\ 0 & \dots & u^n \end{bmatrix}. \tag{2.1}$$

Then, U is a nonzero positive matrix of dimension $n \times m$. Equation (2.1) allows us to prove the theorems below. These results and their proofs are given Cottle and Dantzg [**4**].

THEOREM 2.2. *Let N be an $m \times n, m \geq n$, vertical block matrix of type $(m_1, \dots, m_n)$. If N is a P-matrix, then*

$$M = UN$$

is a P-matrix of size n.

PROOF. Let $(N^j)_i$ be the i-th row of the jth-block, N^j. Then

$$M = UN = \begin{bmatrix} \sum_{i=1}^{m_1} u_i^1 (N^1)_i \\ \vdots \\ \sum_{i=1}^{m_n} u_i^n (N^n)_i \end{bmatrix}.$$

The determinant of M is equal to

$$detM = \sum_{i_1=1}^{m_1} \cdots \sum_{i_n=1}^{m_n} \prod_{j=1}^{n} u_{i_j}^j det \begin{bmatrix} (N^1)_{i_1} \\ \vdots \\ (N^n)_{i_n} \end{bmatrix}. \tag{2.2}$$

For every j, $(j = 1, \dots, n)$, the determinant is positive since the matrix is a representative submatrix of a vertical block P-matrix, and hence, a square P-matrix. For each j, $j_1, \dots, n$, there is an $_{i_j}$ such that $u_{i_j}^j > 0$ by the definion of the vector u^j. Hence, all the terms in (2.2) are positive. Therefore, $M = UN$ is a square P-matrix of size n. □

Gale and Nikaido [**5**] gave a geometric characterization of the set of square P-matrices below.

THEOREM 2.3. *M is an $n \times n$ P-matrix if and only if M reverses the sign of no vector except the zero vector.*

The above result was extended by Szanc [**6**] to the vertical block P-matrix of type $(m_1, \dots, m_n)$ as provided in the next definition and theorem.

DEFINITION 2.4. Let N be a vertical block P-matrix of type $(m_1, \ldots, m_n)$ and x a nonzero $n \times 1$ column vector. The matrix N is said to reverse the sign of x if for any representative submatrix M of N and $b \in R^n$, we have that

$$Mx = b, \tag{2.3}$$

and

$$x_i b_i \leq 0, (i = 1, \ldots, n). \tag{2.4}$$

THEOREM 2.5. *Let N be an $m \times n, m \geq n$, vertical block matrix of type $(m_1, \ldots, m_n)$. N is a P-matrix if and only if N reverses the sign of no nonzero vector.*

PROOF. Let N be an $m \times n, m \geq n$, vertical block P-matrix of type $(m_1, \ldots, m_n)$, and M a representative submatrix of N. If x is an $n \times 1$ column vector such that

$$Mx = b$$

and

$$x_i b_i \leq 0,$$

then x is a zero vector, since M is a square P-matrix. Thus, in this case, N is a vertical block P-matrix of type $(m_1, \ldots, m_n)$.

Conversely, suppose x is a nonzero vector. Assume that N does not reverse the sign of x. Then each representative submatrix of N does not reverse the sign of x. Hence, each representative submatrix of N is not a square P-matrix, by the definition of representative submatrices. Therefore, N is not a vertical block P-matrix of type $(m_1, \ldots, m_n)$. □

The following results and proofs are given in Cottle and Dantzig [**4**].

THEOREM 2.6. *Let N be an $m \times n, m \geq n$, vertical block matrix of type $(m_1, \ldots, m_n)$. If N is a P-matrix, then the system of inequalities*

$$Nz \leq 0, \quad z \geq 0, \tag{2.5}$$

has only the trivial solution, $z = 0$.

PROOF. We prove by contradiction. Suppose that there exists a nontrivial solution, $z > 0$, such that Equation (2.5) is true. Let

$$U = \begin{bmatrix} u^1 & \ldots & 0 \\ \vdots & \ddots & \vdots \\ 0 & \ldots & u^n \end{bmatrix}$$

be defined as in (2.1). By Theorem 2.2, $M = UN$ is a square P-matrix of size n. However,

$$Mz = U(Nz) \leq 0,$$

since $U > 0$. This contradicts the fact that a P-matrix can not reverse the sign a nonzero vector. Hence, system (2.5) has only the trivial solution. □

THEOREM 2.7. *Let N be an $m \times n, m \geq n$, vertical block matrix of type $(m_1, \ldots, m_n)$. If N is a P-matrix, then the system of inequalities*

$$Nz > 0, \quad z > 0, \tag{2.6}$$

has a solution.

PROOF. We prove by contradiction. Let u be an $n \times 1$ column vector. Assume that N is a P-matrix and that

$$Nu = b \leq 0, \tag{2.7}$$

$$u \neq 0, \; u \geq 0, \tag{2.8}$$

where $b \in R^n$. Let M be a representative submatrix of N, then M is an $n \times n$ P-matrix. Let p be an $n \times 1$ vector defined from b such that if row j of M is row i_n of N, then row j of p is from row i_n of b . Then

$$Mu = p \leq 0, \tag{2.9}$$

$$u \neq 0, u \geq 0, \tag{2.10}$$

is true. Since $u_i \geq 0, i = 1, \ldots, n$, we have that

$$u_i p_i \leq 0, i = 1, \ldots, n. \tag{2.11}$$

This implies that M reverses the sign of the nonzero nonnegative vector u, and contradicts the fact that M is a square P-matrix. □

In solving complementarity problems, a principal pivoting operation is used to interchange the basic and nonbasic variables. Consider the system of linear equations below.

$$w - Mz = q, \tag{2.12}$$

where M is $n \times n$, and w, z, and q are in R^n.

A principal pivot operation can be used to interchange the variables w_i and z_i to have the resulting system

$$u - \bar{M}v = \bar{q}. \tag{2.13}$$

Tucker [**7**] showed that if M is a square P-matrix, the resulting matrix $\bar{M}$ is also a P-matrix. We state this result below.

THEOREM 2.8 (Tucker [**7**]). *Let M be an $n \times n$ P-matrix. Every matrix $\bar{M}$ that results from a principal pivot performed on M is also a P-matrix.*

To extend this result to the vertical block P -matrix of type $(m_1, \ldots, m_n)$, we use the definition and procedures below that were first developed in Szanc [**6**].

DEFINITION 2.9. Let N be a vertical block matrix of type $(m_1, \ldots, m_n)$. Consider the system $w = Nz + q$. We say that an $m \times n, m \geq n$, vertical block matrix $\bar{N}$ of type $(m_1, \ldots, m_n)$ is the results of a principal pivot of the jth block in the (t, j) position on N, if for some $j \in \{1, \ldots, n\}$ and some $t \in \{1, \ldots, m_j\}$, the mapping

$$\bar{N}^j_{t.} : R^n \longrightarrow R$$

satisfies

$$\alpha = \bar{N}^j_{t.} z$$

if and only if

$$z_j = N^j_{t.} \bar{z}$$

where

$$\bar{z} = (z_1, \ldots, z_{j-1}, \alpha, z_{j+1}, \ldots, z_n).$$

If $k \neq j$ or if $k = j$ but $i \neq t$, then $\bar{N}^k_{i.} z = N^k_{i.} \bar{z}$.

For an $m \times n$ vertical block matrices of type $(m_1, \dots, m_n)$, principal pivots exists at the (i,j) position with respect to all $j \in \{1, \dots, n\}$ and $i \in \{1, \dots, m_j\}$. This is because $n^j_{ij} > 0$, where $N^j_{i.} = (n^j_{11}, \dots, n^j_{1n})$. That is, $N^j_{i.}$ maps R^n onto R and is an increasing function on its jth component.

Below is a formal description of a principal pivot transform on the jth block, N^j, of N, at the (i,j) position of the system

$$w = Nz + q, \tag{2.14}$$

where N is an $m \times n$ vertical block matrix of type $(m_1, \dots, m_n)$.

(1) Solve the system

$$w^j_i = N^j_{i.} z + q^j_i, \qquad i \in \{1, \dots, m_j\}$$

for the variable z_j in terms of $\{z_1, \dots, z_{j-1}, w^j_i, z_{j+1}, \dots, z_n\}$.

Replace the ith equation of the jth block in equation (2.14) by this equation expressing z_j in terms of the above given variables.

(2) Substitute the expression obtained for z_j in each of the other equations in (2.14).

Observe that a principal pivot transform in the jth block at the (i,j) position is defined if $n_{ij} \neq 0, \quad i \in \{1, \dots, m_j\}$. In the process, we have interchanged the variables w^j_i and z_j in such a way that the jth set of related variables remain unchanged. By Step 1 of the procedure, the resulting system, after the transform operation, is similar to the system in equation (2.14) except that the variables differs in the ith component of the jth block.

After a series of pivot operations have been performed, the system in equation (2.14) is transformed into the system

$$u = \bar{N} v + \bar{q}$$

where $(u^j_i, \dots, u^j_{m_j}, v_j)$ is a permutation of the set of related variables $(w^j_1, \dots, w^j_{m_j}, z_j)$ for each $j \in \{1, \dots, n\}$.

Szanc [**6**] also generalized the result of principal transforms for square matrices to the vertical block P-matrix of type $(m_1, \dots, m_n)$. We state the result and proof below.

THEOREM 2.10. *If N is an $m \times n$ vertical block P-matrix of type $(m_1, \dots, m_n)$, then any matrix $\bar{N}$ that results from a principal pivot is also an $m \times n$ vertical block P-matrix of type $(m_1, \dots, m_n)$.*

PROOF. Consider $\bar{N}$ as the matrix that results from performing a principal pivot on the jth block, located at the (t,j) position in N. Define the function $J : R^n \longrightarrow R^n$ by the following:

$$J(z) = (z_1, \dots, z_{j-1}, \bar{N}^j_t z, z_{j+1}, \dots, z_n)^T.$$

Hence, J is a one-to-one mapping of R^n onto R^n. Let $\bar{M}$ be any representative submatrix of $\bar{N}$. For any $z \in \mathbb{R}^n$, we obtain

$$z_k \bar{M}_{k.} z = z_k \bar{N}^k_{i.} z, \quad \text{for some } i \in \{1, \dots, m_k\}, \quad (k = 1, \dots, n).$$

The following cases arise.

(1) If $k \neq j$, then $z_k \bar{N}^k_{i.} = (J(z)_k (N)^k_{i.} J(z))$

(2) If $k = j$ and $i \neq t$, then $z_k \bar{N}^k_{i.} z = (N^j_{t.} J(z))(N^j_{i.} J(z))$.

(3) If $k = j$ and $i = t$, then $z_k \bar{N}^k_{i.} z = (N^j_{t.} J(z))(J(z))_j$.

Now, assume there exists a representative submatrix $\bar{M}$ of $\bar{N}$ and a vector $z \neq 0$ such that

$$z_k \bar{M}_{k.} z \leq 0, \quad (k = 1, \ldots, n).$$

For $k \neq j$, we have

$$(J(z))_k (N^k_{i.} J(z)) \leq 0, \quad i \in \{i, \ldots, m_k\}$$

- If $k = j$ and $i \neq t$, then

$$(N^j_{t.} J(z))(N^j_i J(z)) \leq 0. \tag{2.15}$$

- If $i = t$ and $k = j$,

$$(J(z))_j N^j_{t.} J(z) \leq 0. \tag{2.16}$$

If the representative submatrix $\bar{M}$ of $\bar{N}$ has its jth row as $N^j_{t.}$, then equations (2.13) through (2.16) indicate that for this particular representative submatrix $\bar{M}$, there exists a corresponding representative submatrix M of N and a vector $J(z) \in R^n$ such that

$$(J(z))_k (M_{k.} J(z)) \leq 0, \quad \forall k \in \{1, \ldots, n\}.$$

Since M is a representative submatrix of an $m \times n$ vertical block P-matrix of type $(m_1, \ldots, m_n)$), it follows that $J(z)$ must be the zero vector as M is known to be a square P-matrix. [**5**].

Consequently, the vector z must be the zero vector. If, however, z is not the zero vector, we must conclude that the jth row of M cannot be $N^j_{t.}$ and that

$$(J(z))_j (N^j_{t.} J(z)) > 0 \tag{2.17}$$

Thus, the jth row of $\bar{M}$ cannot be $\bar{N}^j_{t.}$ and instead must be $\bar{N}^j_{i.}$ for some $i \in \{1, \ldots, m_j\} - \{t\}$. For this particular i, we obtain

$$z_j \bar{M}_{j.} z = (N^j_{t.} J(z))(N^j_{i.} J(z)) \leq 0.$$

This leads to the conclusion that the product

$$((J(z))_j (N^j_{t.} J(z)))((J(z))_j (N^j_{i.} J(z))) \leq 0.$$

Equation (2.17) implies that $(J(z))_j \neq 0$ and that $(J(z))_j (N^j_{i.} J(z)) > 0$. Thus, we conclude that

$$(J(z))_j (N^j_{i.} J(z)) \leq 0.$$

This result implies the existence of a representative submatrix M of N, where the jth row is $N^j_{i.}$, and a vector $J(z) \in R^n$ such that

$$(J(z))_k (M_{k.} J(z)) \leq 0, \quad \text{for } k = 1, \ldots, n.$$

Because M is an $n \times n$ P-matrix, this implies that $J(z)$ must be the zero vector. Therefore, for every representative submatrix $\bar{M}$ of $\bar{N}$ and any nonzero vector $z \in R^n$, there exists an index $k \in \{1, \ldots, n\}$ such that

$$z_k \bar{M}_{k.} z > 0.$$

Thus, every representative submatrix $\bar{M}$ of $\bar{N}$ is a P-matrix. Hence, that $\bar{N}$ is an $m \times n$ vertical block P-matrix of type $(m_1, \ldots, m_n)$. □

2.1. The $2n \times n$ Vertical Block P-matrix. The $2n \times n$ vertical block P-matrix of type $(m_1, \ldots, m_n)$, where $m_1 = 2, (i = 1, \ldots, n)$, is a special case of the vertical block P-matrix. The presentations in this section are similar to the presentations given in Szanc [**6**]. We start with notation and definitions.

Let N be a $2n \times n$ vertical block matrix of type $(m_1, \ldots, m_n)$. We define the $n \times n$ matrices A and B as follows: The jth row of A is from the 1st row of N^j and the jth row of B is from the second row of N^j for $j = 1, \ldots, n$. Hence, the matrix $\bar{N}$, defined by

$$\bar{N} = \begin{bmatrix} A \\ B \end{bmatrix}, \tag{2.18}$$

is a re-arrangement of the rows of N.

Definition 2.11. A representative submatrix of $\bar{N}$ is an $n \times n$ matrix $\bar{M}$ such that the jth row of $\bar{M}$ is either the jth row of A or the jth row of B.

The matrix $\bar{N}$ has 2^n representative submatrices, which are also representative submatrices of N. All the definitions and results established for the representative matrices of N also apply to the representative submatrices of $\bar{N}$.

For example, $\bar{N}$ is a vertical block P-matrix of type $(m_1, \ldots, m_n)$ if each representative submatrix of $\bar{N}$ is a square P-matrix.

Theorem 2.12. *Suppose N is a $2n \times n$ matrix vertical block matrix. Define the matrix $\bar{N}$ by*

$$\bar{N} = \begin{pmatrix} A \\ B \end{pmatrix}, \tag{2.19}$$

where A and B are $n \times n$ matrices as defined in (2.18). *Suppose that A is a P-matrix. Then for each $x \in R^n, x \neq 0$, there exists an $i \in \{1, \ldots, n\}$ that depends on x such that*

$$(Ax)_i(Bx)_i > 0 \tag{2.20}$$

if and only if BA^{-1} is a P-matrix.

Proof. Suppose that for each $x \neq 0$,

$$(Ax)_i(Bx)_i > 0, \tag{2.21}$$

for some $\{i \in 1, \ldots, n\}$. If BA^{-1} is not a P-matrix, there exists some $y \in R^n, y \neq 0$ such that

$$y_i(BA^{-1}y)_i \leq 0 \quad \forall i \in \{1, \ldots, n\} \tag{2.22}$$

Let $x = A^{-1}y$. Then $x \neq 0$ and

$$y = Ax \tag{2.23}$$

$$y_i = (Ax)_i \tag{2.24}$$

$$(Ax)_i(Bx)_i \leq 0 \quad \forall i \in \{1, \ldots, n\} \tag{2.25}$$

This contradiction implies that BA^{-1} is a P-matrix.

Conversely, suppose BA^{-1} is a P-matrix and there exists some $x \in R^n, x \neq 0$ such that

$$(Ax)_i(Bx)_i \leq 0 \quad \forall i \in \{1, \ldots, n\}. \tag{2.26}$$

Let $y = Ax$. Then $y \neq 0$ and

$$y_i(BA^{-1}y)_i \leq 0 \quad \forall i \in \{1, \dots, n\} \tag{2.27}$$

which implies BA^{-1} is not a P-matrix, contradicting our assumption. □

THEOREM 2.13. *Suppose $\bar{N}$ is the $2n \times n$ matrix defined by*

$$\bar{N} = \begin{pmatrix} A \\ B \end{pmatrix}, \tag{2.28}$$

where A and B are as defined in (2.18). *$\bar{N}$ is a $2n \times n$ P-matrix if and only if A is a P-matrix and for any $x \in R^n, x \neq 0$, there exists an $i \in \{1, \dots, n\}$, that depends on x, such that*

$$x_i^2(Ax)_i(Bx)_i > 0 \tag{2.29}$$

PROOF. Suppose $\bar{N}$ is a $2n \times n$ P-matrix. This implies that each of its representative submatrix is a P-matrix. By definition, A is a representative submatrix and therefore a P-matrix. Assume for some $x \in R^n, x \neq 0$ and

$$x_i^2(Ax)_i(Bx)_i \leq 0 \quad \forall i \in \{1, \dots, n\} \tag{2.30}$$

We have the following possibilities:

(1) $x_i(Ax)_i = 0$ and/or $x_i(Bx)_i = 0$
(2) $(x_i(Ax)_i)(x_i(Bx)_i) < 0$

In either case, this implies that we can choose a representative submatrix $\bar{M}$ of $\bar{N}$ such that

$$x_i(\bar{M}x)_i \leq 0, \quad \forall i \in \{1, \dots, n\}. \tag{2.31}$$

This implies that $\bar{M}$ is not a P-matrix [**5**] and contradicts our assumption that $\bar{N}$ is a $2n \times n$ vertical block P-matrix.

Conversely, suppose that A is a P-matrix and that for each $x \in R^n, x \neq 0$, there exists some $i \in \{1, \dots, n\}$, that depends on x, such that

$$x_i^2(Ax)_i(Bx)_i > 0 \tag{2.32}$$

We need to show that all the principal minors of $\bar{N}$ are strictly positive in order to show that $\bar{N}$ is a $2n \times n$ vertical block P-matrix.

Let C be a $k \times k$ principal submatrix obtained from $\bar{N}$. That is, C is a principal submatrix of some representative submatrix $\bar{M}$ of $\bar{N}$. Let $\bar{N}_1$ be the $2k \times k$ matrix where

$$\bar{N}_1 = \begin{pmatrix} A_1 \\ B_1 \end{pmatrix}. \tag{2.33}$$

The matrices A_1 and B_1 are the $k \times k$ matrices that are the result of eliminating the same rows and columns from A and B, respectively, that were eliminated from $\bar{M}$ to create C. C is therefore a representative submatrix of $\bar{N}_1$.

A_1, being a principal submatrix of A, is a $k \times k$ P-matrix and therefore is nonsingular. We can rewrite $\bar{N}_1$ as

$$\bar{N}_1 = \begin{pmatrix} I \\ B_1A_1^{-1} \end{pmatrix} A_1. \tag{2.34}$$

To show that $B_1A_1^{-1}$ is a P-matrix, let $J = \{i_1, \ldots, i_k\}$ be the set of rows involved in creating C. Consider any $x \in R^n, x \neq 0$, such that $x_i = 0$ for $i \notin J$. By our assumption that for all $x \in R^n$, $x \neq 0$, there exists some $i \in \{1, \ldots, n\}$

$$x_i^2(Ax)_i(Bx)_i > 0. \tag{2.35}$$

We observe that $i \neq 0 \in J$. Thus for any $z \in R^k$, $z \neq 0$, there exists an $i \in J$ such that

$$z_i^2(A_1z)_i(B_1z)_i > 0. \tag{2.36}$$

This in turn implies that

$$(A_1z)_i(B_1z)_i > 0 \tag{2.37}$$

Since A_1 is a P-matrix, the previous theorem implies that $B_1A_1^{-1}$ is a P-matrix. This in turn implies that

$$\begin{pmatrix} I \\ B_1A_1^{-1} \end{pmatrix} \tag{2.38}$$

is a $2k \times k$ P-matrix. Since C is a representative submatrix of $\bar{N}_1$, we have $C = DA_1$, where D is a representative submatrix of

$$\begin{pmatrix} I \\ B_1A_1^{-1} \end{pmatrix}. \tag{2.39}$$

So, D must be a P-matrix. The determinant of D is equal to

$$\det C = \det D \det A_1 > 0. \tag{2.40}$$

Since C was an arbitrary principal submatrix of $\bar{N}$, we conclude that $\bar{N}$ is a $2n \times n$ vertical block P-matrix. □

3. The Vertical Block P_o-matrix

Let N be a vertical block matrix of type $(m_1, \ldots, m_n)$. Then N is a P_0-matrix if and only if each representative submatrix is a square P_0-matrix. As we pointed out before, the set of P-matrices is a subset of the set of P_0-matrices. The results provided in this section are from Ebiefung and Kostreva [**8**]. A necessary condition for a vertical block matrix N to be P_0 is provided below.

THEOREM 3.1. *Suppose N is a vertical block matrix of type $(m_1, \ldots, m_n)$. If N is a P_0-matrix, then the elements $a_{i1}, i = 1, \ldots, m_1; a_{i2}, i = m_1 + 1, \ldots, m_2; \ldots; a_{in}, i = m_{n-1} + 1, \ldots, m_n$, must be nonnegative.*

PROOF. By definition, the minors of N are the minors of the representative submatrices. The above listed elements are the main diagonal entries of the representative submatrices of N. Hence, they must be nonnegative if N is P_0-matrix. □

Theorem 3.1 points to the fact that the class to which N belongs depends on the type $(m_1, \ldots, m_n)$. It shows that it is possible to change the class to which N belongs by simply changing its type from $(m_1, \ldots, m_n)$ to $(n_1, \ldots, n_n)$, where $m_j \neq n_j$ for some j, $1 \leq j \leq n$. This concept is demonstrated in the following example.

EXAMPLE 3.2. Let

$$N = \begin{bmatrix} 3 & 1 \\ 4 & 5 \\ -2 & 1 \\ 0 & 2 \end{bmatrix}$$

be a vertical block matrix of type (m_1, m_2).

(1) Suppose $m_1 = m_2 = 2$. That is, N is of type $(2, 2)$. Then N has two blocks N^1, N^2 given by

$$N^1 = \begin{bmatrix} 3 & 1 \\ 4 & 5 \end{bmatrix}, \quad N^2 = \begin{bmatrix} -2 & 1 \\ 0 & 2 \end{bmatrix}.$$

The representative submatrices are:

$$M_1 = \begin{bmatrix} 3 & 1 \\ -2 & 1 \end{bmatrix}, \quad M_2 = \begin{bmatrix} 3 & 1 \\ 0 & 2 \end{bmatrix},$$

$$M_3 = \begin{bmatrix} 4 & 5 \\ -2 & 1 \end{bmatrix}, \quad M_4 = \begin{bmatrix} 4 & 5 \\ 0 & 2 \end{bmatrix}.$$

All the minors of the representative submatrices are nonnegative. So N is a P_0-matrix with respect to the type $(2, 2)$.

(2) Set $m_1 = 3, m_2 = 1$, so that N is a vertical block matrix of type $(3, 1)$. In this case, the blocks of N are

$$N^1 = \begin{bmatrix} 3 & 1 \\ 4 & 5 \\ -2 & 1 \end{bmatrix}, \quad N^2 = \begin{bmatrix} 0 & 2 \end{bmatrix}.$$

The corresponding representative submatrices are

$$M_1 = \begin{bmatrix} 3 & 1 \\ 0 & 2 \end{bmatrix}, \quad M_2 = \begin{bmatrix} 4 & 5 \\ 0 & 2 \end{bmatrix}, \quad M_3 = \begin{bmatrix} -2 & 1 \\ 0 & 2 \end{bmatrix}.$$

Observe that the minors

$$-2 \quad \text{and} \quad \det \begin{bmatrix} -2 & 1 \\ 0 & 2 \end{bmatrix}$$

are negative. Thus N is not a P_0-matrix if it is defined as of type $(3, 1)$.

DEFINITION 3.3. Let N be a vertical block matrix of type $(m_1, \ldots, m_n)$. For each $j (j = 1, \ldots, n)$, let v^j be a nonzero nonnegative row vector of dimension $1 \times m_j$. Define the matrix V by

$$V = \begin{bmatrix} v^1 & \ldots & 0 \\ \vdots & \ddots & \vdots \\ 0 & \ldots & v^n \end{bmatrix}. \tag{3.1}$$

Define a new matrix M by $M = VN$. Then M is of order $n \times n$, and the jth row of M is a weighted sum of the rows in N^j.

THEOREM 3.4. *Let N be a vertical block matrix of type $(m_1, \ldots, m_n)$. If N is a P_0-matrix, then the matrix*

$$M = VN, \tag{3.2}$$

is a P_0-matrix.

PROOF. All the principal minors of N are nonnegative since N is P_0 by assumption. Let N^j be the jth block of N, and $(N^j)_i$ the ith row of N^j. Then

$$M = VN = \begin{bmatrix} \sum_{i=1}^{m_1} v_i^1 N_i^1 \\ \vdots \\ \sum_{i=1}^{m_n} v_i^n N_i^n \end{bmatrix}$$

Moreover,

$$\det M = \sum_{i_1=1}^{m_1} \cdots \sum_{i_n=1}^{m_n} \prod_{j=1}^{n} v_{i_j}^j \det \begin{bmatrix} N_{i_1}^1 \\ \vdots \\ N_{i_n}^n \end{bmatrix}$$

But $v_{i_j}^j \geq 0, j = 1, \ldots, n$, by definition, and

$$\det \begin{bmatrix} N_{i_1}^1 \\ \vdots \\ N_{i_n}^n \end{bmatrix} \geq 0$$

by assumption. Hence, the result follows. □

It is worth noting that the converse of Theorem 3.4 is not necessarily true. This is shown by the example below.

EXAMPLE 3.5. Let

$$N = \begin{bmatrix} -3 & 5 \\ 3 & 1 \\ 0 & 7 \end{bmatrix}$$

be a vertical block matrix of type $(2, 1)$.

The matrix V defined by

$$V = \begin{bmatrix} 0 & 1 & 0 \\ 0 & 0 & 2 \end{bmatrix}$$

satisfies the conditions given in (3.1). By Theorem 3.4,

$$M = VN = \begin{bmatrix} 3 & 1 \\ 0 & 14 \end{bmatrix}$$

is a P_0-matrix. However, by Theorem 3.1, N is not a vertical block P_0-matrix.

If M is a square P_0-matrix, then the system of inequalities

$$Mx \geq 0, \quad x \geq 0$$

has a nonzero solution, according to Eaves [**9**]. This result is extended to the vertical block P_0-matrix in the theorem below.

THEOREM 3.6. *Let N be a vertical block matrix of type $(m_1, \ldots, m_n)$. If N is a P_0-matrix, then the system of linear inequalities*

$$Nx \geq 0, \quad x \geq 0$$

has a nonzero solution.

PROOF. Suppose that N be a vertical block P_o-matrix of type $(m_1, \ldots, m_n)$ and that for each $x \geq 0$, $Nx < 0$. Le M be as defined in Equation (3.2). Then for each $x \geq 0$,

$$Mx = V(Nx) < 0$$

since $V \geq 0$, $V \neq 0$. This contradicts the fact that M is a P_0-matrix. □

THEOREM 3.7. *If N is a vertical block P_0-matrix and the system*

$$Nx = 0, \quad x > 0$$

has a solution, then there exists y in $\mathbb{R}^m$, $y \geq 0$, $y \neq 0$, such that

$$y^t N = 0.$$

PROOF. Assume that N is a vertical block P_o-matrix of type $(m_1, \ldots, m_n)$ and that the system

$$Nx = 0, \quad x > 0$$

has a solution. Suppose that for all y in R^m, $y \geq 0$, $y \neq 0$, we have that

$$y^t N \neq 0.$$

If M is defined by

$$M = VN,$$

then

$$y^t M = y^t VN = \bar{y}^t N \neq 0$$

by hypothesis, where $\bar{y}^t = y^t V$. Since $x > 0$, we have that

$$y^t Mx \neq 0.$$

But

$$y^t Mx = VNx = V(Nx) = 0.$$

This contradicts our assumption, and the result follows. □

The next results provide necessary and sufficient conditions under which N is a vertical block P_0-matrix. The characterizations are generalizations of the results of Fiedler and Pták [**10**] for the square P_0-matrix. These results are provided in [**8**].

THEOREM 3.8. *Let N be a vertical block matrix of type $(m_1, \ldots, m_n)$. Then N is a P_0-matrix if and only if for each z in $\mathbb{R}^n$, $z \neq 0$, there exists an index i such that*

$$z_i \neq 0, \quad z_i(Nz)_i \geq 0.$$

PROOF. Suppose that N is a vertical block P_0-matrix. Let

$$V = \begin{bmatrix} v^1 & & \\ & \ddots & \\ & & v^n \end{bmatrix}$$

be as defined in Equation (3.1). Assume, without loss of generality, that $v^i = (0, \ldots, 0, v_i^i, 0, \ldots, 0)$. By Theorem 3.4, the matrix A defined by

$$A = VN$$

is a square P_0-matrix. Suppose that there exists z in R^n, $z \neq 0$, such that for all i for which $z_i \neq 0$,

$$z_i(Nz)_i < 0.$$

Then

$$z_i(Az)_i = z_i(VNz)_i = v_i^i z_i(Nz)_i < 0,$$

since $v_i^i \geq 0$, $v_i^i \neq 0$, and $z_i(Nz)_i < 0$. This contradicts the fact that A is a P_0-matrix.

Conversely, assume that for each z in R^n, $z \neq 0$, there exists an index i such that if $z_i \neq 0$, $z_i(Nz)i \geq 0$. Then there exists an index j_i, $1 \leq j_i \leq n$, such that $z_{j_i} \neq 0$, $z_{j_i}(N_{j_i})z_{j_i} \geq 0$.

Consequently, if M is a representative submatrix of N, we must have

$$z_{j_i} \neq 0, z_{j_i}(M_{j_i})z_{j_i} \geq 0.$$

So, M is a square P_o-matrix. Because the choice of M was arbitrary, each representative submatrix of N is a P_0-matrix. Therefore, N is a vertical block P_0-matrix. □

COROLLARY 3.9. *Suppose N is a vertical block matrix of type $(m_1, \ldots, m_n)$. Then N is a P_0-matrix if and only if for all z in $\mathbb{R}^n$, $z \neq 0$, there are indices j, $(j = 1, \ldots, n)$ and i_j such that*

$$z_{i_j} \neq 0, \quad z_{i_j}(Nz^j z)_{i_j} \geq 0.$$

PROOF. If N is a P_0-matrix, then by Theorem 3.8 there exists i such that $z_i \neq 0$, $z_i(Nz)_i \geq 0$. Hence, there exists j such that

$$z_i \neq 0, \quad z_i(Nz^j)_i \geq 0. \tag{3.3}$$

The result follows by setting $i = i_j$.

Conversely, if for all z in $\mathbb{R}^n$, $z \neq 0$, there are indices j and i_j, $1 \leq j \leq n$, such that equation (3.3) holds, then for each representative submatrix M of N we must have

$$z_{i_j} \neq 0, \quad z_{i_j}(Mz)_{i_j} \geq 0.$$

Thus every representative submatrix of N is a P_0-matrix. The result follows by the definition of a vertical block P_0-matrix. □

The following result shows that a perturbation of a vertical block P_o-matrix of type $(m_1, \ldots, m_n)$ is P-matrix. First, we state the definition below.

DEFINITION 3.10. Let N is a vertical block matrix of type $(m_1, \ldots, m_n)$. For each j, let e_j be an $m_j \times 1$ column vector of ones. Define a matrix E by

$$E = \begin{bmatrix} e_1 & \ldots & 0 \\ \vdots & \ddots & \vdots \\ 0 & \ldots & e_n \end{bmatrix}. \tag{3.4}$$

Then E is $m \times n$ and nonnegative. For each $\epsilon > 0, N + \epsilon E$ is a vertical block matrix of type $(m_1, \ldots, m_n)$.

THEOREM 3.11. *Suppose N is a vertical block matrix of type $(m_1, \ldots, m_n)$. Then N is a P_0-matrix if and only if for each $\epsilon > 0$, $N + \epsilon E$ is a P-matrix.*

PROOF. Suppose N is a vertical P_o-matrix. Let $\bar{M}$ be a principal submatrix of $N + \epsilon E$. Then $\bar{M} = M + \epsilon I$, where M is a principal submatrix of N and I is the identity matrix. Since N is P_0, we have $\det M \geq 0$ and any eigenvalue of M is nonnegative, say $\lambda \geq 0$. Moreover, $\lambda + \epsilon > 0$ is an eigenvalue of $\bar{M}$. Thus $\det \bar{M} > 0$

for all $\epsilon > 0$. Therefore, $N + \epsilon E$ is a vertical block P-matrix, since $\bar{M}$ is a principal submatrix of $N + \epsilon E$.

For the converse, suppose that $N + \epsilon E$ is a P-matrix for each $\epsilon > 0$. If M is a principal submatrix of N, then $M + \epsilon I$ is a principal submatrix of $N + \epsilon E$ for each $\epsilon > 0$. So $\det(M + \epsilon I) > 0$ for all $\epsilon > 0$. But $\det(M + \epsilon I)$ is a polynomial function in the variable ϵ. Hence

$$\lim_{\epsilon \to 0} \det(M + \epsilon I) = \det M \geq 0.$$

Therefore, N is a P_0-matrix. □

4. The $P \cap Z$ and $P_o \cap Z$ Vertical Block Matrix Classes

In this section, we characterize the vertical block $P \cap Z$ and $P_o \cap Z$ matrix classes. There are existing applications, such as the generalize Leontief Input-Output Production Model [**1**], in which the associated matrix may belong to these matrix classes. These results are the generalizations of the results of Fiedler and Pta'k [**11**]. As in the case of the P-matrix class or the P_o-matrix class, an $m \times n, m \geq n$, matrix N is a vertical block Z-matrix of type $(m_1, \ldots, m_n)$ if every representative matrix of N is a square Z-matrix.

THEOREM 4.1. *Let N be a vertical block Z-matrix of type $(m_1, \ldots, m_n)$. Then the statements below are equivalent:*

(1) *There exists a vector $x \geq 0$ such that $Nx > 0$.*
(2) *N is a vertical block P-matrix of type $(m_1, \ldots, m_n)$.*

PROOF. (1)$\longrightarrow$ (2): Suppose that there is $x \geq 0$ such that $Nx > 0$. Let e be an $n \times 1$ vector with each component equal to 1. For any $\epsilon > 0$, let $y = x + \epsilon e$. Then $Ny = N(x + \epsilon e) = Nx + \epsilon Ne > 0$ for ϵ sufficiently small. Thus there exists $y > 0$ such that $Ny > 0$. Hence, for each representative submatrix M of N, we must have $My > 0, y > 0$. By [**11**], we have that for each $k \neq 0$, there exists an index k such that $x_k \neq 0$ and $x_k(Mx)_k > 0$. This is true for all representatives submatrices of N, since the choice of M is arbitrary. Thus each representative submatrix of N is a square P-matrix. Therefore, N is a vertical block P-matrix of type $(m_1, \ldots, m_n)$.

(2)$\longrightarrow$(1): Let N is a vertical block P-matrix of type $(m_1, \ldots, m_n)$. Then the vertical generalized linear complementarity problem has a solution for all $q \in R^m$, as provided in [**12**]. Thus for any $q < 0, q \in R^m$, there exists $x \geq 0$, such that $Nx + q \geq 0$. Hence, the system

$$Nx > 0,$$
$$x \geq 0,$$

has a solution. □

Next, we provide a characterization theorem that gives the relationships between the vertical block P-, P_o-, and Z-matrix classes. Through out the rest of this section, E will always denote the matrix in equation (3.4).

THEOREM 4.2. *Let N be a vertical block Z-matrix of type $(m_1, \ldots, m_n)$. Then the statements below are equivalent:*

(1) *The eigenvalues of all principal submatrices of N are nonnegative.*
(2) *All principal minors of N are nonnegative.*
(3) *$N + \epsilon E$ is a P-matrix for all $\epsilon > 0$.*
(4) *N is a P_0-matrix of type $(m_1, \ldots, m_n)$.*

PROOF. (1) $\to$ (2): The determinant of a principal submatrix is the product of its eigenvalues. Since all the eigenvalues are nonnegative, the minors are also nonnegative.

(2) $\to$ (3): Let $\bar{M}$ be a principal submatrix of $N + \epsilon E$. Then $\bar{M}$ can be represented as $\bar{M} = M + \epsilon E$, where M is a principal submatrix of N and I is an identity matrix of appropriate size. Since det $M \geq 0$, we have the det $\bar{M} > 0$ for any $\epsilon > 0$. Therefore, $N + \epsilon E$ is a P-matrix for any $\epsilon > 0$.

(3) $\to$ (4): Let M be a representative submatrix of N. Then $M + \epsilon I$ is a representative submatrix of $N + \epsilon E$. So, det $M + \epsilon I \geq 0$ for all $\epsilon > 0$. Hence,

$$detM = \lim_{\epsilon > 0} det(M + \epsilon I) \geq 0.$$

Thus, M is a square P_o-matrix. Therefore, N is a vertical block P_o-matrix.

(4) $\to$ (1): Suppose N is a vertical block P_o-matrix of type $(m_1, \ldots, m_n)$. Let M be a representative submatrix of N. By the definition of N, M is a square P_o-matrix. By Fiedler and Pta'k [**11**], all the eigenvalues of M and of all its principal submatrices are nonnegative. But a principal submatrix of N is a principal submatrix of a representative submatrix. Therefore, all the eigenvalues of all principal submatrices of N are nonnegative. □

COROLLARY 4.3. *Suppose N is a vertical block Z-matrix of type $(m_1, \ldots, m_n)$. If there is an $x > 0$ such that $Nx \geq 0$, then N is a P_o-matrix.*

PROOF. Let N is a vertical block Z-matrix of type $(m_1, \ldots, m_n)$. Suppose there is an $x > 0$ such that $Nx \geq 0$. If M be a representative submatrix of N, then $Mx \geq 0$ □

5. Applications Involving the vertical block P- and P_o-matrices

This section of the chapter is devoted to an economic application of the vertical generalized linear complementarity problem. The generalized inputoutput model was defined and formulated as a vertical generalized linear complementarity problem (VGLCP) in [**1**]. The application of the model to the choice of new technoligies is presented below. We start with some notation and definitions.

Given a vertical block matrix N of type $(m_1, \ldots, m_n)$ and a vector $q \in R^m$, the vertical generalized linear complementarity problem (VGLCP) associated with N and q, denoted VGLCP(q, N), is to find vectors $z \in R^n$ and $w \in R^m$ such that:

$$w = Nz + q \geq 0, \qquad z \geq 0 \tag{5.1}$$

$$z_j \prod_{i}^{m_j} (N^j z + q^j)_i = 0, \qquad j = 1, \ldots, n. \tag{5.2}$$

where N, w, and q are partitioned conformably as in Definition 1.2.

The solution of the VGLCP depends on the matrix class that the associated matrix N belongs to. If the matrix N, that results from a mathematical modelling involving the VGLCP, is a P-matirx or $P \cap Z$ class, then the VGLCP has a unique solution [**12**]. The algorithm for finding this unique solution is provided in [**13**].

5.1. Application of the generalized Leontief Input-Output Linear Model to the Choice of New Technology. Consider n industries each of which wants to choose betwwen an old technology and a new technology. Assume that there is an interaction between the industries in the sense that an output from a given industry can be used as an input for the production of an output in another industry and that an industry can use one of its outputs for the production of another output. Production is restricted to a single period of activities.

Let

g^j_{ik} = units of output i by industry j to sector k using the old technology.

f^j_{ik} = units of output i by industry j to sector k using the new technology.

The technology matrix for the sectors are defined below.

$$A^1 = \begin{bmatrix} g^1_{11} & \cdots & g^1_{1n} \\ f^1_{11} & \cdots & f^1_{1n} \end{bmatrix}, \ldots, A^n = \begin{bmatrix} g^n_{11} & \cdots & g^n_{1n} \\ f^n_{11} & \cdots & f^n_{1n} \end{bmatrix}, \tag{5.3}$$

By combining these sector technology matrices, we have the technology matrix for the economy:

$$A = \begin{bmatrix} A^1 \\ A^2 \\ \vdots \\ A^n \end{bmatrix}$$

Observe that A is a $2n \times n$ vertical block matrix of type $(2, \ldots, 2)$. For the demand parameters, let d^j_i $(i = 1, 2, j = 1, \ldots, n)$ be the total demand vector for sector j's outputs under the old technology if $i = 1$ and under the new technology if $i = 2$. The sector demand quantities are then given by:

$$d^1 = \begin{bmatrix} d^1_1 \\ d^1_2 \end{bmatrix}, \ldots, d^n = \begin{bmatrix} d^n_1 \\ d^n_2 \end{bmatrix}$$

and the demand quantities for the whole economy are represented by the column vector:

$$d = \begin{bmatrix} d^1 \\ d^2 \\ \vdots \\ d^n \end{bmatrix}$$

Next, let's define the production quantities. If we let x_j be the number of units of type j outputs by industry j, then the production vector for the economy is:

$$x = \begin{bmatrix} x_1 \\ x_2 \\ \vdots \\ x_n \end{bmatrix}$$

Finally, we define the vector e_2 by:

$$e_2 = \begin{bmatrix} 1 \\ 1 \end{bmatrix}$$

and the matrix J by:

$$J = \begin{bmatrix} e_2 & 0 & 0 & \cdots & 0 \\ 0 & e_2 & 0 & \cdots & 0 \\ \cdots & \cdots & \cdots & \cdots & 0 \\ 0 & 0 & 0 & \cdots & e_2 \end{bmatrix}$$

Then J is a $2n \times n$ vertical block matrix of type $(2, \ldots, 2)$, when partitioned to conform with the partition of A. Moreover, d is in R^{2n} and x in R^n.

5.1.1. *The Model.* Given the technology matrix A of type $(2,\dots,2)$, and the demand vector d in R^{2n}, find the production vector x in R^n such that:

$$(J-A)x-d\geq 0 \tag{5.4}$$

$$x_j\prod_{i=1}^{2}(J-A)_i^j x-d_i^j=0, \qquad (j=1,\dots,n) \tag{5.5}$$

$$x\geq 0$$

Equations (5.4) and (5.5) define a vertical generalized linear complementarity problem. In practice, the matrix $J-A$ is a vertical block P-matrix, and so the solution can be obtained by the algorithm given in [**13**]. If the $J-A$ is Z-matrix that is not a P-matrix, then the problem can be solved by any of the algorithms provided [**14**] and [**15**].

EXAMPLE 5.1. Consider an economy with three sectors, say 1, 2, and 3. Suppose that the sectors' outputs are land (acres), food (pounds), and capital (billions of dollars), respectively. The sectors' operations are based on a single period of activities. Suppose that under the current operations, the technology matrix and bill of goods for the economy are

$$A=\begin{bmatrix}0.6 & 0.1 & 0.3\\ 0.3 & 0.6 & 0.1\\ .1 & 0.3 & 0.6\end{bmatrix}, \qquad f=\begin{bmatrix}150\\ -130\\ -20\end{bmatrix}$$

respectively, where the columns and rows are for land, food, and capital in that order. Suppose a new technology has recently been introduced in the market. To compare the worth of the new technology over the old, the OR departments for the sectors came up with a corresponding new technology matrix B and demand vector g,

$$B=\begin{bmatrix}0.5 & 0.2 & 0.3\\ 0.4 & 0.2 & 0.4\\ 0.1 & 0.6 & 0.3\end{bmatrix}, \qquad g=\begin{bmatrix}150\\ -190\\ -45\end{bmatrix}$$

The vertical block matrix $J-A$ and the vector d are equal to:

$$J-A=\begin{bmatrix}0.4 & -0.1 & -0.3\\ 0.5 & -0.2 & -0.3\\ -0.3 & 0.4 & -0.1\\ -0.4 & 0.8 & -0.4\\ -0.1 & -0.3 & 0.4\\ -0.1 & -0.6 & 0.7\end{bmatrix} \qquad d=\begin{bmatrix}150\\ 150\\ -130\\ -190\\ -20\\ -45\end{bmatrix}$$

When the model is solved using any of the techniques discussed above, we get the approximate solution x = [415, 0, 54]. The demand vector as modified by the model is

$$d=\begin{bmatrix}150, & 192, & -130, & -188, & -20, & -4\end{bmatrix}$$

to the nearest whole number.

Thus, in this scenerio, the sectors should continue to use the old technology.

6. Conclusion

The vertical block matrix structure, which arises in the analysis of the vertical generalized linear complementarity problem, is studied in the context of the P-, P_o-, $P \cap Z$-, and $P_o \cap Z$-matrix classes. We choose these matrix classes because they arise in many vertical generalized linear complementarity mathematical models, and their potential to further advance areas of applications of linear algebra. An application, the generalization of the linear Leontief input-output model, is provided to show an application instance in which the associated matrix is a vertical block P- or $P \cap Z$-matrix of type $(m_1, \ldots, m_n)$. We hope that a deeper understanding of the mathematical properties of the matrix classes presented in this chapter, will shed more light on their properties and usefulness, and lead to other theoretical results and application areas.

References

[1] Aniekan A. Ebiefung and Michael M. Kostreva, *The generalized Leontief input-output model and its application to the choice of new technology*, Ann. Oper. Res. **44** (1993), no. 1-4, 161–172, DOI 10.1007/BF02061065. Advances in equilibrium modeling, analysis and computation. MR1246841

[2] M. Seetharama Gowda and Roman Sznajder, *A generalization of the Nash equilibrium theorem on bimatrix games*, Internat. J. Game Theory **25** (1996), no. 1, 1–12, DOI 10.1007/BF01254380. MR1375785

[3] George J. Habetler and Caroline N. Haddad, *Global stability of a two-species piecewise linear Volterra ecosystem*, Appl. Math. Lett. **5** (1992), no. 6, 25–28, DOI 10.1016/0893-9659(92)90007-V. MR1345917

[4] Richard W. Cottle and George B. Dantzig, *A generalization of the linear complementarity problem*, J. Combinatorial Theory **8** (1970), 79–90. MR254064

[5] David Gale and Hukukane Nikaidô, *The Jacobian matrix and global univalence of mappings*, Math. Ann. **159** (1965), 81–93, DOI 10.1007/BF01360282. MR204592

[6] Bohdan P. Szanc, *The generalized complementarity problem*, ProQuest LLC, Ann Arbor, MI, 1989. Thesis (Ph.D.)–Rensselaer Polytechnic Institute. MR2637595

[7] A. W. Tucker, *Principal Pivot Transorms of Square Matrices*, SIAM Review. **5**, 305 (1965).

[8] A. A. Ebiefung and M. M. Kostreva, *Generalized P_0- and Z-matrices*, Linear Algebra Appl. **195** (1993), 165–179, DOI 10.1016/0024-3795(93)90262-M. MR1253275

[9] B. Curtis Eaves, *The linear complementarity problem*, Management Sci. **17** (1971), 612–634, DOI 10.1287/mnsc.17.9.612. MR282663

[10] Miroslav Fiedler and Vlastimil Pták, *Some generalizations of positive definiteness and monotonicity*, Numer. Math. **9** (1966), 163–172, DOI 10.1007/BF02166034. MR209309

[11] Miroslav Fiedler and Vlastimil Pták, *On matrices with non-positive off-diagonal elements and positive principal minors* (English, with Russian summary), Czechoslovak Math. J. **12(87)** (1962), 382–400. MR142565

[12] G. J. Habetler and B. P. Szanc, *Existence and uniqueness of solutions for the generalized linear complementarity problem*, J. Optim. Theory Appl. **84** (1995), no. 1, 103–116, DOI 10.1007/BF02191738. MR1312965

[13] A. A. Ebiefung, G. Habetler, M. M. Kostreva, and P. P. Szanc, *A Direct Algorithm for the Vertical Generalized Complementarity Problem Associated with P-Matrices*, Open Journal of Optimization, **6**, 101–114 (2017).

[14] S. R. Mohan and S. K. Neogy, *Algorithms for the generalized linear complementarity problem with a vertical block Z-matrix*, SIAM J. Optim. **6** (1996), no. 4, 994–1006, DOI 10.1137/S1052623494275586. MR1416525

[15] A. A. Ebiefung, M. M. Kostreva, and V. Ramanujam, *An Algorithm to Solve the Generalized Linear Complementarity Problem with a Vertical Block Z-matrix*, Journal of Optimization Methods and Software, **7**, 123–138 (1997).

Department of Mathematics, University of Tennessee at Chattanooga, 615 McCallie Ave, Chattanooga, Tennessee 37403

Email address: aniekan-ebiefung@utc.edu

Contemporary Mathematics
Volume **833**, 2026
https://doi.org/10.1090/conm/833/16663

Some remarks on the reparametrization of linear models

Stephen J. Haslett, Jarkko Isotalo, Augustyn Markiewicz,
and Simo Puntanen

ABSTRACT. We consider the partitioned linear model $\mathbf{y} = \mathbf{X}_1\boldsymbol{\beta}_1+\mathbf{X}_2\boldsymbol{\beta}_2+\boldsymbol{\varepsilon}$, for which the standard notation is the triplet $\mathscr{M} = \{\mathbf{y}, \mathbf{X}\boldsymbol{\beta}, \mathbf{V}\} = \{\mathbf{y}, \mathbf{X}_1\boldsymbol{\beta}_1 + \mathbf{X}_2\boldsymbol{\beta}_2, \mathbf{V}\}$. Vector $\boldsymbol{\beta} \in \mathbb{R}^p$ is a vector of fixed (but unknown) parameters, partitioned as $\boldsymbol{\beta}_1$ and $\boldsymbol{\beta}_2$. Our aim is to estimate parametric functions like $\mathbf{K}'\boldsymbol{\beta}$ under $\mathscr{M}$; here $\mathbf{K}$ is a given matrix of p rows and $'$ denotes the transpose. In particular, we are interested in the best linear unbiased estimators, BLUEs, of $\mathbf{K}'\boldsymbol{\beta}$. Let $\mathbf{X}_*$ be any matrix such that the column spaces of $\mathbf{X}$ and $\mathbf{X}_*$ are equal. Then there exists a matrix $\mathbf{A}$ such that $\mathbf{X}_* = \mathbf{X}\mathbf{A}$ and we get the reparametrized model $\mathscr{T} = \{\mathbf{y}, \mathbf{X}_*\boldsymbol{\gamma}, \mathbf{V}\}$. Models $\mathscr{M}$ and $\mathscr{T}$ are called equivalent in the sense that $\mathrm{BLUE}(\mathbf{X}\boldsymbol{\beta} \mid \mathscr{M}) = \mathrm{BLUE}(\mathbf{X}_*\boldsymbol{\gamma} \mid \mathscr{T})$. In this paper we consider two kinds of choice of $\mathbf{A}$:

$$\mathbf{A} = \begin{pmatrix} \mathbf{I}_{p_1} & -\mathbf{X}_1^+\mathbf{X}_2 \\ \mathbf{0} & \mathbf{I}_{p_2} \end{pmatrix}, \quad \mathbf{A} = (\mathbf{K} : \mathbf{K}^{\perp}),$$

where $\mathbf{K}^{\perp}$ is a matrix spanning the orthocomplement of the column space of $\mathbf{K}$ and the superscript $^+$ refers to the Moore–Penrose inverse.

1. Preliminaries

In this section we introduce the notation to be used and some mathematical preliminaries which prepare the reader for the main material of the paper. The results in this section are not new but we find it helpful to go them through before the newer stuff presented in the later sections.

The main focus of this paper lies in the partitioned linear model $\mathbf{y} = \mathbf{X}_1\boldsymbol{\beta}_1 + \mathbf{X}_2\boldsymbol{\beta}_2 + \boldsymbol{\varepsilon}$, denoted as a triplet

$$\mathscr{M}_{12} = \{\mathbf{y}, \mathbf{X}\boldsymbol{\beta}, \mathbf{V}\} = \{\mathbf{y}, \mathbf{X}_1\boldsymbol{\beta}_1 + \mathbf{X}_2\boldsymbol{\beta}_2, \mathbf{V}\}.$$

Here $\mathbf{y}$ is an n-dimensional observable random vector, and $\boldsymbol{\varepsilon}$ is an unobservable random error vector with a known (possibly singular) covariance matrix $\mathrm{cov}(\boldsymbol{\varepsilon}) = \mathbf{V} = \mathrm{cov}(\mathbf{y})$ and expectation $\mathrm{E}(\boldsymbol{\varepsilon}) = \mathbf{0}$. The matrix $\mathbf{X}$ is a known $n \times p$ matrix, i.e., $\mathbf{X} \in \mathbb{R}^{n\times p}$, partitioned columnwise as $\mathbf{X} = (\mathbf{X}_1 : \mathbf{X}_2)$. Vector $\boldsymbol{\beta} = (\boldsymbol{\beta}_1', \boldsymbol{\beta}_2')' \in \mathbb{R}^p$ is a vector of fixed (but unknown) parameters; here $\boldsymbol{\beta}_i \in \mathbb{R}^{p_i}, i=1,2$, and symbol $'$ stands for the transpose. We will also denote $\boldsymbol{\mu} = \mathbf{X}\boldsymbol{\beta}$, $\boldsymbol{\mu}_i = \mathbf{X}_i\boldsymbol{\beta}_i$, $i = 1,2$.

2020 *Mathematics Subject Classification.* Primary 62J05, 62J10.

Key words and phrases. Best linear unbiased estimator, reparametrized linear model, covariance matrix, misspecified model, partitioned linear model.

Occasionally we use the notations $\mathscr{M}_{12}, \mathscr{M}_{12}(\mathbf{V})$ or $\mathscr{M}$ depending which property of the model we wish to emphasize.

As for notations, the symbols $\mathrm{r}(\mathbf{A})$, $\mathbf{A}^-$, $\mathbf{A}^+$, $\mathscr{C}(\mathbf{A})$, and $\mathscr{C}(\mathbf{A})^\perp$, denote, respectively, the rank, a generalized inverse, the Moore–Penrose inverse, the column space, and the orthogonal complement of the column space of the matrix $\mathbf{A}$. By $\mathbf{A}^\perp$ we denote any matrix satisfying $\mathscr{C}(\mathbf{A}^\perp) = \mathscr{C}(\mathbf{A})^\perp$. We will write $\mathbf{P}_\mathbf{A} = \mathbf{A}\mathbf{A}^+ = \mathbf{A}(\mathbf{A}'\mathbf{A})^-\mathbf{A}'$ to denote the orthogonal projector onto $\mathscr{C}(\mathbf{A})$ and $\mathbf{Q}_\mathbf{A} = \mathbf{I}_a - \mathbf{P}_\mathbf{A}$, where $\mathbf{I}_a$ is the identity matrix of order a. In particular, we denote

$$\mathbf{M} = \mathbf{I}_n - \mathbf{P}_\mathbf{X}\,, \quad \mathbf{M}_i = \mathbf{I}_n - \mathbf{P}_i\,, \quad \mathbf{P}_i = \mathbf{P}_{\mathbf{X}_i}\,, \ i = 1, 2\,.$$

Under the model $\{\mathbf{y}, \mathbf{X}\boldsymbol{\beta}, \mathbf{V}\}$, the statistic $\mathbf{G}\mathbf{y}$, where $\mathbf{G}$ is an $n \times n$ matrix, is the best linear unbiased estimator, BLUE, of $\mathbf{X}\boldsymbol{\beta}$ if $\mathbf{G}\mathbf{y}$ is unbiased, i.e., $\mathbf{G}\mathbf{X} = \mathbf{X}$, and it has the smallest covariance matrix in the Löwner sense among all linear unbiased estimators of $\mathbf{X}\boldsymbol{\beta}$. In other words, $\mathrm{cov}(\mathbf{C}\mathbf{y}) - \mathrm{cov}(\mathbf{G}\mathbf{y})$ is nonnegative definite. The BLUE of an estimable parametric function $\boldsymbol{\eta} = \mathbf{K}'\boldsymbol{\beta}$, where $\mathbf{K} \in \mathbb{R}^{p\times q}$, is defined in the corresponding way. Estimability of $\mathbf{K}'\boldsymbol{\beta}$ means that it has a linear unbiased estimator which happens if and only if $\mathscr{C}(\mathbf{K}) \subseteq \mathscr{C}(\mathbf{X}')$. It is well known that $\boldsymbol{\mu}_1 = \mathbf{X}_1\boldsymbol{\beta}_1$ is estimable in the partitioned model if and only if

$$\mathscr{C}(\mathbf{X}_1) \cap \mathscr{C}(\mathbf{X}_2) = \{\mathbf{0}\}\,, \quad \text{i.e.,} \quad \mathrm{r}(\mathbf{M}_2\mathbf{X}_1) = \mathrm{r}(\mathbf{X}_1)\,. \tag{1.1}$$

For the proof of the following fundamental lemma, see, e.g., [**16**, p. 282].

LEMMA 1.1. *Consider the partitioned linear model* $\mathscr{M}_{12} = \{\mathbf{y}, \mathbf{X}\boldsymbol{\beta}, \mathbf{V}\}$. *Let* $\boldsymbol{\eta} = \mathbf{K}'\boldsymbol{\beta}$ *and* $\boldsymbol{\mu}_1 = \mathbf{X}_1\boldsymbol{\beta}_1$ *be estimable. Then*

(a) $\mathbf{G}\mathbf{y} = \mathrm{BLUE}(\boldsymbol{\mu}) \iff \mathbf{G}(\mathbf{X} : \mathbf{V}\mathbf{M}) = (\mathbf{X} : \mathbf{0})$, *i.e.,* $\mathbf{G} \in \{\mathbf{P}_{\boldsymbol{\mu}\,|\,\mathbf{V}}\}$.
(b) $\mathbf{B}\mathbf{y} = \mathrm{BLUE}(\boldsymbol{\eta}) \iff \mathbf{B}(\mathbf{X} : \mathbf{V}\mathbf{M}) = (\mathbf{K}' : \mathbf{0})$, *i.e.,* $\mathbf{B} \in \{\mathbf{P}_{\boldsymbol{\eta}\,|\,\mathbf{V}}\}$.
(c) $\mathbf{C}\mathbf{y} = \mathrm{BLUE}(\boldsymbol{\mu}_1) \iff \mathbf{C}(\mathbf{X}_1 : \mathbf{X}_2 : \mathbf{V}\mathbf{M}) = (\mathbf{X}_1 : \mathbf{0} : \mathbf{0})$; $\mathbf{C} \in \{\mathbf{P}_{\boldsymbol{\mu}_1\,|\,\mathbf{V}}\}$.

In Lemma 1.1 the set $\{\mathbf{P}_{\boldsymbol{\mu}\,|\,\mathbf{V}}\}$ is defined as

$$\mathbf{G} \in \{\mathbf{P}_{\boldsymbol{\mu}\,|\,\mathbf{V}}\} \iff \mathbf{G}(\mathbf{X} : \mathbf{V}\mathbf{M}) = (\mathbf{X} : \mathbf{0})\,,$$

and $\{\mathbf{P}_{\boldsymbol{\eta}\,|\,\mathbf{V}}\}$ and $\{\mathbf{P}_{\boldsymbol{\mu}_1\,|\,\mathbf{V}}\}$ in the corresponding way.

Solutions to $\mathbf{G}, \mathbf{B}$ and $\mathbf{C}$ in Lemma 1.1 are unique if and only if $\mathscr{C}(\mathbf{X} : \mathbf{V}\mathbf{M}) = \mathbb{R}^n$. It is well known, see, e.g., [**15**, Sec. 4], that solutions for $\mathbf{G}, \mathbf{B}$ and $\mathbf{C}$ in Lemma 1.1 can be expressed, e.g., as

$$\mathbf{G}_1 = \mathbf{X}(\mathbf{X}'\mathbf{W}^-\mathbf{X})^-\mathbf{X}'\mathbf{W}^+ =: \mathbf{P}_{\mathbf{X};\mathbf{W}^+}, \quad \mathbf{B}_1 = \mathbf{K}'(\mathbf{X}'\mathbf{W}^-\mathbf{X})^-\mathbf{X}'\mathbf{W}^+,$$
$$\mathbf{C}_1 = \mathbf{X}_1(\mathbf{X}_1'\dot{\mathbf{M}}_2\mathbf{X}_1)^-\mathbf{X}_1'\dot{\mathbf{M}}_2 =: \mathbf{P}_{\mathbf{X}_1;\dot{\mathbf{M}}_2}\,, \ \mathbf{C}_2 = (\mathbf{X}_1 : \mathbf{0})(\mathbf{X}'\mathbf{W}^-\mathbf{X})^-\mathbf{X}'\mathbf{W}^+,$$

where

$$\dot{\mathbf{M}}_2 = \mathbf{M}_2(\mathbf{M}_2\mathbf{W}\mathbf{M}_2)^+\mathbf{M}_2 = (\mathbf{M}_2\mathbf{W}\mathbf{M}_2)^+, \tag{1.2}$$

and $\mathbf{W}$ belongs to the class of nonnegative definite matrices defined as

$$\mathcal{W} = \{\mathbf{W} : \mathbf{W} = \mathbf{V} + \mathbf{X}\mathbf{U}\mathbf{U}'\mathbf{X}',\ \mathscr{C}(\mathbf{W}) = \mathscr{C}(\mathbf{X} : \mathbf{V})\}\,. \tag{1.3}$$

In (1.3), $\mathbf{U}$ can be any matrix comprising p rows so that $\mathscr{C}(\mathbf{W}) = \mathscr{C}(\mathbf{X} : \mathbf{V})$ is satisfied. It is clear that we can always choose $\mathbf{U} = \alpha\mathbf{I}_n$, where α is an arbitrary nonzero scalar. Moreover, $\mathbf{V} \in \mathcal{W}$ if and only if $\mathscr{C}(\mathbf{X}) \subseteq \mathscr{C}(\mathbf{V})$. If there is a need to emphasize that there is a particular model $\mathscr{A}$, say, under consideration we will use the notation $\mathcal{W}(\mathscr{A})$.

Notice that $\mathbf{G}_1$ and also $\mathbf{B}_1, \mathbf{C}_1$ and $\mathbf{C}_2$ are invariant for the choices of the generalized inverses denoted by "$-$". This is due to [**17**, Lemma 2.2.4] stating that for any nonnull $\mathbf{A}$ and $\mathbf{C}$ the following holds:

$$\text{(1.4)} \qquad \mathbf{AB^-C} = \mathbf{AB^+C} \text{ for all } \mathbf{B}^- \iff \mathscr{C}(\mathbf{C}) \subseteq \mathscr{C}(\mathbf{B}) \;\&\; \mathscr{C}(\mathbf{A}') \subseteq \mathscr{C}(\mathbf{B}').$$

Moreover, the following statements will be useful:

$$\mathbf{AA^-C} = \mathbf{C} \text{ for some (and hence for all) } \mathbf{A}^- \iff \mathscr{C}(\mathbf{C}) \subseteq \mathscr{C}(\mathbf{A}).$$

$$\mathbf{C}'\mathbf{A}^-\mathbf{A} = \mathbf{C}' \text{ for some (and hence for all) } \mathbf{A}^- \iff \mathscr{C}(\mathbf{C}) \subseteq \mathscr{C}(\mathbf{A}').$$

We assume that the model $\mathscr{M} = \{\mathbf{y}, \mathbf{X}\boldsymbol{\beta}, \mathbf{V}\}$ is *consistent* in the sense that the observed value of the random vector $\mathbf{y}$ belongs to $\mathscr{C}(\mathbf{X} : \mathbf{V})$, i.e.,

$$\mathbf{y} \in \mathscr{C}(\mathbf{X} : \mathbf{V}) = \mathscr{C}(\mathbf{X} : \mathbf{VM}) = \mathscr{C}(\mathbf{X} : \mathbf{MV}).$$

Assuming that our model is consistent, we also have, for all $\mathbf{W} \in \mathcal{W}$,

$$\mathbf{X}(\mathbf{X}'\mathbf{W}^-\mathbf{X})^-\mathbf{X}'\mathbf{W}^-\mathbf{y} =: \mathbf{P}_{\mathbf{X};\mathbf{W}^-}\mathbf{y} = \mathrm{BLUE}(\mathbf{X}\boldsymbol{\beta}) \quad \text{for all } \mathbf{W}^-.$$

The matrix $\mathbf{P}_{\mathbf{X};\mathbf{W}^-} \in \{\mathbf{P}_{\boldsymbol{\mu}\,|\,\mathbf{V}}\}$ itself is not necessarily invariant for the choice of generalized inverses involved but the vector $\mathbf{P}_{\mathbf{X};\mathbf{W}^-}\mathbf{y}$ is.

In several connections we can replace $(\mathbf{M}_2\mathbf{W}\mathbf{M}_2)^+$ with any $(\mathbf{M}_2\mathbf{W}\mathbf{M}_2)^-$ but using the Moore–Penrose inverse makes $\mathbf{C}_1$ unique and in particular, we have the equality $\mathbf{C}_1 = \mathbf{C}_2$. Observe that in view of $\mathscr{C}(\mathbf{M}_2\mathbf{X}_1) \subseteq \mathscr{C}(\mathbf{M}_2\mathbf{W}\mathbf{M}_2)$ and (1.4) we have

$$\mathbf{X}_1'\mathbf{M}_2(\mathbf{M}_2\mathbf{W}\mathbf{M}_2)^-\mathbf{M}_2\mathbf{X}_1 = \mathbf{X}_1'\mathbf{M}_2(\mathbf{M}_2\mathbf{W}\mathbf{M}_2)^+\mathbf{M}_2\mathbf{X}_1 \quad \text{for all } (\mathbf{M}_2\mathbf{W}\mathbf{M}_2)^-.$$

As for (1.2), we refer to [**4**, Cor. 2] showing the following:

LEMMA 1.2. *For the matrices* $\mathbf{A}_{n\times b}$ *and* $\mathbf{B}_{n\times n}$ *the following holds:*

$$\mathbf{P}_\mathbf{A}(\mathbf{P}_\mathbf{A}\mathbf{B}\mathbf{P}_\mathbf{A})^+\mathbf{P}_\mathbf{A} = (\mathbf{P}_\mathbf{A}\mathbf{B}\mathbf{P}_\mathbf{A})^+ = \mathbf{P}_\mathbf{A}(\mathbf{P}_\mathbf{A}\mathbf{B}\mathbf{P}_\mathbf{A})^+ = (\mathbf{P}_\mathbf{A}\mathbf{B}\mathbf{P}_\mathbf{A})^+\mathbf{P}_\mathbf{A}.$$

Lemma 1.3 collects together some important properties of the class $\mathcal{W}$; see, e.g., [**14**, Prop. 12.1].

LEMMA 1.3. *Given the model* $\mathscr{M} = \{\mathbf{y}, \mathbf{X}\boldsymbol{\beta}, \mathbf{V}\}$, *the matrix* $\mathbf{W} = \mathbf{V} + \mathbf{XUU}'\mathbf{X}'$ *belongs to* $\mathcal{W}(\mathscr{M})$ *if and only if any of the following equivalent conditions holds:*

(a) $\mathscr{C}(\mathbf{X} : \mathbf{V}) = \mathscr{C}(\mathbf{W})$.

(b) $\mathscr{C}(\mathbf{X}) \subseteq \mathscr{C}(\mathbf{W})$.

(c) $\mathbf{X}'\mathbf{W}^-\mathbf{X}$ *is invariant for any choice of* $\mathbf{W}^-$.

(d) $\mathscr{C}(\mathbf{X}'\mathbf{W}^-\mathbf{X}) = \mathscr{C}(\mathbf{X}')$ *for any choice of* $\mathbf{W}^-$.

(e) $\mathbf{X}(\mathbf{X}'\mathbf{W}^-\mathbf{X})^-\mathbf{X}'\mathbf{W}^-\mathbf{X} = \mathbf{X}$ *for any choices of* $\mathbf{W}^-$ *and* $(\mathbf{X}'\mathbf{W}^-\mathbf{X})^-$.

Moreover, each of these statements is equivalent to (a') $\mathscr{C}(\mathbf{X} : \mathbf{V}) = \mathscr{C}(\mathbf{W}')$, *and hence to the statements* (b')–(e') *obtained from* (b)–(e), *by setting* $\mathbf{W}'$ *in place of* $\mathbf{W}$.

Actually the class $\mathcal{W}$ can be replaced with set

$$\mathcal{W}_\# = \{\mathbf{W} : \mathbf{W} = \mathbf{V} + \mathbf{XTX}',\ \mathscr{C}(\mathbf{W}) = \mathscr{C}(\mathbf{X} : \mathbf{V})\},$$

where $\mathbf{T}$ need not be symmetric nonnegative definite. However, we will use (1.3) for it simplifies some considerations.

Let $\mathbf{G}$ be such a matrix that $\mathbf{Gy}$ is the BLUE for $\mathbf{X}\boldsymbol{\beta}$ under $\mathscr{M}(\mathbf{V}_0)$, Then we say that $\mathbf{Gy}$ remains BLUE under $\mathscr{M}(\mathbf{V})$ if the following implication holds:

$$\mathbf{G}(\mathbf{X} : \mathbf{V}_0\mathbf{M}) = (\mathbf{X} : \mathbf{0}) \implies \mathbf{G}(\mathbf{X} : \mathbf{VM}) = (\mathbf{X} : \mathbf{0}).$$

The general solution for $\mathbf{G}$ satisfying $\mathbf{G}(\mathbf{X} : \mathbf{V}_0\mathbf{M})$ can be expressed as

$$\mathbf{P}_{\mathbf{X};\mathbf{W}_0^+} + \mathbf{E}(\mathbf{I}_n - \mathbf{P}_{\mathbf{W}_0}), \quad \text{where } \mathbf{E}_{n\times n} \text{ is free to vary}$$

and $\mathbf{W}_0 \in \mathcal{W}(\mathscr{M}(\mathbf{V}_0))$. Let the set of all representations of BLUE of $\boldsymbol{\mu}$ under $\mathscr{M}(\mathbf{V}_0)$ be denoted as

$$\begin{aligned}\mathcal{B}(\boldsymbol{\mu} \mid \mathbf{V}_0) &= \{\mathrm{BLUE}(\boldsymbol{\mu} \mid \mathbf{V}_0)\} = \{\mathbf{Gy} : \mathbf{G}(\mathbf{X} : \mathbf{V}_0\mathbf{M}) = (\mathbf{X} : \mathbf{0})\} \\ &= \{\mathbf{Gy} : \mathbf{G} \in \{\mathbf{P}_{\boldsymbol{\mu} \mid \mathbf{V}_0}\}\}.\end{aligned}$$

For the property that *every* representation of the BLUE of $\boldsymbol{\mu}$ under $\mathscr{M}(\mathbf{V}_0)$ remains BLUE of $\boldsymbol{\mu}$ under $\mathscr{M}(\mathbf{V})$ we will use the notation

$$\mathcal{B}(\boldsymbol{\mu} \mid \mathbf{V}_0) \subseteq \mathcal{B}(\boldsymbol{\mu} \mid \mathbf{V}), \quad \text{i.e.,} \quad \{\mathbf{P}_{\boldsymbol{\mu} \mid \mathbf{V}_0}\} \subseteq \{\mathbf{P}_{\boldsymbol{\mu} \mid \mathbf{V}}\}. \tag{1.5}$$

We may consider $\mathscr{M}(\mathbf{V}_0)$ as the original model and $\mathscr{M}(\mathbf{V})$ as the misspecified model; misspecification concerning only the covariance matrix. Some equivalent statements to (1.5) are given as follows.

LEMMA 1.4. *Consider the linear models $\mathscr{M}(\mathbf{V}_0)$ and $\mathscr{M}(\mathbf{V})$. Then the following statements are equivalent:*

(a) $\mathcal{B}(\boldsymbol{\mu} \mid \mathbf{V}_0) \subseteq \mathcal{B}(\boldsymbol{\mu} \mid \mathbf{V})$.
(b) $\mathscr{C}(\mathbf{VM}) \subseteq \mathscr{C}(\mathbf{V}_0\mathbf{M})$.
(c) $\mathbf{V} = \mathbf{XAA}'\mathbf{X}' + \mathbf{V}_0\mathbf{MBB}'\mathbf{MV}_0$, *for some* $\mathbf{A}$ *and* $\mathbf{B}$.
(d) $\mathbf{V} = \mathbf{V}_0 + \mathbf{XCC}'\mathbf{X}' + \mathbf{V}_0\mathbf{MDD}'\mathbf{MV}_0$, *for some* $\mathbf{C}$ *and* $\mathbf{D}$.

For the proof of Lemma 1.4 and related discussion, see, e.g., [**13**, Th. 4.1–4.2], [**15**, Th. 5.2, Th. 5.5], [**16**, p. 289], and [**3**, Th. 3]. Notice that if $\mathbf{V}_0 = \mathbf{I}_n$,

$$\mathrm{BLUE}(\boldsymbol{\mu} \mid \mathbf{V}_0) = \mathbf{P_X y} = \mathrm{OLSE}(\boldsymbol{\mu} \mid \mathbf{V}) = \mathrm{OLSE}(\boldsymbol{\mu} \mid \mathbf{V}_0),$$

where OLSE refers to the ordinary least squares estimator. As a consequence the properties (a) and (b) of Lemma 1.4 give the following:

$$\mathrm{OLSE}(\boldsymbol{\mu} \mid \mathbf{V}) = \mathrm{BLUE}(\boldsymbol{\mu} \mid \mathbf{V}) \iff \mathscr{C}(\mathbf{VM}) \subseteq \mathscr{C}(\mathbf{M}) \iff \mathbf{X}'\mathbf{VM} = \mathbf{0}. \tag{1.6}$$

In view of [**8**, Th. 2.1], and [**6**, Th. 11.4]; see also [**12**, Th. 2.1, Th. 2.4], the following holds:

LEMMA 1.5. *Consider the partitioned linear models $\mathscr{M}_{12}(\mathbf{V}_0)$ and $\mathscr{M}_{12}(\mathbf{V})$, where $\boldsymbol{\mu}_1 = \mathbf{X}_1\boldsymbol{\beta}_1$ is estimable. Then the following statements are equivalent:*

(a) $\mathcal{B}(\boldsymbol{\mu}_1 \mid \mathbf{V}_0) \subseteq \mathcal{B}(\boldsymbol{\mu}_1 \mid \mathbf{V})$.
(b) $\mathscr{C}(\mathbf{M}_2\mathbf{VM}) \subseteq \mathscr{C}(\mathbf{M}_2\mathbf{V}_0\mathbf{M})$.
(c) $\mathscr{C}(\mathbf{VM}) \subseteq \mathscr{C}(\mathbf{X}_2 : \mathbf{V}_0\mathbf{M})$.
(d) *The matrix* $\mathbf{V}$ *can be expressed, for some* $\mathbf{L}_i$, $\mathbf{L}_{ij} = \mathbf{L}_i\mathbf{L}_j'$, *as*

$$\mathbf{V} = \mathbf{V}_0 + \mathbf{X}_1\mathbf{L}_{11}\mathbf{X}_1' + \mathbf{X}_2\mathbf{L}_{22}\mathbf{X}_2' + \mathbf{V}_0\mathbf{ML}_{33}\mathbf{MV}_0 + \mathbf{F} + \mathbf{F}',$$

where $\mathbf{F} = \mathbf{X}_1\mathbf{L}_{12}\mathbf{X}_2' + \mathbf{X}_2\mathbf{L}_{23}\mathbf{MV}_0$.

Let's put here Theorem 2.4 of [**12**]:

LEMMA 1.6. *Consider the linear models $\mathscr{M}(\mathbf{V}_0)$ and $\mathscr{M}(\mathbf{V})$, where $\mathbf{K}'\boldsymbol{\beta}$ is estimable. Then the following statements are equivalent:*

(a) $\mathcal{B}(\mathbf{K}'\boldsymbol{\beta} \mid \mathbf{V}_0) \subseteq \mathcal{B}(\mathbf{K}'\boldsymbol{\beta} \mid \mathbf{V})$.
(b) $\mathscr{C}(\mathbf{Q}_{\mathbf{XK}^\perp}\mathbf{VM}) \subseteq \mathscr{C}(\mathbf{Q}_{\mathbf{XK}^\perp}\mathbf{V}_0\mathbf{M})$.
(c) $\mathscr{C}(\mathbf{VM}) \subseteq \mathscr{C}(\mathbf{XK}^\perp : \mathbf{V}_0\mathbf{M})$.

We will later utilize the following result of [**7**].

LEMMA 1.7. *Consider the partitioned linear models* $\mathscr{M}_{12}(\mathbf{V}_0)$ *and* $\mathscr{M}_{12}(\mathbf{V})$, *where* $\boldsymbol{\mu}_1 = \mathbf{X}_1\boldsymbol{\beta}_1$ *is estimable and denote* $\mathbf{W}_0 = \mathbf{V}_0 + \mathbf{X}\mathbf{X}', \mathbf{W} = \mathbf{V} + \mathbf{X}\mathbf{X}'$. *Suppose that* $\mathcal{B}(\boldsymbol{\mu}_1 \mid \mathbf{V}_0) \subseteq \mathcal{B}(\boldsymbol{\mu}_1 \mid \mathbf{V})$. *Then the following two statements are equivalent:*

(a) $\operatorname{cov}(\tilde{\boldsymbol{\mu}}_1 \mid \mathbf{V}_0) = \operatorname{cov}(\tilde{\boldsymbol{\mu}}_1 \mid \mathbf{V})$, *i.e.*,

$$\operatorname{cov}(\operatorname{BLUE}(\boldsymbol{\mu}_1 \mid \mathbf{V}_0)) = \operatorname{cov}(\operatorname{BLUE}(\boldsymbol{\mu}_1 \mid \mathbf{V})),$$

(b) $(\mathbf{M}_2\mathbf{W}\mathbf{M}_2)(\mathbf{M}_2\mathbf{W}_0\mathbf{M}_2)^{+}\mathbf{X}_1 = \mathbf{M}_2\mathbf{X}_1$.

We will frequently need the following result, known as the Frisch–Waugh–Lovell theorem, see, e.g., [**5**].

LEMMA 1.8. *Consider the partitioned linear model* $\mathscr{M}_{12} = \{\mathbf{y}, \mathbf{X}\boldsymbol{\beta}, \mathbf{V}\}$ *and its reduced form*

$$\mathscr{M}_{12\cdot 2} = \{\mathbf{M}_2\mathbf{y},\ \mathbf{M}_2\mathbf{X}_1\boldsymbol{\beta}_1,\ \mathbf{M}_2\mathbf{V}\mathbf{M}_2\}.$$

Then every BLUE *of* $\mathbf{M}_2\mathbf{X}_1\boldsymbol{\beta}_1$ *under* $\mathscr{M}_{12\cdot 2}$ *remains* BLUE *under* $\mathscr{M}_{12}$ *and vice versa. In other words, let* $\mathbf{L}\boldsymbol{\beta}_1$ *be an estimable parametric function in* $\mathscr{M}_{12}$. *Then every representation of the* BLUE *of* $\mathbf{L}\boldsymbol{\beta}_1$ *under* $\mathscr{M}_{12\cdot 2}$ *remains* BLUE *under* $\mathscr{M}_{12}$ *and vice versa. In particular, this concerns* $\boldsymbol{\mu}_1 = \mathbf{X}_1\boldsymbol{\beta}_1$ *supposing it is estimable under* $\mathscr{M}_{12}$.

2. Reparametrization: case 1

To begin, denote the models under consideration as

$$\mathscr{M}_{12} = \{\mathbf{y}, \mathbf{X}\boldsymbol{\beta}, \mathbf{V}\} = \{\mathbf{y}, (\mathbf{X}_1 : \mathbf{X}_2)\boldsymbol{\beta}, \mathbf{V}\},$$
$$\mathscr{M}_{1R} = \{\mathbf{y}, \mathbf{X}_*\boldsymbol{\beta}, \mathbf{V}\} = \{\mathbf{y}, (\mathbf{X}_1 : \mathbf{M}_1\mathbf{X}_2)\boldsymbol{\beta}, \mathbf{V}\},$$

where we assume that the disjointness condition (1.1) holds so that $\boldsymbol{\mu}_1 = \mathbf{X}_1\boldsymbol{\beta}_1$ is estimable under $\mathscr{M}_{12}$; it is obviously estimable under $\mathscr{M}_{1R}$. Above we have

$$\mathbf{X}_* = \mathbf{X}\mathbf{A}, \quad \text{where } \mathbf{A} = \begin{pmatrix} \mathbf{I}_{p_1} & -\mathbf{X}_1^{+}\mathbf{X}_2 \\ \mathbf{0} & \mathbf{I}_{p_2} \end{pmatrix} \in \mathbb{R}^{p \times p}. \tag{2.1}$$

In view of $\mathscr{C}(\mathbf{X}) = \mathscr{C}(\mathbf{X}_*)$, the models $\mathscr{M}_{12}$ and $\mathscr{M}_{1R}$ are equivalent in the sense that

$$\operatorname{BLUE}(\mathbf{X}\boldsymbol{\beta} \mid \mathscr{M}_{12}) = \operatorname{BLUE}(\mathbf{X}_*\boldsymbol{\beta} \mid \mathscr{M}_{1R}). \tag{2.2}$$

Premultiplying $\mathscr{M}_{1R}$ by the orthogonal projector $\mathbf{M}_1$ and keeping the Frisch–Waugh–Lovell theorem in mind we get the model

$$\mathscr{M}_{1R\cdot 1} = \{\mathbf{M}_1\mathbf{y}, \mathbf{M}_1\mathbf{X}_2\boldsymbol{\beta}_2, \mathbf{M}_1\mathbf{V}\mathbf{M}_1\},$$

which is nothing but the *reduced* version of the model $\mathscr{M}_{12}$ after elimination of $\boldsymbol{\beta}_1$ for which the usual notation is

$$\mathscr{M}_{12\cdot 1} = \{\mathbf{M}_1\mathbf{y}, \mathbf{M}_1\mathbf{X}_2\boldsymbol{\beta}_2, \mathbf{M}_1\mathbf{V}\mathbf{M}_1\}.$$

Thus by Lemma 1.8, we have (with probability 1, as is the case when dealing with the equalities between random vectors)

$$\operatorname{BLUE}(\boldsymbol{\mu}_2 \mid \mathscr{M}_{1R}) = \operatorname{BLUE}(\boldsymbol{\mu}_2 \mid \mathscr{M}_{12}) = \operatorname{BLUE}(\boldsymbol{\mu}_2 \mid \mathscr{M}_{12\cdot 1}),$$

or shortly,

$$\tilde{\boldsymbol{\mu}}_2(\mathscr{M}_{1R}) = \tilde{\boldsymbol{\mu}}_2(\mathscr{M}_{12}) = \tilde{\boldsymbol{\mu}}_2(\mathscr{M}_{12\cdot 1}).$$

Recall that one representation for $\tilde{\boldsymbol{\mu}}_2(\mathscr{M}_{12})$ is

$$\mathbf{P}_{\mathbf{X}_2;\dot{\mathbf{M}}_1}\mathbf{y} = \mathbf{X}_2(\mathbf{X}_2'\dot{\mathbf{M}}_1\mathbf{X}_2)^-\mathbf{X}_2'\dot{\mathbf{M}}_1\mathbf{y}\,,$$

where $\dot{\mathbf{M}}_1 = \mathbf{M}_1(\mathbf{M}_1\mathbf{W}_2\mathbf{M}_1)^+\mathbf{M}_1$ and $\mathbf{W}_2 = \mathbf{V} + \mathbf{X}_2\mathbf{X}_2'$. Correspondingly, one representation for $\tilde{\boldsymbol{\mu}}_1(\mathscr{M}_{12})$ is

$$\mathbf{P}_{\mathbf{X}_1;\dot{\mathbf{M}}_2}\mathbf{y} = \mathbf{X}_2(\mathbf{X}_2'\dot{\mathbf{M}}_2\mathbf{X}_2)^-\mathbf{X}_2'\dot{\mathbf{M}}_2\mathbf{y}\,,$$

where $\dot{\mathbf{M}}_2 = \mathbf{M}_2(\mathbf{M}_2\mathbf{W}_1\mathbf{M}_2)^+\mathbf{M}_2$ and $\mathbf{W}_1 = \mathbf{V} + \mathbf{X}_1\mathbf{X}_1'$.

What about the relation between

$$\mathrm{BLUE}(\boldsymbol{\mu}_1 \mid \mathscr{M}_{1R}) \quad \text{vs} \quad \mathrm{BLUE}(\boldsymbol{\mu}_1 \mid \mathscr{M}_{12})?$$

It is clear that

$$\mathrm{BLUE}(\mathbf{X}\boldsymbol{\beta} \mid \mathscr{M}_{12}) = \mathrm{BLUE}(\mathbf{X}_1\boldsymbol{\beta}_1 \mid \mathscr{M}_{12}) + \mathrm{BLUE}(\mathbf{X}_2\boldsymbol{\beta}_2 \mid \mathscr{M}_{12})\,.$$

Thereby, in light of (2.2) we have

$$\begin{aligned}\tilde{\boldsymbol{\mu}}_1(\mathscr{M}_{12}) + \tilde{\boldsymbol{\mu}}_2(\mathscr{M}_{12}) &= \tilde{\boldsymbol{\mu}}_1(\mathscr{M}_{1R}) + \mathbf{M}_1\tilde{\boldsymbol{\mu}}_2(\mathscr{M}_{1R})\\ &= \tilde{\boldsymbol{\mu}}_1(\mathscr{M}_{1R}) + \mathbf{M}_1\tilde{\boldsymbol{\mu}}_2(\mathscr{M}_{12})\,,\end{aligned}$$

which further implies that $\tilde{\boldsymbol{\mu}}_1(\mathscr{M}_{12}) = \tilde{\boldsymbol{\mu}}_1(\mathscr{M}_{1R}) - \mathbf{P}_1 \cdot \tilde{\boldsymbol{\mu}}_2(\mathscr{M}_{12})$, where $\mathbf{P}_1 = \mathbf{P}_{\mathbf{X}_1}$, so that

$$\tilde{\boldsymbol{\mu}}_1(\mathscr{M}_{1R}) - \tilde{\boldsymbol{\mu}}_1(\mathscr{M}_{12}) = \mathbf{P}_1 \cdot \tilde{\boldsymbol{\mu}}_2(\mathscr{M}_{12})\,. \tag{2.3}$$

Next we prove the following:

$$\tilde{\boldsymbol{\mu}}_1(\mathscr{M}_{12}) = \tilde{\boldsymbol{\mu}}_1(\mathscr{M}_{1R}) \iff \mathbf{X}_1'\mathbf{X}_2 = \mathbf{0}\,. \tag{2.4}$$

This can be seen by considering the equality

$$\begin{aligned}\mathbf{P}_1 \cdot \tilde{\boldsymbol{\mu}}_2(\mathscr{M}_{12}) &= \mathbf{P}_1\mathbf{X}_2(\mathbf{X}_2'\dot{\mathbf{M}}_1\mathbf{X}_2)^-\mathbf{X}_2'\dot{\mathbf{M}}_1\mathbf{y}\\ &= \mathbf{P}_1\mathbf{P}_{\mathbf{X}_2;\dot{\mathbf{M}}_1}\mathbf{y}\\ &= \mathbf{0} \;\; \forall\, \mathbf{y} \in \mathscr{C}(\mathbf{X} : \mathbf{VM})\,.\end{aligned} \tag{2.5}$$

Notice that we assume that the model $\mathscr{M}_{12}$ is consistent in the sense that $\mathbf{y} \in \mathscr{C}(\mathbf{X} : \mathbf{VM})$. Thus (2.5) implies that

$$\mathbf{P}_1\mathbf{X}_2(\mathbf{X}_2'\dot{\mathbf{M}}_1\mathbf{X}_2)^-\mathbf{X}_2'\dot{\mathbf{M}}_1\mathbf{X}_2 = \mathbf{0}\,. \tag{2.6}$$

In view of $\mathscr{C}(\mathbf{X}_2'\dot{\mathbf{M}}_1\mathbf{X}_2) = \mathscr{C}(\mathbf{X}_2')$, which follows from the rule of the rank of matrix product

$$\begin{aligned}\mathrm{r}(\mathbf{X}_2'\dot{\mathbf{M}}_1\mathbf{X}_2) &= \mathrm{r}[\mathbf{X}_2'\mathbf{M}_1(\mathbf{M}_1\mathbf{W}_2\mathbf{M}_1)^+]\\ &= \mathrm{r}(\mathbf{X}_2'\mathbf{M}_1) - \dim \mathscr{C}(\mathbf{M}_1\mathbf{X}_2) \cap \mathscr{C}(\mathbf{M}_1\mathbf{W}_2)^\perp = \mathrm{r}(\mathbf{X}_2'\mathbf{M}_1) = \mathrm{r}(\mathbf{X}_2')\,,\end{aligned}$$

we observe that

$$\mathbf{X}_2(\mathbf{X}_2'\dot{\mathbf{M}}_1\mathbf{X}_2)^-\mathbf{X}_2'\dot{\mathbf{M}}_1\mathbf{X}_2 = \mathbf{X}_2\,. \tag{2.7}$$

Substituting (2.7) into (2.6) confirms that (2.4) indeed holds.

Let us next find an explicit expression for the $\tilde{\boldsymbol{\mu}}_1(\mathscr{M}_{1R})$. This could be done on the basis of (2.3) giving

$$\tilde{\boldsymbol{\mu}}_1(\mathscr{M}_{1R}) = \tilde{\boldsymbol{\mu}}_1(\mathscr{M}_{12}) + \mathbf{P}_1 \cdot \tilde{\boldsymbol{\mu}}_2(\mathscr{M}_{12})\,.$$

An interesting alternative way is to reduce the model $\mathscr{M}_{1R} = \{\mathbf{y}, (\mathbf{X}_1 : \mathbf{M}_1\mathbf{X}_2)\boldsymbol{\beta}, \mathbf{V}\}$ by premultiplying it by

$$\mathbf{Q} := \mathbf{Q}_{\mathbf{M}_1\mathbf{X}_2} = \mathbf{I}_n - \mathbf{P}_{\mathbf{M}_1\mathbf{X}_2} = \mathbf{M} + \mathbf{P}_1 .$$

This gives the model

$$\mathscr{M}_{1R\cdot 2} = \{\mathbf{Qy},\ \mathbf{QX}_1\boldsymbol{\beta}_1,\ \mathbf{QVQ}\} = \{\mathbf{Qy},\ \mathbf{X}_1\boldsymbol{\beta}_1,\ \mathbf{QVQ}\},$$

where we have used $\mathbf{QX}_1 = \mathbf{X}_1$. In view of the Frisch–Waugh–Lovell theorem, we have

$$\tilde{\boldsymbol{\mu}}_1(\mathscr{M}_{1R\cdot 2}) = \tilde{\boldsymbol{\mu}}_1(\mathscr{M}_{1R}) .$$

Moreover, in light of

$$\mathbf{X}_1\mathbf{X}_1' + \mathbf{QVQ} = \mathbf{QW}_1\mathbf{Q} \in \mathcal{W}(\mathscr{M}_{1R\cdot 2}), \quad \mathbf{W}_1 = \mathbf{V} + \mathbf{X}_1\mathbf{X}_1' ,$$

we have

$$\begin{aligned} \tilde{\boldsymbol{\mu}}_1(\mathscr{M}_{1R}) &= \mathbf{X}_1[\mathbf{X}_1'\mathbf{Q}(\mathbf{QW}_1\mathbf{Q})^+\mathbf{QX}_1]^-\mathbf{X}_1'\mathbf{Q}(\mathbf{QW}_1\mathbf{Q})^+\mathbf{Qy} \\ &= \mathbf{X}_1[\mathbf{X}_1'(\mathbf{QW}_1\mathbf{Q})^+\mathbf{X}_1]^-\mathbf{X}_1'(\mathbf{QW}_1\mathbf{Q})^+\mathbf{y} \\ &= \mathbf{P}_{\mathbf{X}_1;(\mathbf{QW}_1\mathbf{Q})^+}\mathbf{y} , \end{aligned} \tag{2.8}$$

where we have used Lemma 1.2. On the other hand, we know that

$$\begin{aligned} \tilde{\boldsymbol{\mu}}_1(\mathscr{M}_{12}) &= \mathbf{X}_1[\mathbf{X}_1'(\mathbf{M}_2\mathbf{W}_1\mathbf{M}_2)^+\mathbf{X}_1]^-\mathbf{X}_1'(\mathbf{M}_2\mathbf{W}_1\mathbf{M}_2)^+\mathbf{y} \\ &= \mathbf{P}_{\mathbf{X}_1;(\mathbf{M}_2\mathbf{W}_1\mathbf{M}_2)^+}\mathbf{y} \\ &= \mathbf{P}_{\mathbf{X}_1;\dot{\mathbf{M}}_2}\mathbf{y} . \end{aligned} \tag{2.9}$$

The difference between (2.8) and (2.9) lies in the "innerproduct matrices" $(\mathbf{QW}_1\mathbf{Q})^+$ and $(\mathbf{M}_2\mathbf{W}_1\mathbf{M}_2)^+$. The multipliers of $\mathbf{y}$ are some kind of generalized projectors with singular innerproduct matrices.

The difference of the BLUEs of $\boldsymbol{\mu}_1$ under $\mathscr{M}_{1R}$ and $\mathscr{M}_{12}$ can be written as

$$\begin{aligned} \tilde{\boldsymbol{\mu}}_1(\mathscr{M}_{1R}) - \tilde{\boldsymbol{\mu}}_1(\mathscr{M}_{12}) &= \mathbf{P}_{\mathbf{X}_1;(\mathbf{QW}_1\mathbf{Q})^+}\mathbf{y} - \mathbf{P}_{\mathbf{X}_1;(\mathbf{M}_2\mathbf{W}_1\mathbf{M}_2)^+}\mathbf{y} \\ &= \mathbf{P}_1 \cdot \tilde{\boldsymbol{\mu}}_2(\mathscr{M}_{12}) = \mathbf{P}_1\mathbf{X}_2(\mathbf{X}_2'\dot{\mathbf{M}}_1\mathbf{X}_2)^-\mathbf{X}_2'\dot{\mathbf{M}}_1\mathbf{y} \\ &= \mathbf{P}_1\mathbf{P}_{\mathbf{X}_2;\dot{\mathbf{M}}_1}\mathbf{y} = \mathbf{P}_1\mathbf{P}_{\mathbf{X}_2;(\mathbf{M}_1\mathbf{W}_2\mathbf{M}_1)^+}\mathbf{y} . \end{aligned} \tag{2.10}$$

The equality (2.10) holds for all $\mathbf{y} \in \mathscr{C}(\mathbf{X} : \mathbf{VM}) = \mathscr{C}(\mathbf{W})$, where $\mathbf{W} = \mathbf{V} + \mathbf{XX}'$. What happens if $\mathbf{y}$ is outside that area? In other words, is the following equality correct:

$$\mathbf{P}_{\mathbf{X}_1;(\mathbf{QW}_1\mathbf{Q})^+} - \mathbf{P}_{\mathbf{X}_1;(\mathbf{M}_2\mathbf{W}_1\mathbf{M}_2)^+} = \mathbf{P}_1\mathbf{P}_{\mathbf{X}_2;(\mathbf{M}_1\mathbf{W}_2\mathbf{M}_1)^+} . \tag{2.11}$$

To prove (2.11) we need to show, in addition to (2.10), that

$$\mathbf{P}_{\mathbf{X}_1;(\mathbf{QW}_1\mathbf{Q})^+}\mathbf{Q}_\mathbf{W} - \mathbf{P}_{\mathbf{X}_1;(\mathbf{M}_2\mathbf{W}_1\mathbf{M}_2)^+}\mathbf{Q}_\mathbf{W} = \mathbf{P}_1\mathbf{P}_{\mathbf{X}_2;(\mathbf{M}_1\mathbf{W}_2\mathbf{M}_1)^+}\mathbf{Q}_\mathbf{W} , \tag{2.12}$$

where $\mathbf{Q}_\mathbf{W} = \mathbf{I}_n - \mathbf{P}_\mathbf{W}$. Now $(\mathbf{M}_2\mathbf{W}_1\mathbf{M}_2)^+\mathbf{Q}_\mathbf{W} = \mathbf{0}$ because

$$\begin{aligned} \mathscr{C}[(\mathbf{M}_2\mathbf{W}_1\mathbf{M}_2)^+] &= \mathscr{C}(\mathbf{M}_2\mathbf{W}_1\mathbf{M}_2) = \mathscr{C}(\mathbf{M}_2\mathbf{W}_1) \\ &\subseteq \mathscr{C}(\mathbf{X}_2 : \mathbf{M}_2\mathbf{W}_1) = \mathscr{C}(\mathbf{X}_2 : \mathbf{W}_1) \subseteq \mathscr{C}(\mathbf{W}) . \end{aligned}$$

In the same way we can conclude that

$$(\mathbf{M}_1\mathbf{W}_2\mathbf{M}_1)^+\mathbf{Q}_\mathbf{W} = (\mathbf{QW}_1\mathbf{Q})^+\mathbf{Q}_\mathbf{W} = \mathbf{0} ,$$

and thus (2.12) and (2.11) indeed hold.

The following theorem summarises our findings; of course the claims (a)–(c) are not "new" findings.

THEOREM 2.1. *Consider the partitioned linear models*

$$\mathscr{M}_{12} = \{\mathbf{y}, \mathbf{X}\boldsymbol{\beta}, \mathbf{V}\} = \{\mathbf{y}, (\mathbf{X}_1 : \mathbf{X}_2)\boldsymbol{\beta}, \mathbf{V}\},$$
$$\mathscr{M}_{1R} = \{\mathbf{y}, \mathbf{X}_*\boldsymbol{\beta}, \mathbf{V}\} = \{\mathbf{y}, (\mathbf{X}_1 : \mathbf{M}_1\mathbf{X}_2)\boldsymbol{\beta}, \mathbf{V}\},$$

where $\boldsymbol{\mu}_1$ *is estimable in* $\mathscr{M}_{12}$ *(and trivially in* $\mathscr{M}_{1R}$*) and denote*

$$\mathbf{Q} = \mathbf{I}_n - \mathbf{P}_{\mathbf{M}_1\mathbf{X}_2}, \dot{\mathbf{M}}_1 = (\mathbf{M}_1\mathbf{W}_2\mathbf{M}_1)^+, \dot{\mathbf{M}}_2 = (\mathbf{M}_2\mathbf{W}_1\mathbf{M}_2)^+, \mathbf{W}_i = \mathbf{V} + \mathbf{X}_i\mathbf{X}_i'.$$

Then the following statements hold:

(a) $\mathrm{BLUE}(\boldsymbol{\mu}_2 \mid \mathscr{M}_{1R}) = \mathrm{BLUE}(\boldsymbol{\mu}_2 \mid \mathscr{M}_{12})$, *or shortly,* $\tilde{\boldsymbol{\mu}}_2(\mathscr{M}_{1R}) = \tilde{\boldsymbol{\mu}}_2(\mathscr{M}_{12})$.
(b) $\tilde{\boldsymbol{\mu}}_1(\mathscr{M}_{12}) = \mathbf{P}_{\mathbf{X}_1;(\mathbf{M}_2\mathbf{W}_1\mathbf{M}_2)^+}\mathbf{y} = \mathbf{P}_{\mathbf{X}_1;\dot{\mathbf{M}}_2}\mathbf{y}$.
(c) $\tilde{\boldsymbol{\mu}}_2(\mathscr{M}_{12}) = \mathbf{P}_{\mathbf{X}_2;(\mathbf{M}_1\mathbf{W}_2\mathbf{M}_1)^+}\mathbf{y} = \mathbf{P}_{\mathbf{X}_2;\dot{\mathbf{M}}_1}\mathbf{y}$.
(d) $\tilde{\boldsymbol{\mu}}_1(\mathscr{M}_{1R}) = \mathbf{X}_1[\mathbf{X}_1'(\mathbf{Q}\mathbf{W}_1\mathbf{Q})^+\mathbf{X}_1]^-\mathbf{X}_1'(\mathbf{Q}\mathbf{W}_1\mathbf{Q})^+\mathbf{y} = \mathbf{P}_{\mathbf{X}_1;(\mathbf{Q}\mathbf{W}_1\mathbf{Q})^+}\mathbf{y}$.
(e) $\tilde{\boldsymbol{\mu}}_1(\mathscr{M}_{1R}) - \tilde{\boldsymbol{\mu}}_1(\mathscr{M}_{12}) = \mathbf{P}_1 \cdot \tilde{\boldsymbol{\mu}}_2(\mathscr{M}_{12})$
$= \mathbf{P}_1\mathbf{P}_{\mathbf{X}_2;\dot{\mathbf{M}}_1}\mathbf{y} = \mathbf{P}_1\mathbf{P}_{\mathbf{X}_2;(\mathbf{M}_1\mathbf{W}_2\mathbf{M}_1)^+}\mathbf{y}$.
(f) $\mathbf{P}_{\mathbf{X}_1;(\mathbf{Q}\mathbf{W}_1\mathbf{Q})^+} - \mathbf{P}_{\mathbf{X}_1;(\mathbf{M}_2\mathbf{W}_1\mathbf{M}_2)^+} = \mathbf{P}_1\mathbf{P}_{\mathbf{X}_2;(\mathbf{M}_1\mathbf{W}_2\mathbf{M}_1)^+}$.
(g) $\tilde{\boldsymbol{\mu}}_1(\mathscr{M}_{1R}) = \tilde{\boldsymbol{\mu}}_1(\mathscr{M}_{12}) \iff \mathbf{X}_1'\mathbf{X}_2 = \mathbf{0}$.

EXAMPLE 2.1. Consider the ultra-simple situation when the models are

$$\mathscr{M}_{12} = \{\mathbf{y},\ \beta_0\mathbf{1} + \beta_1\mathbf{x},\ \sigma^2\mathbf{I}_n\} = \{\mathbf{y},\ (\mathbf{1} : \mathbf{x})\boldsymbol{\beta},\ \sigma^2\mathbf{I}_n\},$$
$$\mathscr{M}_{1R} = \{\mathbf{y},\ \beta_0\mathbf{1} + \beta_1\mathbf{C}\mathbf{x},\ \sigma^2\mathbf{I}_n\} = \{\mathbf{y},\ (\mathbf{1} : \mathbf{C}\mathbf{x})\boldsymbol{\beta},\ \sigma^2\mathbf{I}_n\},$$
$$\mathscr{M}_{12\cdot 1} = \{\mathbf{C}\mathbf{y},\ \beta_1\mathbf{C}\mathbf{x},\ \sigma^2\mathbf{C}\},$$

where $\mathbf{1} \in \mathbb{R}^n$ is a vector of ones, σ^2 is an unknown nonzero scalar and

$$\mathbf{C} = \mathbf{I}_n - \mathbf{P}_{\mathbf{1}} = \text{the centering matrix.}$$

Corresponding to (2.1) we now have

$$\mathbf{X}_* = (\mathbf{1} : \mathbf{C}\mathbf{x}) = (\mathbf{1} : \mathbf{x})\mathbf{A} = \mathbf{X}\mathbf{A}, \text{ where } \mathbf{A} = \begin{pmatrix} 1 & -\mathbf{1}^+\mathbf{x} \\ 0 & 1 \end{pmatrix} = \begin{pmatrix} 1 & -\bar{x} \\ 0 & 1 \end{pmatrix},$$

and

$$\mathbf{1}^+\mathbf{x} = (\mathbf{1}'\mathbf{1})^{-1}\mathbf{1}'\mathbf{x} = (x_1 + \cdots + x_n)/n = \bar{x}.$$

We assume that $\mathrm{r}(\mathbf{1} : \mathbf{x}) = 2$, i.e, $\mathbf{x} \notin \mathscr{C}(\mathbf{1})$; otherwise β_1 would not be estimable in $\mathscr{M}_{12}$.

We observe that in all models $\mathscr{M}_{12}$, $\mathscr{M}_{1R}$ and $\mathscr{M}_{12\cdot 1}$ the BLUEs of the parameters are the same as their ordinary least squares estimators, OLSEs. Using Theorem 2.1 we obtain the following well-known results; see also [**14**, Prop. 3.2]:

$$\tilde{\beta}_1(\mathscr{M}_{12}) = \tilde{\beta}_1(\mathscr{M}_{12\cdot 1}) = \tilde{\beta}_1(\mathscr{M}_{1R}) = \mathbf{x}'\mathbf{C}\mathbf{y}/\mathbf{x}'\mathbf{C}\mathbf{x},$$
$$\tilde{\beta}_0(\mathscr{M}_{12}) = \bar{y} - \bar{x}\tilde{\beta}_1, \quad \tilde{\beta}_0(\mathscr{M}_{1R}) = \bar{y}.$$

Now for all $\mathbf{y} \in \mathbb{R}^n$ we have $\tilde{\beta}_0(\mathscr{M}_{12}) = \tilde{\beta}_0(\mathscr{M}_{1R})$ if and only if

$$n^{-1}\mathbf{1}'\mathbf{x} \cdot \mathbf{x}'\mathbf{C}\mathbf{y}/\mathbf{x}'\mathbf{C}\mathbf{x} = 0 \quad \text{for all } \mathbf{y} \in \mathbb{R}^n. \tag{2.13}$$

Assuming that $\mathbf{x} \notin \mathscr{C}(\mathbf{1})$ holds, (2.13) is obviously equivalent to $\mathbf{1}'\mathbf{x} = 0$. □

3. Reparametrization: case 2

In the context of estimating the estimable parametric function $\mathbf{K}'\boldsymbol{\beta}$, Isotalo and Puntanen in (2009a, 2009b) and Isotalo et al. (2009) considered the following reparametrized version of $\mathscr{M} = \{\mathbf{y}, \mathbf{X}\boldsymbol{\beta}, \mathbf{V}\}$:

$$\mathscr{T} = \{\mathbf{y},\, \mathbf{X}_*\boldsymbol{\gamma},\, \mathbf{V}\} = \{\mathbf{y}, (\mathbf{XK} : \mathbf{XK}^{\perp})\boldsymbol{\gamma}, \mathbf{V}\} = \{\mathbf{y}, \mathbf{X}_{1*}\boldsymbol{\gamma}_1 + \mathbf{X}_{2*}\boldsymbol{\gamma}_2, \mathbf{V}\},$$

where $\mathbf{X}_* = \mathbf{X}(\mathbf{K} : \mathbf{K}^{\perp})$ and so

$$\mathbf{X}_*\boldsymbol{\gamma} = \mathbf{XK}\boldsymbol{\gamma}_1 + \mathbf{XK}^{\perp}\boldsymbol{\gamma}_2 = \mathbf{X}_{1*}\boldsymbol{\gamma}_1 + \mathbf{X}_{2*}\boldsymbol{\gamma}_2\,.$$

Above $\mathbf{K}^{\perp}$ can be any matrix spanning $\mathscr{C}(\mathbf{K})^{\perp}$, and of course one choice is $\mathbf{Q}_{\mathbf{K}} = \mathbf{I}_p - \mathbf{P}_{\mathbf{K}}$. It is clear that $\mathscr{M}$ and $\mathscr{T}$ are equivalent in the sense that

$$\text{BLUE}(\mathbf{X}\boldsymbol{\beta} \mid \mathscr{M}) = \text{BLUE}(\mathbf{X}_*\boldsymbol{\gamma} \mid \mathscr{T})\,. \tag{3.1}$$

What is the benefit in using the reparametrization like in $\mathscr{T}$? Shortly said, the usefulness follows from the Frisch–Waugh–Lovell theorem. Notice that $\mathscr{T}$ is a partitioned linear model and it is just partitioned models where the Frisch–Waugh–Lovell theorem works. It is crucial to observe that when $\mathbf{K}'\boldsymbol{\beta}$ is estimable under $\mathscr{M}$, i.e.,

$$\mathbf{K}' = \mathbf{ZX} \quad \text{for some } \mathbf{Z}\,, \tag{3.2}$$

then the parametric function $\mathbf{X}_{1*}\boldsymbol{\gamma}_1$ is estimable under $\mathscr{T}$ since

$$\mathscr{C}(\mathbf{XK}) \cap \mathscr{C}(\mathbf{XQ_K}) = \{\mathbf{0}\}\,. \tag{3.3}$$

To confirm (3.3), consider the equality

$$\mathbf{XKa} = \mathbf{XQ_K b} \quad \text{for some } \mathbf{a}, \mathbf{b} \in \mathbb{R}^p. \tag{3.4}$$

Premultiplying (3.4) by the matrix $\mathbf{Z}$ and using (3.2) we observe that (3.4) implies that $\mathbf{K}'\mathbf{Ka} = \mathbf{0}$, i.e., $\mathbf{Ka} = \mathbf{0}$, and thereby (3.3) indeed holds.

It appears that the BLUE of $\mathbf{K}'\boldsymbol{\beta}$ is actually related only to the unknown subvector $\boldsymbol{\gamma}_1$ and not to the whole parameter vector $\boldsymbol{\beta}$ or $\boldsymbol{\gamma}_2$. Thus the reparametrized model $\mathscr{T}$ is a partitioned linear model where the part $\mathbf{XK}^{\perp}\boldsymbol{\gamma}_2 = \mathbf{X}_{2*}\boldsymbol{\gamma}_2$ can be treated as a nuisance factor.

Following [**10**, Lemma 1] we can observe that

$$\text{BLUE}(\mathbf{K}'\boldsymbol{\beta} \mid \mathscr{M}) = \text{BLUE}(\mathbf{K}'\mathbf{K}\boldsymbol{\gamma}_1 \mid \mathscr{T})\,. \tag{3.5}$$

Namely, in light of (3.1) and (3.2) we have the following:

$$\begin{aligned}\text{BLUE}(\mathbf{K}'\boldsymbol{\beta} \mid \mathscr{M}) &= \text{BLUE}(\mathbf{ZX}\boldsymbol{\beta} \mid \mathscr{M}) = \mathbf{Z}\,\text{BLUE}(\mathbf{X}\boldsymbol{\beta} \mid \mathscr{M}) \\ &= \mathbf{Z}\,\text{BLUE}(\mathbf{X}_*\boldsymbol{\gamma} \mid \mathscr{T}) = \text{BLUE}[\mathbf{ZX}(\mathbf{K} : \mathbf{K}^{\perp})\boldsymbol{\gamma} \mid \mathscr{T}] \\ &= \text{BLUE}[\mathbf{K}'(\mathbf{K} : \mathbf{K}^{\perp})\boldsymbol{\gamma} \mid \mathscr{T}] = \text{BLUE}(\mathbf{K}'\mathbf{K}\boldsymbol{\gamma}_1 \mid \mathscr{T})\,,\end{aligned}$$

and thus (3.5) is confirmed. Premultiplying the model

$$\mathscr{T}\colon\ \mathbf{y} = \mathbf{XK}\boldsymbol{\gamma}_1 + \mathbf{XK}^{\perp}\boldsymbol{\gamma}_2 + \boldsymbol{\varepsilon} = \mathbf{X}_{1*}\boldsymbol{\gamma}_1 + \mathbf{X}_{2*}\boldsymbol{\gamma}_2 + \boldsymbol{\varepsilon}$$

by $\mathbf{M}_{2*} = \mathbf{I}_n - \mathbf{P}_{\mathbf{X}_{2*}} = \mathbf{Q}_{\mathbf{XK}^{\perp}}$ gives the reduced reparametrized model

$$\begin{aligned}\mathscr{T}_{12\cdot 2} &= \{\mathbf{M}_{2*}\mathbf{y},\ \mathbf{M}_{2*}\mathbf{X}_{1*}\boldsymbol{\gamma}_1,\ \mathbf{M}_{2*}\mathbf{V}\mathbf{M}_{2*}\} \\ &= \{\mathbf{M}_{2*}\mathbf{y},\ \mathbf{M}_{2*}\mathbf{XK}\boldsymbol{\gamma}_1,\ \mathbf{M}_{2*}\mathbf{V}\mathbf{M}_{2*}\}\,.\end{aligned}$$

Thus, in view of the Frisch–Waugh–Lovell theorem and (3.5) the following holds, see also [**11**, Th. 2.1]:

$$\mathrm{BLUE}(\mathbf{K}'\boldsymbol{\beta} \mid \mathscr{M}) = \mathrm{BLUE}(\mathbf{K}'\mathbf{K}\boldsymbol{\gamma}_1 \mid \mathscr{T}) = \mathrm{BLUE}(\mathbf{K}'\mathbf{K}\boldsymbol{\gamma}_1 \mid \mathscr{T}_{12\cdot 2}), \tag{3.6}$$

$$\mathrm{BLUE}(\mathbf{XK}\boldsymbol{\gamma}_1 \mid \mathscr{T}) = \mathrm{BLUE}(\mathbf{XK}\boldsymbol{\gamma}_1 \mid \mathscr{T}_{12\cdot 2}). \tag{3.7}$$

Lemma 1.6 says the following: Every linear representation of the BLUE of estimable parametric function $\mathbf{K}'\boldsymbol{\beta}$ under $\mathscr{M}(\mathbf{V}_0)$ continues to be its BLUE under $\mathscr{M}(\mathbf{V})$, i.e.,

$$\mathcal{B}(\mathbf{K}'\boldsymbol{\beta} \mid \mathscr{M}(\mathbf{V}_0)) \subseteq \mathcal{B}(\mathbf{K}'\boldsymbol{\beta} \mid \mathscr{M}(\mathbf{V})), \tag{3.8}$$

if and only if

$$\mathscr{C}(\mathbf{VM}) \subseteq \mathscr{C}(\mathbf{XK}^{\perp} : \mathbf{V}_0\mathbf{M}). \tag{3.9}$$

Consider then the reparametrized models $\mathscr{T}(\mathbf{V}_0)$ and $\mathscr{T}(\mathbf{V})$ and ask what is needed for

$$\mathcal{B}(\mathbf{XK}\boldsymbol{\gamma}_1 \mid \mathscr{T}(\mathbf{V}_0)) \subseteq \mathcal{B}(\mathbf{XK}\boldsymbol{\gamma}_1 \mid \mathscr{T}(\mathbf{V})). \tag{3.10}$$

We immediately observe that in light of Lemma 1.5 the inclusion (3.10) holds if and only if (3.9) holds. Thus we have proved the equivalence of (3.8) and (3.10). Moreover, using (3.6) we observe that (3.8) is also equivalent to

$$\mathcal{B}(\mathbf{K}'\mathbf{K}\boldsymbol{\gamma}_1 \mid \mathscr{T}(\mathbf{V}_0)) \subseteq \mathcal{B}(\mathbf{K}'\mathbf{K}\boldsymbol{\gamma}_1 \mid \mathscr{T}(\mathbf{V})).$$

We can combine our conclusions as an extended version of Lemma 1.6.

THEOREM 3.1. *Consider the above models $\mathscr{M}$ and $\mathscr{T}$ and let $\mathbf{K}'\boldsymbol{\beta}$ be estimable under $\mathscr{M}$. Then the following statements are equivalent:*

(a) $\mathcal{B}(\mathbf{K}'\boldsymbol{\beta} \mid \mathscr{M}(\mathbf{V}_0)) \subseteq \mathcal{B}(\mathbf{K}'\boldsymbol{\beta} \mid \mathscr{M}(\mathbf{V}))$.
(b) $\mathcal{B}(\mathbf{K}'\mathbf{K}\boldsymbol{\gamma}_1 \mid \mathscr{T}(\mathbf{V}_0)) \subseteq \mathcal{B}(\mathbf{K}'\mathbf{K}\boldsymbol{\gamma}_1 \mid \mathscr{T}(\mathbf{V}))$.
(c) $\mathcal{B}(\mathbf{K}\boldsymbol{\gamma}_1 \mid \mathscr{T}(\mathbf{V}_0)) \subseteq \mathcal{B}(\mathbf{K}\boldsymbol{\gamma}_1 \mid \mathscr{T}(\mathbf{V}))$.
(d) $\mathscr{C}(\mathbf{VM}) \subseteq \mathscr{C}(\mathbf{XK}^{\perp} : \mathbf{V}_0\mathbf{M})$.
(e) $\mathscr{C}(\mathbf{Q}_{\mathbf{XK}^{\perp}}\mathbf{VM}) \subseteq \mathscr{C}(\mathbf{Q}_{\mathbf{XK}^{\perp}}\mathbf{V}_0\mathbf{M})$, *i.e.*, $\mathscr{C}(\mathbf{M}_{2*}\mathbf{VM}) \subseteq \mathscr{C}(\mathbf{M}_{2*}\mathbf{V}_0\mathbf{M})$.

In particular, if $\mathbf{K}$ *has a full column rank, then* (b) *is equivalent to*

(f) $\mathcal{B}(\boldsymbol{\gamma}_1 \mid \mathscr{T}(\mathbf{V}_0)) \subseteq \mathcal{B}(\boldsymbol{\gamma}_1 \mid \mathscr{T}(\mathbf{V}))$.

Another interesting summary is the following theorem where part (e) seems to be a new (or at least less well known) result.

THEOREM 3.2. *Consider the models $\mathscr{M}$, $\mathscr{T}$ and $\mathscr{T}_{12\cdot 2}$ defined above and denote*

$$\mathbf{M}_{2*} = \mathbf{Q}_{\mathbf{XK}^{\perp}}, \quad \dot{\mathbf{M}}_{2*} = (\mathbf{M}_{2*}\mathbf{W}\mathbf{M}_{2*})^{+}, \quad \mathbf{W} = \mathbf{V} + \mathbf{XX}'.$$

Then

(a) $\mathrm{BLUE}(\mathbf{K}'\boldsymbol{\beta} \mid \mathscr{M}) = \mathrm{BLUE}(\mathbf{K}'\mathbf{K}\boldsymbol{\gamma}_1 \mid \mathscr{T}) = \mathrm{BLUE}(\mathbf{K}'\mathbf{K}\boldsymbol{\gamma}_1 \mid \mathscr{T}_{12\cdot 2})$,
(b) $\mathrm{BLUE}(\mathbf{XK}\boldsymbol{\gamma}_1 \mid \mathscr{T}) = \mathrm{BLUE}(\mathbf{XK}\boldsymbol{\gamma}_1 \mid \mathscr{T}_{12\cdot 2})$,
(c) $\mathrm{BLUE}(\mathbf{K}'\boldsymbol{\beta} \mid \mathscr{M}) = \mathbf{K}'\mathbf{K}(\mathbf{K}'\mathbf{X}'\dot{\mathbf{M}}_{2*}\mathbf{XK})^{-}\mathbf{K}'\mathbf{X}'\dot{\mathbf{M}}_{2*}\mathbf{y}$
$= \mathbf{K}'\,\mathrm{BLUE}(\mathbf{K}\boldsymbol{\gamma}_1 \mid \mathscr{T})$,
(d) $\mathrm{BLUE}(\mathbf{XK}\boldsymbol{\gamma}_1 \mid \mathscr{T}) = \mathbf{XK}(\mathbf{K}'\mathbf{X}'\dot{\mathbf{M}}_{2*}\mathbf{XK})^{-}\mathbf{K}'\mathbf{X}'\dot{\mathbf{M}}_{2*}\mathbf{y}$
$= \mathbf{X}\,\mathrm{BLUE}(\mathbf{K}\boldsymbol{\gamma}_1 \mid \mathscr{T})$,
(e) $\mathbf{K}'\mathbf{K}(\mathbf{K}'\mathbf{X}'\dot{\mathbf{M}}_{2*}\mathbf{XK})^{-}\mathbf{K}'\mathbf{X}'\dot{\mathbf{M}}_{2*} = \mathbf{K}'(\mathbf{X}'\mathbf{W}^{-}\mathbf{X})^{-}\mathbf{X}'\mathbf{W}^{+}$.

PROOF. Claims (a) and (b) follow from (3.6) and (3.7), respectively, and (c) is shown in [**11**, Thm. 2.1]. Part (d) follows from the Frisch–Waugh–Lovell theorem. Denoting

$$\mathbf{A} = \mathbf{K}'\mathbf{K}(\mathbf{K}'\mathbf{X}'\dot{\mathbf{M}}_{2*}\mathbf{X}\mathbf{K})^{-}\mathbf{K}'\mathbf{X}'\dot{\mathbf{M}}_{2*}\,, \quad \mathbf{B} = \mathbf{K}'(\mathbf{X}'\mathbf{W}^{-}\mathbf{X})^{-}\mathbf{X}'\mathbf{W}^{+},$$

we know that

$$\mathbf{AW} = \mathbf{BW}.$$

It is clear that $\mathbf{BQ_W} = \mathbf{0}$. Next we prove that also $\mathbf{AQ_W} = \mathbf{0}$ which follows from

$$(\mathbf{M}_{2*}\mathbf{W}\mathbf{M}_{2*})^{+}\mathbf{Q_W} = \mathbf{0}\,.$$

The above equality holds if and only if the following inclusion holds:

$$\mathscr{C}(\mathbf{M}_{2*}\mathbf{W}\mathbf{M}_{2*}) = \mathscr{C}(\mathbf{M}_{2*}\mathbf{W}) = \mathscr{C}(\mathbf{Q}_{\mathbf{XK}^{\perp}}\mathbf{W}) \subseteq \mathscr{C}(\mathbf{W})\,,$$

which indeed holds since

$$\mathbf{P_W}\mathbf{Q}_{\mathbf{XK}^{\perp}}\mathbf{W} = \mathbf{P_W}(\mathbf{I}_n - \mathbf{P}_{\mathbf{XK}^{\perp}})\mathbf{W}] = \mathbf{W} - \mathbf{P}_{\mathbf{XK}^{\perp}}\mathbf{W} = \mathbf{Q}_{\mathbf{XK}^{\perp}}\mathbf{W}.$$

Thus $\mathbf{A}(\mathbf{W} : \mathbf{W}^{\perp}) = \mathbf{B}(\mathbf{W} : \mathbf{W}^{\perp})$, i.e., $\mathbf{A} = \mathbf{B}$. □

4. Example on a linear mixed model

This example illustrates the use of the reparametrization technique presented in Section 3. This is an extended version of example given in [**9**, Sec. 3].

Let us consider the linear mixed model $\mathscr{M}\colon \mathbf{y} = \mathbf{X}\boldsymbol{\beta} + \mathbf{Z}\boldsymbol{\delta} + \boldsymbol{\varepsilon}$, where

$$\mathbf{X} = \begin{pmatrix} \mathbf{1}_2 & \mathbf{0} & \mathbf{0} \\ \mathbf{0} & \mathbf{1}_2 & \mathbf{0} \\ \mathbf{0} & \mathbf{0} & \mathbf{1}_2 \end{pmatrix}, \quad \mathbf{Z} = \begin{pmatrix} \mathbf{1}_3 & \mathbf{0} \\ \mathbf{0} & \mathbf{1}_3 \end{pmatrix}, \quad \mathbf{1}_a = (1, 1, \dots, 1)' \in \mathbb{R}^a.$$

The vector $\boldsymbol{\beta} = (\beta_1, \beta_2, \beta_3)'$ is a vector of the fixed effects, and $\boldsymbol{\delta} = (\delta_1, \delta_2)'$ is an unobservable vector of the random effects with $\mathrm{E}(\boldsymbol{\delta}) = \mathbf{0}$, $\mathrm{cov}(\boldsymbol{\delta}) = \omega_0^2\mathbf{I}_2$, and $\mathrm{cov}(\boldsymbol{\delta}, \boldsymbol{\varepsilon}) = \mathbf{0}$. Furthermore, let $\mathrm{cov}(\boldsymbol{\varepsilon}) = \sigma_0^2\mathbf{I}_6$. Denoting

$$\mathrm{cov}(\mathbf{y}) = \mathrm{cov}(\mathbf{Z}\boldsymbol{\delta} + \boldsymbol{\varepsilon}) = \omega^2\mathbf{Z}\mathbf{Z}' + \sigma^2\mathbf{I}_6 = \mathbf{V},$$

where ω^2 and σ^2 are unknown positive scalars, the mixed model can be written as a triplet

$$\mathscr{M}(\mathbf{V}) = \{\mathbf{y}, \mathbf{X}\boldsymbol{\beta}, \mathbf{V}\}\,.$$

Unless ω^2 and σ^2 are known, the BLUE of $\boldsymbol{\mu} = \mathbf{X}\boldsymbol{\beta}$ under $\mathscr{M}(\mathbf{V})$ is not obtainable since the covariance matrix $\mathbf{V}$ depends on the unknown parameters ω^2 and σ^2. Hence we may consider the covariance structure $\mathbf{V}_0 = \sigma_0^2\mathbf{I}_6$ and the corresponding model $\mathscr{M}(\mathbf{V}_0) = \{\mathbf{y}, \mathbf{X}\boldsymbol{\beta}, \mathbf{V}_0\}$ as an alternative model to $\mathscr{M}(\mathbf{V})$.

Notice that under $\mathscr{M}(\mathbf{V}_0) = \{\mathbf{y}, \mathbf{X}\boldsymbol{\beta}, \mathbf{V}_0\}$, where $\mathbf{V}_0 = \sigma_0^2\mathbf{I}_6$, the BLUE of $\boldsymbol{\mu} = \mathbf{X}\boldsymbol{\beta}$ is the OLSE of $\boldsymbol{\mu}$, i.e.,

$$\mathrm{BLUE}(\boldsymbol{\mu} \mid \mathscr{M}(\mathbf{V}_0)) = \mathbf{X}\mathbf{X}^{+}\mathbf{y} = \mathbf{P_X}\mathbf{y} = \mathrm{OLSE}(\boldsymbol{\mu} \mid \mathscr{M}(\mathbf{V}))\,.$$

However, the condition $\mathcal{B}(\boldsymbol{\mu} \mid \mathscr{M}(\mathbf{V}_0)) \subseteq \mathcal{B}(\boldsymbol{\mu} \mid \mathscr{M}(\mathbf{V}))$, or in other words,

$$\mathrm{OLSE}(\boldsymbol{\mu} \mid \mathscr{M}(\mathbf{V})) = \mathrm{BLUE}(\boldsymbol{\mu} \mid \mathscr{M}(\mathbf{V}))$$

does not hold since

$$\mathbf{X}'\mathbf{V}\mathbf{M} = \omega^2\mathbf{X}'\mathbf{Z}\mathbf{Z}'\mathbf{M} \neq \mathbf{0}$$

for any values of $\omega^2 > 0$; see (1.6). But if we consider the estimation of

$$\alpha = \tfrac{1}{3}(\beta_1 + \beta_2 + \beta_3) = \mathbf{k}'\boldsymbol{\beta}, \quad \text{where } \mathbf{k} = \tfrac{1}{3}\mathbf{1}_3\,,$$

then the condition $\mathcal{B}(\alpha \,|\, \mathscr{M}(\mathbf{V}_0)) \subseteq \mathcal{B}(\alpha \,|\, \mathscr{M}(\mathbf{V}))$ appears to hold for all $\omega^2 > 0$, $\sigma^2 > 0$ and $\sigma_0^2 > 0$. This is next proved using the reparametrization technique presented in Section 3.

Consider the reparametrized linear mixed model

$$\begin{aligned}\mathscr{T}\colon\ \mathbf{y} &= \mathbf{X}(\mathbf{k} : \mathbf{k}^{\perp})\boldsymbol{\gamma} + \mathbf{Z}\boldsymbol{\delta} + \boldsymbol{\varepsilon} = \mathbf{x}_{1*}\gamma_1 + \mathbf{X}_{2*}\boldsymbol{\gamma}_2 + \mathbf{Z}\boldsymbol{\delta} + \boldsymbol{\varepsilon}\\ &= \mathbf{Xk}\gamma_1 + \mathbf{Xk}^{\perp}\boldsymbol{\gamma}_2 + \mathbf{Z}\boldsymbol{\delta} + \boldsymbol{\varepsilon}\,,\end{aligned}$$

where $\boldsymbol{\gamma} = (\gamma_1, \boldsymbol{\gamma}_2')' = (\gamma_1, \gamma_2, \gamma_3)'$ and

$$\mathbf{k}^{\perp} = \begin{pmatrix} 1 & 1/2 \\ -1 & 1/2 \\ 0 & -1 \end{pmatrix}, \quad \mathbf{x}_{1*} = \mathbf{xk} = \tfrac{1}{3}\mathbf{1}_6\,, \quad \mathbf{x}_{2*} = \mathbf{xk}^{\perp} = \begin{pmatrix} 1 & 1/2 \\ 1 & 1/2 \\ -1 & 1/2 \\ -1 & 1/2 \\ 0 & -1 \\ 0 & -1 \end{pmatrix}.$$

The reparametrized linear mixed models can be written as

$$\begin{aligned}\mathscr{T}(\mathbf{V}_0) &= \{\mathbf{y},\, \mathbf{x}_{1*}\gamma_1 + \mathbf{X}_{2*}\boldsymbol{\gamma}_2, \mathbf{V}_0\} = \{\mathbf{y},\, \mathbf{Xk}\gamma_1 + \mathbf{Xk}^{\perp}\boldsymbol{\gamma}_2, \mathbf{V}_0\}\,,\\ \mathscr{T}(\mathbf{V}) &= \{\mathbf{y},\, \mathbf{x}_{1*}\gamma_1 + \mathbf{X}_{2*}\boldsymbol{\gamma}_2, \mathbf{V}\} = \{\mathbf{y},\, \mathbf{Xk}\gamma_1 + \mathbf{Xk}^{\perp}\boldsymbol{\gamma}_2, \mathbf{V}\}\,.\end{aligned}$$

In view of Theorem 3.1, the inclusion

$$\mathcal{B}(\mathbf{k}'\boldsymbol{\beta} \,|\, \mathscr{M}(\mathbf{V}_0)) \subseteq \mathcal{B}(\mathbf{k}'\boldsymbol{\beta} \,|\, \mathscr{M}(\mathbf{V})) \tag{4.1}$$

can be equivalently considered in terms of models $\mathscr{T}(\mathbf{V}_0)$ and $\mathscr{T}(\mathbf{V})$ so that (4.1) is equivalent to $\mathcal{B}(\mathbf{x}_{1*}\gamma_1 \,|\, \mathscr{T}(\mathbf{V}_0)) \subseteq \mathcal{B}(\mathbf{x}_{1*}\gamma_1 \,|\, \mathscr{T}(\mathbf{V}))$ as well as to

$$\mathscr{C}(\mathbf{M}_{2*}\mathbf{VM}) \subseteq \mathscr{C}(\mathbf{M}_{2*}\mathbf{V}_0\mathbf{M}) = \mathscr{C}(\mathbf{M})\,, \quad \text{where } \mathbf{M}_{2*} = \mathbf{I}_6 - \mathbf{P}_{2*}\,, \tag{4.2}$$

for all ω^2, σ^2 and σ_0^2. Notice that for $\mathbf{V}_0 = \sigma_0^2\mathbf{I}_6$ we have

$$\mathbf{M}_{2*}\mathbf{V}_0\mathbf{M} = \sigma_0^2\mathbf{M}_{2*}\mathbf{M} = \sigma_0^2(\mathbf{I}_6 - \mathbf{P}_{\mathbf{Xk}^{\perp}})\mathbf{M} = \mathbf{M}\,.$$

Now we have

$$\begin{aligned}\mathbf{M}_{2*}\mathbf{VM} &= \mathbf{M}_{2*}(\sigma^2\mathbf{I} + \omega^2\mathbf{ZZ}')\mathbf{M} = \sigma^2\mathbf{M} + \omega^2 \begin{pmatrix} \mathbf{0} & \mathbf{0} & \mathbf{0} \\ \mathbf{0} & \mathbf{I}_2 - \frac{1}{2}\mathbf{1}_2\mathbf{1}_2' & \mathbf{0} \\ \mathbf{0} & \mathbf{0} & \mathbf{0} \end{pmatrix}\\ &= \sigma^2\mathbf{M} + \omega^2\mathbf{F} = (\sigma_0\mathbf{M} : \omega\mathbf{F})(\sigma\mathbf{M} : \omega\mathbf{F})'.\end{aligned}$$

Hence $\mathscr{C}(\mathbf{M}) \subseteq \mathscr{C}(\mathbf{M}_{2*}\mathbf{VM})$ for all ω^2, σ^2 and σ_0^2, which further implies that $\mathscr{C}(\mathbf{M}) = \mathscr{C}(\mathbf{M}_{2*}\mathbf{VM})$ showing that (4.2) indeed holds.

Let us also consider when $\operatorname{cov}(\tilde{\boldsymbol{\mu}}_{1*} \,|\, \mathscr{T}(\mathbf{V}_0)) = \operatorname{cov}(\tilde{\boldsymbol{\mu}}_{1*} \,|\, \mathscr{T}(\mathbf{V}))$ holds. Denoting $\mathbf{W}_{0*} = \mathbf{V}_0 + \mathbf{X}_*\mathbf{X}_*'$ and $\mathbf{W}_* = \mathbf{V} + \mathbf{X}_*\mathbf{X}_*'$, the Moore–Penrose inverse of the matrix $\mathbf{M}_{2*}\mathbf{W}_{0*}\mathbf{M}_{2*}$ has the form

$$\begin{aligned}(\mathbf{M}_{2*}\mathbf{W}_{0*}\mathbf{M}_{2*})^+ &= [\sigma^2\mathbf{M}_{2*} + \omega^2(\tfrac{1}{2}\mathbf{1}_6\mathbf{1}_6' + \mathbf{F}) + \tfrac{1}{9}\mathbf{1}_6\mathbf{1}_6']^+\\ &= \tfrac{1}{\sigma^2}\mathbf{M}_{2*} - a\mathbf{1}_6\mathbf{1}_6' - b\mathbf{F},\end{aligned}$$

where

$$a = \frac{9\omega^2 + 2}{6\sigma^2(3\sigma^2 + 9\omega^2 + 2)}\,, \qquad b = \frac{\omega^2}{\sigma^2(\sigma^2 + \omega^2)}\,.$$

Hence, denoting $\mathbf{E} = \mathbf{1}_6\mathbf{1}_6'$,

(4.3)

$$\begin{aligned}(\mathbf{M}_{2*}\mathbf{W}_*\mathbf{M}_{2*})(\mathbf{M}_{2*}\mathbf{W}_{0*}\mathbf{M}_{2*})^+\mathbf{x}_{1*} &= (\sigma_0^2\mathbf{M}_{2*} + \tfrac{1}{9}\mathbf{E})(\tfrac{1}{\sigma^2}\mathbf{M}_{2*} - a\mathbf{E} - b\mathbf{F})\mathbf{x}_{1*} \\ &= [\tfrac{\sigma_0^2}{\sigma^2}\mathbf{M}_{2*} + (\tfrac{1}{9\sigma^2} - \sigma_0^2 a - \tfrac{2a}{3})\mathbf{E} - \sigma_0^2 b\mathbf{F}]\mathbf{x}_{1*} \\ &= \tfrac{\sigma_0^2}{3\sigma^2}\mathbf{1}_6 + (\tfrac{2}{9\sigma^2} - 2\sigma_0^2 a - \tfrac{4a}{3})\mathbf{1}_6 \\ &= (\tfrac{\sigma_0^2}{3\sigma^2} + \tfrac{2}{9\sigma^2} - 2\sigma_0^2 a - \tfrac{4a}{3})\mathbf{1}_6 = c\mathbf{1}_6\,,\end{aligned}$$

since $\mathbf{x}_{1*} = \frac{1}{3}\mathbf{1}_6$. In view of $\mathbf{M}_{2*}\mathbf{x}_{1*} = \frac{1}{3}\mathbf{1}_6$ and part (b) of Lemma 1.7, the equality $\operatorname{cov}(\tilde{\boldsymbol{\mu}}_{1*} \mid \mathbf{V}_0) = \operatorname{cov}(\tilde{\boldsymbol{\mu}}_{1*} \mid \mathbf{V})$ holds when in (4.3) $c = \frac{1}{3}$. Thus given the original model $\mathscr{T}(\mathbf{V}_0)$, the parameter σ_0^2 in the model $\mathscr{T}(\mathbf{V})$ needs to fulfill the following equation

$$\sigma_0^2 = \frac{4a - \frac{2\sigma^2}{3} + 1}{\frac{1}{\sigma^2} - 6a} = \sigma^2 + 3\omega^2$$

in order for the covariance matrix equality $\operatorname{cov}(\tilde{\boldsymbol{\mu}}_{1*} \mid \mathbf{V}_0) = \operatorname{cov}(\tilde{\boldsymbol{\mu}}_{1*} \mid \mathbf{V})$ to hold.

5. Conclusions

In this article we consider the partitioned linear model

$$\mathscr{M} = \mathscr{M}_{12} = \{\mathbf{y},\, \mathbf{X}\boldsymbol{\beta},\, \mathbf{V}\} = \{\mathbf{y},\, \mathbf{X}_1\boldsymbol{\beta}_1 + \mathbf{X}_2\boldsymbol{\beta}_2,\, \mathbf{V}\}\,,$$

and in particular, our aim is to estimate parametric functions like $\mathbf{K}'\boldsymbol{\beta}$ under $\mathscr{M}$; here $\mathbf{K}$ is a given matrix of p rows. In particular, we are interested in the best linear unbiased estimators, BLUEs, of $\mathbf{K}'\boldsymbol{\beta}$. Let $\mathbf{X}_*$ be any matrix such that the column space equality $\mathscr{C}(\mathbf{X}) = \mathscr{C}(\mathbf{X}_*)$ holds. Then it is clear that there exists a matrix $\mathbf{A}$ such that $\mathbf{X}_* = \mathbf{X}\mathbf{A}$ and we get the reparametrized model

$$\mathscr{T} = \{\mathbf{y},\, \mathbf{X}_*\boldsymbol{\gamma},\, \mathbf{V}\} = \{\mathbf{y},\, \mathbf{X}\mathbf{A}\boldsymbol{\gamma},\, \mathbf{V}\}\,.$$

Models $\mathscr{M}$ and $\mathscr{T}$ are called equivalent in the sense that

$$\operatorname{BLUE}(\mathbf{X}\boldsymbol{\beta} \mid \mathscr{M}) = \operatorname{BLUE}(\mathbf{X}_*\boldsymbol{\gamma} \mid \mathscr{T})\,.$$

We consider two kinds of choice of $\mathbf{A}$:

$$\mathbf{A} = \begin{pmatrix} \mathbf{I}_{p_1} & -\mathbf{X}_1^+\mathbf{X}_2 \\ \mathbf{0} & \mathbf{I}_{p_2} \end{pmatrix}, \quad \mathbf{A} = (\mathbf{K} : \mathbf{K}^\perp)\,,$$

where the first choice leads to model

$$\mathscr{M}_{1R} = \{\mathbf{y}, \mathbf{X}_*\boldsymbol{\beta}, \mathbf{V}\} = \{\mathbf{y}, (\mathbf{X}_1 : \mathbf{M}_1\mathbf{X}_2)\boldsymbol{\beta}, \mathbf{V}\}\,,$$

We compare the BLUEs of $\boldsymbol{\mu}_i = \mathbf{X}_i\boldsymbol{\beta}_i$ under $\mathscr{M}_{12}$ and $\mathscr{M}_{1R}$ and summarize our findings (some being less well known) in Theorem 2.1.

We may mention one interesting usage of $\mathscr{M}_{1R}$: to develop a general expression for matrix $\mathbf{V}$ such that every representation of the BLUE of $\boldsymbol{\mu}$ under $\mathscr{M}(\mathbf{V}_0) = \{\mathbf{y}, \mathbf{X}\boldsymbol{\beta}, \mathbf{V}_0\}$ remains BLUE under $\mathscr{M}(\mathbf{V}) = \{\mathbf{y}, \mathbf{X}\boldsymbol{\beta}, \mathbf{V}\}$ or shortly

$$\mathcal{B}(\boldsymbol{\mu} \mid \mathbf{V}_0) \subseteq \mathcal{B}(\boldsymbol{\mu} \mid \mathbf{V})\,. \tag{5.1}$$

Instead of $\mathscr{M}(\mathbf{V}_0)$ and $\mathscr{M}(\mathbf{V})$ we can equivalently consider $\mathscr{M}_{1R}(\mathbf{V}_0)$ and $\mathscr{M}_{1R}(\mathbf{V})$ and thus by Lemma 1.4 the inclusion (5.1) holds if and only if

$$\mathbf{V} = \mathbf{V}_0 + \mathbf{X}_*\mathbf{C}\mathbf{C}'\mathbf{X}_*' + \mathbf{V}_0\mathbf{M}\mathbf{D}\mathbf{D}'\mathbf{M}\mathbf{V}_0 \text{ for some } \mathbf{C} \text{ and } \mathbf{D}\,. \tag{5.2}$$

Representation (5.2) was utilised by Haslett S. J et al. [**6**].

In Section 3 of this article we consider the following reparametrized version of $\mathscr{M}$:

$$\mathscr{T} = \{\mathbf{y}, \mathbf{X}_*\boldsymbol{\gamma}, \mathbf{V}\} = \{\mathbf{y}, (\mathbf{X}\mathbf{K} : \mathbf{X}\mathbf{K}^{\perp})\boldsymbol{\gamma}, \mathbf{V}\} = \{\mathbf{y}, \mathbf{X}_{1*}\boldsymbol{\gamma}_1 + \mathbf{X}_{2*}\boldsymbol{\gamma}_2, \mathbf{V}\}\,.$$

It is clear that $\mathscr{M}$ and $\mathscr{T}$ are equivalent in the sense that

$$\mathrm{BLUE}(\mathbf{X}\boldsymbol{\beta} \mid \mathscr{M}) = \mathrm{BLUE}(\mathbf{X}_*\boldsymbol{\gamma} \mid \mathscr{T})\,.$$

The benefit in using the reparametrization like in $\mathscr{T}$ follows from the Frisch–Waugh–Lovell theorem. The reason is that $\mathscr{T}$ is a partitioned linear model where we can apply the Frisch–Waugh–Lovell theorem. It can be observed that when $\mathbf{K}'\boldsymbol{\beta}$ is estimable under $\mathscr{M}$, then the parametric function $\mathbf{X}_{1*}\boldsymbol{\gamma}_1$ is estimable under $\mathscr{T}$. It appears that the BLUE of $\mathbf{K}'\boldsymbol{\beta}$ is actually related only to the value of unknown subvector $\boldsymbol{\gamma}_1$ and not to the whole parameter vector $\boldsymbol{\beta}$ or $\boldsymbol{\gamma}_2$. Thus the reparametrized model $\mathscr{T}$ is a partitioned linear model where the part $\mathbf{X}\mathbf{K}^{\perp}\boldsymbol{\gamma}_2 = \mathbf{X}_{2*}\boldsymbol{\gamma}_2$ can be treated as a nuisance factor.

[**9**, Thm. 1] utilized the model $\mathscr{T}$ for comparing the OLSE and BLUE of estimable $\mathbf{K}'\boldsymbol{\beta}$ and introduced, in terms of $\mathscr{T}$, a representation for the class of matrices $\mathbf{V}$ such that $\mathrm{OLSE}(\mathbf{K}'\boldsymbol{\beta})$ equals $\mathrm{BLUE}(\mathbf{K}'\boldsymbol{\beta})$ under $\{\mathbf{y}, \mathbf{X}\boldsymbol{\beta}, \mathbf{V}\}$. For the corresponding expression for the class of matrices $\mathbf{V}$ (for a given $\mathbf{V}_0$) to allow $\mathcal{B}(\mathbf{K}'\boldsymbol{\beta} \mid \mathbf{V}_0) \subseteq \mathcal{B}(\mathbf{K}'\boldsymbol{\beta} \mid \mathbf{V})$, see [**12**, Th. 2.4].

The concept of linear sufficiency, originally introduced by Baksalary and Kala [**1**, **2**], was applied to the model $\mathscr{T}$ by Isotalo and Puntanen [**9**, **10**]. To recall briefly, a linear statistic $\mathbf{F}\mathbf{y}$ is called linearly sufficient for estimable $\mathbf{K}'\boldsymbol{\beta}$ under the model $\{\mathbf{y}, \mathbf{X}\boldsymbol{\beta}, \mathbf{V}\}$ if there exists a matrix $\mathbf{A}$ such that $\mathbf{A}\mathbf{F}\mathbf{y}$ is the BLUE for $\mathbf{K}'\boldsymbol{\beta}$. We denote then $\mathbf{F}\mathbf{y} \in \mathcal{S}(\mathbf{K}'\boldsymbol{\beta} \mid \mathbf{V})$. In particular, $\mathbf{F}\mathbf{y}$ is linearly minimal sufficient, denoted as $\mathbf{F}\mathbf{y} \in \mathcal{S}_{\min}(\mathbf{K}'\boldsymbol{\beta} \mid \mathbf{V})$, if for any other linearly sufficient statistics $\mathbf{S}\mathbf{y}$, say, there exists a matrix $\mathbf{B}$ such that $\mathbf{F}\mathbf{y} = \mathbf{B}\mathbf{S}\mathbf{y}$ almost surely.

In view of Lemma 1.1, $\mathbf{F}\mathbf{y}$ is linearly sufficient for estimable $\mathbf{K}'\boldsymbol{\beta}$ if and only if the equation $\mathbf{A}\mathbf{F}(\mathbf{X} : \mathbf{V}\mathbf{M}) = (\mathbf{K}' : \mathbf{0})$ has a solution for $\mathbf{A}$. From the results of [**9**, **10**] we may briefly mention the following:

- $\mathbf{F}\mathbf{y} \in \mathcal{S}(\mathbf{K}'\boldsymbol{\beta} \mid \mathscr{M}) \iff \mathbf{F}\mathbf{y} \in \mathcal{S}(\mathbf{K}\boldsymbol{\gamma}_1 \mid \mathscr{T})$,
- $\mathbf{M}_{2*}\mathbf{y} \in \mathcal{S}(\mathbf{K}'\boldsymbol{\beta} \mid \mathscr{M})$, and $\mathbf{K}'\mathbf{X}'\dot{\mathbf{M}}_{2*}\mathbf{y} \in \mathcal{S}_{\min}(\mathbf{K}'\boldsymbol{\beta} \mid \mathscr{M})$,

where $\mathbf{M}_{2*} = \mathbf{Q}_{\mathbf{X}\mathbf{K}^{\perp}}$, $\dot{\mathbf{M}}_{2*} = (\mathbf{M}_{2*}\mathbf{W}\mathbf{M}_{2*})^{+}$ and $\mathbf{W} = \mathbf{V} + \mathbf{X}\mathbf{X}'$.

References

[1] J. K. Baksalary and R. Kala, *Linear transformations preserving best linear unbiased estimators in a general Gauss-Markoff model*, Ann. Statist. **9** (1981), no. 4, 913–916. MR619297

[2] J. K. Baksalary and R. Kala, *Linear sufficiency with respect to a given vector of parametric functions*, J. Statist. Plann. Inference **14** (1986), no. 3, 331–338, DOI 10.1016/0378-3758(86)90171-0. MR859980

[3] J. K. Baksalary and T. Mathew, *Linear sufficiency and completeness in an incorrectly specified general Gauss-Markov model*, Sankhyā Ser. A **48** (1986), no. 2, 169–180. MR905457

[4] J. K. Baksalary, S. Puntanen, and G. P. H. Styan, *A property of the dispersion matrix of the best linear unbiased estimator in the general Gauss-Markov model*, Sankhyā Ser. A **52** (1990), no. 3, 279–296. MR1178038

[5] J. Groß and S. Puntanen, *Estimation under a general partitioned linear model*, Linear Algebra Appl. **321** (2000), no. 1-3, 131–144, DOI 10.1016/S0024-3795(00)00028-8. Linear algebra and statistics (Fort Lauderdale, FL, 1998). MR1799989

[6] S. J. Haslett, J. Isotalo, A. Markiewicz, and S. Puntanen, *Permissible covariance structures for simultaneous retention of BLUEs in small and big linear models*, Applied linear algebra, probability and statistics—a volume in honour of C. R. Rao and Arbind K. Lal, Indian Stat. Inst. Ser., Springer, Singapore, [2023] ©2023, pp. 197–213, DOI 10.1007/978-981-99-2310-6_11. MR4647558

[7] S. J. Haslett, J. Isotalo, A. Markiewicz, and S. Puntanen, *How data or error covariance can change and still retain BLUEs as well as their covariance or the sum of squares of errors*, Aust. N. Z. J. Stat. **67** (2025), no. 2, 175–201, DOI 10.1111/anzs.70003. MR4931907

[8] S. J. Haslett and S. Puntanen, *Effect of adding regressors on the equality of the BLUEs under two linear models*, J. Statist. Plann. Inference **140** (2010), no. 1, 104–110, DOI 10.1016/j.jspi.2009.06.010. MR2568125

[9] J. Isotalo and S. Puntanen, *A note on the equality of the OLSE and the BLUE of the parametric function in the general Gauss-Markov model*, Statist. Papers **50** (2009), no. 1, 185–193, DOI 10.1007/s00362-007-0055-6. MR2476178

[10] J. Isotalo and S. Puntanen, *Linear sufficiency and completeness in the context of estimating the parametric function in the general Gauss-Markov model*, J. Statist. Plann. Inference **139** (2009), no. 3, 722–733, DOI 10.1016/j.jspi.2007.11.020. MR2479823

[11] J. Isotalo, S. Puntanen, and G. P. H. Styan, *Some comments on the Watson efficiency of the ordinary least squares estimator under the Gauss-Markov model*, Calcutta Statist. Assoc. Bull. **61** (2009), no. 241-244, 1–15. MR2554134

[12] T. Mathew and P. Bhimasankaram, *On the robustness of LRT in singular linear models*, Sankhyā Ser. A **45** (1983), no. 3, 301–312. MR747600

[13] S. K. Mitra and B. J. Moore, *Gauss-Markov estimation with an incorrect dispersion matrix*, Sankhyā Ser. A **35** (1973), 139–152. MR345341

[14] S. Puntanen, G. P. H. Styan, and J. Isotalo, *Matrix tricks for linear statistical models: Our personal top twenty*, Springer, Heidelberg, 2011, DOI 10.1007/978-3-642-10473-2. MR3013662

[15] C. R. Rao, *Unified theory of linear estimation*, Sankhyā Ser. A **33** (1971), 371–394. MR319321

[16] C. R. Rao, *Representations of best linear unbiased estimators in the Gauss-Markoff model with a singular dispersion matrix*, J. Multivariate Anal. **3** (1973), 276–292, DOI 10.1016/0047-259X(73)90042-0. MR341765

[17] C. R. Rao and S. K. Mitra, *Generalized inverse of matrices and its applications*, John Wiley & Sons, Inc., New York-London-Sydney, 1971. MR338013
doi=10.1023/A:1008224418676,

School of Mathematical and Computational Sciences, Massey University, Palmerston North, New Zealand; Research School of Finance, Actuarial Studies and Statistics, The Australian National University, Canberra, Australia; Faculty of Engineering and Information Sciences, University of Wollongong, NSW, Australia

Email address: `s.j.haslett@massey.ac.nz`

Faculty of Information Technology and Communication Sciences, FI-33014 Tampere University, Finland

Email address: `jarkko.isotalo@tuni.fi`

Department of Mathematical and Statistical Methods, Poznań University of Life Sciences, Wojska Polskiego 28, PL-60637 Poznań, Poland

Email address: `amark@up.poznan.pl`

Faculty of Information Technology and Communication Sciences, FI-33014 Tampere University, Finland

Email address: `simo.puntanen@tuni.fi`

Contemporary Mathematics
Volume **833**, 2026
https://doi.org/10.1090/conm/833/16664

Products of idempotents in Banach algebras of operators

S. K. Jain, André Leroy, and Ajit Iqbal Singh

ABSTRACT. Let X be a Banach space and $\mathcal{A}$ be the Banach algebra $\mathcal{B}(X)$ of bounded (i.e. continuous) linear transformations (to be called operators) on X to itself. Let $\mathcal{E}$ be the set of idempotents in $\mathcal{A}$ and $\mathcal{S}$ be the semigroup generated by $\mathcal{E}$ under composition as multiplication. If $T \in \mathcal{S}$ with $0 \neq T \neq I_X$ then T has a local block representation of the form $\begin{pmatrix} T_1 & T_2 \\ 0 & 0 \end{pmatrix}$ on $X = Y \oplus Z$, a topological sum of non-zero closed subspaces Y and Z of X, and any $A \in \mathcal{A}$ has the form $\begin{pmatrix} A_1 & A_2 \\ A_3 & A_4 \end{pmatrix}$ with $T_1, A_1 \in \mathcal{B}(Y)$, $T_2, A_2 \in \mathcal{B}(Z, Y)$, $A_3 \in \mathcal{B}(Y, Z)$, and $A_4 \in \mathcal{B}(Z)$. The purpose of this paper is to study conditions for A to be in $\mathcal{S}$.

1. Introduction

This paper is addressed to a general mathematician interested in Algebra, Linear Algebra and Operator theory. The problem of expressing a non-invertible (i.e. singular) function f on a non-empty set X to itself as a (finite) product of idempotent functions on X to itself was started by Howie [**6**] in 1960's. He proved it for finite sets X and studied the situation for infinite sets in detail. The fact that taking X as a basis of a linear space $\tilde{X}$ with $dim(\tilde{X}) = \#X$ and function f on X to X giving rise to linear transformation T on $\tilde{X}$ to itself led Erdos [**5**] to study the problem for rings of $n \times n$ matrices over fields. Howie's student Dawlings [**3**] studied the question for Banach algebra of bounded (i.e. continuous) linear transformations (to be called operators) on a Hilbert space $\mathcal{H}$ to itself. Several authors studied this further or extended the context to a Banach space X and the Banach algebra $\mathcal{A}$ of operators in X to itself, like Wu [**12**] , Arias, Corach, and Maestripieri [**2**]. As expected , finite rank operators on X to itself are finite products of idempotents in $\mathcal{A}$. Arias, Corach, and Maestripieri [**2**] concentrated on products of two idempotents of $\mathcal{A}$. We find that the technique of local block representation gives several new results related to the expression of operators in certain classes as finite product of idempotents in $\mathcal{A}$ in infinitely many ways. At times, we get a better deal when X is a Hilbert space $\mathcal{H}$. This constitutes the subject matter of this paper.

The paper is divided in four sections. Section 2 is concerned with preliminaries and motivation. Section 3 deals with properties of the set $\mathcal{E}$ of idempotents in

2020 *Mathematics Subject Classification.* Primary 16S50, 16U40, 16W80, 46B28, 47B01.
Key words and phrases. Idempotents, finite products, Banach algebra of operators, Block representation, Local block representation, Conditions for products of two idempotents.

$\mathcal{A}$ and the semigroup $\mathcal{S}$ generated by $\mathcal{E}$ with composition as multiplication. This includes the basic concept of local block representation for members of $\mathcal{S}$, other than 0 and I_X, which coincides with the set $\mathcal{F}$ of finite products of members of $\mathcal{E}$ as well, other than 0 and I_X. Section 4 develops the technique further to obtain more classes of operators expressible as products of two idempotents in infinitely many ways.

2. Preliminaries and motivation

This section collects basics from standard books, monographs or relevant papers such as Howie [**6**], Erdos [**5**], Dawlings [**3**], Lindenstrauss and Tzafriri [**10**], survey article by Jain and Leroy [**7**], Alahmadi, Jain, Leroy, and Sathaye [**1**], and Douglas [**4**] with a tinge of novelty.

2.1. Preliminaries.

(i) Let X be a complex Banach space with norm $\|.\|$. In case X is a Hilbert space $\mathcal{H}$ with inner product $< .,. >$ we will usually write $\mathcal{H}$ in place of X.

(a) Let $\mathcal{A}$ be the Banach algebra $\mathcal{B}(X)$ of bounded (i.e., continuous) linear transformations on X to itself (to be called operators in this paper) with pointwise addition, scalar multiplication and composition as multiplication, and for $T \in \mathcal{A}$ $\|T\| = \sup\{\|Tx\| : \|x\| \leq 1\}$. Let I_X be the identity operator on X to itself.

(b) Let $\mathcal{E}$ be the set of idempotents in $\mathcal{A}$ and $\mathcal{S}$ the semigroup generated by $\mathcal{E}$ with composition as multiplication. Then $\mathcal{S}$ coincides with the set of (finite) products of members of $\mathcal{E}$.
We denote the set of such products other than 0 and I_X by $\mathcal{F}$.

(c) For Banach spaces X_1 and X_2, $\mathcal{B}(X_1, X_2)$ will denote the Banach space of bounded (i.e., continuous) linear transformations (to be called operators) on X_1 to X_2.

Then $\mathcal{B}(X_1, X_2)$ is a left $\mathcal{B}(X_2)$ Banach module and right $\mathcal{B}(X_1)$ Banach module.

We shall use the same symbol 0 for different 0's in respective spaces.

(ii) Let $\widetilde{\mathcal{K}}$ be the set of closed (linear) subspaces of X.

(a) Suppose $\dim X \geq 2$. We write $X = Y \oplus Z$, a topological sum to mean that $Y, Z \in \widetilde{\mathcal{K}}$, $\{0\} \neq Y \neq X$, $\{0\} \neq Z \neq X$, $Y \cap Z = \{0\}, Y + Z = X$, the idempotent surjective linear maps $p_Y : X \to Y$ and $p_Z = I_X - p_Y : X \to Z$ given by $p_Y(x) = y$, $p_Z(x) = z$ for $x = y + z$ with $x \in X$, $y \in Y$, $z \in Z$ are both continuous (i.e., bounded). In this case, Y will be said to be topologically complemented (or in short complemented) in X and Z will be called a topological complement (in short, a complement) of Y in X. The inclusion maps $i_Y : Y \to X$ and $i_Z : Z \to X$ are in $\mathcal{B}(Y, X)$ and $\mathcal{B}(Z, X)$ respectively and indeed isometric and satisfy $p_Y \circ i_Y = I_Y$ and $p_Z \circ i_Z = I_Z$.

(b) Situation in (a) permits us to treat $T \in \mathcal{A}$ as a block matrix

$$T = \begin{bmatrix} T_1 & T_2 \\ T_3 & T_4 \end{bmatrix}$$

in $X = Y \oplus Z$, a topological sum with $T_1 = p_Y \circ T \circ i_Y \in \mathcal{B}(Y) = \mathcal{A}_1$, say, $T_2 = p_Y \circ T \circ i_Z \in \mathcal{B}(Z, Y) = \mathcal{A}_2$, say, $T_3 = p_Z \circ T \circ i_Y \in \mathcal{B}(Y, Z) = \mathcal{A}_3$, say and $T_4 = p_Z \circ T \circ i_Z \in \mathcal{B}(Z) = \mathcal{A}_4$, say. We call it a block representation of T in $X = Y \oplus Z$. We just mention that it is an example of a Morita context [**11**].

Let $\widehat{\mathcal{K}}$ be the subset of $\widetilde{\mathcal{K}}$ consisting of Y's that possess complements in X.

(c) If $X = \mathcal{H}$, a Hilbert space then every closed subspace $\mathcal{K}$ of $\mathcal{H}$ is complemented and we may even have its orthogonal complement $\mathcal{L} = \mathcal{K}^{\perp}$. We write $\mathcal{H} = \mathcal{K} \oplus^{\perp} \mathcal{L}$ or simply $\mathcal{H} = \mathcal{K} \oplus \mathcal{L}$.

(d) An old result of Lindenstrauss and Tzafriri [**10**] says that a Banach space X in which every closed subspace Y is topologically complemented is isomorphic to a Hilbert space $\mathcal{H}$. But in spite of that we may really not worry about such uncomplemented subspaces in $\widetilde{\mathcal{K}}$ in our paper because of the special basic properties of $\mathcal{E}$ and $\mathcal{F}$.

(e) Let G be a Banach space, G_1 a dense linear subspace of G, S_1 a non-empty subset of G_1, $\Gamma(S_1)$ the linear span of S_1 and $\Gamma_q(S_1)$ finite linear combinations of members of S_1 using rational scalars only. We note that if S_1 is infinite, then $\#\Gamma_q(S_1) = \#S_1$. Also, the closures of $\Gamma(S_1)$ and $\Gamma_q(S_1)$ in G_1 coincide and so do the closures of $\Gamma(S_1)$ and $\Gamma_q(S_1)$ in G, let us denote them by $\overline{\Gamma(S_1)}^{G_1}$, $\overline{\Gamma_q(S_1)}^{G_1}$, $\overline{\Gamma(S_1)}$, and $\overline{\Gamma_q(S_1)}$, respectively. Finally, $\overline{\Gamma(S_1)}^{G_1} = G_1 \Leftrightarrow \overline{\Gamma(S_1)} = G$. So, the smallest cardinality of S_1 such that $\overline{\Gamma(S_1)} = G$, equivalently, $\overline{\Gamma(S_1)}^{G_1} = G_1$ coincide. We call it the (topological) **dimension of** G as well as **dimension of** G_1, and denote them by $\dim(G)$ and $\dim(G_1)$, respectively. If G is a Hilbert space, then $\dim(G)$ is the cardinality of any complete orthonormal basis of G.

(f) For a closed subspace Y of X with $\{0\} \neq Y \neq X$, the (topological) dimension of the quotient Banach space X/Y will be called the codimension of Y in X. If $Y \in \widetilde{\mathcal{K}}$ is finite dimensional or has finite codimension, then $Y \in \widehat{\mathcal{K}}$. This gives us the next interspersing proposition.

PROPOSITION 2.1. *Suppose* $\dim X \geq 2$. *If Y is a closed subspace of X with* $\{0\} \neq Y \neq X$, *then there exist Y_0 and $\widehat{Y} \in \widehat{\mathcal{K}}$ that satisfy*

$$\{0\} \neq Y_0 \subset Y \subset \widehat{Y} \neq X.$$

Indeed, Y_0 is available even for any non-zero subspace Y of X. Further, if $Y \notin \widehat{\mathcal{K}}$, then there exist infinitely many choices for Y_0 as well as $\hat{Y}$; indeed, there are a strictly increasing sequence $(Y_n)_{n\in\mathbb{N}}$ in $\hat{\mathcal{K}}$ and a strictly decreasing sequence $(\hat{Y}_n)_{n\in\mathbb{N}}$ in $\hat{\mathcal{K}}$ that satisfy for $n \in \mathbb{N}$, $Y_n \subset Y \subset \hat{Y}_n$, Y_n is complemented in Y_{n+1} and $\hat{Y}_{n+1}$ is complemented in $\hat{Y}_n$.

PROOF. It is enough to prove the last part. We have Y is neither finite dimensional nor finite codimensional. We instantly have a desired sequence $(Y_n)_{n\in\mathbb{N}}$ of finite dimensional spaces contained in Y which is strictly increasing. By Hahn Banach Theorem, there is a continuous linear functional $f \neq 0$ which is zero on Y. We let $\hat{Y}_1 = \mathcal{N}(f)$, the nullspace of f. We can proceed by induction by taking $X = \hat{Y}_n$ and obtain $\hat{Y}_{n+1}$. □

COROLLARY 2.2. *Let X and Y be as in Proposition 2.1 above. Then we have the following.*

(i) *There exists a $0 \neq A \in \mathcal{E}$, which is zero on Y.*
(ii) *There exist $0 \neq T_0$ and $I_X \neq T_1 \in \mathcal{E}$ with $\mathcal{R}(T_0) \subset Y \subset \mathcal{R}(T_1)$, where $\mathcal{R}(B)$ stands for the range of an element B of $\mathcal{A}$.*

PROOF. (i) We may take $A = I_X - p_{\widehat{Y}}$, coming from some complement of $\widehat{Y}$ in X.
(ii) We may take $T_0 = p_{Y_0}$ and $T_1 = p_{\hat{Y}}$ coming from some complements of Y_0 and $\hat{Y}$ in X.

□

DEFINITION 2.3. Lindenstraus [**9**] called an infinitedimensional Banach space X *decomposable* if it is a topological sum $Y \oplus Z$ of some pair (Y, Z) of infinite dimensional closed subspaces Y and Z of itself, and *indecomposable* otherwise. In other words X is indecomposable if $\hat{\mathcal{K}}$ consists only of finite- dimensional or finite-codimensional closed subspaces of X.

REMARK 2.4.

(i) We thank M.A. Sofi for the suggestion that we give an idea about the problem discussed in this paper in the context of indecomposable Banach spaces as well. We do so at appropriate places.
(ii) The concept of indecomposable Banach spaces led to deep work and concepts of hereditarily indecomposable (H.I.) Banach spaces, spaces with a few operators, unconditional basic sequences, infinite Ramsey Theorem etc. We shall not go into that in this short expository research paper. The impressive long list of contributors includes names like S.A. Argysos, J. Bourgain, W.T. Gowers, R.C. Haydon, P. Kuszmider, B. Maurey, H.P. Rosenthatl, V. Zigler.
(iii) There is a large class of indecomposable Banach spaces that are not hereditarily indecomposable Banach spaces [**8**].

2.2. Basic properties of $\mathcal{E}$ and $\mathcal{F}$ and examples. Consider p_Y as in subsection 2.1(ii)(a) above. Then, p_Y is an idempotent in $\mathcal{A}$, $\mathcal{R}(p_Y) = Y$ and $\mathcal{N}(p_Y) = Z = \mathcal{R}(I_X - p_Y)$. Geometrically speaking, p_Y is a projection on Y along Z. They constitute the base for Cartesian co-ordinate system. As already noted in Section 1, Erdos [**5**] showed that they are the building blocks for singular linear transformations of $\mathbb{C}^n$ to itself. But it is not so for infinite-dimensional Banach spaces or even Hilbert spaces. The following lemma and remarks thereafter explain it to begin with.

LEMMA 2.5. (i) *For $Q \in \mathcal{E}$ with $0 \neq Q \neq I_X$. $\mathcal{N}(Q)$, the null space of Q and $\mathcal{R}(Q)$, the range of Q, are both closed subspaces of X that are neither $\{0\}$ nor X. Indeed, $X = \mathcal{R}(Q) \oplus \mathcal{N}(Q)$, a topological sum.*
(ii) *Let $F \in \mathcal{F}$, other than 0 and I_X, say $F = Q_1 \cdots Q_p$ with $p \in \mathbb{N}$, $Q_j \in \mathcal{E}$, $0 \neq Q_j \neq I_X$ for $1 \leq j \leq p$. Then $\mathcal{N}(Q_p) \subset \mathcal{N}(F)$ and $\{0\} \neq \mathcal{R}(F) \subset \mathcal{R}(Q_1)$. Also, $\{0\} \neq \mathcal{N}(F) \neq X$ and $\overline{\mathcal{R}(F)}$, the closure of $\mathcal{R}(F)$, is $\neq X$.*
(iii) *Let $0 \neq T \in \mathcal{A}$ and let $\mathcal{K}_0 = \mathcal{R}(T)$, $\mathcal{K}_1$ its closure in X.*
  (a) *Then T has a non-zero left annihilator if and only if T has a non-zero idempotent left annihilator if and only if $\mathcal{K}_1 \neq X$.*

(b) *Moreover, T has a non-zero right annihilator if and only if T has a non-zero idempotent right annihilator if and only if $\mathcal{N}(T) \neq \{0\}$.*

(c) *As a consequence, any $F \in \mathcal{F} \setminus \{0, I_X\}$ has non-zero idempotent left and right annihilators.*

(iv) *For $X = \mathbb{C}^n$, $n \geq 2$, $T \in \mathcal{A}$, $\mathcal{N}(T) \neq \{0\}$ if and only if T is singular, i.e., not invertible in $\mathcal{A}$ if and only if $\mathcal{R}(T) \neq X$. Furthermore, $\mathcal{R}(T)$ is closed. Also, the concepts of T having a non-zero left annihilator coincides with that of T having a non-zero right annihilator.*

PROOF. Parts (i) and (ii) are immediate from discussion in this section. For (iii), we appeal to Corollary 2.2. To give an idea, we elaborate for (iii) (a), suppose T has a non-zero left annihilator A in $\mathcal{A}$. Then, $AT = 0$. So, A is zero on $\mathcal{K}_0$. But A is continuous, so, A is zero on $\mathcal{K}_1$. Because $A \neq 0$, we have $\mathcal{K}_1 \neq X$. On the other hand, suppose $\mathcal{K}_1 \neq X$. We know $\mathcal{K}_1 \neq \{0\}$. So, by Corollary 2.2 (i), there exists $0 \neq B \in \mathcal{E}$, which is zero on $\mathcal{K}_1$. So, B is a non-zero idempotent left annihilator of T.

Finally, (iv) comes from basic Linear Algebra. □

REMARK 2.6. The statements in Lemma 2.5 (iv) above are not true for the space $\mathcal{H} = l_2 = l_2(\mathbb{N}) = \{x = (x_n)_{n\in\mathbb{N}}, \sum_{n\in\mathbb{N}} |x_n|^2 < \infty\}$ with inner product $< x, y >= \sum_{n\in\mathbb{N}} x_n \bar{y_n}$ for $x, y \in l_2(\mathbb{N})$ as shown below. For $n \in \mathbb{N}$ let e_n be the sequence with 1 at the n-th place and zero elsewhere. The discussion can be adapted to Banach spaces $X = l_p$, $1 \leq p < \infty$ as well.

(i) Let $T \in \mathcal{A}$ be defined by

$$(Tx)_j = \begin{cases} 0, & j = 1 \\ x_{j-1} & j \geq 2 \end{cases}.$$

(a) Then $\mathcal{N}(T) = \{0\}$ and $\mathcal{R}(T) = \{x \in l_2 : x_1 = 0\}$, which is closed but $\neq \mathcal{H}$. So, T has a non-zero left annihilator, viz., the orthogonal projection P_1 given by $P_1(x) = x_1 e_1$, $x \in l_2$ and its non-zero scalar multiples as well. But T has no non-zero right annihilators.

(b) As a consequence, T is not a finite product of idempotents in $\mathcal{A}$, i.e., $T \notin \mathcal{F}$.

(ii) Let $T \in \mathcal{A}$ be defined by $(Tx)_j = x_{j+1}$, for $j \in \mathbb{N}$ and $x \in \mathcal{H}$.

(a) Then $Te_1 = 0$. So, $\mathcal{N}(T) \neq \{0\}$ and T has a non-zero right annihilator viz., P_1 as in (i) (a) above together with its non-zero scalar multiples. But $\mathcal{R}(T) = \mathcal{H}$. So, T has no non-zero left annihilators.

(b) As a consequence, T is not a finite product of idempotents in $\mathcal{A}$, i.e., $T \notin \mathcal{F}$.

(iii) Let $T \in \mathcal{A}$ be given by $(Tx)_j = \frac{1}{j} x_j$ for $j \in \mathbb{N}$ and $x \in l_2$.

(a) Then $\mathcal{N}(T) = \{0\}$ and therefore, T has no non-zero right annihilators. Moreover, for $j \in \mathbb{N}$, $e_j \in \mathcal{R}(T)$ simply because $T(je_j) = e_j$. So, $\mathcal{R}(T)$ is dense in $\mathcal{H}$, i.e., $\overline{\mathcal{R}(T)} = \mathcal{H}$. So, T has no non-zero left annihilators either.

(b) Further, $\mathcal{R}(T) \neq \mathcal{H}$ because $x = (x_j)_{j\in\mathbb{N}}$ given by $x_j = \frac{1}{j}$ is in $\mathcal{H}$ but not in $\mathcal{R}(T)$. This gives another instance of $T \in \mathcal{A}$ with $\mathcal{N}(T) = \{0\}$, but $\mathcal{R}(T) \neq \mathcal{H}$ in contrast to the situation in $\mathbb{C}^n$, $n \in \mathbb{N}$.

(c) T is not a regular element of $\mathcal{A}$. Indeed, let, if possible, $G \in \mathcal{A}$ be such that $TGT = T$. Then $T(GT - I_{\mathcal{H}}) = 0$. But $\mathcal{N}(T) = \{0\}$. So,

$GT = I_{\mathcal{H}}$. So, for $j \in \mathbb{N}$, $je_j = GT(je_j)$. So, $\|G(e_j)\| = j$ whereas $\|e_j\| = 1$. So, G is not bounded, a contradiction.
Hence $\mathcal{A}$ is not von Neumann regular. This demands individual study of properties of $\mathcal{F}$ as we cannot use the literature on von Neumann regular rings.
We do that in the next section.

The equivalence of statements in the last part of (iii) (a) and (b) may not hold for all closed subalgebras of $\mathcal{A} = \mathcal{B}(X)$.

EXAMPLES 2.7. Let $\mathfrak{C}$ be the Banach algebra $C[0,1]$ of continuous scalar valued functions on the closed interval $[0,1]$ with the sup norm. Let $f, g \in \mathfrak{C}$ be given by $f(t) = 2t - 1$ for $0 \leq t \leq \frac{1}{2}$ and $f(t) = 0$ for $\frac{1}{2} \leq t \leq 1$, whereas $g(t) = 0$ for $0 \leq t \leq \frac{1}{2}$ and $2t - 1$ for $\frac{1}{2} \leq t \leq 1$. Then $fg = 0$. Now, let $X = \mathfrak{C}^2$. Then $\mathcal{A} = \mathcal{B}(X)$ can be identified with $M_2(\mathcal{B}(\mathfrak{C}))$. We consider the closed subalgebra $\mathcal{A}_1 = M_2(\mathfrak{M}(\mathfrak{C}))$, where $\mathfrak{M}(\mathfrak{C})$ is the algebra of multipliers of $\mathfrak{C}$ identified with $\mathfrak{C}$ itself. Then $T = \begin{bmatrix} 1 & 0 \\ 0 & f \end{bmatrix}$ has $A = \begin{bmatrix} 0 & 0 \\ 0 & g \end{bmatrix}$ as a non-zero left annihilator but T has no non-zero idempotent left annihilator. To elaborate, the left annihilators of T in $\mathcal{A}_1$ are given by $\begin{bmatrix} 0 & b \\ 0 & d \end{bmatrix}$ with $bf = 0 = df$. But $\begin{bmatrix} 0 & b \\ 0 & d \end{bmatrix}$ is an idempotent if and only if $bd = b$ and $d^2 = d$. The only idempotents in $\mathfrak{C}$ are 0 and 1. Now, $d = 0$ forces $b = 0$ whereas $d = 1$ does not satisfy $df = 0$. So, T has no non-zero idempotent left annihilators in $\mathcal{A}_1$.

REMARK 2.8. Let X be an infinite-dimensional indecomposable Banach space.

(i) Let Q be an idempotent in $\mathcal{A}$ other than 0 and I_X. The either Q has finite rank or $\mathcal{N}(Q)$ has finite codimension.
(ii) Let $F \in \mathcal{F}$ be as in Lemma 2.5 (ii). Then either F has finite rank or each Q_j there has finite -dimensional null-space. As a consequence, in the latter case, $\mathcal{N}(F)$ is finite-dimensional.

EXAMPLES 2.9. (i) We consider the algebra $M_2(\mathbb{R})$ and express some nonzero singular matrices T in $M_2(\mathbb{R})$ as finite products of idempotents on the right.

(a) For $0 \neq a \in \mathbb{R}$,
$$\begin{bmatrix} a & 0 \\ 0 & 0 \end{bmatrix} = \begin{bmatrix} 1 & a-1 \\ 0 & 0 \end{bmatrix} \begin{bmatrix} 1 & 0 \\ 1 & 0 \end{bmatrix}.$$
(b) For $0 \neq b$, $0 \neq c \in \mathbb{R}$,
$$\begin{bmatrix} bc & 0 \\ 0 & 0 \end{bmatrix} = \begin{bmatrix} 1 & b \\ 0 & 0 \end{bmatrix} \begin{bmatrix} 0 & 0 \\ 0 & 1 \end{bmatrix} \begin{bmatrix} 1 & 0 \\ c & 0 \end{bmatrix}.$$
(c) For $0 \neq a$, $b \in \mathbb{R}$
$$\begin{bmatrix} a & ab \\ 0 & 0 \end{bmatrix} = \begin{bmatrix} 1 & a-1+b \\ 0 & 0 \end{bmatrix} \begin{bmatrix} 1-b & b(1-b) \\ 1 & b \end{bmatrix}.$$

(ii) For any ring R with identity 1 and $n = 2r$, $r \in \mathbb{N}$, $\mathbb{R}$ in (i) above can be replaced by $M_r(R)$ giving us factorizations in analogy with (i).
(iii) For any ring R with identity 1, $n \in \mathbb{N}$, $n > 2r, r$ in $\mathbb{N}$, any $a \in M_r(R)$ can be written as bc with $b \in M_{r,n-r}(R)$ and $c \in M_{n-r,r}(R)$ such as

$b = \begin{bmatrix} a & 0 \end{bmatrix}$, $c = \begin{bmatrix} I_r \\ 0 \end{bmatrix}$. So, we can have an analogue to (i)(b) by replacing 1 in the first row by I_r and second row by I_{n-r}.

(iv) For a Hilbert space $\mathcal{H} = \mathcal{K} \oplus \mathcal{L}$ with $\mathcal{K} \perp \mathcal{L}$, $0 < \dim \mathcal{K} \leq \dim \mathcal{L}$, we can have a $\mathcal{J} : \mathcal{K} \to \mathcal{L}_1 \subset \mathcal{L}$ an isometric isomorphism with $\mathcal{L}_1$ a closed subspace of $\mathcal{L}$. Then $\mathcal{J}^*\mathcal{J} = I_{\mathcal{K}}$. So, we may have the following analogue of (i).

(a)
$$\begin{bmatrix} a & 0 \\ 0 & 0 \end{bmatrix} = \begin{bmatrix} I_{\mathcal{K}} & (a - I_{\mathcal{K}})\mathcal{J}^* \\ 0 & 0 \end{bmatrix} \begin{bmatrix} I_{\mathcal{K}} & 0 \\ \mathcal{J} & 0 \end{bmatrix},$$
for $a \in \mathcal{B}(\mathcal{K})$.

(b) For $0 \neq b \in \mathcal{B}(\mathcal{L}, \mathcal{K})$, $0 \neq c \in \mathcal{B}(\mathcal{K}, \mathcal{L})$,
$$\begin{bmatrix} bc & 0 \\ 0 & 0 \end{bmatrix} = \begin{bmatrix} I_{\mathcal{K}} & b \\ 0 & 0 \end{bmatrix} \begin{bmatrix} 0 & 0 \\ 0 & I_{\mathcal{L}} \end{bmatrix} \begin{bmatrix} I_{\mathcal{K}} & 0 \\ c & 0 \end{bmatrix}.$$

(c) For $0 \neq a \in \mathcal{B}(\mathcal{K}), b \in \mathcal{B}(\mathcal{L}, \mathcal{K})$,
$$\begin{bmatrix} a & ab \\ 0 & 0 \end{bmatrix} = \begin{bmatrix} I_{\mathcal{K}} & (a - I_{\mathcal{K}})\mathcal{J}^* + b \\ 0 & 0 \end{bmatrix} \begin{bmatrix} I_{\mathcal{K}} - b\mathcal{J} & b - b\mathcal{J}b \\ \mathcal{J} & \mathcal{J}b \end{bmatrix}.$$

(v) Indeed, in Corollary 3.11 in [**2**], the conditions given there on T force T to satisfy $\mathcal{R}(T)^{\perp} = \mathcal{N}(T) = \mathcal{R}(T^*)^{\perp} = \mathcal{N}(T)$ and $T = 0$ on $\mathcal{R}(T)^{\perp}$ and therefore, have the form $\begin{bmatrix} T_1 & 0 \\ 0 & 0 \end{bmatrix}$ with T_1 invertible in $\mathcal{B}(\mathcal{K})$. Then (ii)(a) above renders an easy proof of a generalization of Corollary 3.11 [**2**].

We give a useful factorization theorem, due to Douglas [**4**, Theorem 1 and the last paragraph]. It will be referred to as Theorem D.

THEOREM 2.10. *Let X_1, X_2, X_3 be Banach spaces and $U \in \mathcal{B}(X_1, X_3)$, $V \in \mathcal{B}(X_2, X_3)$. Then $\mathcal{R}(U) \subset \mathcal{R}(V)$ if and only if $U = VW$ for some $W \in \mathcal{B}(X_1, X_2)$. In this case, there exists a unique $W_0 \in \mathcal{B}(X_1, X_2)$ that satisfies $U = VW_0$ and $\mathcal{N}(U) = \mathcal{N}(W_0)$.*

3. Properties of idempotents in $\mathcal{A}$ and their products

3.1. Basic concepts and set-up. Let us first consider $0 \neq T \in \mathcal{A}$ that has a nonzero left annihilator as well as a non-zero right annihilator. The subject of this paper is to see if we can decompose such an operator $T \in \mathcal{A}$ into (finite) product of idempotents. We start with the following rewording of a part of Lemma 2.5 for use here.

LEMMA 3.1. *Let $0 \neq T \in \mathcal{A}$ have a non-zero left annihilator and denote by $\mathcal{K}_0$ its range and $\mathcal{K}_1$, the closure of $\mathcal{K}_0$. Then there exists $\mathcal{K}$ in $\hat{\mathcal{K}}$ that satisfies $\{0\} \neq \mathcal{K}_0 \subset \mathcal{K}_1 \subset \mathcal{K} \subsetneq X$.*

This motivates us to work in the scenario set up in the previous section especially 2.1 and 2.2. We shall write $\mathcal{K}$ in place of Y and $\mathcal{L}$ in place of Z with the understanding that $\mathcal{L}$ will be an arbitrary but then fixed complement of $\mathcal{K}$ in X, and that in case $X = \mathcal{H}$, a Hilbert space, $\mathcal{L} = \mathcal{K}^{\perp}$ as indicated in 2.1 (ii) (c) above.

For subspaces G and G' of $\mathcal{K}$ and $\mathcal{L}$ respectively, we may take the liberty of writing $G \oplus G'$ and we do so. Let us first make some observations.

LEMMA 3.2. *Let* $T = \begin{pmatrix} T_1 & T_2 \\ T_3 & T_4 \end{pmatrix}$ *be an operator in its block representation according to a decomposition* $X = \mathcal{K} \oplus \mathcal{L}$.

(i) *T is an idempotent if and only if* $T_1^2 + T_2T_3 = T_1, T_1T_2 + T_2T_4 = T_2, T_3T_1 + T_4T_3 = T_3$, *and* $T_3T_2 + T_4^2 = T_4$.
(ii) *Let T be an idempotent with* $\mathcal{K} \subseteq \mathcal{R}(T)$. *Then* $T_1 = I_{\mathcal{K}}$. $T_2T_3 = 0, T_2T_4 = 0, T_4T_3 = 0$, *and* $T_3T_2 = T_4 - T_4^2$.
(iii) (a) $\mathcal{R}(T) \subseteq \mathcal{K}$ *if and only if* $T_3 = 0$ *and* $T_4 = 0$.
  (b) *In case (a) happens, then:*
  - $\mathcal{R}(T) = \mathcal{R}(T_1) + \mathcal{R}(T_2)$.
  - $\mathcal{N}(T) = (\mathcal{N}(T_1) \oplus \mathcal{N}(T_2)) + \{k + l : k \in \mathcal{K}, l \in \mathcal{L}, (k,l) = (0,0)$ *or* $T_1k = -T_2l \neq 0\}$.

(iv) *T is an idempotent with* $\mathcal{R}(T) \subset \mathcal{K}$ *if and only if it is of the form* $\begin{pmatrix} T_1 & T_2 \\ 0 & 0 \end{pmatrix}$ *with*
  (a) $T_1^2 = T_1$ *and*
  (b) $T_1T_2 = T_2$.
  Under (a), the condition (b) can be replaced by any of the following:
  (b') $T_2 = T_1B$ *for some* $B \in \mathcal{B}(\mathcal{L}, \mathcal{K})$.
  (b'') $\mathcal{R}(T_2) \subset \mathcal{R}(T_1)$.
  (b''') $\mathcal{R}(T) = \mathcal{R}(T_1)$.
(v) *T is an idempotent with* $\mathcal{R}(T) = \mathcal{K}$ *if and only if* $T_3 = 0, T_4 = 0$, *and* $T_1 = I_{\mathcal{K}}$, *i.e.*, $T = \begin{bmatrix} I_{\mathcal{K}} & T_2 \\ 0 & 0 \end{bmatrix}$.

PROOF. (i) Direct computations give that T is an idempotent if and only if $T_1^2 + T_2T_3 = T_1, T_1T_2 + T_2T_4 = T_2, T_3T_1 + T_4T_3 = T_3$, and $T_3T_2 + T_4^2 = T_4$.
(ii) Since $\mathcal{K} \subseteq \mathcal{R}(T)$ and $T^2 = T$, we immediately obtain that $T_1 = I_{\mathcal{K}}$ and hence thanks to the equations above, the other equalities in (ii) follow.
(iii) is clear.
(iv) Follows from (ii) and Theorem D.

□

Let $\mathcal{E}_{\mathcal{K}}$ be the set of idempotents in $\mathcal{A}$ with range $\mathcal{K}$ and $\widetilde{\mathcal{E}_{\mathcal{K}}}$, the set of idempotents Q in $\mathcal{A}$ with range $\mathcal{R}(Q) \subset \mathcal{K}$.

With this notation we easily get the following theorem.

THEOREM 3.3. (i) $\mathcal{E}_{\mathcal{K}}$ *is a non-commutative semigroup.*
(ii) $\widetilde{\mathcal{E}_{\mathcal{K}}}$ *is left as well as right* $\mathcal{E}_{\mathcal{K}}$*-module.*

PROOF. (i) Let $T = \begin{bmatrix} T_1 & T_2 \\ T_3 & T_4 \end{bmatrix} \in \mathcal{E}_{\mathcal{K}}$. Since the range of T is $\mathcal{K}$, the statement *(iv)* in Lemma 3.2 shows that T is of the form $T = \begin{bmatrix} I_{\mathcal{K}} & T_2 \\ 0 & 0 \end{bmatrix}$. It is then obvious that these elements form a noncommutative semigroup.
(ii) The elements of $\widetilde{\mathcal{E}_{\mathcal{K}}}$ and $\mathcal{E}_{\mathcal{K}}$ are described in the statement (iii) and (iv) of the above lemma. Let T_1, T_1' be idempotents in $\mathcal{B}(\mathcal{K})$ and $T_2, T_2' \in \mathcal{B}(\mathcal{L}, \mathcal{K})$

with $T_1T_2 = T_2$ and $T_1'T_2' = T_2'$. We have

$$\begin{bmatrix} T_1 & T_1T_2 \\ 0 & 0 \end{bmatrix} \begin{bmatrix} I_{\mathcal{K}} & T_2' \\ 0 & 0 \end{bmatrix} = \begin{bmatrix} T_1 & T_1T_2' \\ 0 & 0 \end{bmatrix},$$

$$\begin{bmatrix} I_{\mathcal{K}} & T_2' \\ 0 & 0 \end{bmatrix} \begin{bmatrix} T_1 & T_1T_2 \\ 0 & 0 \end{bmatrix} = \begin{bmatrix} T_1 & T_1T_2 \\ 0 & 0 \end{bmatrix}.$$

This leads to the fact that $\widetilde{\mathcal{E}_{\mathcal{K}}}$ is left as well as right $\mathcal{E}_{\mathcal{K}}$-module

□

DEFINITION 3.4. Let $T \in \mathcal{F}$ with $T \notin \mathcal{E}$. Then there is a smallest $t \geq 2$ such that $T = Q_1 \cdots Q_t$ with each $Q_j \in \mathcal{E}$. We shall call t the *idempotent index* of T, in short $ii(T)$. For T having idempotent index t a factorization of T into t idempotents, say $E_1 \cdots E_t$ will be said to be a *minimal idempotent factorization.*

REMARKS 3.5. (i) The idempotent index of $T \in \mathcal{E}$ will be taken to be 1 and for $T \in \mathcal{A} \setminus \mathcal{S}$, it will be taken to be ∞.

At times, we write $\mathcal{F}_t$ for $\mathcal{E}^t$ for $t \geq 2$.

(ii) For $t \in \mathbb{N}$, let $\mathcal{S}_t$ be the subset of $\mathcal{S}$ consisting of T with idempotent index t. Then $\mathcal{S}_t$'s are mutually disjoint and have union equal to $\mathcal{S}$.

(iii) For $s, t \in \mathbb{N}$, $\mathcal{S}_{s+t} \subset \mathcal{S}_s\mathcal{S}_t$.

(iv) Clearly $\mathcal{S}_t \subseteq \mathcal{E}^t$. Equality may not occur.

  (a) Examples 2.9 give a good idea to begin with. See Example 2.9 (i) (b) for $\mathcal{E}^3$ to be different from $\mathcal{S}_3$.

  (b) Another instance is provided by non-zero mutually orthogonal projections in a Hilbert space $\mathcal{H}$.

(v) Several other results in literature, for instance, [**7**] and [**1**] can be adapted to this general situation. We just give one more which helps in understanding idempotent index better, viz., part (vi) below and Theorem 3.6 below which is Theorem 3 (1) in [**7**] adapted to our set-up.

Indeed, part (i) of Theorem 3.6 below with $\mathcal{K} = \mathcal{R}(T)$ combined with the old result of Erdos [**5**] for matrices gives the known result that if $T \in \in\mathcal{A}$ has finite rank then $T \in \mathcal{S}$.

(vi) Let R be a ring with identity 1. Let $0 \neq a \in R$ and $0 \neq q \neq 1$ an idempotent in R and $q' = 1 - q$. Then $qa = 0 \Leftrightarrow a = q'a$. So, in this case, for $b \in R$, $a = q'b \Leftrightarrow q'(b - a) = 0$, i.e., $c = b - a$ is a right annihilator of q'. So, to begin with we can target elements like $b = a + c$ to be expressible as finite products of idempotents in R. Indeed, we also get that $ii(a) \leq 1 + ii(b)$.

(vii) Minimal representations for members of $\mathcal{S} \setminus \mathcal{E}$ are not unique. We have seen this before like in Example 2.9 (iv)(a) where we can have several maps J. Section 4 will give several Theorems and Examples for $t = 2$.

THEOREM 3.6. *Let $T \in \mathcal{A}$ with block representation $T = \begin{bmatrix} T_1 & T_2 \\ 0 & 0 \end{bmatrix}$ in $X = \mathcal{K} \oplus \mathcal{L}$, a topological sum.*

(i) *If T_1 is a product of s idempotents in $\mathcal{B}(\mathcal{K})$, then T is a product of $s+1$ idempotents in $\mathcal{A}$. In this case, idempotent index of $T \leq 1+$ idempotent index of T_1, in short $ii(T) \leq ii(T_1) + 1$.*

(ii) *Inequality in (i) above can be strict.*

PROOF. (i) We follow the proof of Theorem 3 (1) in [7]. We have $T_1 = E_1 \cdots E_s$ with each E_j an idempotent in $\mathcal{B}(\mathcal{K})$.

Let for $1 \leq j \leq s$, $Q_j = \begin{bmatrix} E_j & 0 \\ 0 & I_{\mathcal{L}} \end{bmatrix}$ in $X = \mathcal{K} \oplus \mathcal{L}$ and let $Q_0 = \begin{bmatrix} I_{\mathcal{K}} & T_2 \\ 0 & 0 \end{bmatrix}$. Then for $0 \leq j \leq s$, $Q_j \in \mathcal{E}$ and $0 \neq Q_j \neq I_X$. Now $T = Q_0 Q_1 \cdots Q_s$. So, $T \in \mathcal{E}^{s+1}$. By Remark 3.5 (iii), we have $ii(T) \leq ii(T_1) + 1$.

(ii) We give some instances of the strict inequality.

(a) In Example 2.9 (i) (a), for $a \neq 1$, $ii(a) = \infty$ whereas $ii \begin{bmatrix} a & 0 \\ 0 & 0 \end{bmatrix} = 2$.

(b) We consider Example 2.9 (ii) and take $r \geq 3$. It is well-known ([5], [7], [1]) there is an $a \in M_r(\mathbb{R})$ with $ii(a) \geq 3$. By Example 2.9 (ii), $ii \begin{bmatrix} a & 0 \\ 0 & 0 \end{bmatrix} = 2$.

(c) We consider Example 2.9 (iv) and take $\mathcal{K} = l_2 = \mathcal{L}$ and $\mathcal{H} = l_2 \oplus l_2$. Next, we take T_1 to be any of the T's in Remark 2.6 above. Then $ii(T_1) = \infty$. But by Example 2.9 (iv) (a), $ii \begin{bmatrix} T_1 & 0 \\ 0 & 0 \end{bmatrix} = 2$.

□

LEMMA 3.7. *If $T \in \mathcal{F} \setminus \mathcal{E}$ is written as $T = Q_1 \ldots Q_t$ where $Q_1, \ldots, Q_t$ are idempotents and t is the idempotent index of T, then for any $1 \leq j < t$ we must have*

(i) *Q_j and Q_{j+1} cannot be in any semigroup $\mathcal{S}' \subset \mathcal{E}$.*
(ii) *$K_{j+1} \setminus K_j \neq \emptyset$ and $K_j \setminus K_{j+1} \neq \emptyset$, where K_s denotes $\mathcal{R}(Q_s)$, $1 \leq s \leq t$.*

PROOF. Indeed, it follows from Theorem 3.3 (ii). But we also give an idea of a direct proof. It is enough to consider the case when $T = Q_1 Q_2$, i.e. when $t = 2$. If we have $K_1 \subset K_2$, then for any $x \in X$ we have that $Q_2 Q_1(x) = Q_1(x)$ and hence $Q_1 Q_2 Q_1 Q_2(x) = Q_1 Q_1 Q_2(x) = Q_1 Q_2(x)$, so that $Q_1 Q_2$ is indeed an idempotent. It is then clear that $Q_1 Q_2 \in \mathcal{E}_{K_1}$, but this contradicts the fact that $t = 2$ is the idempotent index of T. One can show that if $K_2 \subseteq K_1$, then $Q_1 Q_2 = Q_2$, this contradicts again the fact that $t = 2$. □

REMARK 3.8. The converse of the result in Lemma 3.7 (ii) is not true as can be seen by taking $Q_1 = \begin{bmatrix} 1 & a \\ 0 & 0 \end{bmatrix}$ and $Q_2 = \begin{bmatrix} 0 & 0 \\ b & 1 \end{bmatrix}$ with $ab = 1$.

REMARK 3.9. It will be helpful to list more idempotents in $\mathcal{A}$ in the form as in Lemma 3.2 (i) simply by interchanging the roles of $\mathcal{K}$ and $\mathcal{L}$, T_1 and T_4, T_2 and T_3 in the statements in the discussion above.

(i) Idempotents Q in $\mathcal{A}$ with $\mathcal{R}(Q) \supset \mathcal{L}$, are given by $Q = \begin{bmatrix} T_1 & T_2 \\ T_3 & I_{\mathcal{L}} \end{bmatrix}$ that satisfy $T_2 T_3 = T_1(I_{\mathcal{K}} - T_1) = (I_{\mathcal{K}} - T_1)T_1$, $T_1 T_2 = 0$, $T_3 T_1 = 0$, and $T_3 T_2 = 0$.

(ii) $Q \in \mathcal{E}$ with $\mathcal{R}(Q) \subset \mathcal{L}$ are given by $Q = \begin{bmatrix} 0 & 0 \\ DC & D \end{bmatrix}$ with $D \in \mathcal{B}(\mathcal{L})$ an idempotent, $C \in \mathcal{B}(\mathcal{K}, \mathcal{L})$ i.e., $Q = \begin{bmatrix} 0 & 0 \\ C_1 & D \end{bmatrix}$ with $D \in \mathcal{B}(\mathcal{L})$ an idempotent and $C_1 \in \mathcal{B}(\mathcal{K}, \mathcal{L})$ with $\mathcal{R}(C_1) \subset \mathcal{R}(D)$; further, $\mathcal{R}(Q) = \mathcal{R}(D)$.

(iii) Any $Q \in \mathcal{E}$ with $\mathcal{R}(Q) = \mathcal{L}$ is given by $Q = \begin{bmatrix} 0 & 0 \\ C & I_{\mathcal{L}} \end{bmatrix}$ with $C \in \mathcal{B}(\mathcal{K}, \mathcal{L})$.

We note that all these form a non-commutative semigroup with product $Q_1Q_2 = Q_2$, viz.,

$$\begin{bmatrix} 0 & 0 \\ C & I_{\mathcal{L}} \end{bmatrix} \begin{bmatrix} 0 & 0 \\ C' & I_{\mathcal{L}} \end{bmatrix} = \begin{bmatrix} 0 & 0 \\ C' & I_{\mathcal{L}} \end{bmatrix}$$

in line with Theorem 3.3 (i).

3.2. Concepts of local block representation and "essentially contained in". We recast these concepts in the way we like for further use.

DEFINITION 3.10. Let $0 \neq T \in \mathcal{A}$. Assume that T has a non-zero left annihilator. Let $\mathcal{K}_0 = \mathcal{R}(T)$ and $\mathcal{K}_1$ its closure in X. Lemma 3.1 and Lemma 3.2 above give us a $\mathcal{K} \in \hat{\mathcal{K}}$ with $\mathcal{K}_1 \subseteq \mathcal{K}$ and enable us to express $X = \mathcal{K} \oplus \mathcal{L}$, a topological sum for any complement $\mathcal{L}$ of $\mathcal{K}$ in X and have the block, representation of T with respect to this decomposition as $T = \begin{bmatrix} T_1 & T_2 \\ 0 & 0 \end{bmatrix}$ in $X = \mathcal{K} \oplus \mathcal{L}$. We call it the **Local Block representation** of T with respect to the decomposition $\mathcal{K} \oplus \mathcal{L}$ of X, or simply a local block representation if no confusion can arise.

Notice that such an operator T has a non-zero right annihilator if and only if at least one of the following conditions is satisfied:

(a) $\mathcal{N}(T_1) \neq \{0\}$,
(b) $\mathcal{N}(T_2) \neq \{0\}$,
(c) $\mathcal{R}(T_1) \cap \mathcal{R}(T_2) \neq \{0\}$.

We assume that it is so.

If $\mathcal{K}_1 \in \hat{\mathcal{K}}$, then we may take $\mathcal{K} = \mathcal{K}_1$. If $\mathcal{K}_1 \notin \hat{\mathcal{K}}$, then by Proposition 2.1, there are infinitely many $\mathcal{K}$'s in $\hat{\mathcal{K}}$ available. We can carry out our discussion in any of them. There will be some interrelations which we do not go into at this moment. Usefulness of different $\mathcal{K}$'s will become clear as we go along and is indicated towards the end of the paper as well.

REMARK 3.11. (i) We work in the context of 2.1(ii)(e) above. For Hilbert spaces G and G' we have that if $\dim G \leq \dim G'$, then there exists an isometric linear map J on G onto a closed subspace $G'' \subset G'$ and the adjoint $J^* : G' \to G$ satisfies $J^*J = I_G$. Indeed, $J^* = J^{-1} \circ P_{G''}$, where $P_{G''}$ is the (orthogonal) projection on G' onto G''.

(ii) To have a map like J in the context of Banach spaces, we can impose the condition that $G'' \in \hat{\mathcal{K}}$ for G' replacing X. To elaborate, we say that G is **essentially contained in** G', in short $G \subset_e G'$, if there exists a complemented closed subspace G'' of G' and a bicontinuous linear bijection $J : G \to G''$ and for notational convenience, denote $J^{-1} \circ p_{G''}$ by J^*, where $p_{G''}$ is an idempotent in $\mathcal{B}(G')$ with range G'' coming from some complement of G'' in G'. Then $J^*J = I_G$. We note that in this case, $\dim G = \dim G'' \leq \dim G'$. The converse is true if G is finite-dimensional.

(iii) We next note that if $A \in \mathcal{B}(G, G')$, then $\dim \mathcal{R}(A) \leq \dim G$.

THEOREM 3.12. *We shall freely use Remarks and results above. Let* $t \in \mathbb{N}$ *with* $t \geq 2$. *Let* $T \in \mathcal{S}_t$, *for* $1 \leq j \leq t$, E_j *in* $\mathcal{E}$ *with* $T = E_1 \dots E_t$ *a minimal decomposition of* T. *Let* $\mathcal{K} = \mathcal{R}(E_1)$, $\mathcal{L}$, *a complement of* $\mathcal{K}$, *for instance* $\mathcal{L} = \mathcal{N}(E_1)$, *in* X. *Then we have the following.*

(i) $E_1 = \begin{bmatrix} I_{\mathcal{K}} & B \\ 0 & 0 \end{bmatrix}$ *for some* $B \in \mathcal{B}(\mathcal{L}, \mathcal{K})$, *and* $T = \begin{bmatrix} T_1 & T_2 \\ 0 & 0 \end{bmatrix}$ *in* $X = \mathcal{K} \oplus \mathcal{L}$, *a topological sum for some* $T_1 \in \mathcal{B}(\mathcal{K})$, $T_2 \in \mathcal{B}(\mathcal{L}, \mathcal{K})$, *its local block representation with respect to* $X = \mathcal{K} \oplus \mathcal{L}$.

(ii) $S = E_2 \cdots E_t = \begin{bmatrix} T_1 - BC & T_2 - BD \\ C & D \end{bmatrix} = T_{B,C,D}$, *say, for some* $C \in \mathcal{B}(\mathcal{K}, \mathcal{L})$ *and* $D \in \mathcal{B}(\mathcal{L})$.

(iii) $ii(\mathcal{S}) = t - 1$.

(iv) $\dim \mathcal{R}(S) \geq \dim \mathcal{K}_1$, *where* $\mathcal{K}_1$ *is the closure of* $\mathcal{R}(T)$ *in* X.

PROOF.

(i) it is immediate from Lemma 3.2 (v) and Definition 3.10.

(ii) Let $S = \begin{bmatrix} S_1 & S_2 \\ S_3 & S_4 \end{bmatrix}$ in $X = \mathcal{K} \oplus \mathcal{L}$. We use (i) and try to solve the equation $T = E_1 S$. Direct computations give the solutions as stated in (ii). An alternative argument is given in Remark 3.13 (i) below.

(iii) To see this, the definition of S gives that it has idempotent index $\leq t-1$, if its idempotent index is $< t - 1$ then that makes idempotent index of $T \leq t - 1$, a contradiction.

(iv) Because $T = E_1 S$, we have that $\dim \mathcal{R}(T) \leq \dim \mathcal{R}(S)$. But $\dim \mathcal{R}(T) = \dim \mathcal{K}_1$. So, the result follows.

□

REMARK 3.13. Let $0 \neq T \in \mathcal{A}$ have a non-zero left annihilator and $T = \begin{bmatrix} T_1 & T_2 \\ 0 & 0 \end{bmatrix}$, a local block representation for T in $X = \mathcal{K} \oplus \mathcal{L}$ with $\mathcal{K} \supset \mathcal{K}_1 = \overline{\mathcal{K}_0}$, where $\mathcal{K}_0 = \mathcal{R}(T)$. We intend to determine if T is in $\mathcal{S}$ or not, if yes then we would like to obtain $ii(T)$ and factorizations of T into idempotents in $\mathcal{A}$. We display our line of thought and approach. We demonstrate it concretely for $\mathcal{F}_2$ in the next section.

(i) For any $B \in \mathcal{B}(\mathcal{L}, \mathcal{K})$, $Q_B = \begin{bmatrix} I_{\mathcal{K}} & B \\ 0 & 0 \end{bmatrix}$ satisfies $Q_B T = T$ and $Q_B \in \mathcal{E}_{\mathcal{K}}$. Right annihilators of Q_B are given by $\begin{bmatrix} -BC & -BD \\ C & D \end{bmatrix}$ in $X = \mathcal{K} \oplus \mathcal{L}$ with $C \in \mathcal{B}(\mathcal{K}, \mathcal{L})$ and $D \in \mathcal{B}(\mathcal{L})$. So, in view of Remark 3.5 (vi) $Q_B T_{B,C,D} = T$, where $T_{B,C,D} = \begin{bmatrix} T_1 - BC & T_2 - BD \\ C & D \end{bmatrix}$. This can be obtained by direct computation as well. So, we target $T_{B,C,D}$ for being in $\mathcal{S}$ or not, if yes, then we determine its idempotent index $t_{B,C,D}$ and conclude that $T \in \mathcal{S}$ with $ii(T) \leq t_{B,C,D} + 1$.

(ii) $C = 0, D = 0$ gives $S = \begin{bmatrix} T_1 & T_2 \\ 0 & 0 \end{bmatrix} = T$, which we do not want.

We would like to figure out $B \in \mathcal{B}(\mathcal{L}, \mathcal{K})$ and $(C, D) \neq (0, 0)$ in $\mathcal{B}(\mathcal{K}, \mathcal{L}) \times \mathcal{B}(\mathcal{L})$, for which $T_{B,C,D} \in \mathcal{F}$. As already noted in Lemma

2.5 (ii), this necessitates $\overline{\mathcal{R}(T_{B,C,D})} \neq X$, and $\mathcal{N}(T_{B,C,D}) \neq \{0\}$ and $\mathcal{N}(T) \neq \{0\}$.

(iii) Indeed, if $\mathcal{K}_1' = \overline{\mathcal{R}(T_{B,c,D})} \neq \mathcal{X}$, then we may repeat the discussion for T for this $T_{B,C,D}$ and have an idempotent Q' with range $\mathcal{K}'$ containing $\mathcal{K}_1'$ as a factor of $T_{B,C,D}$. By Lemma 3.7, we would like $\mathcal{K}' \setminus \mathcal{K} \neq \emptyset \neq \mathcal{K} \setminus \mathcal{K}'$. This leads to an iteration process, which may end or just continue.

$\mathcal{K}' \setminus \mathcal{K} \neq \emptyset$ can be ensured by $(C, D) \neq (0,0)$. For $\mathcal{K} \setminus \mathcal{K}' \neq \emptyset$ we must have $(\mathcal{R}(T_1 - BC) + \mathcal{R}(T_2 - BD)) \neq \mathcal{K}$.

(iv) Now,

$$\begin{aligned}
\mathcal{N}(T_{B,C,D}) \neq 0 \Leftrightarrow & \text{ for some } k \in \mathcal{K} \text{ and } l \in \mathcal{L}, 0 \neq k + l \in \mathcal{N}(T_{B,C,D}) \\
\Leftrightarrow & (T_1 - BC)k + (T_2 - BD)l = 0 \text{ and } Ck + Dl = 0 \\
\Leftrightarrow & T_1 k + T_2 l - B(Ck + Dl) = 0 \text{ and } Ck + Dl = 0, \\
\Leftrightarrow & T_1 k + T_2 l = 0 \text{ and } Ck + Dl = 0.
\end{aligned}$$

We may state that any one of the following conditions is sufficient, viz., $\mathcal{N}(T_1) \cap \mathcal{N}(C) \neq \{0\}$ or $\mathcal{N}(T_2) \cap \mathcal{N}(D) \neq \{0\}$ or , $\exists\ k, l$ with $0 \neq T_1 k = -T_2 l, 0 \neq Ck = -Dl$.

This implies $\mathcal{N}(T_1) \neq \{0\}$, or $\mathcal{N}(T_2) \neq \{0\}$,
or $\mathcal{R}(T_1) \cap \mathcal{R}(T_2) \neq \{0\}$ together with $\mathcal{R}(C) \cap \mathcal{R}(D) \neq \{0\}$.

4. New Classes of Elements of $\mathcal{F}_2$, the products of two idempotents in $\mathcal{A}$

We shall give some general results and some special examples of operators in $\mathcal{A}$ that are products of two idempotents. We begin with some direct applications of discussion in Sections 2 and 3 above. We freely use the notation, terminology, and results therein. For this section, $0 \neq T \neq I_X \in \mathcal{A}$ has non-zero left and right annihilators and $T = \begin{bmatrix} T_1 & T_2 \\ 0 & 0 \end{bmatrix}$ in $X = \mathcal{K} \oplus \mathcal{L}$, is a local block representation for T, unless otherwise stated. We will simply write $T = \begin{bmatrix} T_1 & T_2 \\ 0 & 0 \end{bmatrix}$, if no confusion can arise, take $\mathcal{N}(T) \neq 0$ and keep condition set (a), (b), (c) in definition 3.10 in mind. Because $T_1 = I_{\mathcal{K}}$ forces T to be an idempotent with range equal to $\mathcal{K}$, we usually avoid repetition and take $T_1 \neq I_{\mathcal{K}}$.

We emphasize the two idempotents can be chosen in infinitely many ways and we will not worry much about the rank of T being finite or not and the dimension of $\mathcal{K}$ or $\mathcal{L}$ being finite or not.

4.1. Products of two idempotents by inspection.

LEMMA 4.1. *Let $B \in \mathcal{B}(\mathcal{L}, \mathcal{K})$, and $C \in \mathcal{B}(\mathcal{K}, \mathcal{L})$. Then*

$$\begin{bmatrix} BC & B \\ 0 & 0 \end{bmatrix} \text{ and } \begin{bmatrix} 0 & 0 \\ C & CB \end{bmatrix}$$

are products of two idempotents in $\mathcal{A}$.

PROOF. Using Lemma 3.2 (v) and Remark 3.9 (iii),

$$E_1 = \begin{bmatrix} I_{\mathcal{K}} & B \\ 0 & 0 \end{bmatrix} \text{ and } E_2 = \begin{bmatrix} 0 & 0 \\ C & I_{\mathcal{L}} \end{bmatrix}$$

are idempotents in $\mathcal{A}$. Also,

$$\begin{bmatrix} BC & B \\ 0 & 0 \end{bmatrix} = E_1E_2 \quad \text{and} \quad \begin{bmatrix} 0 & 0 \\ C & CB \end{bmatrix} = E_2E_1.$$

□

THEOREM 4.2. *Let* $T = \begin{bmatrix} T_1 & T_2 \\ 0 & 0 \end{bmatrix}$ *with* $T_1 \in \mathcal{B}(\mathcal{K})$, $T_2 \in \mathcal{B}(\mathcal{L}, \mathcal{K})$, *etc. be as in Lemma 3.2 (iii) and* $\sigma = \begin{bmatrix} 0 & 0 \\ \sigma_3 & \sigma_4 \end{bmatrix}$ *with* $\sigma_3 \in \mathcal{B}(\mathcal{K}, \mathcal{L})$, $\sigma_4 \in \mathcal{B}(\mathcal{L})$.

(i) *If* $\mathcal{R}(T_1) \subset \mathcal{R}(T_2)$ *then we have the following.*
  (a) T *can be expressed as the product* E_1E_2 *of two idempotents* $E_1 = \begin{bmatrix} I_{\mathcal{K}} & T_2 \\ 0 & 0 \end{bmatrix}$, $E_2 = \begin{bmatrix} 0 & 0 \\ C & I_{\mathcal{L}} \end{bmatrix}$, *where* C *is given by Theorem D to satisfy* $T_1 = T_2C$.
  (b) $\dim \mathcal{K}_1 \leq \dim \mathcal{L}$.

(ii) *If* $\mathcal{R}(\sigma_4) \subset \mathcal{R}(\sigma_3)$, *then we have the following.*
  (a) σ *can be expressed as the product* F_1F_2 *of the idempotents* F_1 *and* F_2 *in* $\mathcal{A}$ *given by* $F_1 = \begin{bmatrix} 0 & 0 \\ \sigma_3 & I_{\mathcal{L}} \end{bmatrix}$, $F_2 = \begin{bmatrix} I_{\mathcal{K}} & B \\ 0 & 0 \end{bmatrix}$, *where* $B \in \mathcal{B}(\mathcal{L}, \mathcal{K})$ *is given by Theorem D to satisfy* $\sigma_4 = \sigma_3 B$.
  (b) *If* $\mathcal{L}_1$ *is the closure of* $\mathcal{R}(\sigma)$, *then* $\dim \mathcal{L}_1 \leq \dim \mathcal{K}$.

(iii) *If* X *is an indecomposable (infinite-dimension)al Banach space then under the conditions specified in (i) and (ii) above,* T *and* σ *have finite rank.*

PROOF. (i) (a) and (ii) (a) come from direct application of Theorem D and Lemma 4.1 above.

We only have to prove (ii) (b) and note that (i) (b) is similar. Indeed, $\mathcal{R}(\sigma_3) = \mathcal{R}(\sigma)$ is dense in $\mathcal{L}_1$. But $\mathcal{R}(\sigma_3)$ has dimension less than or equal to that of $\mathcal{K}$. So, $\dim \mathcal{L}_1 \leq \dim \mathcal{K}$. □

Example 2.9 (iv) (c) may be adapted to lead us to the following Theorem.

THEOREM 4.3. *Let* $T = \begin{bmatrix} T_1 & T_2 \\ 0 & 0 \end{bmatrix}$ *in* $X = \mathcal{K} \oplus \mathcal{L}$ *in some local representation. Suppose*

(a) $\mathcal{R}(T) = \mathcal{R}(T_1)$ *i.e.,* $\mathcal{R}(T_2) \subset \mathcal{R}(T_1)$ *and*
(b) $\mathcal{K} \subset_e \mathcal{L}$ *so that there exists an injective* $J \in \mathcal{B}(\mathcal{K}, \mathcal{L})$ *with a closed range* $\mathcal{L}_1 \in \hat{\mathcal{K}}$ *and a corresponding* J^* *in* $\mathcal{B}(\mathcal{L}, \mathcal{K})$ *with* $J^*J = I_{\mathcal{K}}$.

Then T *is product of two idempotents in* $\mathcal{A}$. *Moreover, these factorizations are distinct for distinct* J*'s. Indeed, absolute value 1 scalar multiples of* J *give infinitely many new choices to replace* (J, J^*).

PROOF. Because $\mathcal{R}(T_2) \subset \mathcal{R}(T_1)$, by Theorem D, there exists $B \in \mathcal{B}(\mathcal{L}, \mathcal{K})$ that satisfies $T_2 = T_1 B$. Let

$$U = \begin{bmatrix} I_{\mathcal{K}} & (T_1 - I_{\mathcal{K}})J^* + B \\ 0 & 0 \end{bmatrix}$$

and let

$$V = \begin{bmatrix} I_{\mathcal{K}} - BJ & B - BJB \\ J & JB \end{bmatrix}.$$

Then T is the product of idempotents U and V. □

REMARK 4.4.

(i) Taking $T_2 = 0$ in Theorem 4.3, and $\mathcal{R}(B) \subset \mathcal{N}(T_1)$, for instance, $B = 0$ in its proof, we arrive at ways to express $\begin{bmatrix} T_1 & 0 \\ 0 & 0 \end{bmatrix}$ as a product of two idempotents.

(ii) Conditions (a) and (b) in Theorem 4.3 above are sort of trade-off in the sense that (a) hints at $\mathcal{K}$ large enough and $\mathcal{L}$ small enough wehereas (b) indicates that $\mathcal{K}$ is small enough and $\mathcal{L}$ is large enough.

(iii) If X is indecomposable (infinite-dimensional) Banach space then the condition (b) in Theorem 4.3 above forces $\mathcal{K}$ to be finite-dimensional , and consequently T to be of finite rank.

4.2. Products of two idempotents via Consistency condions. Let us first make a few observations some of which look familiar or related to contents of Section 3, particularly Remark 3.5 (vi), Theorem 3.12 and Remark 3.13.

THEOREM 4.5. *Let $0 \neq T \in \mathcal{A}$ with a non-zero left annihilator.*

(α) *Let $T = \begin{bmatrix} T_1 & T_2 \\ 0 & 0 \end{bmatrix}$ in $X = \mathcal{K} \oplus \mathcal{L}$ in a local block representation form as in Definition 3.10, which is not an idempotent with range $\mathcal{K}$ (i.e., $T_1 \neq I_{\mathcal{K}}$). Then, T is product of two idempotents if there exist $B \in \mathcal{B}(\mathcal{L}, \mathcal{K})$, $C \in \mathcal{B}(\mathcal{K}, \mathcal{L})$, $D \in \mathcal{B}(\mathcal{L})$, that satisfy any of the following equivalent sets (i), (ii), (iii), (iv) of conditions.*

(i) $\mathcal{C}_1$: $S = T_{B,C,D} = \begin{bmatrix} T_1 - BC & T_2 - BD \\ C & D \end{bmatrix}$ *is an idempotent.*

(ii) $\mathcal{C}_2$: $\mathcal{C}_1$ *together with* $TS = T$.

(iii) $\mathcal{C}_3$: *Let* (4.1) *to* (4.6) *be as below.*

(a) The set $\{(4.1), (4.2), (4.3), (4.4)\}$ *holds, or*

(b) The set $\{(4.1), (4.2), (4.3), (4.4), (4.5), (4.6)\}$ *holds.*

$$(T_1 - BC)^2 + (T_2 - BD)C = T_1 - BC \tag{4.1}$$

$$(T_1 - BC)(T_2 - BD) + (T_2 - BD)D = T_2 - BD \tag{4.2}$$

$$C(T_1 - BC) + DC = C \tag{4.3}$$

$$C(T_2 - BD) + D^2 = D \tag{4.4}$$

$$T_1(T_1 - BC) + T_2C = T_1 \tag{4.5}$$

and

$$T_1(T_2 - BD) + T_2D = T_2 \tag{4.6}$$

(iv) $\mathcal{C}_4$: *Let* $U = T_1 - BC$, $V = T_2 - BD$ *and* (4.1′) *to* (4.6′) *be as below.*

(a) the set of conditions $\{(4.1'), (4.2'), (4.3'), (4.4')\}$ *holds or*

(b) The set of conditions $\{(4.1'), (4.2'), (4.3'), (4.4'), (4.6'), (4.6')\}$ *holds*

$$VC = (I_{\mathcal{K}} - U)U \equiv U(I_{\mathcal{K}} - U) \tag{4.1′}$$

$$UV = V(I_{\mathcal{L}} - D) \tag{4.2′}$$

$$CU = (I_{\mathcal{L}} - D)C \tag{4.3′}$$

$$CV = D(I_{\mathcal{L}} - D) \equiv (I_{\mathcal{L}} - D)D \tag{4.4′}$$

$$T_2C = T_1(I_{\mathcal{K}} - U) \tag{4.5′}$$

$$T_1V = T_2(I_{\mathcal{L}} - D) \tag{4.6′}$$

(β) *For the only if part, there exists some decomposition $X = \mathcal{K} \oplus \mathcal{L}$, a topological sum, with $\mathcal{K} \supset \mathcal{K}_0$ and conditions stated in (α) above.*

PROOF. (α) (i) and (ii) follow from Remark 3.13.

(iii) [(a)] For the set {(4.1), (4.2), (4.3), (4.4)}, it follows from (i) on expanding $S^2 = S$. (4.5) and (4.6) follow when we expand $TS = T$.

(iv) The equation (4.1) can be re-written as $U^2 + VC = U$, i.e., $VC = U - U^2 = (I_{\mathcal{K}} = U)U \equiv U(I_{\mathcal{K}} - U)$ which becomes (4.1′).

Similar computations give the remaining part.

(β) Suppose $T = QQ'$ for some Q, Q' in $\mathcal{E}$. Let $\mathcal{K} = \mathcal{R}(Q)$ and $\mathcal{L} = \mathcal{N}(Q)$. □

COROLLARY 4.6. *Following the notation of Theorem 4.5 above, we have more sets of equivalent conditions:*

(i) *Any one of the sets* $\{(4.3), (4.4), (4.5), (4.6)\}$, $\{(4.3), (4.4), (4.1), (4.6)\}$, $\{(4.3), (4.4), (4.2), (4.5)\}$ *holds.*

(ii) *Any of the sets* {(4.3′), (4.4′), (4.5′), (4.6′)}, {(4.3′), (4.4′), (4.1′), (4.6′)}, {(4.3′), (4.4′), (4.2′), (4.5′)} *holds.*

PROOF. For convenience of the reader we will break the propof into parts $(a), (b), (c)$, and (d).

(a) We rewrite equations (4.1) to (4.6) by making needed shifting to make R.H.S zero in each and call the new equations by the same numbering for notational convenience. We note that L.H.S of (4.5)−L.H.S of (4.1)= $B[C(T_1 - BC) + DC - C]$, So, if (4.3) holds then (4.1) and (4.5) become equivalent.

(b) Now, L.H.S of (4.6)−(4.2)= $B[C(T_2 - BD) + D^2 - D]$. So if (4.4) holds then (4.2) and (4.6) become equivalent.

(c) We come to (i). From (a) and (b) we get that $\{(4.1), (4.2), (4.3), (4.4)\}$ is equivalent to the system where (4.1) is replaced by (4.5) or (4.2) is replaced by (4.6). This gives that the remaining three systems in (i) are as good as the first one.

(d) Now (ii) follows from (i).

□

REMARK 4.7. We now proceed to identify T's which satisfy the conditions set out in Theorem 4.5 or Corollary 4.6 and figure out their factorizations into two idempotents. We work out details for some specific cases. For this purpose, we have already discarded $(C, D) = (0, 0)$ in Remark 3.13 (ii). Also we have seen a situation in Theorem 4.2 above, viz., $\mathcal{R}(T_1) \subset \mathcal{R}(T_2)$ for which $B = T_2$, $D = I_{\mathcal{L}}$, and C is such that $T_1 = T_2C$ does work. We give more factoriztions in the results that follow. We do not aim at most general results. Instead we give some sufficient conditions.

The proof of the following theorem follows from Remark 3.13 and direct computations as indicated in Remark 4.9.

THEOREM 4.8. *Let* $T = \begin{bmatrix} T_1 & T_2 \\ 0 & 0 \end{bmatrix}$ *in $X = \mathcal{K} \oplus \mathcal{L}$ in some local block representation with $T_1 \neq I_{\mathcal{K}}$ and $\mathcal{N}(T_1) \neq \{0\}$.*

(i) *Suppose* $\mathcal{R}(T_1) \subset \mathcal{R}(T_2)$ *and* $C \in \mathcal{B}(\mathcal{L},\mathcal{K})$ *as given by Theorem D to satisfy* $T_1 = T_2C$ *and* $\mathcal{N}(C) = \mathcal{N}(T_1)$. *Let* $V \in \mathcal{B}(\mathcal{L},\mathcal{K})$ *with* $\mathcal{R}(V) \subset \mathcal{N}(T_1)$. *Then* T *can be expressed as the product* E_1E_2 *of two idempotents* $E_1 = \begin{bmatrix} I_{\mathcal{K}} & T_2 - V \\ 0 & 0 \end{bmatrix}$, $E_2 = \begin{bmatrix} VC & V \\ C & I_{\mathcal{L}} \end{bmatrix}$. *Furthermore, there are uncountably many* V*'s and therefore, uncountably many factorizations corresponding to these* V*'s.*

(ii) *If* T_1 *is an idempotent, then* T *can be expressed as a product of two idempotents in uncountable number of ways:* $T = E_1E_2$, $E_1 = \begin{bmatrix} I_{\mathcal{K}} & T_2 - V \\ 0 & 0 \end{bmatrix}$, *and* $E_2 = \begin{bmatrix} T_1 + VC & V \\ C & I_{\mathcal{L}} \end{bmatrix}$, *where* $V \in \mathcal{B}(\mathcal{L},\mathcal{K})$ *and* $C \in \mathcal{B}(\mathcal{K},\mathcal{L})$ *satisfy* $\mathcal{R}(V) \subset \mathcal{N}(T_1), (\mathcal{R}(V) \cup \mathcal{R}(T_1)) \subset \mathcal{N}(C)$ *and* $\mathcal{R}(C) \subset \mathcal{N}(T_2)$.

REMARK 4.9. We proceed further with the case $D = I_{\mathcal{L}}$ and we look for more solutions envisaged in Theorem 4.5 (iv), which becomes

(1) $V = T_2 - BD = T_2 - B$. So, $B = T_2 - V$.

(2)

$$\begin{aligned} U &= T_1 - BC \\ &= T_1 - (T_2 - V)C \\ &= (T_1 - T_2C) + VC \\ &= W + VC, \end{aligned}$$

where $W = T_1 - T_2C$.

(3) The system (4.1′) to (4.6′) becomes

$$VC = U(I_{\mathcal{K}} - U) \tag{4.1''}$$

$$WV + VCV = 0 \tag{4.2''}$$

$$CW + CVC = 0 \tag{4.3''}$$

$$CV = 0 \tag{4.4''}$$

$$T_1W + T_1VC = W \tag{4.5''}$$

$$T_1V = 0. \tag{4.6''}$$

(4) Now (4.3″) and (4.4″) can be replaced by (4.3‴) and (4.4″), with (4.3‴) as below.

$$CW = 0. \tag{4.3'''}$$

Also, (4.5″) and (4.6″) can be replaced by (4.5‴) and (4.6″) with (4.5‴) as below.

$$T_1W = W, \text{ i.e., } (I_{\mathcal{K}} - T_1)W = 0. \tag{4.5'''}$$

(5) We may concentrate on (4.3‴), (4.4″), (4.5‴), and (4.6″).

Let $V \in \mathcal{B}(\mathcal{L},\mathcal{K})$ with $\mathcal{R}(V) \subset \mathcal{N}(T_1)$ keeping (4.6″) in mind. We note that $V = 0$ if $\mathcal{N}(T_1) = \{0\}$. Next let $C \in \mathcal{B}(\mathcal{K},\mathcal{L})$ that satisfies $\mathcal{R}(V) \subset \mathcal{N}(C)$ keeping (4.4″) in mind. The conditions (4.3‴) and (4.5‴) can be written as

$$C(T_1 - T_2C) = 0 \tag{C1}$$

$$(I_{\mathcal{K}} - T_1)(T_1 - T_2C) = 0 \tag{C2}$$

and will be called consistency conditions.

Suppose (C1) and (C2) are satisfied. Then, using the fact that $CV = 0$ and $T_1V = 0$, we get that $T = E_1E_2$, a product of two idempotents $E_1 = \begin{bmatrix} I_{\mathcal{K}} & T_2 - V \\ 0 & 0 \end{bmatrix}$, $E_2 = \begin{bmatrix} T_1 - T_2C + VC & V \\ C & I_{\mathcal{L}} \end{bmatrix}$, which are distinct if V's are so. For that $\mathcal{N}(T_1) \neq \{0\}$ is necessary.

(6) We record suitable obvious subcases for which the consistency conditions (C1) and (C2) are satisfied.
   (a) $T_1 - T_2C = 0$
   (b) $T_2C = 0, CT_1 = 0$, and T_1 is an idempotent.

(7) Now (vi) (a) holds for some $C \Leftrightarrow \mathcal{R}(T_1) \subset \mathcal{R}(T_2)$ by applying Theorem D. In this case, there exists a unique C satisfying $\mathcal{N}(T_1) = \mathcal{N}(C)$. We fix such a C now. So, all we have to do is to choose any $V \in \mathcal{B}(\mathcal{L}, \mathcal{K})$ with $\mathcal{R}(V) \subset \mathcal{N}(T_1)$.

   We note that all this is possible only if $\dim \mathcal{K}_1 \leq \dim \mathcal{L}$. As already noted $V = 0$ if $\mathcal{N}(T_1) = \{0\}$ and there are uncountably many V's available if $\mathcal{N}(T_1) \neq \{0\}$. (This validates Theorem 4.8(i))

(8) Next (6) (b) just requires extra care in (4) while choosing C, we have to make sure that $\mathcal{N}(C)$ contains $\mathcal{R}(V) \cup \mathcal{R}(T_1)$ and $\mathcal{R}(C) \subset \mathcal{N}(T_2)$.

(9) Now we come to the case when T_1 is an idempotent. So, $\mathcal{N}(T_1) = \{0\} \Leftrightarrow T_1 = I_{\mathcal{K}}$. But if $T_1 = I_{\mathcal{K}}$ then T is an idempotent and we may ignore this case and ensure $\mathcal{N}(T_1) \neq \{0\}$.

   We note that if $\mathcal{N}(T_2) = \{0\}$, then $C = 0$.

THEOREM 4.10. *Let* $T = \begin{bmatrix} T_1 & T_2 \\ 0 & 0 \end{bmatrix}$ *in* $X = \mathcal{K} \oplus \mathcal{L}$ *in some local block representation. Suppose that*

(a) T_1 *is an idempotent* $\neq I_{\mathcal{K}}$ *and*
(b) *Either* $\mathcal{N}(T_2) \neq \{0\}$ *or* $\mathcal{R}(T_1) \cap \mathcal{R}(T_2) \neq \{0\}$.
(c) $\dim \mathcal{L} \geq 2$.

Then we have the following.

(i) *There exist uncountably many idempotents* D *with* $\{0\} \neq \mathcal{N}(D) \neq \mathcal{L}$ *and* $\mathcal{N}(D) \subset \mathcal{N}(T_2 - T_1T_2)$.
(ii) *Let* D *be as in (i) above. Then* T *can be expressed as a product of two idempotents as follows.*
   (a) $T = E_1E_2$,
   (b) $T = E_1'E_2$,
   where

$$E_1 = \begin{bmatrix} I_{\mathcal{K}} & T_2 \\ 0 & 0 \end{bmatrix}, \quad E_2 = \begin{bmatrix} T_1 & T_2(I_{\mathcal{L}} - D) \\ 0 & D \end{bmatrix} \quad \text{and } E_1' = \begin{bmatrix} I_{\mathcal{K}} & T_2D \\ 0 & 0 \end{bmatrix}.$$

(iii) $E_1 = E_1' \Leftrightarrow \mathcal{N}(D) \subset \mathcal{N}(T_2) \implies \mathcal{N}(T_2) \neq \{0\}$.
(iv) *For distinct* D*'s as in (i) above, the corresponding* E_2*'s are distinct.*
(v) T *can be expressed as a product of two idempotents in uncountably many ways.*

PROOF. Once again direct computations give the results. □

In the next remark we give some more explanations related to the above results.

REMARK 4.11. We consider T defined on the topological sum $X = \mathcal{K} \oplus \mathcal{L}$, $T_1 \neq I_{\mathcal{K}}$ so that T is not an idempotent with range equal to $\mathcal{K}$ and consider the case $C = 0$, $D \neq 0$ other than those considered above.

(i) So $0 \neq D \neq I_{\mathcal{L}}$.

(ii) We work in the context of Theorem 4.5 (iv). So, $U = T_1$ and the system (4.1′)–(4.6′) becomes

$$T_1(I_{\mathcal{K}} - U) = 0, \tag{I}$$

$$T_1 V = V(I_{\mathcal{L}} - D), \tag{II}$$

(4.3′) is satisfied,

$$(I_{\mathcal{L}} - D)D = 0 \tag{IV}$$

$$T_1(I_{\mathcal{K}} - T_1) = 0 \tag{V$\equiv$I}$$

$$T_1 V = T_2(I_{\mathcal{L}} - D). \tag{VI}$$

(iii) (I) holds $\Leftrightarrow$ T_1 is an idempotent. In view of $T_1 \neq I_{\mathcal{K}}$, either T_1 is zero or $\dim(\mathcal{K}) \geq 2$. Also, (IV) holds $\Leftrightarrow$ D is an idempotent. In this case, (i) forces $\dim(\mathcal{L}) \geq 2$.

(VI) can be rewritten as

$$T_1T_2 - T_2 = (T_1B - T_2)D. \tag{VI′}$$

In other words, the system gives T_1 and D are idempotents, $T_1 \neq I_{\mathcal{K}}$, $0 \neq D \neq I_{\mathcal{L}}$ and (VI′) holds. We may call (VI′) as consistency condition C3.

(iv) Let T, B, D be as in (iii) above satisfying C3. Then T is the product E_1E_2 of two idempotents

$$E_1 = \begin{bmatrix} I_{\mathcal{K}} & B \\ 0 & 0 \end{bmatrix}, \quad E_2 = \begin{bmatrix} T_1 & T_2 - BD \\ 0 & D \end{bmatrix}.$$

(v) For $B = T_2$ or T_2D, we have $BD = T_2D$. So, RHS of (VI′) $= T_1T_2D - T_2D = (T_1T_2 - T_2)D$. So, (VI′) is satisfied if and only if

$$(I_{\mathcal{K}} - T_1)T_2(I_{\mathcal{L}} - D) = 0 \Leftrightarrow$$

$$\mathcal{N}(D) = \mathcal{R}(I_{\mathcal{L}} - D) \subset \mathcal{N}((I_{\mathcal{K}} - T_1)T_2). \tag{VI‴}$$

Because $\{0\} \neq \mathcal{N}(D) \neq \mathcal{L}$, this requires

$$\begin{aligned} &\mathcal{N}((I_{\mathcal{K}} - T_1)T_2) \neq \{0\} \\ \Leftrightarrow &\exists x \neq 0 \text{ with } T_2x = T_1T_2x \\ \Leftrightarrow &\exists x \neq 0 \text{ with } T_2x \in \mathcal{R}(T_1) \\ \Leftrightarrow &\mathcal{N}(T_2) \neq \{0\} \text{ or } \mathcal{R}(T_2) \cap \mathcal{R}(T_1) \neq \{0\}. \end{aligned}$$

Also, if this holds we may consider an idempotent D that satisfies (VI‴).

There are uncountably many such D's. The corresponding idempotents are:

$$E_1 = \begin{bmatrix} I_{\mathcal{K}} & T_2 \\ 0 & 0 \end{bmatrix}, \quad E_2 = \begin{bmatrix} T_1 & T_2 - T_2D \\ 0 & D \end{bmatrix}$$

and

$$E_1' = \begin{bmatrix} I_{\mathcal{K}} & T_2D \\ 0 & 0 \end{bmatrix}, \quad E_2' = \begin{bmatrix} T_1 & T_2 - T_2D \\ 0 & D \end{bmatrix}.$$

So, $E_2 = E_2'$ but $E_1 = E_1' \Leftrightarrow T_2 = T_2 D$ i.e., $T_2(I_{\mathcal{L}} - D) = 0$ i.e., $\mathcal{N}(D) = \mathcal{R}(I_{\mathcal{L}} - D) \subset \mathcal{N}(T_2)$. In any case, E_2's are different for different D's.

EXAMPLE 4.12. We continue with our discussion in the context of Remarks 4.9.and 4.11. We now come to the case $C \neq 0$, $D \neq 0$, $D \neq I_{\mathcal{L}}$, C and D both invertible.

(i) Invertibility of C forces $\dim \mathcal{K} = \dim \mathcal{L}$ and if $\dim X = n$ then this forces n even.

(ii) We work in the context of Theorem 4.5 (iv). Then (4.1′) to (4.4′) become equivalent to

$$(\alpha) \qquad V = U(I_{\mathcal{K}} - U)C^{-1}$$

$$(\beta) \qquad V = (I_{\mathcal{K}} - U)VD^{-1}$$

$$(\gamma) \qquad U = C^{-1}(I_{\mathcal{L}} - D)C = I_{\mathcal{K}} - C^{-1}DC$$

$$(\delta) \qquad V = C^{-1}D(I_{\mathcal{L}} - D).$$

Let us assume (γ) and (δ).

Then RHS of (α) is

$$(I_{\mathcal{K}} - C^{-1}DC)C^{-1}DCC^{-1} = C^{-1}D - C^{-1}D^2 = C^{-1}D(I_{\mathcal{L}} - D) = V = \text{LHS of } (\alpha).$$

So, (α) holds.

$$\text{RHS of } (\beta) = (C^{-1}DC)C^{-1}D(I_{\mathcal{L}} - D)D^{-1} = C^{-1}D(I_{\mathcal{L}} - D) = V = \text{LHS of } (\beta).$$

So, (β) is satisfied.

Thus the system can be replaced by (γ) and (δ).

(iii) Let us assume (γ) and (δ) hold. Then $T_1 - BC = U = I_{\mathcal{K}} - C^{-1}DC$. So, $BC = T_1 - I_{\mathcal{K}} + C^{-1}DC$. Thus $B = T_1C^{-1} - C^{-1} + C^{-1}D = T_1C^{-1} - C^{-1}(I_{\mathcal{L}} - D)$. Now, $T_2 - BD = V = C^{-1}D(I_{\mathcal{L}} - D)$. So, $BD = T_2 - C^{-1}(I_{\mathcal{L}} - D)D$. Thus, $B = T_2D^{-1} - C^{-1}(I_{\mathcal{L}} - D)$. Hence, we get a consistency condition $T_1C^{-1} - C^{-1}(I_{\mathcal{L}} - D) = T_2D^{-1} - C^{-1}(I_{\mathcal{L}} - D)$ i.e., $T_1C^{-1} = T_2D^{-1}$.

In particular, $\mathcal{R}(T_1) = \mathcal{R}(T_2)$.

Hence we may state the consistency condition as:

For some invertible operators $C' \in \mathcal{B}(\mathcal{L}, \mathcal{K})$ and $D' \in \mathcal{B}(\mathcal{L})$,

$$\text{(C5)} \qquad T_1C' = T_2D'.$$

Then we may choose $C = {C'}^{-1}$, $D = {D'}^{-1}$, $B = T_1C^{-1} - C^{-1}(I_{\mathcal{L}} - D)$. Thus, we almost have the following theorem.

THEOREM 4.13. *Let* $T = \begin{bmatrix} T_1 & T_2 \\ 0 & 0 \end{bmatrix}$ *in* $X = \mathcal{K} \oplus \mathcal{L}$, *and* $T_1 \neq I_{\mathcal{K}}$. *Suppose* $T_1C^{-1} = T_2D^{-1}$ *for some invertible operators* $C \in \mathcal{B}(\mathcal{K}, \mathcal{L})$, $D \in \mathcal{B}(\mathcal{L})$. *Then we have the following.*

(i) $\dim \mathcal{K} = \dim \mathcal{L}$ *and* $\mathcal{R}(T_1) = \mathcal{R}(T_2)$.

(ii) *(a) T is expressible as a product of two idempotents: $T = E_1E_2$,*

$$E_1 = \begin{bmatrix} I_{\mathcal{K}} & T_1C^{-1} - C^{-1}(I_{\mathcal{L}} - D) \\ 0 & 0 \end{bmatrix},$$

$$E_2 = \begin{bmatrix} I_{\mathcal{K}} - C^{-1}DC & C^{-1}(I_{\mathcal{L}} - D)D \\ C & D \end{bmatrix}.$$

(b) If X is finite-dimensional, then Dim$X$ is even, and if X is infinite-dimensional, then X is decomposable.

(iii) *If $\dim \mathcal{L}$ $(= \dim \mathcal{K}) \geq 2$ then (ii) can be done in uncountably many ways.*

PROOF. In view of Example 4.12 above, we have only to prove (iii). Indeed, there exist uncountably many invertible $D' \in \mathcal{B}(\mathcal{L})$. We may put $D'' = D'D$, $C'' = D'C$. Then, $T_1C''^{-1} = T_1C^{-1}D'^{-1} = T_2D^{-1}D'^{-1} = T_2D''^{-1}$. So, the idempotents corresponding to (C'', D'') in place of (C, D) do have their product as T only. □

We conclude the discussion of operators expressible as product of two idempotents even though some more cases remain to be seen.

Acknowledgments

Authors thank M. A. Sofi for useful discussion. We also thank Umashankara Kelathaya for his unusual help in putting the manuscript in tex form.

References

[1] A. Alahmadi, S. K. Jain, A. Leroy, and A. Sathaye, *Decompositions into products of idempotents*, Electron. J. Linear Algebra **29** (2015), 74–88, DOI 10.13001/1081-3810.2948. MR3414586

[2] M. Laura Arias, Gustavo Corach, and Alejandra Maestripieri, *Products of idempotent operators*, Integral Equations Operator Theory **88** (2017), no. 2, 269–286, DOI 10.1007/s00020-017-2363-0. MR3669130

[3] R. J. H. Dawlings, *The idempotent generated subsemigroup of the semigroup of continuous endomorphisms of a separable Hilbert space*, Proc. Roy. Soc. Edinburgh Sect. A **94** (1983), no. 3-4, 351–360, DOI 10.1017/S0308210500015717. MR709728

[4] R. G. Douglas, *On majorization, factorization, and range inclusion of operators on Hilbert space*, Proc. Amer. Math. Soc. **17** (1966), 413–415, DOI 10.2307/2035178. MR203464

[5] J. A. Erdos, *On products of idempotent matrices*, Glasgow Math. J. **8** (1967), 118–122, DOI 10.1017/S0017089500000173. MR220751

[6] J. M. Howie, *The subsemigroup generated by the idempotents of a full transformation semigroup*, J. London Math. Soc. **41** (1966), 707–716, DOI 10.1112/jlms/s1-41.1.707. MR219649

[7] S. K. Jain and A. Leroy, *Decomposition of singular elements of an algebra into product of idempotents, a survey*, Contributions in algebra and algebraic geometry, Contemp. Math., vol. 738, Amer. Math. Soc., [Providence], RI, [2019] ©2019, pp. 57–74, DOI 10.1090/conm/738/14878. MR4014964

[8] Piotr Koszmider, *On large indecomposable Banach spaces*, J. Funct. Anal. **264** (2013), no. 8, 1779–1805, DOI 10.1016/j.jfa.2013.01.015. MR3022726

[9] Joram Lindenstrauss, *Decomposition of Banach spaces*, Indiana Univ. Math. J. **20** (1971), no. 10, 917–919, DOI 10.1512/iumj.1971.20.20079. MR405074

[10] J. Lindenstrauss and L. Tzafriri, *On the complemented subspaces problem*, Israel J. Math. **9** (1971), 263–269, DOI 10.1007/BF02771592. MR276734

[11] Kiiti Morita, *Duality for modules and its applications to the theory of rings with minimum condition*, Sci. Rep. Tokyo Kyoiku Daigaku Sect. A **6** (1958), 83–142. MR96700

[12] Pei Yuan Wu, *The operator factorization problems*, Linear Algebra Appl. **117** (1989), 35–63, DOI 10.1016/0024-3795(89)90546-6. MR993030

Ohio University, Athens, Ohio
Email address: jain@ohio.edu
URL: sites.ohio.edu/jain

Université d'Artois, France
Email address: leroyandre7555@gmail.com
URL: http://leroy.perso.math.cnrs.fr/

INSA Emeritus Scientist, The Indian National Science Academy, Bahadur Shah Zafar Marg, New Delhi, 110 002, India
Email address: ajitis@gmail.com

Contemporary Mathematics
Volume **833**, 2026
https://doi.org/10.1090/conm/833/16665

Edge sudoku number of a graph—I

J. Maria Jeyaseeli, Jay S. Bagga, and S. Arumugam

Abstract. Let $G = (V, E)$ be a finite undirected connected simple graph of order n and size m with chromatic index $\chi'(G) \geq 2$. For any subset S of E, an edge coloring $\mathscr{C}_0$ of the edge induced subgraph $G[S]$ is called a partial edge coloring of G. The coloring $\mathscr{C}_0$ is called an edge sudoku coloring of G if $\mathscr{C}_0$ can be uniquely extended to a χ'-coloring of G. The minimum size of such an induced subgraph $G[S]$ of G which admits an edge sudoku coloring is called the edge sudoku number of G and is denoted by $esn(G)$. In this paper we present several fundamental results on this parameter.

1. Introduction

By a graph $G = (V, E)$ we mean a finite undirected connected simple graph of order $n = |V|$ and size $m = |E|$. For notations and concepts not defined here we refer to the book [**2**].

A proper edge coloring of a graph G is an assignment of colors to the edges of G such that adjacent edges are assigned distinct colors. The chromatic index $\chi'(G)$ of a graph G is the least positive integer k such that G admits a proper edge coloring with k colors. Vizing [**5**] established the following remarkable result on the chromatic index of a graph.

Theorem 1.1. [**5**] *Let G be a nonempty graph with maximum degree $\Delta(G)$. Then $\chi'(G) \leq 1 + \Delta(G)$.*

Trivially $\chi'(G) \geq \Delta(G)$ and hence it follows from Vizing's theorem that $\chi'(G) = \Delta(G)$ or $\Delta(G) + 1$.

Definition 1.2. A graph G is said to be of class one if $\chi'(G) = \Delta(G)$ and is of class two if $\chi'(G) = \Delta(G) + 1$.

The following theorem is due to König.

Theorem 1.3. ([**2**], Page 461) *Every nonempty bipartite graph is of class one.*

We observe that the chromatic index $\chi'(G)$ is the chromatic number of its line graph $L(G)$. We will need this connection with line graphs again in Section 2 when we define the edge sudoku number. For an extensive description of line graphs we refer to [**1**].

2020 *Mathematics Subject Classification.* Primary 05C78; Secondary 05C69.
Key words and phrases. Edge coloring, chromatic index, line graph, edge sudoku coloring, edge sudoku number.

Motivated by the popular sudoku puzzle, Maria et al. [**4**] introduced the concept of sudoku number of a graph.

DEFINITION 1.4. [**4**] Let G be a connected graph of order n with chromatic number $\chi(G) = k \geq 2$. Let $S \subseteq V$ and let $G[S]$ be the induced subgraph of G. Let $\mathscr{C}_0$ be a proper k-coloring of $G[S]$. We also call $\mathscr{C}_0$ a partial coloring of G. The coloring $\mathscr{C}_0$ is called an extendable coloring of $G[S]$ if $\mathscr{C}_0$ can be extended to a k-coloring for G. $\mathscr{C}_0$ is called a sudoku coloring of G if $\mathscr{C}_0$ can be uniquely extended to a proper k-coloring of G. The smallest order of such an induced subgraph $G[S]$ of G which admits a sudoku coloring is called the sudoku number of G and is denoted by $sn(G)$.

The concept of the sudoku number of a graph is motivated by the popular sudoku puzzle, which consists of 81 cells arranged in 9 rows and 9 columns, which are subdivided into 9 subsquares of order 3×3 each. Some cells are filled with distinct integers 1 through 9. A solution of the puzzle consists of all 81 cells being filled with integers 1 through 9

Thus solving a sudoku puzzle is simply the extension of a partial coloring to a proper χ-coloring of G. In [**4**] the authors formulated the sudoku coloring as a list coloring problem. The concept of list coloring was independently studied by Erdös et al. [**3**] and Vizing [**6**].

DEFINITION 1.5. Given a graph G and a set $L(v)$ of colors for each vertex v, which is called a color list of v, a coloring of G is called a list coloring if the vertex v is colored by a color in the list $L(v)$. If G admits a list coloring for a given list L, then G is called list colorable or L-colorable.

If $\mathscr{C}_0$ is a sudoku coloring of G, then the set of all possible colors that can be assigned to an uncolored vertex is a color list of v and the problem of extending $\mathcal{C}_0$ to a k-coloring of G is equivalent to a list coloring problem.

We need the following theorems.

THEOREM 1.6. [**4**] *Let G be a connected graph of order n. Then $sn(G) = 1$ if and only if G is a bipartite graph.*

THEOREM 1.7. [**4**] *Let G be a connected graph of order n. Then $sn(G) = n - 1$ if and only if $G = K_n$.*

THEOREM 1.8. [**4**] *For $n \geq 3$,*

$$sn(C_n) = \begin{cases} 1 & \text{if } n \text{ is even} \\ \frac{n+1}{2} & \text{if } n \text{ is odd.} \end{cases}$$

In this paper we extend the notion of the sudoku number of a graph in the context of edge-coloring and initiate a study of this parameter.

2. Edge Sudoku Number of Graph

Unless mentioned otherwise, in this paper G denotes a connected graph of order n and size $m \geq 2$. It follows that $\Delta(G) \geq 2$.

DEFINITION 2.1. Let G be a connected graph of order n with chromatic index $\chi'(G) = k \geq 2$. Let $S \subseteq E(G)$ and let $G[S]$ be the edge-induced subgraph of G. Let $\mathscr{C}_0$ be an edge-coloring of $G[S]$. We call $\mathscr{C}_0$ a partial edge-coloring of G. The coloring $\mathscr{C}_0$ is called an extendable edge-coloring of G if $\mathscr{C}_0$ can be extended to a k-edge-coloring for G. We say that $\mathscr{C}_0$ is an edge sudoku coloring of G if $\mathscr{C}_0$ can be uniquely extended to a k-edge-coloring of G. The smallest number of edges in such an edge induced subgraph $G[S]$ of G which admits an edge sudoku coloring is called the edge-sudoku number of G and is denoted by $esn(G)$.

OBSERVATION 2.2.

(1) For a graph with $\chi'(G) = k$, any partial edge-coloring of G using at most $k-2$ colors admits two different extensions. Hence, when G is of class one, $esn(G) \geq \chi'(G) - 1 = \Delta(G) - 1$, and when G is of class two, $esn(G) \geq \chi'(G) - 1 = \Delta(G)$.

(2) We note that $esn(G) = sn(L(G))$, where $L(G)$ is the line graph of G.

Our first result below presents several basic properties of $esn(G)$ that follow from the definition.

THEOREM 2.3. *Let G be a connected graph of order n and size $m \geq 2$.*

(1) *If G is of class one, then $\Delta(G) - 1 = \chi'(G) - 1 \leq esn(G) \leq m - 1$.*

(2) *If G is of class two, then $\Delta(G) = \chi'(G) - 1 \leq esn(G) \leq m - 1$.*

(3) *$esn(G) = 1$ if and only if G is a path or an even cycle.*

(4) *$esn(G) = m - 1$ if and only if $G = K_{1,m}$ or K_3.*

PROOF. (1) and (2) follow from the definition of $esn(G)$ and from Observation 2.2.

(3) The line graph $L(G)$ is bipartite if and only if G is a path or an even cycle. Hence the result follows from Theorem 1.6.

(4) The line graph $L(G)$ is complete if and only if $G = K_{1,m}$ or K_3. Hence the result follows from Theorem 1.7. □

EXAMPLE 2.4.

(i) Let $G = K_{1,4} + e$. Then $esn(G) = 3$. Clearly $\Delta(G) = \chi'(G) = 4$. It follows from Theorem 2.3 that $esn(G) \geq 3$. Also the partial coloring $\mathscr{C}_0$ given in Figure 1 can be uniquely extended to a 4-edge coloring of G. Hence $esn(G) = 3$. Thus $esn(G) = \Delta(G) - 1 = \chi'(G) - 1$ and this shows that the lower bound in (2) of Theorem 2.3 is sharp.

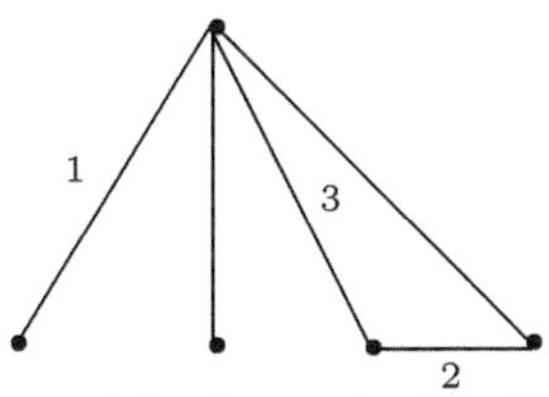

FIGURE 1. A partial edge coloring $\mathscr{C}_0$ of $G = K_4 + e$

(ii) $esn(K_4 - e) = 2$. Clearly $\chi'(K_4 - e) = \Delta(K_4 - e) = 3$. Also the partial edge coloring $\mathscr{C}_0$ given in Figure 2 can be uniquely extended to a 3-edge coloring of $K_4 - e$. Hence it follows that $esn(K_4 - e) = 2$.

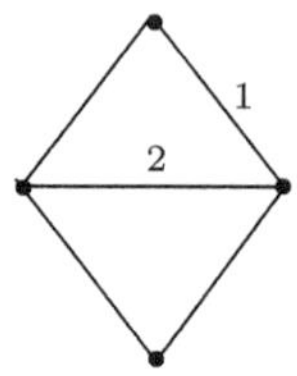

FIGURE 2. A partial edge coloring $\mathscr{C}_0$ of $G = K_4 - e$

(iii) For the hypercube Q_3, a partial edge coloring $\mathscr{C}_0$ and its unique extension to a 3-edge coloring are given in Figure 3. Hence $esn(Q_3) = 2$.

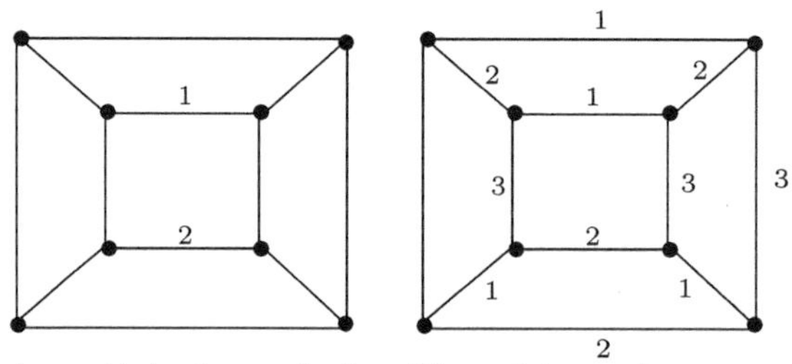

FIGURE 3. A partial edge coloring $\mathscr{C}_0$ and its unique extension for Q_3

(iv) A graph G, a partial edge coloring $\mathscr{C}_0$ and its unique extension to a 3-edge coloring are given in Figure 4. Also if $\mathscr{C}_0$ is any partial coloring of G in which all the three edges of a triangle are not assigned a color, then $\mathscr{C}_0$ has two extensions to a 3-edge coloring of G. Hence $esn(G) = 3$.

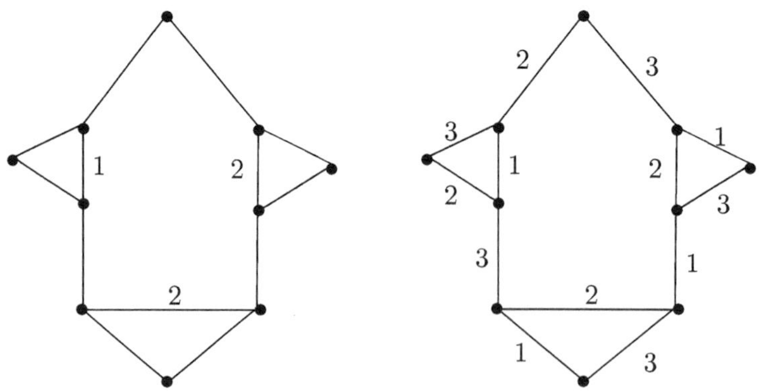

FIGURE 4. A partial edge coloring $\mathscr{C}_0$ and its unique extension for G.

The following theorem gives an infinite family of graphs with $esn(G) = 2$.

THEOREM 2.5. *Let $G = P_r \Box K_2$. Then $esn(G) = 2$*

PROOF. Let $(v_1, v_2, \cdots, v_r)$ and $(w_1, w_2, \cdots, w_r)$ be two copies of P_r in G, where $v_i w_i$ are edges in G. Let $\mathscr{C}_0 = \{E_1, E_2\}$, where $E_1 = \{v_1 v_2\}$ and $E_2 = \{w_1 w_2\}$. Then $\mathscr{C}_0$ is a partial edge-coloring of G and $\mathscr{C} = \{E_1, E_2, E_3\}$, where

$E_1 = \{v_iv_{i+1} : i\ odd, 1 \leq i \leq r\} \cup \{w_iw_{i+1} : i\ even, 1 \leq i \leq r\}$, $E_2 = \{v_iv_{i+1} : i\ even, 1 \leq i \leq r\} \cup \{w_iw_{i+1} : i\ odd\}$, and $E_3 = \{v_iw_i : 1 \leq i \leq r\}$ is the unique extension of $\mathscr{C}_0$ to a 3-edge-coloring of G. Hence $esn(G) \leq 2$. It follows from (2) of Theorem 2.3 that $esn(G) = 2$. □

COROLLARY 2.6. *The graph $G = P_r \square K_2$ is of class one.*

THEOREM 2.7. *Let $G = C_r \square K_2$. Then* $esn(G) = \begin{cases} 2 & \text{if } r \text{ is even} \\ \frac{r+1}{2} & \text{if } r \text{ is odd.} \end{cases}$

PROOF. Let $H_1 = (v_1, v_2, \ldots, v_r, v_1)$ and $H_2 = (w_1, w_2, \ldots, w_r, w_1)$ be two copies of C_r in G and let $v_iw_i \in E(G)$. Let $M = \{v_iw_i : 1 \leq i \leq r\}$. We consider two cases.

CASE 1 (r is even). Let $\mathscr{C}_0 = \{E_1, E_2\}$, where $E_1 = \{v_1v_2\}$ and $E_2 = \{w_1w_2\}$. Then $\mathscr{C}_0$ is a partial edge-coloring of G and $\mathscr{C} = \{E_1, E_2, E_3\}$, where $E_1 = \{v_iv_{i+1} | i\ odd, 1 \leq i < r\} \cup \{w_iw_{i+1} | i\ even, 2 \leq i \leq r\}$, $E_2 = \{v_iv_{i+1} | i\ even, 2 \leq i \leq r\} \cup \{w_iw_{i+1} | i\ odd, 1 \leq i < r\}$, and $E_3 = \{v_iw_i | 1 \leq i \leq r\}$, and the addition of the subscripts is taken modulo r, is the unique extension of $\mathscr{C}_0$ to a 3-edge-coloring of G. Hence $esn(G) \leq 2$. It follows from (2) of Theorem 2.3 that $esn(G) = 2$.

CASE 2 (r is odd). Let $e_i = v_iv_{i+1}$ where addition in the suffix is taken modulo r. Let $\mathscr{C}_0 = \{E_1, E_2, E_3\}$ where $E_1 = \{e_i : i \equiv 1 (mod\ 4)\}, E_2 = \{e_i : i \equiv 3(mod\ 4)\}$ and $E_3 = \{e_{r-1}\}$ be the 3-edge coloring defined on the induced subgraph $H_1[S]$. Then $\mathscr{C}_0$ is an edge sudoku coloring of H_1 and let $\mathscr{C}_1$ be the unique extension of $\mathscr{C}_0$ to a 3-edge coloring of H_1. The case for $r = 9$ is given in Figure 5.

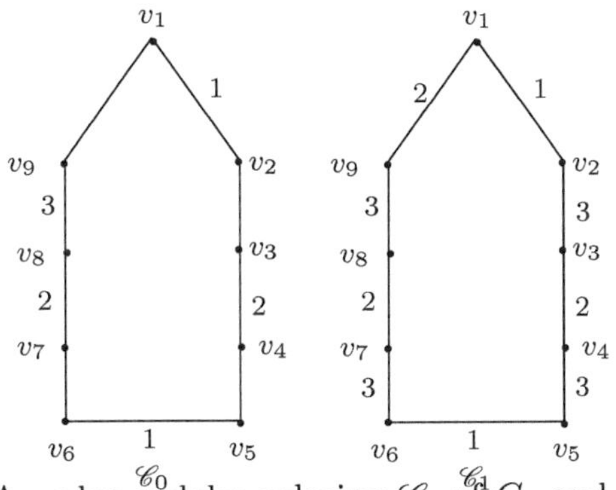

FIGURE 5. An edge sudoku coloring $\mathscr{C}_0$ of C_9 and its unique extension $\mathscr{C}_1$.

Since e_1 and e_r have colors 1 and 2, the edge v_1w_1 is forced to have color 3. Now v_2w_2 gets color 2 and hence w_1w_2 gets color 1. Proceeding like this we get a unique extension of $\mathscr{C}_0$ to a 3-edge coloring $\mathscr{C}$ of G. This is illustrated in Figure 6. Hence $\mathscr{C}_0$ is an edge sudoku coloring of G and $|S| = \frac{r+1}{2}$. Thus $esn(G) \leq \frac{r+1}{2}$.

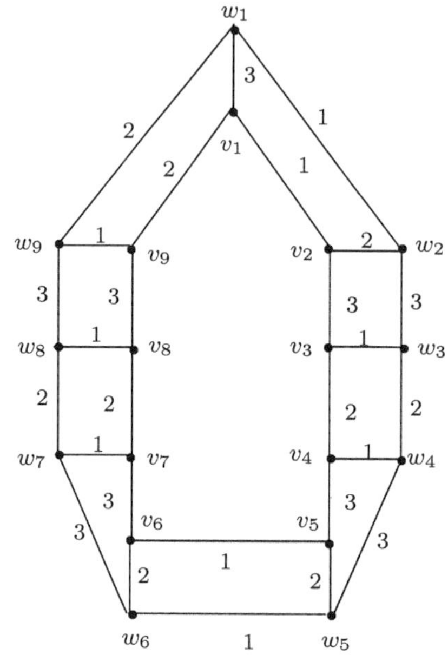

FIGURE 6. The unique extension $\mathscr{C}$ of $\mathscr{C}_0$ to $G = C_9 \Box K_2$.

Now, let $\mathscr{C}_0 = \{E_1, E_2, E_3\}$ be any 3-edge sudoku coloring defined on an edge induced subgraph $G[S]$. If $S \subseteq E(H_1)$ or $S \subseteq E(H_2)$, then $\mathscr{C}_0$ is an edge sudoku coloring the cycle C_r. Hence by Theorem 1.8, $|S| \geq \frac{r+1}{2}$. Now, suppose $S \subseteq M$. Without loss of generality, let $v_1w_1 \in S \cap E_1$. If $v_2w_2 \in S \cap E_2$, then the edges which are uniquely colored are shown in Figure 7.

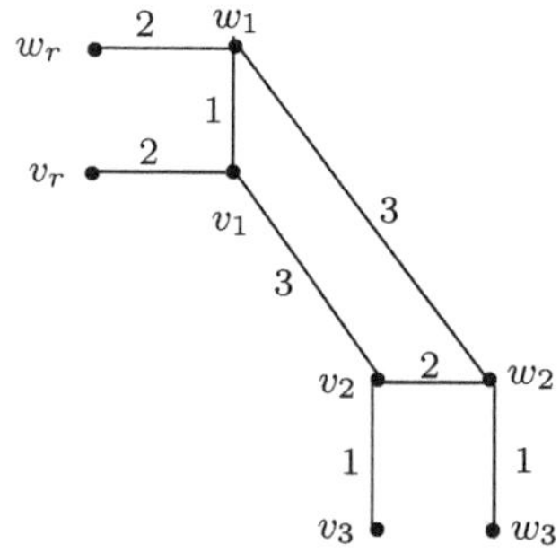

FIGURE 7. The extension process when $v_1w_1 \in E_1$ and $v_2w_2 \in E_2$.

Since $|N\{v_1w_1, v_2w_2\}| = 6$, it follows if two consecutive edges $v_iw_i, v_{i+1}w_{i+1}$ of M are in two different color classes in $\mathscr{C}_0$, then exactly six edges receive unique color in the extension process and these edges are in $E(H_1) \cup E(H_2)$. Hence it follows that two consecutive edges, say v_1w_1 and v_2w_2 and at least one of the edges $v_iw_i, v_{i+1}w_{i+1}$ must be in S for each $i, 3 \leq i \leq r$. Hence $|S| \geq \frac{r+1}{2}$. Now, let S be any subset of $E(G)$. Since $\mathscr{C}_0$ is a sudoku coloring, there exist two edges x_1 and x_2 in S such that $N(x_1) \cap N(x_2) \neq \emptyset$. The number of edges in G whose color is uniquely determined by the edges x_1 and x_2 is at most 6. In the extension process, any edge e of S determines uniquely the color of an edge if at least one edge of $N(e)$ has already received a color and the number of such edges is at most 3. $\square$

COROLLARY 2.8. *The graph* $G = C_r \square K_2$ *is of class one.*

We now proceed to determine the edge sudoku number of a a few classes of trees.

DEFINITION 2.9. Let $G = (V, E)$ be graph of order n with $\delta(G) = 1$. A vertex v is called a support vertex if there exists a pendent vertex adjacent to v. A support vertex v is called a strong support vertex if the number of pendent vertices adjacent to v is at least 2.

LEMMA 2.10. *Let G be a graph with $\chi'(G) \geq 3$. Let $\mathscr{C}_0$ be an edge sudoku coloring defined on $G[S]$. Let v be a strong support vertex and let A be the set of all pendent edges incident with v.*

(i) *If $deg(v) \leq \Delta(G) - 2$, then $A \subseteq S$.*
(ii) *If $deg(v) \leq \Delta(G) - 1$ and G is of class two, then $A \subseteq S$.*
(iii) *If $deg(v) \leq \Delta(G)$, then S contains all edges of A except for one edge.*

PROOF.

(i) Let $L(e)$ denote the list of colors that can be assigned to the edge e.
If there exists $e \in A$ such that $e \notin S$, then $|L(e)| \geq 2$ and hence we get two extensions of $\mathscr{C}_0$, which is a contradiction. Hence $A \subseteq S$.
(ii) Since G is of class two, $\chi'(G) = \Delta + 1$ and as in (i) we get $A \subseteq S$.
(iii) Let $deg(v) = \Delta$. Suppose there exist two edges e_1 and e_2 in $A - S$. Since $\chi' \geq 3$, it follows that $|L(e_1)| \geq 2$ and $|L(e_2)| \geq 2$ and we get two extensions of $\mathscr{C}_0$. Hence at most one edge of A is not in S. $\square$

THEOREM 2.11. *Let $G = B_{r,s}$ be a bistar. Then $esn(G) = r + s - 1$.*

PROOF. Let $V(G) = \{u, v, u_1, u_2, \ldots, u_r, v_1, v_2, \ldots, v_s\}$ and

$$E(B_{r,s}) = \{uv, uu_1, \ldots, uu_r, vv_1, vv_2, \ldots, vv_s\}.$$

Let $A_1 = \{uu_1, uu_2, \ldots, uu_r\}$ and $A_2 = \{vv_1, vv_2, \ldots, vv_s\}$. It follows from Theorem 1.3 that G is of class one.

We consider three cases.

CASE 1 ($r = s = 1$). In this case $G = P_4 = (u_1, u, v, v_1)$. Now if $\mathscr{C}_0 = \{E_1\}$ where $E_1 = \{uv\}$, then $\mathscr{C} = \{E_1, E(G) - E_1\}$ is the unique extension of $\mathscr{C}_0$ to a χ'-edge coloring of G. Hence $esn(G) = 1 = r + s - 1$.

CASE 2 ($r < s$). Let $\mathscr{C}_0$ be an edge sudoku coloring defined on an edge induced subgraph $G[S]$. It follows from Lemma 2.10 that $A_1 \subseteq S$ and $A_2 - \{e\} \subseteq S$. Thus $|S| \geq r + s - 1$ and hence $esn(G) \geq r + s - 1$. Also $\mathscr{C}_0 = \{E_1, E_2\}$ where $E_1 = A_1$ and $E_2 = A_2 - \{vv_1\}$ is an edge sudoku coloring of G. Hence $esn(G) \leq r + s - 1$ and so $esn(G) = r + s - 1$.

CASE 3 ($r = s$ and $r \geq 2$). In this case $deg(u) = deg(v) = r + 1 = \Delta$.

Let $\mathscr{C}_0 = \{E_1, E_2, E_3, \ldots, E_{r-1}, E_r\}$ where $E_i = \{uu_i, vv_i\}, 1 \leq i \leq r - 1$ and $E_r = \{uv\}$.

Clearly $\mathscr{C}_0$ is an edge sudoku coloring of G and hence $esn(G) \leq r + s - 1$. Also it follows from Lemma 2.10 that $esn(G) \geq (r - 1) + (s - 1) = r + s - 2$. Also if three edges uu_r, uv and vv_r are uncolored edges, then the color list of all these three edges is $\{r, r+1\}$ and we can get two extensions of G. Hence $esn(G) \geq r+s-1$. Thus $esn(G) = r+s-1$. □

We now proceed to determine the edge sudoku number of complete binary trees.

THEOREM 2.12. *Let B_h be the complete binary tree of height h. Then*

$$esn(B_h) = \begin{cases} 1 & \text{if } h = 1 \\ 3 & \text{if } h = 2 \\ 3(2^{h-2}) & \text{if } h \geq 3. \end{cases}$$

PROOF. Let $\{e_{i_1}, e_{i_2}, \ldots, e_{i_{2^i}}\}$ be the set of edges of B_h joining a vertex at level $i-1$ to a vertex at level i, where the edges are listed in an order from left to right. Clearly $B_1 = P_3$ and $esn(B_1) = 1$. Now, let $h \geq 2$. The proof is by induction on h. Clearly $\Delta' = \chi' = 3$ and B_h is of class one. We consider two cases.

CASE 1 ($h = 2$). An edge sudoku coloring $\mathscr{C}_0$ of B_2 and its unique extension to a 3-edge coloring are given in Figure 8.

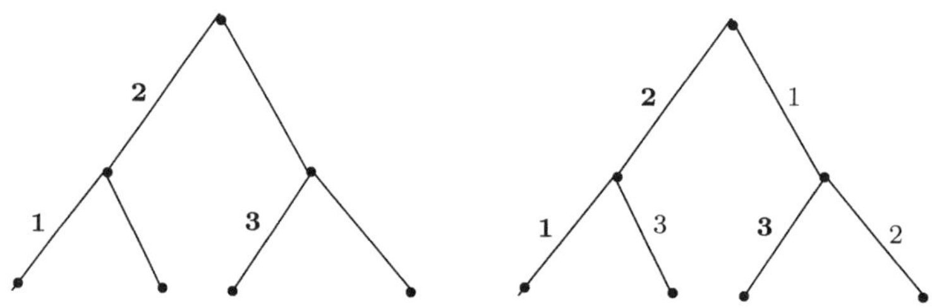

FIGURE 8. Edge sudoku coloring and its unique extension to B_2

Hence $esn(B_2) \leq 3$. Also it follows from Lemma 2.10 that $esn(B_2) \geq 2$. Now let $\mathscr{C}_0$ be any partial coloring such that only two edges are colored. If both pendent edges incident at a vertex are not colored, then $\mathscr{C}_0$ has two different extensions. Hence we may assume without loss of generality that $\mathscr{C}_0 = \{E_1, E_2\}$ where $E_1 = \{e_{2_1}\}$ and $E_2 = \{e_{2_3}\}$. Again $\mathscr{C}_0$ admits two different extensions. Hence $esn(B_2) \geq 3$ and thus $esn(B_2) = 3$.

CASE 2 ($h \geq 3$). Let v_1 be the root of B_h. Let v_2 and v_3 denote respectively the left and right child of v_1 respectively. Let T_1 and T_2 be the subtrees of B_h rooted at v_2 and v_3 respectively. Since T_1 and T_2 are complete binary trees of height $h-1$, by induction hypothesis, there exist edge sudoku colorings $\mathscr{C}_1$ for T_1 defined on $T_1[S_1]$ and $\mathscr{C}_2$ for T_2 defined on $T_2[S_2]$ such that $|S_1| = |S_2| = 3(2^{h-3})$.

We may assume without loss of generality, that in the unique extension of $\mathscr{C}_1$ to T_1 the two edges incident at v_1 receive colors 1 and 3 and in the unique extension of $\mathscr{C}_2$ to T_2, the two edges incident at v_2 receive colors 2 and 3. Now the partial edge coloring $\mathscr{C} = \mathscr{C}_1 \cup \mathscr{C}_2$ is an edge sudoku coloring of B_h and $|\mathscr{C}| = 2 \cdot 3 \cdot 2^{h-3} = 3 \cdot 2^{h-2}$. Hence $esn(B_h) \leq 3 \cdot 2^{h-2}$.

Now let $\mathscr{C} = \{E_1, E_2, E_3\}$ be an edge sudoku coloring of B_h such that $|E_1 \cup E_2 \cup E_3|$ is minimum. Let $e_1 = v_1v_2$ and $e_2 = v_1v_3$ be the edges of B_h incident with the root vertex v_1 of B_h. We consider two subcases.

SUBCASE 2a ($e_1, e_2 \notin E_1 \cup E_2 \cup E_3$). In this case $\mathscr{C} = \mathscr{C}_1 \cup \mathscr{C}_2$ where $\mathscr{C}_1$ and $\mathscr{C}_2$ are partial edge colorings of T_1 and T_2 respectively. Since $\mathscr{C}$ is an edge sudoku coloring of B_h, it follows that $\mathscr{C}_1$ and $\mathscr{C}_2$ are edge sudoku colorings of T_1 and T_2 respectively. By induction hypothesis $|\mathscr{C}_1| \geq 3(2^{h-3})$ and $|\mathscr{C}_2| \geq 3(2^{h-3})$. Hence $|\mathscr{C}| \geq 3(2^{h-2})$. Thus $esn(B_h) \geq 3(2^{h-2})$ and so $esn(B_h) = 3 \cdot (2^{h-2})$.

SUBCASE 2b (One of e_1, e_2 is in $E_1 \cup E_2 \cup E_3$). Without loss of generality we may assume that $e_1 \in E_1$. Let $e_3 = v_2v_4$ and $e_4 = v_2v_5$ be the edges of T_1 incident at its root vertex v_2. If $e_3, e_4 \in E_2 \cup E_3$, then $\mathscr{C}_1 = \{E_1 - \{e_1\}, E_2, E_3\}$ is a sudoku edge coloring of B_h and $|\mathscr{C}_1| < |\mathscr{C}|$ which is a contradiction. If exactly one of the edge e_3, e_4, say $e_3 \in E_2$, then in the extension process, e_4 receives color 3. Now $\mathscr{C}_1 = \{E_1 - \{e_1\}, E_2, E_3 \cup \{e_4\}\}$ is a sudoku edge coloring of B_h and it follows from Subcase 2a that $|\mathscr{C}| = |\mathscr{C}_1| \geq 3(2^{h-2})$.

Now suppose $e_3, e_4 \notin E_2 \cup E_3$. Clearly $L(e_3) = L(e_4) = \{2, 3\}$. Since $\mathscr{C}$ is an edge sudoku coloring of B_h, for forcing the colors for e_3 and e_4, we may assume without loss of generality that one edge incident at v_4 is in E_3 and one edge incident at v_5 is in E_2. This forces e_3 to receive color 2 and e_4 to receive color 3. Noe $\mathscr{C}_1 = \{E_1 - \{e_1\}, E_2 \cup \{e_3\}, E_3\}$ is an edge sudoku coloriing of B_h and it follows from subcase 2a that $|\mathscr{C}_1| = |\mathscr{C}| = 3(2^{h-2})$. Thus $esn(B_h) = 3(2^{h-2})$ and the proof is complete. □

3. Conclusion and Scope

One of the intriguing conjectures on sudoku number of a graph is that $sn(K_n \square K_n) = \left\lfloor \frac{n^2}{4} \right\rfloor$, which is mentioned in [**7**]. Since $L(K_{r,s}) \cong K_r \square K_s$, the above conjecture is equivalent to the statement that $esn(K_{n,n}) = \left\lfloor \frac{n^2}{4} \right\rfloor$. Hence the computation of $esn(K_{r,s})$ is a significant problem for further investigation. We list below some other problems for further investigation.

PROBLEM 3.1. Compute the edge sudoku number of classes of trees such as caterpillar and lobster.

PROBLEM 3.2. Characterize graphs G with $esn(G) = 2$.

PROBLEM 3.3. Characterize the class of graphs for which $esn(G) = \chi'(G) - 1$.

PROBLEM 3.4. Given three positive integers a, b and c with $3 \leq a \leq b \leq c$, does there exist a graph G with $\chi'(G) = a - 1, esn(G) = b$ and $|E(G)| = c - 1$?

References

[1] L. W. Beineke and J. S. Bagga, *Line graphs and line digraphs*, Developments in Mathematics, vol. 68, Springer, Cham, 2021. With a foreword by Gary Chartrand, DOI 10.1007/978-3-030-81386-4. MR4381305

[2] G. Chartrand, L. Lesniak, and P. Zhang, *Graphs and digraphs*, 6th ed., Textbooks in Mathematics, CRC Press, Boca Raton, FL, 2016. MR3445306

[3] P. Erdős, A. L. Rubin, and H. Taylor, *Choosability in graphs*, Proceedings of the West Coast Conference on Combinatorics, Graph Theory and Computing (Humboldt State Univ., Arcata, Calif., 1979), Congress. Numer., XXVI, Utilitas Math., Winnipeg, MB, 1980, pp. 125–157. MR593902

[4] J. M. Jeyaseeli, G. C. Lau, W. C. Shiu, and S. Arumugam, *Sudoku number of graphs*, AKCE Int. J. Graphs Comb. **20** (2023), no. 2, 209–216, DOI 10.1080/09728600.2023.2218917. MR4637193

[5] V. G. Vizing, *On an estimate of the chromatic class of a p-graph* (Russian), Diskret. Analiz **3** (1964), 25–30. MR180505

[6] V.G. Vizing, *Vertex colorings with given colors*, Metody Diskret, Analiz (Russian), **29** (1976), 3–10.

[7] J. A. Bate and G. H. J. van Rees, *The size of the smallest strong critical set in a Latin square*, Ars Combin. **53** (1999), 73–83. MR1724489

National Centre for Advanced Research in Discrete Mathematics (n-CARDMATH), Kalasalingam Academy of Research and Education, Anand Nagar, Krishnankoil-626126, Tamilnadu, India.

Email address: mariyajeyaseeli@gmail.com

Department of Computer Science, Ball State University, Muncie, Indiana 47306, USA

Email address: jbagga@bsu.edu

National Centre for Advanced Research in Discrete Mathematics (n-CARDMATH), Kalasalingam Academy of Research and Education, Anand Nagar, Krishnankoil-626126, Tamilnadu, India.

Email address: s.arumugam.klu@gmail.com

Contemporary Mathematics
Volume **833**, 2026
https://doi.org/10.1090/conm/833/16666

On subspaces of Banach spaces which admit unique norm preserving extensions under various embeddings

Sainik Karak, Tanmoy Paul, and T. S. S. R. K. Rao

Abstract. We study Banach spaces X, that under the canonical embedding in the bidual X^{**}, have unique norm-preserving extensions for norm attaining functionals. This property is called weak Hahn-Banach smoothness of a Banach space. Our study also extend to the case, when the sequence space c_0 can be embedded into classical Banach spaces like $\mathcal{B}(\ell^p), \mathcal{K}(\ell^p)$ for $1 < p < \infty$, and to $C(\Omega)$ spaces, with uniqueness of extensions for norm attaining functionals (when this happens such a subspace is called a (wU) subspace). On the other hand by using matrix representation of compact operators in $\mathcal{K}(\ell^1)$, we show that the canonical embedding of c_0 fails this property. We show that weak Hahn-Banach smoothness is separably determined. It is demonstrated that the property (wU), under some additional hypothesis on the component spaces, is preserved in spaces of Bochner integrable spaces and tensor product spaces.

1. Introduction

Let X be a real Banach space and let $Y \subseteq X$ be a closed subspace. In this paper we study uniqueness of norm-preserving extensions of functionals in Y^* that attain their norm on Y, to X^*. A starting point of the literature in this area is a seminal paper by Robert Phelps ([**17**]) where he investigates the so called *property* (U) for the first time. Another land mark work by A. Lima ([**13**]) characterizes uniqueness of norm-preserving extensions from various geometric viewpoints.

Definition 1.1. Let Y be a subspace of a Banach space X. Y is said to have,

(1) *property* (U) if every linear functional on Y has unique norm preserving extension over X.
(2) *property* (wU) if every linear functional on Y which attains its norm has unique norm preserving extension over X.

The classical space of null sequences, c_0 is an appropriate space to study uniqueness of norm preserving Hahn-Banach extensions to a super space, as it occurs as

2020 *Mathematics Subject Classification.* Primary 47L05, 40H05, 46A22, 46B20, 46E15; Secondary 46B22, 46M05, 41A52.

Key words and phrases. Hahn-Banach smooth, weak Hahn-Banach smooth, spaces of operators, matrix representation of operators, unique best approximations, Bochner integrable functions, tensor product spaces, spaces of operators.

a subspace, in a canonical fashion in several spaces of functions and spaces of operators. In this paper we investigate properties (U) and (wU), for the sequence space c_0, under various embeddings into function spaces and spaces of compact and bounded operators, with surprisingly different consequences. When a Banach space X has property (U) ((wU)) in its canonical embedding X^{**}, X is said to be Hahn-Banach smooth (weakly Hahn-Banach smooth). For a Banach space X, by $NA(X)$ we denote the set of functionals in X^* that attain their norm on the unit ball of X. When Y is a subspace of X and $y^* \in Y^*$, by $HB(y^*)$ we denote the set $\{x^* \in X^* : x^*|_Y = y^*, \|y^*\| = \|x^*\|\}$. $S(X), X_1$ stand for the closed unit sphere and closed unit ball of X respectively. For $x \in X$ and $r > 0$, $B(x,r)$ represents the closed ball in X centered at x and radius r.

DEFINITION 1.2. A Banach space X is said to be *Hahn-Banach smooth* if every linear functional on X has unique norm preserving extension over X^{**}.

In [**20**], Sullivan studied Hahn-Banach smoothness of Banach spaces. It is well-known that spaces which are Hahn-Banach smooth share nice geometric properties; viz. they are Asplund spaces and the w^* and w-topologies coincide on the unit sphere $S(X^*)$ of X^* (see [**10**, Lemma 3.2.4]). See also Chapter III in [**10**] for several examples of Hahn-Banach smooth spaces from among function spaces and spaces of compact operators. In particular, the sequence space c_0 and its non-commutative analogue, space of compact operators, $\mathcal{K}(H)$ for an infinite dimensional Hilbert space H, are Hahn-Banach smooth spaces. These spaces play a significant role in understanding the geometric properties of spaces of compact operators on Banach spaces. It was shown in [**2**] that this property is preserved under c_0-sums.

Motivated by Phelps result that $Y \subseteq X$ is a U subspace, if and only if $Y^\perp$ is Chebyshev subspace of X^* (i.e every $x^* \in X^*$ that is not in $Y^\perp$ has unique best approximation in $Y^\perp$), in Section 2 we show that $Y \subseteq X$ is a (wU) subspace if and only if for every $x^* \in X^*$ that attains its norm on Y, 0 is the unique best approximation from $Y^\perp$. Analogous to the work in [**2**], we show that weakly Hahn-Banach smoothness is also a separably determined property, in the sense that if every separable subspace of X is weakly Hahn-Banach smooth, then so is X.

The space c_0 also has other interesting properties like, it embeds by naturally defined into isometries, in function spaces and spaces of operators. In Section 3, we show that for $1 < p < \infty$, the canonical embedding of c_0 as multiplication operators, in the space of compact operators $\mathcal{K}(\ell^p)$ has property (wU) where as this fails in $\mathcal{K}(\ell^1)$.

In Section 4 we give some topological conditions on a compact set Ω so that a well studied embedding of c_0 in $C(\Omega)$ has unique Hahn-Banach extensions. In the complex scalar case, the corresponding result for function algebras (i.e., closed subalgebras of $C(\Omega)$ containing constants and separating points) are still open. See the recent article [**9**] and its zbMATH Open, review by Rao.

Stability properties of (wU) subspaces are considered in Section 5. In spaces of Bochner integrable functions, we show that if X^* has the Radon-Nikodým property (RNP, see [**4**] for this and for the general theory of spaces of Bochner integrable functions), if $Y \subseteq X$ is a (wU) subspace, then so is $L^p(\mu, Y)$ in $L^p(\mu, X)$ for $1 < p < \infty$.

2. Various aspects of property (wU) in Banach spaces

We have the following characterization of property (wU). For a subspace W of a Banach space Z, we recall the metric projection map is defined by $P_W : Z \to 2^W$ by $P_W(z) = \{w \in W : \|w - z\| = d(z, W)\}$. Here 2^W represents the set of all closed and bounded subsets of W. Note that $P_W(z)$ may be empty, we call W proximinal if $P_W(z) \neq \emptyset$ for all z. Clearly a w^*-closed subspace of a dual space is proximinal.

THEOREM 2.1. *Let Y be a subspace of X then the following are equivalent.*

(1) *Y has property (wU) in X.*
(2) *For every proximinal subspace $Z \subseteq Y$, Y/Z has property (wU) in X/Z.*
(3) *Let $x^* \in X^*$ be such that for some $y \in S(Y)$, $x^*(y) = \|x^*\|$. Then $P_{Y^\perp}(x^*) = \{0\}$.*
(4) *For $i = 1, 2$, let $x_i^* \in NA(X)$ be such that $x_i^*(y_i) = 1$ for some $y_i \in S(Y)$ and also $x_1^* - x_2^* \in Y^\perp$ then $x_1^* - x_2^* = 0$.*

PROOF. (1) $\iff$ (2) $\iff$ (3) follows from [**7**, Theorem 2.1].

(1) $\Rightarrow$ (4). Suppose that $y^* = x_1^*|_Y = x_2^*|_Y$. Then $y^* \in NA(Y)$ and $y^* \in S(Y^*)$. By (1), y^* has unique norm preserving extension to X. Hence $x_1^* = x_2^*$.

(4) $\Rightarrow$ (3). Suppose that, $x^* \in S(X^*)$ such that $x^*(y) = 1$ for some $y \in S(Y)$. Let $y^* = x^*|_Y$. Then $y^* \in S(Y^*)$. Suppose that there exists $z^* \in P_{Y^\perp}(x^*)$ such that $z^* \neq 0$. Now $\|x^* - z^*\| = d(x^*, Y^\perp) = \|x^*|_Y\| = 1$. Hence $x^* - z^* \in NA(X)$, as $(x^* - z^*)(y) = 1$. Now from (4) we have, $z^* = x^* - (x^* - z^*) = 0$, a contradiction. This completes the proof. □

Let us recall the following from Theorem 3.1 in [**16**].

PROPOSITION 2.2. Let X be a subspace of Y, let $x^* \in S(X^*)$. Suppose $HB(x^*)$ is contained in the set of points of weak*-weak continuity on $S(Y^*)$, then x^* is a point of weak*-weak continuity of $S(X^*)$.

THEOREM 2.3. *Let Y be Hahn-Banach (weak Hahn-Banach) smooth and X be a subspace of Y. Then X is also Hahn-Banach (weak Hahn-Banach) smooth.*

PROOF. We only prove the case for weak Hahn-Banach smoothness. Let (x_α^*) be a net in S_{X^*} and $x^* \in S(X^*) \cap NA(X)$. Let $x_\alpha^* \to x^*$ in w^*-sense in X^*. Let $\tilde{x}^* \in HB(x^*)$ be an extension over Y. Since $\tilde{x}^* \in S(Y^*) \cap NA(Y)$, $\tilde{x}^*$ is a point of $w^* - w$-continuity of Y^*. Hence from Proposition 2.2 it follows x^* is a point of $w^* - w$-continuity of of X^*. This completes the proof. □

We now prove that the property weak *Hahn-Banach smoothness* is separably determined. For $x \in S(X)$, we denote $D(x) = \{x^* \in S(X^*) : x^*(x) = 1\}$. Let us recall the following.

DEFINITION 2.4 ([**1**]). X is said to have property $P_1^* - III$, if for any convergent sequence (x_n) in $S(X)$ and any $x_n^* \in D(x_n)$ where $n \in \mathbb{N}$, we have $\cap_{n\geq 1}\overline{conv}\{x_k^* : k \geq n\} \neq \emptyset$.

THEOREM 2.5. *The property weak Hahn-Banach smoothness is separably determined.*

PROOF. It remains to prove that X has property $P_1^* - III$ (see [**1**, Remark 3.3]). Let (x_n) be any convergent sequence in S_X and for $n \geq 1$, let $x_n^* \in D(x_n)$. We show that $\cap_{n\geq 1}\overline{conv}\{x_k^* : k \geq n\} \neq \emptyset$. Choose $Y = \overline{span}\{x_n : n \in \mathbb{N}\}$. Clearly,

Y is a separable subspace of X and x_n^* attains its norm on Y_1, for all $n \in \mathbb{N}$. Now from [**19**, Proposition 2] there exists a separable $Y' \supseteq Y$ and a linear mapping $T : (Y')^* \to X^*$ such that for each $f \in (Y')^*$, $T(f)$ is a norm preserving extension of f and $T(Y')^* \supseteq \overline{span}\{x_n^* : n \in \mathbb{N}\}$, the norm closure of $span\{x_n^* : n \in \mathbb{N}\}$. Now since Y' is separable so it has property $P_1^* - III$. Now let $y_n^* = x_n^*|_{Y'}$, hence $T(y_n^*) = x_n^*$ and $y_n(= x_n) \in S(Y')$ for all $n \in \mathbb{N}$. Finally, we get a convergent sequence (y_n) in $S(Y')$ and $y_n^* \in D(y_n)$, so $\cap_{n\geq 1}\overline{conv}\{y_n^* : k \geq n\} \neq \emptyset$ and hence the result follows. □

REMARK 2.6. A simple way of getting multiple norm-preserving extensions is to consider a proper subspace Y of X and take Z be any non zero Banach space, then $Y \oplus_1 Z$ (ℓ^1-sum) does not have property (wU) in $X \oplus_1 Z$.

We recall that a linear projection $P : X \to X$ is called a L-projection, if $\|P(x)\| + \|x - P(x)\| = \|x\|$ for all $x \in X$. The following corollary is now easy to see. The following are some more easy consequences of Theorem 2.3.

COROLLARY 2.7. *If X is non-reflexive and weakly Hahn-Banach smooth, then X has no nontrivial L-projection.*

PROPOSITION 2.8. *If for a Banach space X, every equivalent renorming is weakly Hahn-Banach smooth, then X is reflexive.*

PROOF. Let $X = Y \oplus Z$ be a nontrivial direct sum, then the norm $\|x\|_1 = \|y\| + \|z\|$, where $y \in Y$ and $z \in Z$ is an equivalent norm on X. Clearly this new norm has a non-trivial L-projection. Therefore every non-reflexive space can be renormed to fail weakly Hahn-Banach smooth. □

REMARK 2.9. It is easy to see that weakly Hahn-Banach smoothness is not a three space property, for example c_0 which is a Hahn-Banach smooth space is a hyperplane in c which is not Hahn-Banach smooth.

In the following Proposition, we will use the fact that for an infinite compact set K, $C(K)$ fails to be Hahn-Banach smooth. The detailed proof of this will be supplied in Section 4, Proposition 4.4

PROPOSITION 2.10. *Let X be a Banach space which is weakly Hahn-Banach smooth. Then $C(K, X)$ is weakly Hahn-Banach smooth if and only if K is finite.*

PROOF. From Corollary 5.6 we get the sufficiency. We now prove the condition is necessary. Fixing a unit vector, $x_0 \in X$, it is clear that $C(K) \subseteq C(K, X)$ as functions valued in $span\{x_0\}$. Since $C(K, X)$ is weakly Hahn-Banach smooth, so is $C(K)$. From Proposition 4.4 it follows that K is discrete. As K is compact, it is a finite set. □

PROBLEM 2.11. We do not know analogues of Phelps [**17**] results for weak*-closed subspaces. In particular if every weak*-closed subspace of X^* is a (wU) subspace, is X^{**} a strictly convex space? Since any finite dimensional subspace is weak*-closed, such an assumption in particular implies that X^* is a smooth space, there by X is a strictly convex space.

3. Property (wU) in operator spaces

We first consider the natural embedding of the Banach space c_0 as multiplication operators into $\mathcal{K}(\ell^p), 1 < p < \infty$. Let us recall that the isometric identification of the dual space as the projective tensor product space, $\mathcal{K}(\ell^p)^* \cong \ell^p \hat{\otimes}_\pi \ell^q$, where $\frac{1}{p} + \frac{1}{q} = 1$ (see pg.114 in [**18**]). For a Banach space X, in this paper we follow the notation $x \otimes x^*$ for $x \in X$ and $x^* \in X^*$ to denote the functional on space of bounded operators, $\mathcal{B}(X)$ defined by , $(x \otimes x^*)(T) = x^*(T(x))$ for $T \in \mathcal{B}(X)$.

It is known the set of extreme points of the dual unit ball is given by

$$ext\ \mathcal{K}(X)_1^* = \{\Lambda \otimes x^* : \Lambda \in ext\ \mathcal{B}(X^{**}),\quad x^* \in ext\ \mathcal{B}(X^*)\}.$$

See Theorem VI.1.3 in [**10**].

In what follows we use that fact that in $c_0^* = \ell^1$, the set of norm attaining functionals is the subspace spanned by the coordinate functionals $e_n \in ext\ \ell_1^1$. Also by an application of the Krein-Milman theorem, an extreme point of the dual unit ball of a subspace $Y \subseteq X$, has a norm-preserving extension to an extreme point of the dual unit ball of X. We also use the same coordinate functional notation in all sequence spaces in this sequel.

THEOREM 3.1. *The subspace c_0 has property (wU) in $\mathcal{K}(\ell^p)$ where $1 < p < \infty$.*

PROOF. We will first prove that if $\Lambda \in ext\ \ell_1^1$ then Λ has a unique norm preserving extension to $\mathcal{K}(\ell^p)$. Hence it is enough to prove that for a fixed n, for $e_n \in \ell^1$, the set of norm-preserving extension $HB(e_n) \subseteq \mathcal{K}(\ell^p)_1^*$, which is always a weak*-closed extreme convex set, is singleton.

We recall that $x \otimes y \in \mathcal{B}(\ell^p)$, for $x \in \ell^p, y \in \ell^q$ is defined by $x \otimes y(T) = \langle Tx, y\rangle$. Now for $\alpha \in c_0$, the multiplication operator $M_\alpha \in \mathcal{K}(\ell^p)$ and clearly $\langle M_\alpha(\zeta), e_n\rangle = \alpha_n\zeta_n$, for all $(\zeta_k) \in \ell^p$. Hence $e_n \otimes e_n \in HB(e_n)$ and is one extreme point of this set. Now suppose $\Lambda \in ext\ HB(e_n)$ is another extreme point, then $\Lambda \in ext\ B_{\mathcal{K}(\ell^p)^*}$ and hence there exist $x \in \ell^p, y \in \ell^q$ where $\|x\|_p = 1 = \|y\|_q$ such that $\Lambda = x \otimes y$. Therefore as $\Lambda|_{c_0} = e_n$ we have for any $M_\alpha \in \mathcal{K}(\ell^p)$, $\Lambda(M_\alpha) = \langle M_\alpha x, y\rangle = \alpha_n x_n y_n = \alpha_n$. This concludes that $x = e_n = y$. Hence $ext\ HB(e_n) = \{e_n \otimes e_n\}$ and consequently $HB(e_n)$ is singleton.

As remarked before if $\Lambda \in \ell^1$ is a norm-attaining unit vector, we may assume that $\Lambda = \sum_1^n \alpha_i e_i$ where $\alpha_i \in (0, 1]$ and $\sum_1^n \alpha_i = 1$. It is easy to verify that $\sum_1^n \alpha_i(e_i \otimes e_i)$ which is a norm preserving extension of Λ is unique by showing that it is the only extreme point of $HB(\Lambda)$. □

Let $1 < p < \infty$. We next consider ℓ^∞ as canonically embedded into $\mathcal{B}(\ell^p)$ by means of multiplication operators. We recall the natural decompositions $(\ell^\infty)^* \cong \ell^1 \oplus_{\ell^1} c_0^\perp$ and that of $\mathcal{B}(\ell^p)^* \cong \mathcal{K}(\ell^p)^* \oplus_{\ell^1} \mathcal{K}(\ell^p)^\perp$, see [**10**] Chapter VI. These decompositions show that c_0 $(\mathcal{K}(\ell^p))$ is a U subspace of ℓ^∞ $(\mathcal{B}(\ell^p))$. Thus it is natural to study uniqueness of extensions under these embeddings.

COROLLARY 3.2. As the space of multiplication operators, c_0 has property (wU) in $\mathcal{B}(\ell^p)$ for $1 < p < \infty$.

PROOF. We show that if $\Lambda \in ext\ \ell_1^1$ then Λ has unique norm preserving extension over $\mathcal{B}(\ell^p)$. From the proof of Theorem 3.1 it is clear that if $\Lambda = e_n$ then it has unique norm preserving extension to $\mathcal{K}(\ell^p)$. In view of the decomposition noted before this Corollary, we see that the extension is also unique in $\mathcal{B}(\ell^p)^*$. Now the conclusion follows as in the proof of Theorem 3.1. □

We next recall a result from [**13**] that we will be using below.

THEOREM 3.3 ([**13**]). *Let M be a closed subspace of X. Then the following are equivalent.*

(1) *M has property (wU) in X.*
(2) *If $x \in M, y \in X$ with $\|x\| = 1 = \|y\|$ and $\varepsilon > 0$ then there exists $r \geq 1$ and $z \in M$ such that $B(y - z + rx, r + \varepsilon) \cap B(y - z - rx, r + \varepsilon) \neq \emptyset$.*

In the following theorem, we use the well known infinite matrix representation of elements of $\mathcal{B}(\ell^1)$. See [**14**].

FACT 3.4. Let $M = (k_{i,j})$ be an infinite matrix with scalar entries. Then M defines a bounded linear map $M : \ell^1 \to \ell^1$, provided $\alpha = \sup_{i\geq 1}\{\sum_{j=1}^{\infty}|k_{i,j}|\} < \infty$, and in this case $\|M\|_1 = \alpha$.

We now derive the following as a consequence of Theorem 3.3.

PROPOSITION 3.5. The canonical embedding of c_0 as multiplication operators in $\mathcal{K}(\ell^1)$ does not have property (wU) in $\mathcal{K}(\ell^1)$.

PROOF. Let $\mathcal{C}$ be the image of the canonical embedding of c_0 in $\mathcal{K}(\ell^1)$.

Choose $T = \begin{pmatrix} 1 & 1 & \cdot & 0 & \cdot \\ 0 & 0 & \cdot & 0 & \cdot \\ \cdot & \cdot & \cdot & \cdot & \cdot \\ \cdot & \cdot & \cdot & \cdot & \cdot \end{pmatrix} \in \mathcal{K}(\ell^1)$ with $\|T\| = 1$ and

$$S = \begin{pmatrix} 0 & 0 & \cdot & 0 & \cdot \\ 0 & 1 & \cdot & 0 & \cdot \\ \cdot & \cdot & \cdot & \cdot & \cdot \\ \cdot & \cdot & \cdot & \cdot & \cdot \end{pmatrix} \in \mathcal{C}, \text{ where } \|S\| = 1.$$

Choose $\varepsilon = \frac{1}{2}$. If possible assume that there exists $r \geq 1$ such that

$$B(T + rS, r + \varepsilon) \cap B(T - rS, r + \varepsilon) \cap \mathcal{C} \neq \emptyset.$$

Now if $A = \begin{pmatrix} a & 0 & \cdot & 0 & \cdot \\ 0 & b & \cdot & 0 & \cdot \\ \cdot & \cdot & c & \cdot & \cdot \\ \cdot & \cdot & \cdot & \cdot & \cdot \end{pmatrix} \in B(T + rS, r + \frac{1}{2}) \cap B(T - rS, r + \frac{1}{2})$ for some $(a, b, c, \dots) \in c_0$. Then,

$$\|T + rS - A\| \leq r + \frac{1}{2} \text{ and } \|T - rS - A\| \leq r + \frac{1}{2}$$

Now $T + rS - A = \begin{pmatrix} 1-a & 1 & \cdot & 0 & \cdot \\ 0 & r-b & \cdot & 0 & \cdot \\ \cdot & \cdot & -c & \cdot & \cdot \\ \cdot & \cdot & \cdot & \cdot & \cdot \end{pmatrix}$ and

$$T - rS - A = \begin{pmatrix} 1-a & 1 & \cdot & 0 & \cdot \\ 0 & -r-b & \cdot & 0 & \cdot \\ \cdot & \cdot & -c & \cdot & \cdot \\ \cdot & \cdot & \cdot & \cdot & \cdot \end{pmatrix}.$$

Now,
$\|T + rS - A\| \leq r + \frac{1}{2}$ implies $\max\{|1 - a|, 1 + |r - b|, |c|, |d|, \dots\} \leq r + \frac{1}{2}$ and
$\|T - rS - A\| \leq r + \frac{1}{2}$ implies $\max\{|1 - a|, 1 + |r + b|, |c|, |d|, \dots\} \leq r + \frac{1}{2}$

Then the above two inequalities imply there exists b and $r \geq 1$ such that

$$r + \frac{1}{2} \geq \max\{1 + |r+b|, 1 + |r-b|\}$$

But for any b and any $r \geq 1$ we have $r + \frac{1}{2} < \max\{1 + |r+b|, 1 + |r-b|\}$.

Hence $B(T + rS, r + \frac{1}{2}) \cap B(T - rS, r + \frac{1}{2}) \cap \mathcal{C} = \emptyset$ and hence $\mathcal{C}$ can not have property (wU) in $\mathcal{K}(\ell^1)$. □

As a consequence of Proposition 3.5 we have the following.

COROLLARY 3.6. The canonical image of ℓ^∞ in $\mathcal{B}(\ell^1)$ can not have property (wU) in $\mathcal{B}(\ell^1)$.

We next consider these questions between spaces of operators. In the following proposition we assume the approximation property (AP) on Y. Similar conclusion also holds when one assumes the AP on X^*. See [**4**] Chapter VIII.

PROPOSITION 3.7. Let $\mathcal{K}(X, Y)$ be Hahn-Banach smooth and suppose Y has the approximation property. Then $\mathcal{K}(X, Y)$ has property (U) in $\mathcal{B}(X, Y)$.

PROOF. It is well known that if Y has the approximation property (AP), then $\mathcal{K}(X, Y) = X^* \otimes_\varepsilon^\vee Y$ [**8**, Theorem 16.35(6)]. So from this identification, we have X^*, Y are subspaces of $\mathcal{K}(X, Y)$. Since Hahn-Banach smoothness is hereditary property, so X^* and Y are Hahn Banach smooth. This implies X is reflexive and Y^* has RNP, see [**20**].

Now as Y has the AP and Y^* has RNP, so by [**8**, Theorem 16.40], we have, $(X^* \otimes_\varepsilon^\vee Y)^* = X^{**} \otimes_\pi^\wedge Y^* = X \otimes_\pi^\wedge Y^*$. Hence we get, $\mathcal{K}(X, Y)^* = X \otimes_\pi^\wedge Y^*$.

Now $\mathcal{K}(X, Y)^{**} = (X \otimes_\pi^\wedge Y^*)^* = \mathcal{B}(X, Y^{**})$ by [**8**, Proposition 16.16].

Finally we have, $\mathcal{K}(X, Y) \subseteq \mathcal{B}(X, Y) \subseteq \mathcal{B}(X, Y^{**})$ and hence $\mathcal{K}(X, Y)$ has property (U) in $\mathcal{B}(X, Y)$. □

4. Embedding into space of type $C(\Omega)$

In this section we consider embeddings of c_0 into continuous function spaces, it also illustrates how the nature of the embedding is important for properties considered in this paper. We refer to a recent paper [**3**] where the authors also study embeddings into spaces of continuous functions. See zbMATH Open review by Rao.

THEOREM 4.1. *Let K be a compact Hausdorff space with countably many isolated points. Then there exists an embedding of c_0, $T : c_0 \to C(K)$ such that $T(c_0)$ is a U subspace of $C(K)$*

PROOF. Let us assume that $(t_n) \subseteq K$ be a countable family of isolated points. Then define $T : c_0 \to C(K)$ by the following.

For $(\alpha_n) \in c_0$, define $T((\alpha_n)) = \sum_n \alpha_n \chi_{\{t_n\}}$. Clearly T is an isometry into. We now prove that T is a (U) embedding.

STEP 1. Consider $e_n \in S(\ell^1)$ and let $\hat{e}_n$ be the corresponding linear functional on $Im(T)$. Now $HB(\hat{e}_n)$ is a w^*-closed face of $C(K)^*$. Now $ext\ HB(\hat{e}_n) \subseteq ext\ C(K)^*_1$. Let us assume that there exist δ_p, δ_q, where $p \neq q$ and $\delta_p, \delta_q \in HB(\hat{e}_n)$. Now we have $T^*(\delta_p) = T^*(\delta_q) = e_n$, and this ensures $\delta_p(T((\alpha_n))) = \delta_q(T((\alpha_n))) = \alpha_n$. Hence $\pm\hat{e}_n$ have unique Hahn-Banach extension to $C(K)$, for all n. Consequently if $(\zeta_n) \in \ell^1$ which is finitely supported then the corresponding linear functional in $Im(T)$ has unique Hahn-Banach extension to $C(K)$.

STEP 2. Now choose $\zeta = (\zeta_n) \in \ell^1$. Consider $\hat{\zeta} \in Im(T)$. If there exist $p, q \in K$ such that $\delta_p, \delta_q \in ext\ HB(\hat{\zeta})$ then $T^*(\delta_p) = T^*(\delta_q) = \zeta$. From the action of the linear functionals on c_0 it is clear that $p = q = \zeta_n$ for some $n \in \mathbb{N}$. This completes the proof.

□

THEOREM 4.2. *Let K be an infinite compact Hausdorff space which contains a clopen set E with more than one point. Then there exists an embedding $T : c_0 \to C(K)$ such that $T(c_0)$ is a (wU) subspace of $C(K)$.*

PROOF. Let $(U_n)_{n\geq 1}$ be a pairwise disjoint family of open sets such that $\cup_n U_n \subseteq K \setminus E$. Now by Uryshon's lemma there exists a sequence $(f_n)_{n\geq 1}$ where $0 \leq f_n \leq 1$ and $S(f_n) \subseteq U_n$, for all n. We now define $T : c_0 \to C(K)$ by the following.

For $(\alpha_n) \in c_0$ define $T((\alpha_n)) = \alpha_1 \chi_E + \sum_{i\geq 1} \alpha_i f_i$.

Clearly $T((\alpha_n)) \in C(K)$ and also $\|T((\alpha_n))\| = \|(\alpha_n)\|_\infty$. Now according to our assumption there exists distinct $p, q \in E$ such that the extreme functionals $\delta_p, \delta_q \in B_{M(K)}$. Now as T is defined it follows that $T^* : C(K)^* \to c_0^*$ where $T^*(\delta_p) = T^*(\delta_q) = e_1$. Which in other words both δ_p and δ_q are norm preserving extensions of e_1 on $Im(T)$. □

Next corollary deals with compact extremally disconnected spaces. See [**11**] for their properties.

COROLLARY 4.3.

(1) Let Ω be an infinite extremally disconnected compact Hausdorff space then there exists an embedding $T : c_0 \to C(\Omega)$ such that the range is not a (wU) subspace.
(2) Let K be a compact Hausdorff space, then corresponding to each clopen set E where E contains more than one point, there exists a non (wU) embedding of c_0 in $C(\Omega)$.

PROOF. (1) As Ω is an infinite extremally disconnected space there must be a component C of Ω which contains more than one point. The result now follows from Theorem 4.2.

(2) Directly follows from Theorem 4.2 □

We next give the detailed proof of a result which we have already used in the case of compact spaces, in Section 2. For a discrete set, using the identification $c_0(S)^* = \ell^1(S)$ and $\ell^\infty(S)^* = \ell^1(S) \oplus_{\ell^1} c_0(S)^\perp$, we have that $c_0(S)$ is a Hahn-Banach smooth space.

PROPOSITION 4.4. Let S be a locally compact Hausdorff space. Then $C_0(S)$ is weakly Hahn-Banach smooth if and only if S is a discrete space.

PROOF. We only need to prove that the condition is necessary.

If possible assume that S has a limit point s_0 and (s_α) be a net of distinct points in S such that $s_\alpha \to s_0$. Consider the Dirac functionals $\delta_{s_\alpha}, \delta_{s_0}$ in $C_0(S)^*$. It is clear that $\delta_{s_\alpha} \to \delta_{s_0}$ in w^*-sense. Now the functional $\delta_{s_0} \in NA(C_0(S))$, we have $\delta_{s_\alpha} \to \delta_{s_0}$ in weak sense. Let us identify $C_0(S)^*$ as the space of measures, $M(S)$ on S . We now consider the functional $\chi_{s_0} \in M(S)^*$. It is clear that $\chi_{s_0}(\delta_{s_\alpha}) = 0$, $\alpha \neq 0$ although $\chi_{s_0}(\delta_{s_0}) = 1$. This is a contradiction. This follows S cannot have a limit point. This completes the proof. □

5. Stability results of the property (wU)

In this section we follow the notations and terminology from [**4**] for matters related to spaces of Bochner integrable functions and tensor product spaces. See also [**12**]. The following serves as a counter example which shows that if Y has property (U) in X then $L^1(\mu, Y)$ may not have property (wU) in $L^1(\mu, X)$.

EXAMPLE 5.1. Let $X = (\mathbb{R}^2, \|.\|_\infty)$ and $Y = span\{(1,0)\}$. Then Y has property (U) in X. Let $([0,1], \mathcal{B}, \lambda)$ be the Lebesgue measure space. We now show that $L^1(\lambda, Y)$ can not have property (wU) in $L^1(\lambda, X)$. Suppose $\Phi \in L^1(\lambda, X)^*$ be such that $\Phi = (2,0).\chi_{(1/2,1]}$. Clearly $\|\Phi\|_\infty = 2$ and Φ attains its norm at some $f \in L^1(\lambda, Y)$, where $\|f\|_1 = 1$. In fact one can choose $f = (2,0).\chi_{[1/2,1]}$. Now it is clear that for any $\Lambda \in L^\infty(\lambda, Y^\perp)$, $\|\Phi - \Lambda\| \geq 2$ and hence $d(\Phi, L^\infty(\lambda, Y^\perp)) = 2$. Let us define $\Theta = (0,1).\chi_{[0,1/4]}$. Then $\Theta \in L^1(\lambda, Y^\perp)$ and $\|\Theta - \Phi\|_\infty = 2$, hence $\Theta \in P_{L^\infty(\lambda, Y^\perp)}(\Phi)$ and our claim follows from Theorem 2.1.

In what follows we need the concept of the Radon-Nikodým property (in short RNP) in Banach spaces. [**4**] is a standard reference for this property. For a Banach space X and $1 < p < \infty$, when μ is a non-atomic measure, a necessary and sufficient condition that $L^p(\mu, X)^* \cong L^q(\mu, X^*)$, where $\frac{1}{p} + \frac{1}{q} = 1$ is that X^* has the RNP (see [**4**, Pg.98]).

THEOREM 5.2. *Let (Ω, Σ, μ) be a probability space and Y be a subspace of X. Suppose X^* has the RNP, then Y has property (wU) in X if and only if $L^p(\mu, Y)$ has property (wU) in $L^p(\mu, X)$ for $1 < p < \infty$.*

PROOF. Let $\Lambda \in L^p(\mu, X)^* \cong L^q(\mu, X^*)$ with $\|\Lambda\| = 1$ be such that $\Lambda \in NA(L^p(\mu, X))$. Suppose that there exists $f \in S_{L^p(\mu,Y)}$ such that $\Lambda(f) = 1$. Then for almost all $t \in \Omega$, either $\Lambda(t) = 0$ or $\frac{\Lambda(t)}{\|\Lambda(t)\|}$ attains its norm on B_Y (see [**5**, Lemma 5]).

CLAIM. $P_{L^p(\mu,Y)^\perp}(\Lambda) = \{0\}$.

From [**6**, Proposition 3.2], it suffices to prove that $P_{L^q(\mu,Y^\perp)}(\Lambda) = \{0\}$, where $\frac{1}{p} + \frac{1}{q} = 1$. As Y has property (wU) in X, so $P_{Y^\perp}\left(\frac{\Lambda(t)}{\|\Lambda(t)\|}\right) = \{0\}$, whenever $\Lambda(t) \neq 0$. Which in other words $\|\Lambda(t)\| = d\left(\Lambda(t), Y^\perp\right)$ a.e. $[\mu]$. This implies $\|\Lambda\|_q = d\left(\Lambda, L^q(\mu, Y^\perp)\right)$ [**12**, Theorem 5]. Therefore we get $0 \in P_{L^q(\mu,Y^\perp)}(\Lambda)$. Let us assume that, $\Theta \in P_{L^q(\mu,Y^\perp)}(\Lambda)$. Then $\|\Theta - \Lambda\|_q = d(\Lambda, L^q(\mu, Y^\perp))$. Hence $\|\Theta(t) - \Lambda(t)\| = d\left(\Lambda(t), Y^\perp\right)$ a.e.$[\mu]$ and consequently $\Theta(t) \in P_{Y^\perp}(\Lambda(t))$ a.e. $[\mu]$. Which in other words $\Theta(t) = 0$ a.e.$[\mu]$. Hence the CLAIM follows.

It is obvious that Y has property (wU) in X, whenever $L^p(\mu, Y)$ has property (wU) in $L^p(\mu, X)$. We do not need the assumption of the RNP for this case. □

PROBLEM 5.3. Can we relax the condition RNP in [**6**, Theorem 3.4] or in Theorem 5.2?

We recall from [**11**] that a Banach space X is a L^1-predual space, if X^* is isometric to $L^1(\mu)$ for some positive measure μ. Spaces of continuous functions satisfy this property, and these spaces also have isometric copies of c_0, see [**11**]. Another property that we will be using is that both X and X^* have the metric approximation property (MAP). Similar techniques used in [**6**, Theorem 3.6] also lead to the following. We will again be using the relation between vector-valued and point-wise best approximations from [**12**],[**15**]. See also [**15**] for the difficulty involved in considering the essentially bounded functions case.

THEOREM 5.4. *Let X be an L^1-predual and Z be a Banach space with Z having RNP. Let Y be a subspace of Z. Then Y has property (wU) in X if and only if $X \otimes_\varepsilon^\vee Y$ has property (wU) in $X \otimes_\varepsilon^\vee Z$.*

PROOF. We again use the duality between injective and projective tensor product spaces, under appropriate additional conditions, from Chapter VIII of [**4**]. From our assumption we have $(X \otimes_\varepsilon^\vee Z)^* \cong X^* \otimes_\pi^\wedge Z^* \cong L^1(\mu, Z^*)$, for some positive measure space (Ω, Σ, μ). On the other hand one can easily show that $(X \otimes_\varepsilon^\vee Y)^\perp \cong X^* \otimes_\pi^\wedge Y^\perp \cong L^1(\mu, Y^\perp)$. It now follows from the relation between the vector-valued best approximations and point-wise ones from [**12**], any $\Lambda \in L^1(\mu, Z^*)$, $\Theta \in P_{L^1(\mu, Y^\perp)}(\Lambda)$ if and only if $\Theta(t) \in P_{Y^\perp}(\Lambda(t))$ a.e. $[\mu]$. The result now follows from Theorem 2.1. □

The following stability results are easy to see following the arguments from [**2**].

PROPOSITION 5.5. For $i \in \Gamma$, let Y_i be a subspace with property (wU) of the Banach space X_i. Then $\oplus_{c_0} Y_i$ is a subspace with property (wU) of $\oplus_{c_0} X_i$.

As a consequence of Proposition 3.5 and with similar arguments stated in [**2**, Corollary 2.8] we have the following.

COROLLARY 5.6. For $i \in \Gamma$, let X_i be a family of Banach spaces which are weakly Hahn-Banach smooth. Then $\oplus_{c_0(\Gamma)} X_i$ is also weakly Hahn-Banach smooth.

Our next counter examples prove that in general property (wU) is not stable under tensor product.

EXAMPLE 5.7.
(1) We have observed that if X is weakly Hahn-Banach smooth, then $C(K, X) = C(K) \otimes_\varepsilon^\vee X$ is weakly Hahn-Banach smooth if and only if K is finite.
(2) Now choose X, which is Hahn-Banach smooth, and then consider $X \otimes_\varepsilon c = X \otimes_\varepsilon C(\alpha\mathbb{N})$, which is not weakly Hahn-Banach smooth as $\alpha\mathbb{N}$ is not finite. Here $\alpha\mathbb{N}$ is one point compactification of $\mathbb{N}$.

Acknowledgments

Part of this work was done when the second author was visiting Shiv Nadar University during March 2024. He takes this opportunity to acknowledge the hospitality that was received during his visit. The aforesaid visit was supported by the Collaboration Travel Grant from IITH and he extends his gratitude to the Director @IITH for the approval of this grant. Rao's work is supported by a 3-year project 'Classification of Banach spaces using differentiability' funded by the Core Research Grant of the Anusandhan National Research Foundation (ANRF)-CRG-2023-000595.

References

[1] Pradipta Bandyopadhyay and Sudeshna Basu, *On a new asymptotic norming property*, Indag. Math. (N.S.) **10** (1999), no. 1, 15–23, DOI 10.1016/S0019-3577(99)80002-0. MR1691471

[2] Sudeshna Basu and T. S. S. R. K. Rao, *Some stability results for asymptotic norming properties of Banach spaces*, Colloq. Math. **75** (1998), no. 2, 271–284, DOI 10.4064/cm-75-2-271-284. MR1490693

[3] Christian Cobollo, Antonio José Guirao, and Vicente Montesinos, *Some remarks on Phelps property U of a Banach space into $C(K)$ spaces*, Rev. R. Acad. Cienc. Exactas Fís. Nat. Ser. A Mat. RACSAM **118** (2024), no. 1, Paper No. 6, 23, DOI 10.1007/s13398-023-01504-9. MR4644455

[4] J. Diestel and J. J. Uhl Jr., *Vector measures*, Mathematical Surveys, No. 15, American Mathematical Society, Providence, RI, 1977. With a foreword by B. J. Pettis. MR453964

[5] S. Dutta and T. S. S. R. K. Rao, *Norm-to-weak upper semi-continuity of the pre-duality map*, J. Anal. **12** (2004), 115–124. MR2142968

[6] Soumitra Daptari, Tanmoy Paul, and T. S. S. R. K. Rao, *Stability of unique Hahn-Banach extensions and associated linear projections*, Linear Multilinear Algebra **70** (2022), no. 21, 6067–6083, DOI 10.1080/03081087.2021.1945526. MR4568539

[7] Soumitra Daptari and Tanmoy Paul, *Uniqueness of Hahn-Banach extensions and some of its variants*, Adv. Oper. Theory **7** (2022), no. 3, Paper No. 36, 19, DOI 10.1007/s43036-022-00201-5. MR4434276

[8] Marián Fabian, Petr Habala, Petr Hájek, Vicente Montesinos, and Václav Zizler, *Banach space theory: The basis for linear and nonlinear analysis*, CMS Books in Mathematics/Ouvrages de Mathématiques de la SMC, Springer, New York, 2011, DOI 10.1007/978-1-4419-7515-7. MR2766381

[9] J. F. Feinstein and Alexander J. Izzo, *Weak sequential completeness of uniform algebras*, Irish Math. Soc. Bull. **92** (2023), 27–32, DOI 10.33232/bims.0092.27.32. MR4693564

[10] P. Harmand, D. Werner, and W. Werner, *M-ideals in Banach spaces and Banach algebras*, Lecture Notes in Mathematics, vol. 1547, Springer-Verlag, Berlin, 1993, DOI 10.1007/BFb0084355. MR1238713

[11] H. Elton Lacey, *The isometric theory of classical Banach spaces*, Die Grundlehren der mathematischen Wissenschaften, Band 208, Springer-Verlag, New York-Heidelberg, 1974. MR493279

[12] W. A. Light, *Proximinality in $L_p(S, Y)$*, Rocky Mountain J. Math. **19** (1989), no. 1, 251–259, DOI 10.1216/RMJ-1989-19-1-251. Constructive Function Theory—86 Conference (Edmonton, AB, 1986). MR1016178

[13] Åsvald Lima, *Uniqueness of Hahn-Banach extensions and liftings of linear dependences*, Math. Scand. **53** (1983), no. 1, 97–113, DOI 10.7146/math.scand.a-12019. MR733942

[14] Ivor J. Maddox, *Infinite matrices of operators*, Lecture Notes in Mathematics, vol. 786, Springer, Berlin, 1980. MR568707

[15] José Mendoza, *Proximinality in $L_p(\mu, X)$*, J. Approx. Theory **93** (1998), no. 2, 331–343, DOI 10.1006/jath.1997.3163. MR1616801

[16] Tanmoy Paul, T.S.S.R.K. Rao, *Order preserving unique Hahn-Banach extensions*, preprint.

[17] R. R. Phelps, *Uniqueness of Hahn-Banach extensions and unique best approximation*, Trans. Amer. Math. Soc. **95** (1960), 238–255, DOI 10.2307/1993289. MR113125

[18] Raymond A. Ryan, *Introduction to tensor products of Banach spaces*, Springer Monographs in Mathematics, Springer-Verlag London, Ltd., London, 2002, DOI 10.1007/978-1-4471-3903-4. MR1888309

[19] Brailey Sims and David Yost, *Linear Hahn-Banach extension operators*, Proc. Edinburgh Math. Soc. (2) **32** (1989), no. 1, 53–57, DOI 10.1017/S0013091500006908. MR981992

[20] Francis Sullivan, *Geometrical peoperties determined by the higher duals of a Banach space*, Illinois J. Math. **21** (1977), no. 2, 315–331. MR458124

DEPARTMENT OF MATHEMATICS, INDIAN INSTITUTE OF TECHNOLOGY HYDERABAD, HYDERABAD, INDIA
Email address: ma22resch11001@iith.ac.in

DEPARTMENT OF MATHEMATICS, INDIAN INSTITUTE OF TECHNOLOGY HYDERABAD, HYDERABAD, INDIA
Email address: tanmoy@math.iith.ac.in

DEPARTMENT OF MATHEMATICS, SHIV NADAR INSTITUTION OF EMINENCE, GAUTAM BUDDHA NAGAR, UP-201314 INDIA
Email address: srin@fulbrightmail.org

Contemporary Mathematics
Volume **833**, 2026
https://doi.org/10.1090/conm/833/16667

Some new characterizations of core-EP inverse

Umashankara Kelathaya, Savitha Varkady,
and Manjunatha Prasad Karantha

ABSTRACT. In this article, several new characterizations of core-EP inverse of an element from an involutory associative ring are given. The core-EP inverse is identified as a particular (b, c)-inverse and an expression for the same is obtained. Further, it is established that an element having core-EP inverse is a strongly π-regular element and the core-EP inverse of an element is a strongly regular element. Motivated by the characterizations of generalized inverses and outer inverses using inverse complements given in the literature, a characterization of core-EP inverse of unit regular elements in a ring with unity is provided using inverse complements and minus partial order.

1. Introduction

For a square matrix of index one over complex field, Rao-Mitra [**16**] discussed a generalized inverse $A^{\#}AA^{\dagger}$, where $A^{\#}$ is the group inverse and $A^{\dagger}$ is the Moore-Penrose inverse of A, and this inverse is found to have typical property of column space and row space. In 2010, Baksalary and Trenkler [**2**] independently reintroduced the same inverse and named it as core inverse. Extending this notion of core inverse to a larger class of matrices, Prasad and Mohana [**15**] introduced the notion of core-EP inverse of a square matrix, which thereafter has been studied by several researchers in the context of matrices, operators, and ring elements. Gao and Chen [**10**], in 2018, introduced the notion of pseudo core inverse of an element from an involutory associative ring with unity and showed that it coincides with the core-EP inverse.

In this article, our aim is to provide new characterizations of core-EP inverse in the case of elements from an associative ring, not necessarily with multiplicative identity. The core-EP inverse and the elements having it appear to have several noteworthy properties associated with the notion of π-regularity, which are not discussed in the literature. Having explored these properties, in this article, the core-EP inverse is observed to be a particular (b, c)-inverse, and this observation led to provide a new characterization and an expression for the core-EP inverse of an element from an associative ring. Further, motivated by the characterization of different generalized inverses and outer inverses given by Bhat, Karantha, and

2020 *Mathematics Subject Classification.* Primary 15A09, 16U90, 16D25, 06A06, 16E50.

Key words and phrases. Core-EP inverse, (b, c)-inverse, strongly regular element, strongly π-regular element, minus partial order, inverse complement.

Eagambaram [6] and Kelathaya, Varkady, and Karantha [13], a characterization of core-EP inverse of unit regular elements using inverse complements is presented.

In Section 2, some of the notations and terminologies are provided as a prerequisite for the further discussion. In Section 3, different characterizations of core-EP inverse are provided in the context of associative rings which need not have unity and several interesting properties of the core-EP inverse as well as of the elements having core-EP inverse are discussed. In Section 4, the core-EP inverse is identified as a particular (b, c)-inverse and a new characterization and an expression for the core-EP inverse are provided. In Section 5, the last section, the core-EP inverse of a unit regular element is characterized using inverse complements in an associative ring with unity.

2. Preliminaries

In this section, we introduce some of the notations and terminologies necessary for the further discussion. For all terminologies in the matrix context, we refer to [16] and [5], and for the terminologies in the context of rings, we refer to [1], [7], [3], and [17].

Throughout this paper, $\mathcal{A}$ denotes an associative ring (need not be with multiplicative identity, unless indicated otherwise) with an involution '$*$'. The notations $a, b, c, \ldots$ are used to denote the elements of a ring, and $A, B, C, \ldots$ are used to denote the matrices under discussion.

Given an element $a \in \mathcal{A}$, consider the following set of equations, called Penrose conditions.

$$(1)\ axa = a \quad (2)\ xax = x \quad (3)\ (ax)^* = ax \quad (4)\ (xa)^* = xa.$$

In addition to Penrose conditions, we consider the following equations:

$$(5)\ ax = xa \quad (6)\ ax^2 = x \quad (1^{\mathrm{k}})xa^{k+1} = a^k \quad ({}^{k}1)a^{k+1}x = a^k,$$

where k is a positive integer. Now, we recall some of the basic definitions in the following.

(i) The element a is said to be *regular (von Neumann regular)* if there exists an element g in $\mathcal{A}$ satisfying (1) of Penrose conditions i.e., $aga = a$. Such an element g, if exists, is called a *generalized inverse* (or *g-inverse* or *inner inverse* or *pseudo inverse*) of a. An arbitrary generalized inverse of a, if exists, is denoted by a^- and the class of all generalized inverses of a is denoted by $\{a^-\}$. Note that an element need not belong to the ideal generated by it, unless the element is regular or the ring has multiplicative identity.

(ii) An element g satisfying (2) is called an *outer inverse* of a. An arbitrary outer inverse of a is denoted by $a^=$ and the class of all outer inverses of a is denoted by $\{a^=\}$.

(iii) An element g satisfying both (1) and (2) is called a *reflexive generalized inverse* of a. An arbitrary reflexive generalized inverse of a, if exists, is denoted by a_r^- and the class of all reflexive generalized inverses of a is denoted by $\{a_r^-\}$.

(iv) An element g satisfying (2), (5), and (1^k) (or equivalently, (2), (5), and $({}^k1)$) for some positive integer k is called *Drazin inverse* of a. The Drazin inverse of a, if exists, is unique and is denoted by a^D. The least positive integer k, for which a satisfies the above conditions is called the *index (Drazin index)* of a and the same is denoted by $\mathrm{ind}(a)$.

(v) An element g satisfying the conditions in a subset $\{i, j, k, \dots\}$ of $\{1, 2, 3, 4, 5, 6, 1^k, {}^k1\}$ is called an *$\{i, j, k, \dots\}$-inverse* of a, and the class of all $\{i, j, k, \dots\}$-inverses of a is denoted by $\{a^{\{i,j,k,\dots\}}\}$.

(vi) The element a is said to be a right (left) regular element if there exists an element $x \in \mathcal{A}$ such that $a^2x = a$ $(xa^2 = a)$, and a is said to be a strongly regular element if it is both right regular element and left regular element

(vii) The element a is said to be a right (left) π-regular element if there exists an element $x \in \mathcal{A}$ such that $a^{n+1}x = a^n$ $(xa^{n+1} = a^n)$ for some positive integer n and, in this case, x is called a right (left) π-inverse of a. The least positive integer n for which a is right (left) π-regular element is called the right (left) index of a

(viii) The element a is said to be a strongly π-regular element if it is both right π-regular element and left π-regular element, in which case, the right index and left index coincide and the common value equals the index of a (see, [**7**]). Also, Drazin [**7**, Theorem 4] proved that an element has Drazin inverse if and only if it is a strongly π-regular element.

In 2012, Drazin [**8**] introduced a general class of outer inverses called (b, c)-inverses, and most of the outer inverses such as Moore-Penrose inverse, Bott-Duffin inverse, Drazin inverse etc., are found to be the special cases of (b, c)-inverses. Bapat et al. [**3**] provided an equivalent definition for the (b, c)-inverse of an element as in the following.

DEFINITION 2.1 ((b, c)-inverse, [**3**]). For $a, b, c \in \mathcal{A}$, an element g is called (b, c)-inverse of a if g is an outer inverse of a such that $g\mathcal{A} = b\mathcal{A}$ and $\mathcal{A}g = \mathcal{A}c$.

The (b, c)-inverse of a, whenever exists, is unique.

The notion of core-EP inverse was introduced by [**15**] for square matrices over real or complex field as in the following.

DEFINITION 2.2 (Core-EP inverse, [**15**]). A matrix G is a core-EP inverse of a if G is an outer inverse satisfying

$$\mathcal{C}(G) = \mathcal{R}(G) = \mathcal{C}(A^d),$$

where d is the index of A.

The authors of the same paper provided the following characterization of core-EP inverse of a square matrix.

LEMMA 2.3 (Lemma 3.3, [**15**]). *Given a square matrix A with index d, the following statements are equivalent.*

(i) *G is the core-EP inverse of a.*

(ii) *G is a matrix satisfying (2), (3), (1^k) for some positive integer k and $\mathcal{C}(G) \subseteq \mathcal{C}(A^k)$.*

Further, the authors characterized the elements having core-EP inverse, and provided an expression for the same as in the following.

THEOREM 2.4 (Theorem 3.5, [**15**]). *Given a square matrix A of index d, the core-EP inverse exists if and only if* $\mathrm{rank}(A^{T^d}A^{d+1}) = \mathrm{rank}(A^d)$. *Further, core-EP* *inverse is unique whenever it exists and is given by*

$$A^d(A^{T^d}A^{d+1})^-A^{T^d}.$$

The definition of core-EP inverse given in Definition 2.2 can be extended to the case of elements from an associative ring as follows.

DEFINITION 2.5 (Core-EP inverse). Given an element $a \in \mathcal{A}$, an element $g \in \mathcal{A}$ is said to be the core-EP inverse of a if g satisfies

$$gag = g \quad \text{and} \quad g\mathcal{A} = g^*\mathcal{A} = a^k\mathcal{A}, \tag{2.1}$$

for some positive integer k.

Equivalently, g is the core-EP inverse of a if g is an outer inverse of a satisfying

$$\mathcal{A}g = \mathcal{A}g^* = \mathcal{A}a^{k^*}, \tag{2.2}$$

for some positive integer k.

Gao and Chen [**10**], in 2018, introduced the notion of pseudo core inverse of a ring element and showed that this notion coincides with the core-EP inverse in an associative ring with unity.

DEFINITION 2.6 (Pseudo core inverse, [**10**]). Given an element $a \in \mathcal{A}$, an element $g \in \mathcal{A}$ is said to be pseudo core inverse of a if it satisfies the following conditions:

$$\text{(i) } ag^2 = g \quad \text{(ii) } (ag)^* = ag \quad \text{(iii) } ga^{k+1} = a^k, \tag{2.3}$$

for some positive integer k.

In the present paper, we provide some new characterizations of core-EP inverse, and give an expression for the same.

3. Characterizations and properties of core-EP inverse

In this section, we provide several characterizations of core-EP inverse. We prove that core-EP inverse is a special case of (b, c)-inverse, and provide an expression for the core-EP inverse in the same terms. To begin with, we recall the following lemma by Bapat et al. [**4**], which is useful for further discussions.

LEMMA 3.1 ([**4**]). *Let $a, b \in \mathcal{A}$ such that $a, b \in a\mathcal{A} = b\mathcal{A}$. Then, a is regular if and only if b is also regular.*

The following theorem gives different characterizations of core-EP inverse. The equivalence of statements (i) and (ii) extends the equivalence of core-EP inverse and the pseudo core inverse, proved by Gao and Chen [**10**], to the case of an associative ring not having multiplicative identity.

THEOREM 3.2. *Given an element $a \in \mathcal{A}$, the following statements are equivalent.*

(i) *g is the core-EP inverse of a.*
(ii) *g satisfies the equations*

$$ag^2 = g, \quad (ag)^* = ag, \quad \text{and} \quad ga^{m+1} = a^m,$$

for some positive integer m.

(iii) *g satisfies*

$$gag = g, \quad (ag)^* = ag, \quad ga^{m+1} = a^m, \quad \text{and} \quad g\mathcal{A} \subseteq g^*\mathcal{A}, \tag{3.1}$$

for some positive integer m.

PROOF. (i)$\Longleftrightarrow$(ii): Consider an element $a \in \mathcal{A}$ with the core-EP inverse g as given in (i). In other words, g is an outer inverse of a satisfying

$$g\mathcal{A} = g^*\mathcal{A} = a^k\mathcal{A},$$

for some positive integer k. To prove (i) $\Longrightarrow$ (ii), first note that

$$(ag)^*\mathcal{A} = g^*a^*\mathcal{A} = g^*\mathcal{A} = a^k\mathcal{A} \supseteq a^{k+1}\mathcal{A} = ag\mathcal{A},$$

which shows that

$$ag\mathcal{A} \subseteq (ag)^*\mathcal{A}. \tag{3.2}$$

Since ag is an idempotent element, we have $ag \in ag\mathcal{A}$, and hence from Eq. (3.2) we obtain that

$$\begin{aligned} ag &= (ag)^*r \quad \text{for some } r \in \mathcal{A} \\ &= (ag)^*(ag)^*r \quad (\because (ag)^* \text{ is idempotent}) \\ &= (ag)^*ag, \end{aligned}$$

proving

$$(ag)^* = ag. \tag{3.3}$$

Now, by $g = gag \in g\mathcal{A} = g^*\mathcal{A}$, we get $g = g^*a^*g$, and hence, using Eq. (3.3), we obtain

$$g = g^*a^*g = (ag)^*g = ag^2. \tag{3.4}$$

Further, for $m = k+1$, we have $a^m \in a^k\mathcal{A} = g\mathcal{A}$. So,

$$a^m \in g\mathcal{A}, \tag{3.5}$$

and therefore,

$$ga^{m+1} = gaa^m = a^m. \tag{3.6}$$

Now, statement (ii) follows from Eqs. (3.3), (3.4), and (3.6).

To prove (ii) $\Longrightarrow$ (i), consider an element g satisfying the conditions as given in (ii), i.e.,

$$ag^2 = g, \quad (ag)^* = ag, \quad \text{and} \quad ga^{m+1} = a^m$$

for some positive integer m. Using $g = ag^2$ repeatedly, we have

$$g = ag^2 = a(ag^2)g = a^2g^3 = \cdots = a^ng^{n+1} \tag{3.7}$$

and, premultiplying $g = ag^2$ by a, we obtain

$$ag = a^2g^2 = a^2(ag^2)g = a^3g^3 = \cdots = a^ng^n \tag{3.8}$$

for every positive integer n. Therefore,

$$\begin{aligned} gag &= ga^{m+1}g^{m+1} \quad (\text{using Eq. (3.8)}) \\ &= a^mg^{m+1} \quad (\because ga^{m+1} = a^m) \\ &= g, \quad (\text{using Eq. (3.7)}) \end{aligned}$$

which proves that g is an outer inverse of a.

To complete the proof of (ii) $\Longrightarrow$ (i), we need to show that

$$gA = g^*A = a^kA \tag{3.9}$$

for some positive integer k. From $ga^{m+1} = a^m$ as given in (ii), and from Eq. (3.7), we have $a^m\mathcal{A} \subseteq g\mathcal{A}$ and $g\mathcal{A} \subseteq a^m\mathcal{A}$, respectively. Therefore, $g\mathcal{A} = a^m\mathcal{A}$. Since g^* is outer inverse of a^*, it follows that

$$\begin{aligned} g^*\mathcal{A} = g^*a^*\mathcal{A} = (ag)^*\mathcal{A} & \\ &= ag\mathcal{A} \quad (\because (ag)^* = ag) \\ &= a^{m+1}\mathcal{A} \quad (\because g\mathcal{A} = a^m\mathcal{A}) \\ &\subseteq a^m\mathcal{A} = g\mathcal{A} = ag^2\mathcal{A} = g^*a^*g\mathcal{A} \subseteq g^*\mathcal{A}, \end{aligned}$$

proving that $g\mathcal{A} = g^*\mathcal{A} = a^m\mathcal{A}$. Thus, the outer inverse g satisfies Eq. (3.9) with $k = m$, proving that g is the core-EP inverse of a.

(ii)$\Longleftrightarrow$(iii): Consider an element g satisfying the conditions as given in (ii), i.e.,

$$ag^2 = g, \quad (ag)^* = ag, \quad \text{and} \quad ga^{m+1} = a^m$$

for some positive integer m. To prove (ii) $\Longrightarrow$ (iii), we need only to show that g is an outer inverse of a with $g\mathcal{A} \subseteq g^*\mathcal{A}$. Following the proof of first part of (ii) $\Longrightarrow$ (i) using Eqs. (3.7), (3.8), and $ga^{m+1} = a^m$, we have that g is an outer inverse of a. Now, using $g = ag^2$, we have

$$\begin{aligned} g\mathcal{A} &= ag^2\mathcal{A} \\ &= (ag)^*g\mathcal{A} \quad (\because (ag)^* = ag) \\ &= g^*a^*g\mathcal{A} \\ &\subseteq g^*\mathcal{A}, \end{aligned}$$

proving (ii) $\Longrightarrow$ (iii).

To prove (iii) $\Longrightarrow$ (ii), we need only to prove that the element g, as given in (iii), satisfies $ag^2 = g$. Since $g = gag \in g\mathcal{A} \subseteq g^*\mathcal{A}$, we have $g \in g^*\mathcal{A}$ and hence, using $(ag)^* = ag$, we get

$$ag^2 = (ag)^*g = g^*a^*g = g,$$

proving (iii) $\Longrightarrow$ (ii). Hence the theorem. □

The condition $ag^2 = g$ in (ii) of the above theorem shows that g is a left regular element. In the following corollary we see that g, in fact, is also a strongly regular element.

COROLLARY 3.3. *If g is the core-EP inverse of a, then g is a strongly regular element.*

PROOF. If g is the core-EP inverse of a, then following (i) $\Longrightarrow$ (ii) of Theorem 3.2, g satisfies $ag^2 = g$, and therefore, g is a left regular element. To prove that g is a right regular element, first note that

$$g\mathcal{A} = g^*\mathcal{A} = g^*a^*\mathcal{A} = (ag)^*\mathcal{A} = ag\mathcal{A}. \tag{3.10}$$

Hence,

$$g\mathcal{A} = gag\mathcal{A} = g^2\mathcal{A}. \tag{3.11}$$

Since $g = gag$ is regular, it follows that $g \in g\mathcal{A}$. Therefore, from Eq. (3.11), we have $g = g^2r$ for some $r \in \mathcal{A}$, showing that g is a right regular element. Thus, g is a strongly regular element, proving the corollary. □

REMARK 3.4. From (i) $\Longrightarrow$ (ii) of Theorem 3.2, we have that an element having core-EP inverse is a left π-regular element with left index $m = k+1$. So, it may be noted that the positive integer k appearing in the Definition 2.5 of core-EP inverse need not be equal to the pseudo core index $I(a)$ defined by Gao and Chen in [**10**], unless a^k is regular or the ring has a multiplicative identity, in which case, $a^k \in a^k\mathcal{A} = g\mathcal{A}$ and hence

$$ga^{k+1} = gaa^k = a^k,$$

showing that $m = k$.

The following corollary shows that every element having core-EP inverse is a strongly π-regular element, and therefore, has Drazin inverse.

COROLLARY 3.5. *If a has core-EP inverse, then we have the following.*

(i) *a is a strongly π-regular element.*
(ii) *a^m is regular for some m.*

PROOF. Consider an element a with core-EP inverse g satisfying

$$gag = g \quad \text{and} \quad g\mathcal{A} = g^*\mathcal{A} = a^k\mathcal{A},$$

for some positive integer k. By referring to (i) $\Longrightarrow$ (ii) of Theorem 3.2, we have $(ag)^* = ag$ and hence,

$$a^k\mathcal{A} = g^*\mathcal{A} = g^*a^*\mathcal{A} = (ag)^*\mathcal{A} = ag\mathcal{A} = a^{k+1}\mathcal{A}. \tag{3.12}$$

Since $a^{k+1} \in a^k\mathcal{A}$ and using Eq. (3.12), we get $a^{k+1} \in a^{k+1}\mathcal{A}$. Therefore, for $m = k+1$, we have

$$a^m \in a^m\mathcal{A} = a^{k+1}\mathcal{A} = a^{k+2}\mathcal{A} = a^{m+1}\mathcal{A}, \tag{3.13}$$

which shows that a is a right π-regular element with right index at most m. Further, from (i) $\Longrightarrow$ (ii) of Theorem 3.2, we have that $ga^{m+1} = a^m$ and hence

$$a^m \in \mathcal{A}a^{m+1}, \tag{3.14}$$

proving that a is a left π-regular element with left index at most m. Thus, a is strongly π-regular with $\operatorname{ind}(a) \leq m$, and this proves part (i).

Now, for $m = k+1$, we have $a^m \in a^k\mathcal{A} = g\mathcal{A}$. Also, from Eq. (3.12), we have that $a^k\mathcal{A} = a^m\mathcal{A}$. Therefore,

$$a^m \in a^m\mathcal{A} = g\mathcal{A}.$$

Since $g = gag$ is regular, by referring to Lemma 3.1, we get that a^m is regular, proving part (ii). Hence the corollary. □

4. Core-EP inverse as a (b, c)-inverse

Observing that core-EP inverse is a particular (b, c)-inverse, in this section, we provide another characterization and an expression for the core-EP inverse of a ring element.

In the following theorem, we shall show that the core-EP inverse is a special case of (b, c)-inverse.

THEOREM 4.1. *Given an element $a \in \mathcal{A}$, the core-EP inverse of a, if exists, is an (a^m, a^{m*})-inverse of a for some positive integer m.*

PROOF. Consider an element $a \in \mathcal{A}$ with core-EP inverse g, i.e., g is an outer inverse of a satisfying

$$g\mathcal{A} = g^*\mathcal{A} = a^k\mathcal{A}, \quad \text{or equivalently,} \quad \mathcal{A}g = \mathcal{A}g^* = \mathcal{A}a^{k^*}$$

for some positive integer k. From Eq. (3.12), we have

$$a^k\mathcal{A} = a^{k+1}\mathcal{A}, \quad \text{or equivalently,} \quad \mathcal{A}a^{k^*} = \mathcal{A}a^{k+1^*}. \tag{4.1}$$

Thus, for $m = k + 1$, the outer inverse g satisfies

$$g\mathcal{A} = a^k\mathcal{A} = a^m\mathcal{A} \tag{4.2}$$

and

$$\mathcal{A}g = \mathcal{A}a^{k^*} = \mathcal{A}a^{m*}, \tag{4.3}$$

proving that g is an (a^m, a^{m*})-inverse of a, and hence the theorem. □

The following theorem provides another new characterization of the elements having core-EP inverse in an associative ring.

THEOREM 4.2. *Given an element $a \in \mathcal{A}$, the following statements are equivalent.*

(i) *a has core-EP inverse.*
(ii) *a^m and $a^{m*}a^{m+1}$ are regular, and*

$$a^{m*}\mathcal{A} = a^{m*}a^{m+1}\mathcal{A} \quad \text{and} \quad \mathcal{A}a^m = \mathcal{A}a^{m*}a^{m+1}$$

for some positive integer m.

In the above case, the core-EP inverse of a is given by $a^m(a^{m*}a^{m+1})^-a^{m*}$ for any generalized inverse $(a^{m*}a^{m+1})^-$ of $a^{m*}a^{m+1}$.

PROOF. (i) $\Longrightarrow$ (ii): Consider an element a having core-EP inverse g, i.e., g is an outer inverse of a with

$$g\mathcal{A} = g^*\mathcal{A} = a^k\mathcal{A}, \quad \text{or equivalently,} \quad \mathcal{A}g = \mathcal{A}g^* = \mathcal{A}a^{k^*}$$

for some positive integer k. From Eqs. (4.2) and (4.3), we have

$$g\mathcal{A} = a^m\mathcal{A} \tag{4.4}$$

and

$$\mathcal{A}g = \mathcal{A}a^{m*}, \tag{4.5}$$

where $m = k + 1$. Further, from Corollary 3.5, we have a^m is regular, in which case, $a^m \in a^m\mathcal{A}$. Now, using Eq. (4.4), we get $ga^{m+1} = gaa^m = a^m$. Similarly, using Eq. (4.5) and $a^{m*} \in \mathcal{A}a^{m*}$, we get $a^{m*}ag = a^{m*}$. Therefore,

$$\begin{aligned} a^{m*}\mathcal{A} &= a^{m*}ag\mathcal{A} \\ &= a^{m*}a^{m+1}\mathcal{A} \quad (\because g\mathcal{A} = a^m\mathcal{A}) \end{aligned}$$

and

$$\begin{aligned} \mathcal{A}a^m &= \mathcal{A}gaa^m \\ &= \mathcal{A}a^{m*}a^{m+1}. \quad (\because \mathcal{A}g = \mathcal{A}a^{m*}) \end{aligned}$$

Further,

$$a^{m*}a^{m+1} \in a^{m*}\mathcal{A} = a^{m*}a^{m+1}\mathcal{A}.$$

Now, the regularity of $a^{m*}a^{m+1}$ follows by regularity of a^{m*} following Lemma 3.1.

(ii) $\implies$ (i): Consider an element a for which a^{m*}, $a^{m*}a^{m+1}$ are regular, and

$$a^{m*}\mathcal{A} = a^{m*}a^{m+1}\mathcal{A} \quad \text{and} \quad \mathcal{A}a^m = \mathcal{A}a^{m*}a^{m+1} \tag{4.6}$$

for some positive integer m, as given in (ii). Set

$$g = a^m(a^{m*}a^{m+1})^- a^{m*}, \tag{4.7}$$

where $(a^{m*}a^{m+1})^-$ is an arbitrary generalized inverse of $a^{m*}a^{m+1}$. Using the regularity of a^m and Eq. (4.6), we get

$$a^m \in \mathcal{A}a^{m*}a^{m+1} \quad \text{and} \quad a^{m*} \in a^{m*}a^{m+1}\mathcal{A}. \tag{4.8}$$

Now, the uniqueness of expression for g given in Eq. (4.7) under the choice of generalized inverse of $a^{m*}a^{m+1}$ follows from above Eq. (4.8). Again using Eq. (4.8), Eq. (4.6), and substituting for gag, we get

$$\begin{aligned} gag &= a^m(a^{m*}a^{m+1})^- a^{m*}aa^m(a^{m*}a^{m+1})^- a^{m*} \\ &= a^m(a^{m*}a^{m+1})^- a^{m*} = g, \end{aligned}$$

proving that g is an outer inverse of a. From the definition of g, it is clear that

$$g, g^* \in a^m\mathcal{A}. \tag{4.9}$$

Using Eq. (4.8), we get

$$a^m = a^m(a^{m*}a^{m+1})^- a^{m*}a^{m+1} = ga^{m+1}$$

and

$$a^{m*} = a^{m*}a^{m+1}(a^{m*}a^{m+1})^- a^{m*} = a^{m*}ag,$$

which proves that

$$a^m \in g\mathcal{A} \cap g^*\mathcal{A}. \tag{4.10}$$

Using Eq. (4.9) and Eq. (4.10), we have

$$g\mathcal{A} = g^*\mathcal{A} = a^m\mathcal{A},$$

and therefore g is a core-EP inverse of a, proving (ii) $\implies$ (i). Hence the theorem. □

5. Characterization of core-EP inverse using inverse complements

In this section, we present another characterization of core-EP inverse using inverse complements and the minus partial order. Throughout this section, unless indicated otherwise, $\mathcal{A}$ denotes an associative ring with unity and with an involution '$*$'. Before proceeding to the main results of this section, we shall quickly recall some of the useful definitions and results from the literature. For more details, we refer to Bhat, Karantha, and Eagambaram [**6**] and Kelathaya, Varkady, and Karantha [**13**] and the references therein.

The minus partial order, introduced by Hartwig [**11**] and Nambooripad [**14**], is one of the important $\mathcal{G}$-based relations which is helpful in obtaining several interesting results in the theory of generalized inverses. The same is defined for ring elements as in the following.

DEFINITION 5.1 (Minus partial order [**11**, **14**]). Given any two elements a, b from an associative ring $\mathcal{A}$ (need not have unity), b is less than a under the relation '$\leq^-$', written as $b \leq^- a$, if there exists a generalized inverse g of b such that

$$bg = ag \quad \text{and} \quad gb = ga. \tag{5.1}$$

The relation $\leq^-$ is a partial order on the class of all regular elements of $\mathcal{A}$, and the same is called the minus partial order.

Given an associative ring $\mathcal{A}$ not having unity, it is easily verified that the relation '$\leq^-$' is a preorder on entire $\mathcal{A}$, and is a partial order on the class of all regular elements of $\mathcal{A}$.

The following theorem, describing the properties of minus partial order, plays a key role in obtaining many of the subsequent results. For the proof of the theorem, we refer to [**12**] and [**3**].

THEOREM 5.2 (Right-left symmetry [**3**, **12**]). *Let $\mathcal{A}$ be an associative ring with unity and let $a, b, c \in \mathcal{A}$ such that $a = b+c$ is regular. Then the following statements are equivalent.*

(i) $b \leq^- a$
(ii) $b\mathcal{A} \oplus c\mathcal{A} = a\mathcal{A}$ *($\oplus$ refers to direct sum)*
(iii) $b\mathcal{A} \cap c\mathcal{A} = \{0\} = \mathcal{A}b \cap \mathcal{A}c$ *(we say b and c are disjoint)*
(iv) $\mathcal{A}b \oplus \mathcal{A}c = \mathcal{A}a$
(v) $c \leq^- a$
(vi) $b \in a\mathcal{A} \cap \mathcal{A}a$ *and* $\{a^-\} \subseteq \{b^-\}$
(vii) $c \in a\mathcal{A} \cap \mathcal{A}a$ *and* $\{a^-\} \subseteq \{c^-\}$.

The above result can be extended to any associative ring, need not be with unity, provided all of a, b, and c are regular.

In 2021, using the properties of minus partial order, Eagambaram, Bhat, and Karantha [**6**] reestablished the inverse complemented matrix method, which was initially introduced by Eagambaram [**9**]. Inspired by this work, Kelathaya, Varkady, and Karantha [**13**] extended the notion of inverse complement to the case of ring elements and introduced the notion of strongly unit regular element in an associative ring. Recall that two elements a and b are *disjoint* if $a\mathcal{A} \cap b\mathcal{A} = \{0\}$ and $\mathcal{A}a \cap \mathcal{A}b = \{0\}$, in which case, the sum $a + b$ is written as $a \oplus b$.

DEFINITION 5.3 (Inverse complement of a ring element [**13**]). Given an element a from an associative ring $\mathcal{A}$ with unity, an element $c \in \mathcal{A}$ disjoint from a is called an *inverse complement* of a if $a \oplus c$ is a unit element in $\mathcal{A}$. In such a case, $a \oplus c$ is called an *inverse complemented element* of a.

The class of all inverse complements of a is denoted by $\mathfrak{K}(a)$ and the class of all inverse complemented elements of a is denoted by $\mathfrak{C}(a)$.

Noting that the existence of inverse complement for a given element is not guaranteed even if the element is regular, the authors of [**13**] introduced the notion of strongly unit regular element as in the following.

DEFINITION 5.4 (Strongly unit regular element [**13**]). An element $a \in \mathcal{A}$ is called a *strongly unit regular element* if for any $g \in \{a_r^-\}$, there exists an element $c \in \mathcal{A}$ such that $gc = cg = 0$ and $a + c$ is invertible.

Further, the authors proved the following.

THEOREM 5.5 (Theorem 3.2 [**13**]). *For a unit regular element $a \in \mathcal{A}$ with a unit generalized inverse v, and $c = u - a$, where $u = v^{-1}$, there exists a unique reflexive generalized inverse g of a such that $gc = cg = 0$, in which case, $g \leq^- u^{-1}(= v)$.*

In fact,

$$g = vav = (a+c)^{-1}a(a+c)^{-1}$$

is the unique reflexive generalized inverse of a such that $gc = cg = 0$.

It is shown in the same paper that the notions of 'unit regular element' and 'strongly unit regular elements are one and the same (see [**13**, Theorem 3.4]).

It is noted that given a reflexive generalized inverse g of a, if c is an inverse complement of a, then

$$ag = (a+c)g \quad \text{and} \quad ga = g(a+c).$$

Therefore, $a \leq^- a + c$ and hence, by (i) $\implies$ (vi) of Theorem 5.2, we get that $(a+c)^{-1}$ is a generalized inverse of a.

In the subsequent part of this section, we characterize the core-EP inverse of a unit regular element (equivalently, strongly unit regular element) using inverse complements and the minus partial order. To begin with, we have the following theorem which characterizes the class of outer inverses of a unit regular element in terms of its inverse complemented elements. This result extends the characterization of class of outer inverses of square matrix given in [**6**] to the case of ring elements.

THEOREM 5.6. *Given a unit regular element $a \in \mathcal{A}$, we have that*

$$\{a^=\} = \{a_c^- a_1 a_c^- : c \in \mathfrak{K}(a), a_1 \leq^- a\}, \tag{5.2}$$

where $a_c^- = (a+c)^{-1}a(a+c)^{-1}$.

PROOF. Consider a unit regular element a and an element $g = a_c^- a_1 a_c^-$ from the set appearing in the right hand side of Eq. (5.2), where $a_c^- = (a+c)^{-1}a(a+c)^{-1}$. Since $(a+c)^{-1}$ is a generalized inverse of a it is easily verified that a_c^- is also a generalized inverse of a. Now using $a_1 \leq^- a$, we have that $a_1 a_c^- a = a_1$ and $a_1 a_c^- a_1 = a_1$. Therefore,

$$\begin{aligned} gag &= a_c^- a_1 a_c^- a a_c^- a_1 a_c^- \\ &= a_c^- a_1 a_c^- a_1 a_c^- \\ &= a_c^- a_1 a_c^- \\ &= g. \end{aligned}$$

Hence, g is an outer inverse of a, which proves that

$$\{a^=\} \supseteq \{a_c^- a_1 a_c^- : c \in \mathfrak{K}(a), a_1 \leq^- a\}.$$

To prove the reverse inclusion, consider any outer inverse g_1 of a. Set $a_1 = ag_1a$ and $a_2 = a - a_1$. By direct substitution, it is verified that g_1 is a reflexive generalized inverse of a_1 such that

$$g_1 a_2 = 0 = a_2 g_1.$$

Also, we have that

$$a_1 g_1 = a g_1 \quad \text{and} \quad g_1 a_1 = g_1 a,$$

and therefore, $a_1 \leq^- a$. Now, by (i)$\iff$(v) of Theorem 5.2, it follows that $a_2 \leq^- a$ and that there exists a reflexive generalized inverse g_2 of a_2 such that

$$g_2 a_1 = 0 = a_1 g_2.$$

Now, by direct substitution, it is easily verified that

$$g = g_1 + g_2$$

is a reflexive generalized inverse of a. Since a is a unit regular element, there exists a $c \in \mathfrak{K}(a)$ such that $gc = cg = 0$ and $a+c$ is invertible, in which case, by Theorem 5.5 the reflexive generalized inverse g is given by

$$\begin{aligned} g = a_c^- &= (a+c)^{-1}a(a+c)^{-1} \\ &= (a+c)^{-1}a_1(a+c)^{-1} \oplus (a+c)^{-1}a_2(a+c)^{-1} \quad (\because a = a_1 \oplus a_2). \end{aligned}$$

Set $f_1 = (a+c)^{-1}a_1(a+c)^{-1}$ and $f_2 = (a+c)^{-1}a_2(a+c)^{-1}$. As $a_1, a_2 \leq^- a$ and $(a+c)^{-1} \in \{a^-\}$, by (i) $\Longrightarrow$ (vi) of Theorem 5.2 we have $(a+c)^{-1} \in \{a_1^-\}$ and $(a+c)^{-1} \in \{a_2^-\}$. Now, it is easily verified by substitution that f_1 is a reflexive generalized inverse of a_1. Also, using $a_2 \leq^- a$ and $a_1 = a - a_2$, we get

$$f_2a_1 = (a+c)^{-1}a_2(a+c)^{-1}a_1 = 0 = a_1(a+c)^{-1}a_2(a+c)^{-1} = a_1f_2. \tag{5.3}$$

Now,

$$\begin{aligned} g_1 &= g_1a_1g_1 \quad (\because g_1 \in \{a_{1_r}^-\}) \\ &= ga_1g \quad (\because g = g_1 + g_2 \text{ and } g_2a_1 = 0 = a_1g_2) \\ &= f_1a_1f_1 \quad (\because g = f_1 + f_2 \text{ and } f_2a_1 = 0 = a_1f_2) \\ &= f_1 \\ &= (a+c)^{-1}a_1(a+c)^{-1} \\ &= (a+c)^{-1}a(a+c)^{-1}a_1(a+c)^{-1}a(a+c)^{-1} \quad (\because a_1 \leq^- a) \\ &= a_c^-a_1a_c^-. \end{aligned}$$

Therefore $g_1 = a_c^-a_1a_c^-$, where $c \in \mathfrak{K}(a)$ and $a_1 \leq^- a$, proving the reverse inclusion and hence the theorem. ☐

Note that for any generalized inverse $(a_c^-)^-$ of a_c^-, we have

$$a_c^-a_1a_c^-(a_c^-)^-a_c^-a_1a_c^- = a_c^-a_1a_c^-.$$

Hence, by (i)$\Longleftrightarrow$(vi) of Theorem 5.2, we get $a_c^-a_1a_c^- \leq^- a_c^-$. Now, the following corollary is immediate from Theorem 5.6 by noting that a_c^- is a reflexive generalized inverse of a.

COROLLARY 5.7. For every outer inverse g of a unit regular element a, there exists a reflexive generalized inverse h of a such that $g \leq^- h$.

The following theorem characterizes the core-EP inverse of a unit regular element in terms of its inverse complemented element. To avoid notational discrepancy, we use $a^{\textcircled{\dagger}}$ to denote the core-EP inverse of a.

THEOREM 5.8. *Given a unit regular element $a \in \mathcal{A}$, we have that*

$$a^{\textcircled{\dagger}} = a_c^-a_1a_c^-,$$

where $a_1 \leq^- a$ and $c \in \mathfrak{K}(a)$ are such that

(i) $a_1^*(a+c) = a_1^*a_1$
(ii) $(a+c)a^k = a_1a^k$, *for some positive integer k*
(iii) $(a+c)^{-1}a_1\mathcal{A} = (a+c)^{-1*}a_1^*\mathcal{A}$.

PROOF. Consider a unit regular element $a \in \mathcal{A}$. To prove that the core-EP inverse $a^{\textcircled{\dagger}}$ of a, satisfying

$$a^{\textcircled{\dagger}}aa^{\textcircled{\dagger}} = a^{\textcircled{\dagger}} \quad \text{and} \quad a^{\textcircled{\dagger}}\mathcal{A} = a^{\textcircled{\dagger}^*}\mathcal{A} = a^k\mathcal{A}$$

for some positive integer k, is of the form $a_c^- a_1 a_c^-$, where a_1 and c satisfy the conditions given in (i), (ii), and (iii), first set

$$a_1 = aa^{\textcircled{\dagger}}a.$$

Then, we have

$$\begin{aligned} a_1 a^{\textcircled{\dagger}} a_1 &= aa^{\textcircled{\dagger}}aa^{\textcircled{\dagger}}aa^{\textcircled{\dagger}}a \\ &= aa^{\textcircled{\dagger}}aa^{\textcircled{\dagger}}a \\ &= aa^{\textcircled{\dagger}}a \\ &= a_1, \\ a^{\textcircled{\dagger}}a_1a^{\textcircled{\dagger}} &= a^{\textcircled{\dagger}}aa^{\textcircled{\dagger}}aa^{\textcircled{\dagger}} \\ &= a^{\textcircled{\dagger}}aa^{\textcircled{\dagger}} \\ &= a^{\textcircled{\dagger}}, \end{aligned}$$

and

$$\begin{aligned} (a_1a^{\textcircled{\dagger}})^* &= (aa^{\textcircled{\dagger}}aa^{\textcircled{\dagger}})^* \\ &= (aa^{\textcircled{\dagger}})^* \\ &= aa^{\textcircled{\dagger}} \\ &= a_1a^{\textcircled{\dagger}}, \end{aligned}$$

which show that $a^{\textcircled{\dagger}}$ is a $\{1, 2, 3\}$-inverse of a_1. Thus, $\mathcal{A}a^{\textcircled{\dagger}} = \mathcal{A}a_1^*$. Also, note that

$$a_1a^{\textcircled{\dagger}} = aa^{\textcircled{\dagger}}aa^{\textcircled{\dagger}} = aa^{\textcircled{\dagger}} \quad \text{and} \quad a^{\textcircled{\dagger}}a_1 = a^{\textcircled{\dagger}}aa^{\textcircled{\dagger}}a = a^{\textcircled{\dagger}}a$$

Since $a^{\textcircled{\dagger}}$ is a generalized inverse of a_1, it follows that $a_1 \leq^- a$. Hence, for $a_2 = a - a_1$, we have $a^{\textcircled{\dagger}}a_2 = a^{\textcircled{\dagger}}(a - a_1) = a^{\textcircled{\dagger}}a - a^{\textcircled{\dagger}}a = 0$ and similarly, $a_2a^{\textcircled{\dagger}} = 0$.

By Corollary 5.7, there exists a reflexive generalized inverse h of a such that $a^{\textcircled{\dagger}} \leq^- h$. Since a is unit regular, there exists $c \in \mathfrak{K}(a)$ with $hc = ch = 0$, $h = a_c^- = (a + c)^{-1}a(a + c)^{-1}$, such that

$$a_c^- a_1 a_c^- = ha_1h = haa^{\textcircled{\dagger}}ah \tag{5.4}$$

$$= a^{\textcircled{\dagger}}ah \tag{5.5}$$

$$= a^{\textcircled{\dagger}} \quad (\because hah = h \text{ and } a^{\textcircled{\dagger}} \leq^- h). \tag{5.6}$$

Further, $a^{\textcircled{\dagger}} \leq^- h$ and $hc = ch = 0$ imply that $a^{\textcircled{\dagger}}c = ca^{\textcircled{\dagger}} = 0$. Now, it follows from $a^{\textcircled{\dagger}}a_2 = a_2a^{\textcircled{\dagger}} = 0$ that $a^{\textcircled{\dagger}}(a_2 + c) = (a_2 + c)a^{\textcircled{\dagger}} = 0$. Therefore, using $\mathcal{A}a^{\textcircled{\dagger}} = \mathcal{A}a_1^*$, we get

$$a_1^*(a + c) = a_1^*(a_1 + a_2 + c) = a_1^*a_1. \tag{5.7}$$

Note that

$$\begin{aligned} a_2a_1 &= (a - aa^{\textcircled{\dagger}}a)aa^{\textcircled{\dagger}}a \\ &= a^2a^{\textcircled{\dagger}}a - aa^{\textcircled{\dagger}}a^2a^{\textcircled{\dagger}}a \\ &= a^2a^{\textcircled{\dagger}}a - a^2a^{\textcircled{\dagger}}a \quad (\because a^{\textcircled{\dagger}}\mathcal{A} = a^k\mathcal{A} \text{ and } a^{\textcircled{\dagger}}a^{k+2} = a^{k+1}) \\ &= 0 \end{aligned}$$

and hence, we have

$$\begin{aligned} a_2^k &= (a - a_1)^k \\ &= a^k - a_1a^k \\ &= a^k - aa^{\textcircled{\dagger}}a^k \\ &= a^k - a^k \quad (\because a^k \in a^k\mathcal{A} = a^{\textcircled{\dagger}}\mathcal{A} \text{ and } a{a^{\textcircled{\dagger}}}^2 = a^{\textcircled{\dagger}}) \\ &= 0. \end{aligned}$$

Therefore,

$$a^k = a_1^k + a_1^{k-1}a_2 + \cdots + a_1a_2^{k-1} \tag{5.8}$$

and hence

$$\begin{aligned} (a+c)a^k &= a^{k+1} \quad (\because a^k\mathcal{A} = a^{\textcircled{\dagger}}\mathcal{A} \text{ and } ca^{\textcircled{\dagger}} = 0) \\ &= aa^k \\ &= (a_1 + a_2)a^k \\ &= a_1a^k. \quad (\because \text{Using Eq. (5.8) and } a_2a_1 = 0) \end{aligned} \tag{5.9}$$

Lastly, from (i)$\Longleftrightarrow$(iii) of Theorem 3.2, we have $a^{\textcircled{\dagger}}\mathcal{A} \subseteq {a^{\textcircled{\dagger}}}^*\mathcal{A}$. Since $a^{\textcircled{\dagger}} = ha_1h = (a+c)^{-1}a_1(a+c)^{-1}$, we have that

$$(a+c)^{-1}a_1\mathcal{A} = a^{\textcircled{\dagger}}\mathcal{A} \subseteq {a^{\textcircled{\dagger}}}^*\mathcal{A} = (a+c)^{-1*}a_1^*\mathcal{A}. \tag{5.10}$$

Thus, from Eqs. (5.4), (5.7), (5.9), and (5.10), it follows that $a^{\textcircled{\dagger}} = a_c^-a_1a_c^-$, where $a_1 \leq^- a$ and $c \in \mathfrak{K}(a)$ satisfy conditions given in (i), (ii), and (iii).

On the other hand, consider $g = a_c^-a_1a_c^-$, where $a_1 \leq^- a$ and $c \in \mathfrak{K}(a)$ satisfy conditions given in (i), (ii), and (iii). Then, by Theorem 5.6, g is an outer inverse of a. Note that

$$g\mathcal{A} = (a+c)^{-1}a_1\mathcal{A} = a^k\mathcal{A} \quad (\because \text{using condition (iii)}). \tag{5.11}$$

From condition (iii), we have

$$g\mathcal{A} = (a+c)^{-1}a_1\mathcal{A} \subseteq (a+c)^{-1*}a_1^*\mathcal{A} = g^*\mathcal{A}. \tag{5.12}$$

Further, from condition (i), we have

$$a_1(a+c)^{-1} = (a+c)^{-1^*}a_1^*a_1(a+c)^{-1},$$

which is symmetric. Therefore,

$$\begin{aligned}
g^*\mathcal{A} &= g^*a^*\mathcal{A} \quad (\because gag = g)\\
&= (a+c)^{-1^*}a_1^*(a+c)^{-1^*}a^*\mathcal{A}\\
&= (a+c)^{-1^*}a_1^*\mathcal{A} \quad (\because a_1 \leq^- a)\\
&= a_1(a+c)^{-1}\mathcal{A} \quad (\because a_1(a+c)^{-1} \text{ is symmetric})\\
&= a(a+c)^{-1}a_1(a+c)^{-1}\mathcal{A} \quad (\because a_1 \leq^- a)\\
&= ag\mathcal{A}\\
&= a^{k+1}\mathcal{A} \quad (\because g\mathcal{A} = a^k\mathcal{A})\\
&\subseteq a^k\mathcal{A} = g\mathcal{A}. \qquad (5.13)
\end{aligned}$$

Now, from Eqs. (5.11), (5.12) and (5.13), we get

$$g\mathcal{A} = g^*\mathcal{A} = a^k\mathcal{A}, \qquad (5.14)$$

and hence, g is the core-EP inverse of a, which proves the theorem. □

6. Conclusion

In this paper, several new characterizations of core-EP inverse of an element from an associative ring are given. The core-EP inverse is identified as a particular (b, c)-inverse, and an expression for core-EP inverse in the same terms is obtained. In the process, it is established that every element having core-EP inverse is a strongly π-regular element and the core-EP inverse of an element is a strongly regular element. Lastly, the core-EP inverse of a unit regular element from an associative ring with unity is characterized using inverse complements and minus partial order.

References

[1] Gorô Azumaya, *Strongly π-regular rings*, J. Fac. Sci. Hokkaido Univ. Ser. I. **13** (1954), 34–39. MR67864

[2] Oskar Maria Baksalary and Götz Trenkler, *Core inverse of matrices*, Linear Multilinear Algebra **58** (2010), no. 5-6, 681–697, DOI 10.1080/03081080902778222. MR2722752

[3] Ravindra B. Bapat, Surender Kumar Jain, K. Manjunatha Prasad Karantha, and M. David Raj, *Outer inverses: characterization and applications*, Linear Algebra Appl. **528** (2017), 171–184, DOI 10.1016/j.laa.2016.06.045. MR3652844

[4] Ravindra B. Bapat, Surender Kumar Jain, K. Manjunatha Prasad Karantha, and M. David Raj, *Outer inverses: characterization and applications*, Linear Algebra Appl. **528** (2017), 171–184, DOI 10.1016/j.laa.2016.06.045. MR3652844

[5] Adi Ben-Israel and Thomas N. E. Greville, *Generalized inverses*, 2nd ed., CMS Books in Mathematics/Ouvrages de Mathématiques de la SMC, vol. 15, Springer-Verlag, New York, 2003. Theory and applications. MR1987382

[6] K. Nayan Bhat, Manjunatha Prasad Karantha, and N. Eagambaram, *Inverse complements of a matrix and applications*, J. Algebra Appl. **20** (2021), no. 8, Paper No. 2150144, 12, DOI 10.1142/S0219498821501449. MR4297328

[7] M. P. Drazin, *Pseudo-inverses in associative rings and semigroups*, Amer. Math. Monthly **65** (1958), no. 7, 506–514, `http://www.jstor.org/stable/2308576`.

[8] M. P. Drazin, *A class of outer generalized inverses*, Linear Algebra Appl. **436** (2012), no. 7, 1909–1923, `http://www.sciencedirect.com/science/article/pii/S0024379511006367`.

[9] N. Eagambaram, *$(i, j, \cdots, k)$-inverses via bordered matrices*, Sankhyā Ser. A **53** (1991), no. 3, 298–308. MR1189773

[10] Yuefeng Gao and Jianlong Chen, *Pseudo core inverses in rings with involution*, Comm. Algebra **46** (2018), no. 1, 38–50, DOI 10.1080/00927872.2016.1260729. MR3764841

[11] Robert E. Hartwig, *How to partially order regular elements*, Math. Japon. **25** (1980), no. 1, 1–13. MR571255

[12] S. K. Jain and K. Manjunatha Prasad, *Right-left symmetry of $aR \oplus bR = (a+b)R$ in regular rings*, J. Pure Appl. Algebra **133** (1998), no. 1-2, 141–142, DOI 10.1016/S0022-4049(97)00189-8. Ring theory (Miskolc, 1996). MR1653702

[13] Umashankara Kelathaya, Savitha Varkady, and Manjunatha Prasad Karantha, *Inverse complements and strongly unit regular elements*, J. Algebra Appl. **21** (2022), no. 10, Paper No. 2250190, 11, DOI 10.1142/S0219498822501900. MR4477150

[14] K. S. Subramonian Nambooripad, *The natural partial order on a regular semigroup*, Proc. Edinburgh Math. Soc. (2) **23** (1980), no. 3, 249–260, DOI 10.1017/S0013091500003801. MR620922

[15] K. Manjunatha Prasad and K. S. Mohana, *Core-EP inverse*, Linear Multilinear Algebra **62** (2014), no. 6, 792–802, DOI 10.1080/03081087.2013.791690. MR3195967

[16] C. Radhakrishna Rao and Sujit Kumar Mitra, *Generalized inverse of matrices and its applications*, John Wiley & Sons, Inc., New York-London-Sydney, 1971. MR338013

[17] Savitha Varkady, Umashankara Kelathaya, and Manjunatha Prasad Karantha, *Drazin inverse and generalization of core-nilpotent decomposition*, J. Algebra Appl. **23** (2024), no. 2, Paper No. 2450041, 12, DOI 10.1142/S0219498824500415. MR4688808

Department of Applied Statistics and Data Science, Prasanna School of Public Health, Manipal Academy of Higher Education, Manipal 576 104, Karnataka, India
Email address: `usksullia@gmail.com; umashankar.kelathaya@learner.manipal.edu`

Department of Applied Statistics and Data Science, Prasanna School of Public Health, Manipal Academy of Higher Education, Manipal 576 104, Karnataka, India
Email address: `savithavarkady@yahoo.co.in; savitha.v@learner.manipal.edu`

Department of Applied Statistics and Data Science, Prasanna School of Public Health, Manipal Academy of Higher Education, Manipal 576 104, Karnataka, India
Email address: `kmprasad63@gmail.com; km.prasad@manipal.edu`

Contemporary Mathematics
Volume **833**, 2026
https://doi.org/10.1090/conm/833/16668

Discrete multilinear time-invariant systems based on t-product

Wanli Ma and Yimin Wei

ABSTRACT. In this paper, we introduce a new class of discrete multilinear time-invariant system under the t-product (tDMLTI system). The t-product is a tensor multiplication which allows a third-order tensor to be written as a product of third-order tensors. The definition of t-product can be extended to higher order tensors through recursion and we focus on third-order tensors. Basic definitions about matrices can be extended to tensors using the t-product, such as the transpose, the identity, and the inverse. Using these definitions, we provide the tensor algebraic conditions for the stability, the reachability, and the observability of the tDMLTI systems, which are analogous to the conditions for traditional linear time-invariant systems. We also discuss the balanced realization and give an error bound for the balanced truncation. Besides, a theorem about the pole assignment under the state feedback control is proved. Finally, we mention the definition of the higher order t-product and some numerical experiments are given to illustrate our theories.

1. Introduction

The theory of linear system has been well studied in the past decades. On one hand, it plays an important role in the application of process control, aerospace and aviation, and many other fields. On the other hand, it provides a foundation for other branches of modern control theory. In recent years, as the applications of high-dimensional data and models have become more common, a batch of research concerning multilinear systems has emerged. Rogers, Li, and Russell [**25**] proposed a novel model for tensor time series. Surana, Patterson, and Rajapakse [**26**] introduced a model reduction and identification approach for multilinear dynamical system driven by noise. Ding *et al.* [**8**] proposed a tensor-based linear dynamical system (tLDS) for modeling tensor observations in time series and employed the Tucker decomposition to estimate the parameters of the tLDS model as action descriptors. These multilinear system representations are limited because the tensor operators are all assumed to be formed from the Tucker decomposition.

2020 *Mathematics Subject Classification.* Primary 15A69, 93B35; Secondary 93C05, 93D15.
Key words and phrases. Discrete time-invariant system, model reduction, multilinear control system, pole assignment, tensor t-product.
The first author is supported by the National Natural Science Foundation of China under grant U24A2001. The second author is supported by the Science and Technology Commission of Shanghai Municipality under grant 23JC1400501 and the National Natural Science Foundation of China under grant 12271108.

Using the Einstein product, Chen *et al.* [**3**] generalized the notion of multilinear time-invariant (MLTI) system to normal even-order tensors. They developed tensor algebraic conditions for the stability, the reachability, and the observability for generalized MLTI systems. In [**4**], Chen *et al.* developed a framework for model reduction as well as providing computational and memory complexity analysis of MLTI systems.

In [**13**] and [**12**], Kilmer *et al.* defined a new kind of multiplication operation called t-product which is closed on the set of third-order tensors. Based on discrete Fourier transform and the properties of the block circulant matrices, the definations of tensor factorizations under t-product such as T-SVD [**12**], T-QR [**11**], T-UTV [**2**], T-Schur [**6**], T-CUR [**5, 27**] and T-eigendecomposition [**19**] have been introduced, which are analogous to matrix factoratizations. Tensor t-product has been proved to be of great use in applications, such as the image processing [**11**], the multilinear principle component analysis [**23**] and the tensor total least squares [**9**]. In [**17**], Martin, Shafer, and Larue extended the definition of the tensor t-product to order-p ($p \geq 4$) tensors and the corresponding notions of order-p tensor identity, the inverse, the orthogonality, and the SVD were also provided.

To take advantage of the properties of the t-product and extend the multilinear system theory to tensors of arbitrary order, we introduce the discrete multilinear time-invariant system under the tensor t-product in this paper. We call it "tDMLTI system" in the following contents. We express and prove the stability and asymptotic stability conditions of the tDMLTI system using T-eigendecomposition and T-Jordan canonical form [**19**]. We describe the reachability, the controllability, and the observability of tDMLTI systems and establish results about the conditions for these properties. In addition, we discuss the balanced realization and the model truncation of tDMLTI systems based on T-SVD. The pole assignment problem under the state feedback control is also involved.

The remainders of this paper is organized as follows. In Section 2, we introduce basic definitions and notations about tensor t-product. Then we give the definition of tDMLTI systems and the transfer function $G(z)$ in Section 3. In Section 4, we provide the tensor algebraic conditions for the stability, the reachability, and the observability of tDMLTI systems. In Section 5, we consider the balanced realization of tDMLTI systems and give an error bound for the balanced truncation. Besides, we discuss the pole assignment problem for tDMLTI systems under the state feedback control in Section 6. Numerical experiments of our theories are given in Section 7.

2. Preliminaries

A tensor $\mathcal{A} = (a_{i_1 \dots i_m})$ is a multi-array of entries $a_{i_1 \dots i_m} \in \mathbb{F}$, where $i_j = 1, 2, \dots, n_j$ for $j = 1, 2, \dots, m$ and $\mathbb{F}$ is a field. Here, m is called the order of $\mathcal{A}$ and $(n_1, n_2, \dots, n_m)$ is the dimension of $\mathcal{A}$, or simply, we call $n_1 \times n_2 \times \dots \times n_m$ the size of $\mathcal{A}$. Throughout this paper, we only consider real or complex tensors, that is, $\mathbb{F} = \mathbb{R}$ or $\mathbb{F} = \mathbb{C}$, and $n_1, n_2, \dots, n_m, m$ are integers. We use the notions a, $\mathbf{a}$, A, and $\mathcal{A}$ to represent scalar, vector, matrix, and tensor, respectively. For simplicity of notations, we use 0 to represent the zero vector, matrix, and tensor. The notation $\imath$ denotes the imaginary unit.

Given a third order tensor $\mathcal{A} = (a_{i_1 i_2 i_3}) \in \mathbb{C}^{m \times n \times p}$, we denote the jth *frontal slice* of $\mathcal{A}$ as $A^{(j)}$, that is, the (i_1, i_2)-th element of the matrix $A^{(j)}$ is equal to $a_{i_1 i_2 j}$. In terms of MATLAB indexing notation, this means $A^{(j)} = \mathcal{A}(:, :, j)$. Similarly, we call $\mathcal{A}(i_1, i_2, :) \in \mathbb{C}^{1\times 1\times p}$ the (i_1, i_2)-th *tubal* of $\mathcal{A}$. The *t-product* between two third-order tensors was first introduced by Kilmer *et al.* in [**13**]. They apply the following operators about converting tensor $\mathcal{A} \in \mathbb{C}^{m\times n\times p}$ into matrix.

$$\mathsf{bcirc}(\mathcal{A}) = \begin{bmatrix} A^{(1)} & A^{(p)} & \dots & A^{(2)} \\ A^{(2)} & A^{(1)} & \ddots & \vdots \\ \vdots & \ddots & \ddots & A^{(p)} \\ A^{(p)} & \dots & A^{(2)} & A^{(1)} \end{bmatrix}, \tag{2.1}$$

$$\mathsf{unfold}(\mathcal{A}) = [(A^{(1)})^\top (A^{(2)})^\top \cdots (A^{(p)})^\top]^\top \tag{2.2}$$

and the fold operator undoes the unfold operation, that is, $\mathsf{fold}(\mathsf{unfold}(\mathcal{A})) = \mathcal{A}$.

DEFINITION 2.1 ([**11**, Definition 2.5]). Let $\mathcal{A}$ be a tensor in $\mathbb{R}^{m\times l\times p}$ and $\mathcal{B}$ be a tensor in $\mathbb{R}^{l\times n\times p}$. Then the t-product $\mathcal{A} * \mathcal{B}$ is the $m\times n\times p$ tensor

$$\mathcal{A} * \mathcal{B} = \mathsf{fold}\left(\mathsf{bcirc}(\mathcal{A}) \cdot \mathsf{unfold}(\mathcal{B})\right), \tag{2.3}$$

where '$\cdot$' denotes the matrix multiplication.

The *transpose* $\mathcal{A}^\top$ is obtained by transposing the frontal slices $A^{(1)}, A^{(2)}, \dots, A^{(p)}$ and then reversing the order of the frontal slices $(A^{(2)})^\top, (A^{(3)})^\top, \dots, (A^{(p)})^\top$. The *conjugate transpose* $\mathcal{A}^{\mathrm{H}}$ is obtained by conjugate transposing all the frontal slices and then reversing the order of $(A^{(2)})^{\mathrm{H}}, (A^{(2)})^{\mathrm{H}}, \dots, (A^{(p)})^{\mathrm{H}}$. The *identity tensor* $\mathcal{I}$ is defined as the tensor whose first frontal slice is the identity matrix and the other frontal slices are all zeros. A tensor $\mathcal{P} \in \mathbb{R}^{n\times n\times p}$ is *orthogonal* if $\mathcal{P}^\top * \mathcal{P} = \mathcal{P} * \mathcal{P}^\top = \mathcal{I}$. A tensor $\mathcal{Q} \in \mathbb{C}^{n\times n\times p}$ is *unitary* if $\mathcal{Q}^{\mathrm{H}} * \mathcal{Q} = \mathcal{Q} * \mathcal{Q}^{\mathrm{H}} = \mathcal{I}$. Tensor $\mathcal{B} \in \mathbb{C}^{n\times n\times p}$ is called the *inverse* of $\mathcal{A} \in \mathbb{C}^{n\times n\times p}$ if $\mathcal{A} * \mathcal{B} = \mathcal{I}$ and $\mathcal{B} * \mathcal{A} = \mathcal{I}$. Lund [**16**] proved the following properties.

LEMMA 2.2 ([**16**, Lemma 1]). *Let tensors $\mathcal{A} \in \mathbb{C}^{m\times l\times p}$ and $\mathcal{B} \in \mathbb{C}^{l\times n\times p}$, then*
(1) $\mathsf{bcirc}(\mathcal{A} * \mathcal{B}) = \mathsf{bcirc}(\mathcal{A})\mathsf{bcirc}(\mathcal{B})$,
(2) $\mathsf{bcirc}(\mathcal{A})^j = \mathsf{bcirc}(\mathcal{A}^j)$, *for all* $j = 0, 1, \dots,$
(3) $(\mathcal{A} * \mathcal{B})^{\mathrm{H}} = \mathcal{B}^{\mathrm{H}} * \mathcal{A}^{\mathrm{H}}$,
(4) $\mathsf{bcirc}(\mathcal{A}^\top) = \mathsf{bcirc}(\mathcal{A})^\top$, $\mathsf{bcirc}(\mathcal{A}^{\mathrm{H}}) = \mathsf{bcirc}(\mathcal{A})^{\mathrm{H}}$.

We define the *q-norm* $(q \geq 1, q \in \mathbb{Z})$ of a third-order tensor $\mathcal{X}$ as

$$\| \mathcal{X} \|_q = \|\mathsf{bcirc}(\mathcal{X})\|_q. \tag{2.4}$$

Here "$\|\mathsf{bcirc}(\mathcal{X})\|_q$" represents the q-norm of matrix $\mathsf{bcirc}(\mathcal{X})$. Using Lemma 2.2, we can prove the following inequality of the tensor q-norm.

LEMMA 2.3. *For any third-order tensors $\mathcal{A} \in \mathbb{C}^{m\times l\times p}$, $\mathcal{B} \in \mathbb{C}^{l\times n\times p}$, and integer $q \geq 1$, we have the following inequality: $\| \mathcal{A} * \mathcal{B} \|_q \leq \| \mathcal{A} \|_q \| \mathcal{B} \|_q$.*

We call a third-order tensor $\mathcal{D} \in \mathbb{C}^{n\times n\times p}$ an *F-diagonal tensor* if all of its frontal slices $D^{(1)}, D^{(2)}, \dots, D^{(p)}$ are diagonal matrices. For two tensors $\mathcal{A}, \mathcal{B} \in \mathbb{C}^{n\times n\times p}$, we say $\mathcal{B}$ is *similar* to $\mathcal{A}$ if there exists an invertible tensor $\mathcal{P} \in \mathbb{C}^{n\times n\times p}$ satisfying $\mathcal{B} = \mathcal{P}^{-1} * \mathcal{A} * \mathcal{P}$. Miao, Qi, and Wei [**19**] proved the following lemma.

LEMMA 2.4 ([**19**, Lemma 4]). *Suppose $A^{(1)}, A^{(2)}, \dots, A^{(p)}$, $\tilde{A}^{(1)}, \tilde{A}^{(2)}, \dots, \tilde{A}^{(p)} \in \mathbb{C}^{n\times n}$ are complex matrices satisfying*

$$\mathsf{bcirc}(\mathcal{A}) = (F_p \otimes I_n)\,\mathrm{diag}(\tilde{A}^{(1)}, \tilde{A}^{(2)}, \cdots, \tilde{A}^{(p)})\left(F_p^{\mathrm{H}} \otimes I_n\right),$$

where F_p is the normalized discrete Fourier transform matrix of size $p \times p$ and "$\otimes$" denotes the Kronecker product. Then, $\tilde{A}^{(1)}, \tilde{A}^{(2)}, \ldots, \tilde{A}^{(p)}$ are diagonal (sub-diagonal, upper-triangular, lower-triangular) matrices if and only if $A^{(1)}, A^{(2)}, \ldots, A^{(p)}$ are diagonal (sub-diagonal, upper-triangular, lower-triangular) matrices.

Using MATLAB notation, we define $\tilde{\mathcal{A}} = \mathsf{ifft}(\mathcal{A}, [], 3)$ as the tensor obtained by applying inverse fast Fourier transform (IFFT) along each tubal element of $\mathcal{A}$ and then $\tilde{A}^{(i)} = \tilde{\mathcal{A}}(:, :, i)$. In the following contents of this paper, the tilde notation remains the same meaning as we have introduced here.

DEFINITION 2.5. $\mathcal{A} \in \mathbb{C}^{n \times n \times p}$ is *Hermitian positive definite* if all the matrices $\tilde{A}^{(i)}$ are Hermitian positive definite.

DEFINITION 2.6. The *multirank* of a tensor $\mathcal{A} \in \mathbb{C}^{n \times n \times p}$ is a vector $\mathbf{r} = (r_1, r_2, \ldots, r_p)$ such that r_i is the rank of $\tilde{A}^{(i)}$.

Based on Lemma 2.4, Miao, Qi, and Wei [**19**] deduced the *T-Jordan canonical form* of third-order tensors.

LEMMA 2.7 ([**19**, Theorem 1]). *Let $\mathcal{A} \in \mathbb{C}^{n \times n \times p}$ be a complex tensor. Then there exists an invertible tensor $\mathcal{P} \in \mathbb{C}^{n \times n \times p}$ and an F-upper-bi-diagonal tensor $\mathcal{J} \in \mathbb{C}^{n \times n \times p}$ such that*

$$\mathcal{A} = \mathcal{P}^{-1} * \mathcal{J} * \mathcal{P}.$$

Here "F-upper-bi-diagonal tensor" means that all the frontal slices are upper-bi-diagonal matrices. The diagonal elements of $\tilde{J}^{(i)} (i = 1, 2, \ldots, p)$ are called the T-eigenvalues *of $\mathcal{A}$.*

Liu and Jin [**14**] also defined the T-eigenvalue and T-eigenvector of a third-order tensor $\mathcal{A}$. This definition is equivalent to the one given in [**19**]. Perturbation analysis on T-eigenvalues of third-order tensors can be found in [**21**].

DEFINITION 2.8 ([**14**, Definition 2.5]). Let $\mathcal{A} \in \mathbb{C}^{n \times n \times p}$. Suppose that $\mathcal{X} \in \mathbb{C}^{n \times 1 \times p}$ and $\mathcal{X} \neq 0$. If

$$\mathcal{A} * \mathcal{X} = \lambda \mathcal{X}, \quad \lambda \in \mathbb{C},$$

then λ is called a *T-eigenvalue* of $\mathcal{A}$ and $\mathcal{X}$ is a *T-eigenvector* of $\mathcal{A}$ associated to λ.

Kilmer, Martin, and Perrone extended the definition of singular value decomposition to third-order tensors using t-product in [**13**].

DEFINITION 2.9 ([**13**, Theorem 5.1]). Let $\mathcal{A}$ be an $n \times n \times p$ complex-valued tensor, then $\mathcal{A}$ can be factorized as

$$\mathcal{A} = \mathcal{U} * \mathcal{S} * \mathcal{V}^{\mathrm{H}}, \tag{2.5}$$

where $\mathcal{U}, \mathcal{V} \in \mathbb{C}^{n \times n \times p}$ are unitary tensors and $\mathcal{S} \in \mathbb{R}^{n \times n \times p}$ is an F-diagonal tensor. The factorization (2.5) is called T-SVD.

According to Lemma 2.4, $\tilde{S}^{(1)}, \tilde{S}^{(2)}, \ldots, \tilde{S}^{(p)}$ are diagonal matrices, their diagonal elements are called the singular values of $\mathcal{A}$. T-SVD has been widely used in many high-dimensional problems, such as low-rank tensor completion [**28**], tensor principal component analysis [**15**] and the tensor function [**18**].

3. Discrete multilinear time-invariant systems based on t-product

Using the Einstein product, Chen *et al.* [**3**] generalized the notion of multilinear systems introduced in [**25, 26**] to general even-order tensors. To take advantage of the properties of tensor t-product and extend the definition of multilinear systems to tensors of arbitrary order, we develop the multilinear time-invariant system based on t-product. Recently, Hoover, Caudle, and Braman [**10**] extended the results of [**16**] and discussed some basic theories of continuous multilinear time-invariant systems. We focus on discrete-time systems in this paper. In addition, for the convenience of description, our discussion is restrict to the situation of third-order tensors. The results in the remainders of this paper can be extended to order-p ($p \geq 4$) tensors easily using the higher order t-product in [**17**].

DEFINITION 3.1. The discrete multilinear time-invariant system based on t-product (tDMLTI system) is given by

$$\begin{cases} \mathcal{X}_{t+1} = \mathcal{A} * \mathcal{X}_t + \mathcal{B} * \mathcal{U}_t, \\ \mathcal{Y}_t = \mathcal{C} * \mathcal{X}_t, \end{cases} \tag{3.1}$$

where $\mathcal{A}, \mathcal{B}, \mathcal{C} \in \mathbb{C}^{n \times n \times p}$ are complex time-invariant state tensor, input tensor, and output tensor, respectively; $\mathcal{U}_t, \mathcal{X}_t, \mathcal{Y}_t \in \mathbb{C}^{n \times n \times p}$ are time-dependent input tensor, state tensor, and output tensor, respectively.

LEMMA 3.2. *For an unforced tDMLTI system $\mathcal{X}_{t+1} = \mathcal{A} * \mathcal{X}_t$, the solution for $\mathcal{X}$ at time t, given initial condition $\mathcal{X}_0$, is $\mathcal{X}_t = \mathcal{A}^t * \mathcal{X}_0$, where $\mathcal{A}^t = \mathcal{A} * \mathcal{A} * \cdots * \mathcal{A}$.*

The proof is straightforward using the definition of t-product. According to Lemma 3.2, we can express the explicit solution of (3.1) as

$$\mathcal{X}_t = \mathcal{A}^t * \mathcal{X}_0 + \sum_{i=0}^{t-1} \mathcal{A}^{t-i-1} * \mathcal{B} * \mathcal{U}_j. \tag{3.2}$$

DEFINITION 3.3. The transfer function $G(z)$ of (3.1) is given by

$$G(z) = C * (zI - A)^{-1} * B, \tag{3.3}$$

where z is a complex variable.

4. tDMLTI systems theory

In this section, we introduce the stability, reachability, controllability, and observability of tDMLTI systems. We provide tensor algebra-based criteria for these properties and discuss the difference between them.

4.1. Stability. Generalized from LTI systems, the equilibrium point $\mathcal{X} = 0$ of an unforced tDMLTI system is called *stable* if $\| \mathcal{X}_t \|_2 \leq \gamma \| \mathcal{X}_0 \|_2$ for some $\gamma > 0$, *asymptotically stable* if $\| \mathcal{X}_t \|_2 \to 0$ as $t \to \infty$, and *unstable* if it is not stable.

LEMMA 4.1. *Let λ_j be the T-eigenvalues of $\mathcal{A}$ for $j = 1, 2, \ldots, np$. For an unforced tDMLTI system, the equilibrium point $\mathcal{X} = 0$ is*

1. *stable if and only if all T-eigenvalues of $\mathcal{A}$ have magnitude no larger than 1 and for those equal to 1, the algebraic and geometry multiplicities must be equal;*
2. *asymptotically stable if all T-eigenvalues of $\mathcal{A}$ have magnitude strictly less than 1.*

PROOF. According to Lemma 3.2 and Lemma 2.7, we have

$$\mathcal{X}_t = \mathcal{A}^t * \mathcal{X}_0 = \mathcal{P}^{-1} * \mathcal{J}^t * \mathcal{P} * \mathcal{X}_0,$$

thus

$$\begin{aligned}\| \mathcal{X}_t \|_2 &= \|\mathcal{P}^{-1} * \mathcal{J}^t * \mathcal{P} * \mathcal{X}_0 \|_2 \le \|\mathcal{P}^{-1}\|_2 \| \mathcal{J}^t \|_2 \|\mathcal{P}\| \| \mathcal{X}_0 \|_2 \\ &\le c \| \mathcal{J}^t \|_2 \| \mathcal{X}_0 \|_2 = c \|\mathsf{bcirc}(\mathcal{J}^t)\|_2 \| \mathcal{X}_0 \|_2 = c \|\mathsf{bcirc}(\mathcal{J})^t\|_2 \| \mathcal{X}_0 \|_2,\end{aligned}$$

where c is a constant. From Lemma 2.4, there exist $\tilde{J}^{(i)}$ $(i = 1, 2, \dots, p)$ such that

$$\mathsf{bcirc}(\mathcal{J}) = (F_p \otimes I_n)\mathrm{diag}(\tilde{J}^{(1)}, \tilde{J}^{(2)}, \cdots, \tilde{J}^{(p)})(F_p^{\mathrm{H}} \otimes I_n),$$

where $\tilde{J}^{(1)}, \tilde{J}^{(2)}, \dots, \tilde{J}^{(p)}$ are upper-bi-diagonal matrices and the diagonal elements are the T-eigenvalues of $\mathcal{X}_t$. Therefore,

$$\begin{aligned}\| \mathcal{X}_t \|_2 &\le c \|\mathsf{bcirc}(\mathcal{J})^t\|_2 \| \mathcal{X}_0 \|_2 \\ &= c \| \mathcal{X}_0 \|_2 \left\| (F_p \otimes I_n)\mathrm{diag}(\tilde{J}^{(1)}, \tilde{J}^{(2)}, \cdots, \tilde{J}^{(p)})^t (F_p^{\mathrm{H}} \otimes I_n) \right\|_2 \\ &= c \left\| \mathrm{diag}(\tilde{J}^{(1)}, \tilde{J}^{(2)}, \cdots, \tilde{J}^{(p)})^t \right\|_2 \| \mathcal{X}_0 \|_2.\end{aligned}$$

Denote $J_1, J_2, \dots, J_m$ as all the Jordan blocks in the set $\{\tilde{J}^{(1)}, \tilde{J}^{(2)}, \dots, \tilde{J}^{(p)}\}$. For those J_i with magnitude of diagonal elements less than 1, we have $J_i^t \to 0$ when $t \to +\infty$. For those J_i with magnitude of diagonal elements equal to 1, $\lim\limits_{t \to +\infty} J_i^t$ are finite if and only if the block size is 1, that is, the algebraic multiplicity equals to the geometry multiplicity. If all the eigenvalues have magnitude less than 1, then $\|\mathsf{bcirc}(\mathcal{J})^t\| \to 0$ when $t \to +\infty$. The conclusion follows immediately. □

COROLLARY 4.2. *For an unforced tDMLTI system, the equilibrium point $\mathcal{X} = 0$ is asymptotically stable if the largest singular value of $\mathcal{A}$ is less than 1.*

PROOF. Equation (2.4) and Lemma 2.2 imply that

$$\begin{aligned}\| \mathcal{A}^t \|_2 &= \|\mathsf{bcirc}(\mathcal{A})^t\|_2 \le \|\mathsf{bcirc}(\mathcal{A})\|_2^t = \|\mathsf{bcirc}(\mathcal{U} * \mathcal{S} * \mathcal{V}^{\mathrm{H}})\|_2^t \\ &= \|\mathsf{bcirc}(\mathcal{U})\mathsf{bcirc}(\mathcal{S})\mathsf{bcirc}(\mathcal{V}^{\mathrm{H}})\|_2^t = \|\mathsf{bcirc}(\mathcal{S})\|_2^t \\ &= \left\| (F_p \otimes I_n)\mathrm{diag}(\tilde{S}^{(1)}, \tilde{S}^{(2)}, \cdots, \tilde{S}^{(p)})(F_p^{\mathrm{H}} \otimes I_n) \right\|_2^t \\ &= \left\| \mathrm{diag}(\tilde{S}^{(1)}, \tilde{S}^{(2)}, \cdots, \tilde{S}^{(p)}) \right\|_2^t = \sigma_{\max}(\mathcal{A})^t.\end{aligned}$$

□

4.2. Reachability and Controllability. In this section, we introduce the definition of reachability and controllability of tDMLTI systems, and then establish the sufficient and necessary conditions for these properties. In addition, we discuss the relation between reachability and controllability for tDMLTI systems.

DEFINITION 4.3 (Reachability). A state $\mathcal{X}_1$ is *reachable* from state $\mathcal{X}_0$ in the time interval $[t_0, t_1]$ if there exists a sequence of inputs $\mathcal{U}_t$ that steers the state of the system from $\mathcal{X}_{t_0} = \mathcal{X}_0$ to $\mathcal{X}_{t_1} = \mathcal{X}_1$. The tDMLTI system (3.1) is *reachable* on $[t_0, t_1]$ if, given initial condition $\mathcal{X}_0 = 0$ and any final state $\mathcal{X}_1$, there exists a sequence of inputs $\mathcal{U}_t$ that steers the state of the system from $\mathcal{X}_{t_0} = \mathcal{X}_0$ to $\mathcal{X}_{t_1} = \mathcal{X}_1$.

THEOREM 4.4. *The pair $(\mathcal{A}, \mathcal{B})$ is reachable on $[t_0, t_1]$ if and only if the reachability Gramian*

$$\mathcal{W}_r(t_0, t_1) = \sum_{t=t_0}^{t_1-1} \mathcal{A}^{t_1-t-1} * \mathcal{B} * \mathcal{B}^{\mathrm{H}} * (\mathcal{A}^{\mathrm{H}})^{t_1-t-1} \tag{4.1}$$

is positive definite.

PROOF. "$\Leftarrow$": Suppose $\mathcal{W}_r(t_0, t_1)$ is positive definite. Denote $\mathcal{X}_0 = 0$ as the initial state and $\mathcal{X}_1$ as the desired state. Let $\mathcal{U}_t = \mathcal{B}^{\mathrm{H}} * (\mathcal{A}^{\mathrm{H}})^{t_1-t-1} * \mathcal{W}_r^{-1}(t_0, t_1) * \mathcal{V}$ for some constant tensor $\mathcal{V}$. According to (3.2),

$$\mathcal{X}_{t_1} = \mathcal{A}^{t_1} * \mathcal{X}_0 + \sum_{j=t_0}^{t_1-1} \mathcal{A}^{t_1-j-1} * \mathcal{B} * \mathcal{U}_t = \mathcal{W}_r(t_0, t_1) * \mathcal{W}_r^{-1}(t_0, t_1) * \mathcal{V} = \mathcal{V}.$$

Let $\mathcal{V} = \mathcal{X}_1$, we have $\mathcal{X}_{t_1} = \mathcal{X}_1$.

"$\Rightarrow$": Suppose that $\mathcal{W}_r(t_0, t_1)$ is not positive definite, then there exist $\mathbf{x} \neq 0 \in \mathbb{C}^n$ and $j \in \{1, 2, \ldots, p\}$ such that $\mathbf{x}^{\mathrm{H}} \tilde{W}^{(j)} \mathbf{x} = 0$. Without loss of generality, we suppose $j = 1$. Let $\mathcal{X} \in \mathbb{C}^{n \times n \times p}$ be the tensor with $\tilde{X}^{(2)} = \tilde{X}^{(3)} = \cdots \tilde{X}^{(p)} = 0$ and $\tilde{X}^{(1)} = [\mathbf{x}, 0, \ldots, 0]$, then $\mathcal{X}^{\mathrm{H}} * \mathcal{W} * \mathcal{X} = 0$ hence $\mathcal{X}^{\mathrm{H}} * \sum_{j=t_0}^{t_1-1} \mathcal{A}^{t_1-j-1} * \mathcal{B} = 0$. Take $\mathcal{X}_0 = 0$ and $\mathcal{X}_1 = \mathcal{X}$, we have $\mathcal{X} = \sum_{j=t_0}^{t_1-1} \mathcal{A}^{t_1-j-1} * \mathcal{B} * \mathcal{U}_j$. Multiplying $\mathcal{X}^{\mathrm{H}}$ from the left on both sides of the equation shows that

$$\mathcal{X}^{\mathrm{H}} * \mathcal{X} = \left(\mathcal{X}^{\mathrm{H}} * \sum_{j=t_0}^{t_1-1} \mathcal{A}^{t_1-j-1} * \mathcal{B} \right) * \mathcal{U}_j = 0,$$

hence $\mathcal{X} = 0$, which leads to a contradiction. □

LEMMA 4.5. *The pair $(\mathcal{A}, \mathcal{B})$ is reachable if and only if the multirank of the reachability tensor*

$$\mathcal{R} = \begin{bmatrix} \mathcal{B} & \mathcal{A} * \mathcal{B} & \cdots & \mathcal{A}^{np-1} * \mathcal{B} \end{bmatrix} \tag{4.2}$$

equals to $[n, n, \ldots, n] \in \mathbb{R}^{1 \times p}$.

PROOF. From (3.2) we can see that

$$\begin{bmatrix} \mathcal{B} & \mathcal{A} * \mathcal{B} & \cdots & \mathcal{A}^{t_1-t_0-1} * \mathcal{B} \end{bmatrix} * \begin{bmatrix} \mathcal{U}_{t_1-t_0-1} \\ \mathcal{U}_{t_1-t_0-2} \\ \vdots \\ \mathcal{U}_{t_0} \end{bmatrix} = \mathcal{X}_{t_1}.$$

Therefore reachability on $[t_0, t_1]$ depends on $t_1 - t_0$. The Cayley-Hamilton theorem for the T-eigenvalue problem [**19**, Theorem 6] implies that consideration of $t_1 - t_0 > np$ is superfluous to the rank condition. Hence we only need to consider the tensor $\mathcal{R}$. Suppose that the T-SVD of $\mathcal{R}$ is $\mathcal{R} = \mathcal{U} * \mathcal{S} * \mathcal{V}^{\mathrm{H}}$, if $\operatorname{rank}(\mathcal{R}) = [n, n, \ldots, n]$, then we have $\operatorname{rank}(\mathsf{bcirc}(\mathcal{S})) = np$, which means the matrix $\mathsf{bcirc}(\mathcal{S})\mathsf{bcirc}(\mathcal{S})^{\mathrm{H}}$ is invertible. As a consequence, the tensor $\mathcal{R} * \mathcal{R}^{\mathrm{H}}$ is invertible, so for given $\mathcal{X}_{t_1}$ we can input sequence by setting $\begin{bmatrix} \mathcal{U}_{t_1-t_0-1} \\ \mathcal{U}_{t_1-t_0-2} \\ \vdots \\ \mathcal{U}_{t_0} \end{bmatrix} = \mathcal{R}^{\mathrm{H}} * (\mathcal{R} * \mathcal{R}^{\mathrm{H}})^{-1} * \mathcal{X}_{t_1} = \mathcal{R}^{\dagger} * \mathcal{X}_{t_1}$.

On the other hand, if the rank condition fails, then there exists $\mathbf{x} \neq 0 \in \mathbb{C}^{np}$ such that $\mathsf{bcirc}(\mathcal{R}^{\mathrm{H}})\mathbf{x} = 0$. Denote $\mathsf{unfold}(\mathcal{X}) = \mathbf{x}$ and let $\mathcal{X}_0 = 0$, $\mathcal{X}_{np} = \mathcal{X}$, then if we premultiplying $\mathcal{X}^{\mathrm{H}}$ on both sides of $\mathcal{X} = \mathcal{R} * \begin{bmatrix} \mathcal{U}_{np-1} \\ \mathcal{U}_{np-2} \\ \vdots \\ \mathcal{U}_0 \end{bmatrix}$, we get $\mathcal{X}^{\mathrm{H}} * \mathcal{X} = 0$, which leads to a contradiction. □

DEFINITION 4.6. Suppose that system (3.1) is stable, then $\sum_{t=0}^{\infty} \mathcal{A}^t * \mathcal{B} * \mathcal{B}^{\mathrm{H}} * (\mathcal{A}^{\mathrm{H}})^t$ converges as $t \to +\infty$. Then the *infinite horizon reachability Gramian* can be defined as $\mathcal{W}_r = \sum_{t=0}^{\infty} \mathcal{A}^t * \mathcal{B} * \mathcal{B}^{\mathrm{H}} * (\mathcal{A}^{\mathrm{H}})^t$. It can be computed by the Lyapunov equation $\mathcal{A} * \mathcal{W}_r * \mathcal{A}^{\mathrm{H}} - \mathcal{W}_r = -\mathcal{B} * \mathcal{B}^{\mathrm{H}}$.

If the pair $(\mathcal{A}, \mathcal{B})$ is reachable and all the T-eigenvalues of $\mathcal{A}$ have magnitude less than one, then there exists a unique Hermitian positive definite solution $\mathcal{W}_r$. This can be proved by consider the equations $\tilde{A}^{(i)} \tilde{W}_r^{(i)} (\tilde{A}^{(i)})^{\mathrm{H}} - \tilde{W}_r^{(i)} = -\tilde{B}^{(i)} (\tilde{B}^{(i)})^{\mathrm{H}}$ $(i = 1, 2, \ldots, p)$ immediately.

DEFINITION 4.7 (Controllability). A state $\mathcal{X}_0$ is *controllable* to state $\mathcal{X}_1$ in the time interval $[t_0, t_1]$ if there exists a sequence of inputs $\mathcal{U}_t$ that steers the state of the system from $\mathcal{X}_{t_0} = \mathcal{X}_0$ to $\mathcal{X}_{t_1} = \mathcal{X}_1$. The tDMLTI system (3.1) is *controllable* on $[t_0, t_1]$ if, given any initial condition $\mathcal{X}_0$, there exists a sequence of inputs $\mathcal{U}_t$ that steers the state of the system from $\mathcal{X}_{t_0} = \mathcal{X}_0$ to $\mathcal{X}_{t_1} = 0$.

LEMMA 4.8. *The pair* $(\mathcal{A}, \mathcal{B})$ *is controllable if and only if* $\mathrm{Im}(\mathcal{A}^{np}) \subseteq \mathrm{Im}(\mathcal{R})$.

PROOF. "$\Rightarrow$": Suppose that $(\mathcal{A}, \mathcal{B})$ is controllable, then for any initial state $\mathcal{X}_{t_0} = \mathcal{X}_0$, there exists a sequence of $\mathcal{U}_t$ such that

$$0 = \mathcal{X}_{t_1} = \mathcal{A}^{t_1 - t_0} \mathcal{X}_0 + \sum_{t=t_0}^{t_1-1} \mathcal{A}^{t_1-t-1} * \mathcal{B} * \mathcal{U}_t,$$

i.e., $-\mathcal{A}^{t_1-t_0} \mathcal{X}_0 = \sum_{t=t_0}^{t_1-1} \mathcal{A}^{t_1-t-1} * \mathcal{B} * \mathcal{U}_t$. Therefore,

$$\mathcal{A}^{t_1-t_0} \mathcal{X}_0 \in \mathrm{Im}\left(\begin{bmatrix} \mathcal{B} & \mathcal{A} * \mathcal{B} & \cdots & \mathcal{A}^{t_1-t_0-1} * \mathcal{B} \end{bmatrix} \right)$$

for any $\mathcal{X}_0$. Let $t_1 - t_0 = np$, we get $\mathrm{Im}(\mathcal{A}^{np}) \subseteq \mathrm{Im}(\mathcal{R})$.

"$\Leftarrow$": Suppose that $\mathrm{Im}(\mathcal{A}^{np}) \subseteq \mathrm{Im}(\mathcal{R})$, then for any initial state $\mathcal{X}_{t_0} = \mathcal{X}_0$, there exists a sequence of $\mathcal{U}_t$, $t = t_0, t_0 + 1, \ldots, t_0 + np - 1$ such that $-\mathcal{A}^{np} \mathcal{X}_0 = \sum_{t=t_0}^{t_0+np-1} \mathcal{A}^{t_0+np-t-1} * \mathcal{B} * \mathcal{U}_t$, which means $(\mathcal{A}, \mathcal{B})$ is controllable. □

COROLLARY 4.9. *For tDMLTI system* (3.1), *the reachability implies the controllability but the controllability does not imply the reachability.*

PROOF. According to Lemma 4.5, the reachability implies that $\mathsf{rank}(\mathcal{R}) = np$, from which it follows that $\mathrm{Im}(\mathcal{A}^{np}) \subseteq \mathrm{Im}(\mathcal{R}) = \mathbb{C}^{n\times n\times p}$, then the system is surely controllable.

On the contrary, if $\mathcal{A} = 0$ and $\mathsf{rank}(\mathcal{B}) < [n, n, \ldots, n]$, then $\mathsf{rank}(\mathcal{R}) < [n, n, \ldots, n]$. Therefore the system is not reachable even though it is controllable. □

4.3. Observability.

DEFINITION 4.10 (Observability). The tDMLTI system (3.1) is observable on $[t_0, t_1]$ if any initial state $\mathcal{X}_{t_0} = \mathcal{X}_0$ can be uniquely determined by $\mathcal{Y}_t$ on $[t_0, t_1]$.

THEOREM 4.11. *The pair $(\mathcal{A}, \mathcal{C})$ is observable on $[t_0, t_1]$ if and only if the observability Gramian*

$$\mathcal{W}_o(t_0, t_1) = \sum_{t=t_0}^{t_1-1} (\mathcal{A}^{\mathrm{H}})^{t-t_0} * \mathcal{C}^{\mathrm{H}} * \mathcal{C} * \mathcal{A}^{t-t_0}, \tag{4.3}$$

is positive definite.

PROOF. "$\Leftarrow$": Suppose that $\mathcal{W}_o(t_0, t_1)$ is positive definite. Let $\mathcal{X}_0$ be the initial state and $\mathcal{Y}_t = \mathcal{C} * \mathcal{X}_t = \mathcal{C} * \mathcal{A}^{t-t_0} * \mathcal{X}_0$. Multiplying $(\mathcal{A}^{\mathrm{H}})^{t-t_0} * \mathcal{C}^{\mathrm{H}}$ from the left on both sides of the equation, we have $(\mathcal{A}^{\mathrm{H}})^{t-t_0} * \mathcal{C}^{\mathrm{H}} * \mathcal{Y}_t = (\mathcal{A}^{\mathrm{H}})^{t-t_0} * \mathcal{C}^{\mathrm{H}} * \mathcal{C} * \mathcal{A}^{t-t_0} * \mathcal{X}_0$, which implies that $\sum_{t=t_0}^{t_1-1} (\mathcal{A}^{\mathrm{H}})^{t-t_0} * \mathcal{C}^{\mathrm{H}} * \mathcal{Y}_t = \mathcal{W}_o(t_0, t_1) * \mathcal{X}_0$. Since $\mathcal{W}_o(t_0, t_1)$ is positive definite, this equation has a unique solution $\mathcal{X}_0 = \mathcal{W}_o^{-1}(t_0, t_1) \sum_{t=t_0}^{t_1-1} (\mathcal{A}^{\mathrm{H}})^{t-t_0} * \mathcal{C}^{\mathrm{H}} * \mathcal{Y}_t$. Therefore, $(\mathcal{A}, \mathcal{C})$ is observable on $[t_0, t_1]$.

"$\Rightarrow$": Suppose that $\mathcal{W}_o(t_0, t_1)$ is not positive definite, then there exists $\mathbf{x} \neq 0 \in \mathbb{C}^n$ and $j \in \{1, 2, \ldots, p\}$ such that $\mathbf{x}^{\mathrm{H}} \tilde{W}^{(j)} \mathbf{x} = 0$. Suppose that $j = 1$, let $\mathcal{X}$ be the tensor in $\mathbb{C}^{n\times n\times p}$ with $\tilde{X}^{(2)} = \cdots \tilde{X}^{(p)} = 0$ and $\tilde{X}^{(1)} = [\mathbf{x}, 0, \ldots, 0]$. From (4.3) we can see that $\mathcal{X}^{\mathrm{H}} * \mathcal{W}_o(t_0, t_1) * \mathcal{X} = 0$ and hence $\mathcal{C} * \mathcal{A}^{t-t_0} * \mathcal{X} = 0$. Take $\mathcal{X}_{t_0} = \mathcal{X}_0 + \mathcal{X}$, then $\mathcal{Y}_{t_1} = \mathcal{C} * \mathcal{A}^{t-t_0} * (\mathcal{X}_0 + \mathcal{X}) + \sum_{t=t_0}^{t_1-1} \mathcal{C} * \mathcal{A}^{t-t_0} * \mathcal{B} * \mathcal{U}_t = \mathcal{C} * \mathcal{A}^{t-t_0} * \mathcal{X}_0 + \sum_{t=t_0}^{t_1-1} \mathcal{C} * \mathcal{A}^{t-t_0} * \mathcal{B} * \mathcal{U}_t$. This means that the two states $\mathcal{X}_0$ and $\mathcal{X}_0 + \mathcal{X}$ produce the same output, which leads to a contradiction. □

LEMMA 4.12. *The pair $(\mathcal{A}, \mathcal{C})$ is observable if and only if the multirank of the observability tensor*

$$\mathcal{O} = \begin{bmatrix} \mathcal{C} \\ \mathcal{C} * \mathcal{A} \\ \vdots \\ \mathcal{C} * \mathcal{A}^{np-1} \end{bmatrix} \tag{4.4}$$

equals to $[n, n, \ldots, n] \in \mathbb{R}^{1\times p}$.

PROOF. The proof is similar to Lemma 4.5. □

DEFINITION 4.13. The infinite horizon observability Gramian of a given tDMLTI system (3.1) is defined as $\mathcal{W}_o = \sum_{t=0}^{\infty}(\mathcal{A}^{\mathrm{H}})^t * \mathcal{C}^{\mathrm{H}} * \mathcal{C} * \mathcal{A}^t$. It can be computed by the Lyapunov equation $\mathcal{A}^{\mathrm{H}} * \mathcal{W}_o * \mathcal{A} - \mathcal{W}_o = -\mathcal{C}^{\mathrm{H}} * \mathcal{C}$.

Similarly, if the pair $(\mathcal{A}, \mathcal{C})$ is observable and all the T-eigenvalues of $\mathcal{A}$ have magnitude less than one, then there exists a unique Hermitian positive definite solution $\mathcal{W}_o$.

5. Model reduction for tDMLTI systems

The aim of model order reduction problem is to simplify a high order model for easier implementation. Plenty of reduction methods have been developed for classical control systems. Using t-SVD and the idea of balanced realization, we propose a model reduction method for tDMLTI system. Proposed by Moore [**22**], the idea of balanced realization was derived from the concept of principal component analysis in statistical methods. Similar to the matrix case, the key property of the balanced realization of a tDMLTI system is to find a state coordinate basis such that the reachability and observability Gramians are both equal to some F-diagonal tensor $\mathcal{S}$. If $(\bar{\mathcal{A}}, \bar{\mathcal{B}}, \bar{\mathcal{C}})$ is a balanced realization, then the reachability and observability Gramian $\mathcal{S}$ must satisfy the following two Lyapunov equations,

$$\begin{cases} \bar{\mathcal{A}} * \mathcal{S} * \bar{\mathcal{A}}^{\mathrm{H}} - \mathcal{S} = -\bar{\mathcal{B}} * \bar{\mathcal{B}}^{\mathrm{H}}, \\ \bar{\mathcal{A}}^{\mathrm{H}} * \mathcal{S} * \bar{\mathcal{A}} - \mathcal{S} = -\bar{\mathcal{C}}^{\mathrm{H}} * \bar{\mathcal{C}}. \end{cases} \tag{5.1}$$

In this section, the initially given system is assumed to be asymptotically stable, reachable, and observable.

5.1. Balanced realization for tDMLTI system. For the tDMLTI system (3.1), the main step of a balanced realization is to find a state coordinate basis such that the infinite horizon reachability Gramian and observability Gramian are both equal to some F-diagonal tensor $\mathcal{S}$. The state space representation is then called a balanced realization. Let $\mathcal{W}_r$ and $\mathcal{W}_o$ be the infinite horizon reachability Gramian and observability Gramian, respectively. Since $\mathcal{W}_r$ and $\mathcal{W}_o$ are Hermitian, there exist unitary tensors $\mathcal{U}_r$ and $\mathcal{U}_o$ such that

$$\begin{cases} \mathcal{W}_r = \mathcal{U}_r * \mathcal{S}_r * \mathcal{U}_r^{\mathrm{H}}, \\ \mathcal{W}_o = \mathcal{U}_o * \mathcal{S}_o * \mathcal{U}_o^{\mathrm{H}}, \end{cases}$$

where $\mathcal{S}_r$ and $\mathcal{S}_o$ are F-diagonal tensors.

The tensor $\mathcal{H} = \mathcal{S}_o^{1/2} * \mathcal{U}_o^{\mathrm{H}} * \mathcal{U}_r * \mathcal{S}_r^{1/2}$ is constructed, and the t-SVD of $\mathcal{H}$ is denoted as $\mathcal{H} = \mathcal{U}_H * \mathcal{S}_H * \mathcal{V}_H^{\mathrm{H}}$. Using these tensors, the balancing transformation is given by $\mathcal{T} = \mathcal{U}_o * \mathcal{S}_o^{-1/2} * \mathcal{U}_H * \mathcal{S}_H^{1/2}$.

Then the balanced realization is $\bar{\mathcal{A}} = \mathcal{T}^{-1} * \mathcal{A} * \mathcal{T}$, $\bar{\mathcal{B}} = \mathcal{T}^{-1} * \mathcal{B}$, $\bar{\mathcal{C}} = \mathcal{C} * \mathcal{T}$ and the following formula is established,

$$\mathcal{T}^{\mathrm{H}} * \mathcal{W}_o * \mathcal{T} = \mathcal{T}^{-1} * \mathcal{W}_r * (\mathcal{T}^{-1})^{\mathrm{H}} = \mathcal{S}_H. \tag{5.2}$$

Therefore $\mathcal{S}_H = \mathcal{S}$.

5.2. Balanced truncation. For simplicity of notation, we assume that $(\mathcal{A},\mathcal{B},\mathcal{C})$ is a balanced realization for a given transfer function $G(z)$ such that the infinite horizon reachability Gramian and observability Gramian equal to the same F-diagonal tensor $\mathcal{S}$. Then we can truncate $\mathcal{A},\mathcal{B},\mathcal{C},\mathcal{S}$ into multirank-$\mathbf{r}$ tensors, where $\mathbf{r}=(r,r,\ldots,r)$. Specifically, suppose that

$$\mathsf{bcirc}(\mathcal{S})=(F_p\otimes I_n)\operatorname{diag}(\tilde{S}^{(1)},\tilde{S}^{(2)},\cdots,\tilde{S}^{(p)})(F_p^{\mathrm{H}}\otimes I_n),$$

where $\tilde{S}^{(i)}, i=1,2,\ldots,p$ are diagonal matrices and we assume that the diagonal elements $\sigma_1^{(i)},\sigma_2^{(i)},\ldots,\sigma_n^{(i)}$ of $\tilde{S}^{(i)}$ are arranged in desending order, that is, $\sigma_1^{(i)}\geq\sigma_2^{(i)}\geq\cdots\geq\sigma_n^{(i)}\geq 0$. Then the truncation $\mathcal{S}_{\mathbf{r}}$ has the form

$$\mathcal{S}_{\mathbf{r}}=\mathsf{bcirc}^{-1}\left((F_p\otimes I_n)\operatorname{diag}(\tilde{S}_r^{(1)},\tilde{S}_r^{(2)},\cdots,\tilde{S}_r^{(p)})(F_p^{\mathrm{H}}\otimes I_n)\right),\tag{5.3}$$

where $\tilde{S}_r^{(i)}$ is the rank-r truncation of the diagonal matrix $\tilde{S}^{(i)}$. Similarly, we truncate $\mathcal{A},\mathcal{B},\mathcal{C}$ into multirank-$\mathbf{r}$ tensors $\mathcal{A}_{\mathbf{r}},\mathcal{B}_{\mathbf{r}},\mathcal{C}_{\mathbf{r}}$, respectively. If we denote

$$\mathsf{bcirc}(\mathcal{A})=(F_p\otimes I_n)\operatorname{diag}(\tilde{A}^{(1)},\tilde{A}^{(2)},\cdots,\tilde{A}^{(p)})(F_p^{\mathrm{H}}\otimes I_n)$$

and $\tilde{A}^{(i)}=\begin{bmatrix}\tilde{A}_{11}^{(i)} & \tilde{A}_{12}^{(i)}\\ \tilde{A}_{21}^{(i)} & \tilde{A}_{22}^{(i)}\end{bmatrix}$, where $\tilde{A}_{11}^{(i)}\in\mathbb{C}^{r\times r}$. Then the truncation $\mathcal{A}_{\mathbf{r}}$ is formed as

$$\mathcal{A}_{\mathbf{r}}=\mathsf{bcirc}^{-1}\left((F_p\otimes I_n)\operatorname{diag}(\tilde{A}_{11}^{(1)},\tilde{A}_{11}^{(2)},\cdots,\tilde{A}_{11}^{(p)})(F_p^{\mathrm{H}}\otimes I_n)\right).$$

Similarly, $\mathcal{B},\mathcal{C}$ are partitioned as $\tilde{B}^{(i)}=\begin{bmatrix}\tilde{B}_1^{(i)}\\ \tilde{B}_2^{(i)}\end{bmatrix}, \tilde{C}^{(i)}=\begin{bmatrix}\tilde{C}_1^{(i)} & \tilde{C}_2^{(i)}\end{bmatrix}$.

Consequently, the truncated tensors $\mathcal{B}_{\mathbf{r}}$ and $\mathcal{C}_{\mathbf{r}}$ are formed as

$$\begin{cases}\mathcal{B}_{\mathbf{r}}=\mathsf{bcirc}^{-1}\left((F_p\otimes I_n)\operatorname{diag}(\tilde{B}_1^{(1)},\tilde{B}_1^{(2)},\cdots,\tilde{B}_1^{(p)})(F_p^{\mathrm{H}}\otimes I_n)\right)\\ \mathcal{C}_{\mathbf{r}}=\mathsf{bcirc}^{-1}\left((F_p\otimes I_n)\operatorname{diag}(\tilde{C}_1^{(1)},\tilde{C}_1^{(2)},\cdots,\tilde{C}_1^{(p)})(F_p^{\mathrm{H}}\otimes I_n)\right).\end{cases}$$

LEMMA 5.1. *The balanced truncation of the tDMLTI system* (3.1) *is*

$$\begin{cases}(\mathcal{X}_{t+1})_{\mathbf{r}}=\mathcal{A}_{\mathbf{r}}*(\mathcal{X}_t)_{\mathbf{r}}+\mathcal{B}_{\mathbf{r}}*(\mathcal{U}_t)_{\mathbf{r}},\\ \quad(\mathcal{Y}_t)_{\mathbf{r}}=\mathcal{C}_{\mathbf{r}}*(\mathcal{X}_t)_{\mathbf{r}},\end{cases}\tag{5.4}$$

where $\mathcal{A}_{\mathbf{r}},\mathcal{B}_{\mathbf{r}},\mathcal{C}_{\mathbf{r}}$ denote the balanced truncation of $\mathcal{A},\mathcal{B},\mathcal{C}$, respectively.

The truncated system in (5.4), derived from an asymptotically stable balanced realization, is also asymptotically stable. Following [**24**, Theorem 4.2], an even stronger result holds.

THEOREM 5.2. *Assume that a realization $(\mathcal{A},\mathcal{B},\mathcal{C})$ is asymptotically stable, reachable, observable, and that either the reachability Gramian or the observability Gramian is nonsingular and F-diagonal. Then every subsystem $(\mathcal{A}_{\mathbf{r}},\mathcal{B}_{\mathbf{r}},\mathcal{C}_{\mathbf{r}})$ is asymptotically stable.*

PROOF. Assume that the reachability Gramian is nonsingular, F-diagonal, and equals to $\mathcal{S}$, then the following Lyapunov equation holds, $\mathcal{A}*\mathcal{S}*\mathcal{A}^{\mathrm{H}}-\mathcal{S}+\mathcal{B}*\mathcal{B}^{\mathrm{H}}=0$.

Applying operator bcirc on both sides of the equation results

$$\mathsf{bcirc}(\mathcal{A})\mathsf{bcirc}(\mathcal{S})\mathsf{bcirc}(\mathcal{A})^{\mathrm{H}}-\mathsf{bcirc}(\mathcal{S})+\mathsf{bcirc}(\mathcal{B})\mathsf{bcirc}(\mathcal{B})^{\mathrm{H}}=0.$$

According to Lemma 2.4, this implies $\tilde{A}^{(i)}\tilde{S}^{(i)}(\tilde{A}^{(i)})^{\mathrm{H}} - \tilde{S}^{(i)} + \tilde{B}^{(i)}(\tilde{B}^{(i)})^{\mathrm{H}} = 0$, for $i = 1, 2, \ldots, p$. Hence

$$\begin{aligned} &\begin{bmatrix} \tilde{A}_{11}^{(i)} & \tilde{A}_{12}^{(i)} \\ \tilde{A}_{21}^{(i)} & \tilde{A}_{22}^{(i)} \end{bmatrix} \begin{bmatrix} \tilde{S}_r^{(i)} & \\ & \tilde{S}_{n-r}^{(i)} \end{bmatrix} \begin{bmatrix} (\tilde{A}_{11}^{(i)})^{\mathrm{H}} & (\tilde{A}_{21}^{(i)})^{\mathrm{H}} \\ (\tilde{A}_{12}^{(i)})^{\mathrm{H}} & (\tilde{A}_{22}^{(i)})^{\mathrm{H}} \end{bmatrix} \\ &- \begin{bmatrix} \tilde{S}_r^{(i)} & \\ & \tilde{S}_{n-r}^{(i)} \end{bmatrix} + \begin{bmatrix} \tilde{B}_1^{(i)} \\ \tilde{B}_2^{(i)} \end{bmatrix} \begin{bmatrix} (\tilde{B}_1^{(i)})^{\mathrm{H}} & (\tilde{B}_2^{(i)})^{\mathrm{H}} \end{bmatrix} = 0, \end{aligned} \tag{5.5}$$

for $i = 1, 2, \ldots, p$. According to Lemma 4.1, the truncated system is asymptotically stable if and only if all the T-eigenvalues of $\mathcal{A}_{\mathbf{r}}$ have magnitude less than 1. From the construction of $\mathcal{A}_{\mathbf{r}}$, we only need to prove that all the eigenvalues of $\tilde{A}_{11}^{(i)}$ have magnitude less than 1, for $i = 1, 2, \ldots, p$.

Consider the upper left corner of (5.5).

$$\tilde{A}_{11}^{(i)}\tilde{S}_r^{(i)}(\tilde{A}_{11}^{(i)})^{\mathrm{H}} + \tilde{A}_{12}^{(i)}\tilde{S}_{n-r}^{(i)}(\tilde{A}_{12}^{(i)})^{\mathrm{H}} - \tilde{S}_r^{(i)} = -\tilde{B}_1^{(i)}(\tilde{B}_1^{(i)})^{\mathrm{H}}. \tag{5.6}$$

Let $\mathbf{v}$ be an eigenvector of $(\tilde{A}_{11}^{(i)})^{\mathrm{H}}$ and let λ be the corresponding eigenvalue, that is, $(\tilde{A}_{11}^{(i)})^{\mathrm{H}}\mathbf{v} = \lambda\mathbf{v}$. Multiplying (5.6) from the right by $\mathbf{v}$ and from the left by $\mathbf{v}^{\mathrm{H}}$, then $(|\lambda|^2 - 1)\mathbf{v}^{\mathrm{H}}\tilde{S}_r^{(i)}\mathbf{v} = -(\mathbf{v}^{\mathrm{H}}\tilde{A}_{12}^{(i)}\tilde{S}_{n-r}^{(i)}(\tilde{A}_{12}^{(i)})^{\mathrm{H}} + \mathbf{v}^{\mathrm{H}}\tilde{B}_1^{(i)}(\tilde{B}_1^{(i)})^{\mathrm{H}}\mathbf{v})$.

Since the right-hand side is non-positive and $\mathbf{v}^{\mathrm{H}}\tilde{S}_{r_i}^{(i)}\mathbf{v}$ is positive, it follows that $|\lambda| \le 1$. If $|\lambda| = 1$, then from the positive definiteness of $\tilde{S}_{n-r}^{(i)}$ we get $\mathbf{v}^{\mathrm{H}}\tilde{A}_{12}^{(i)}\mathbf{v} = 0$ and $\mathbf{v}^{\mathrm{H}}\tilde{B}_1^{(i)} = 0$. Therefore,

$$\begin{bmatrix} \mathbf{v}^{\mathrm{H}} & 0 \end{bmatrix} \begin{bmatrix} \tilde{A}_{11}^{(i)} & \tilde{A}_{12}^{(i)} \\ \tilde{A}_{21}^{(i)} & \tilde{A}_{22}^{(i)} \end{bmatrix} = \bar{\lambda} \begin{bmatrix} \mathbf{v}^{\mathrm{H}} & 0 \end{bmatrix}, \qquad \begin{bmatrix} \mathbf{v}^{\mathrm{H}} & 0 \end{bmatrix} \begin{bmatrix} \tilde{B}_1^{(i)} \\ \tilde{B}_2^{(i)} \end{bmatrix} = 0,$$

which contradict the reachability of the origin system. Hence $|\lambda| < 1$ and the subsystem is asymptotically stable. □

LEMMA 5.3. *For the balanced realization* $(\mathcal{A}, \mathcal{B}, \mathcal{C})$, *we have* $\|\mathsf{bcirc}(\mathcal{A})\|_2 \le 1$. *If* $\tilde{S}^{(i)}$ *has distinct diagonal entries, for* $i = 1, 2, \ldots, p$, *then the strict inequality holds.*

PROOF. Multiplying (5.1) from the left by $\mathcal{A}^{\mathrm{H}}$ and from the right by $\mathcal{A}$ and adding the result to (5.1), we get

$$\mathcal{A}^{\mathrm{H}} * \mathcal{A} * \mathcal{S} * \mathcal{A}^{\mathrm{H}} * \mathcal{A} - \mathcal{S} = -(\mathcal{A}^{\mathrm{H}} * \mathcal{B} * \mathcal{B}^{\mathrm{H}} * \mathcal{A} + \mathcal{C}^{\mathrm{H}} * \mathcal{C}), \tag{5.7}$$

which implies

$$(\tilde{A}^{(i)})^{\mathrm{H}}\tilde{A}^{(i)}\tilde{S}^{(i)}(\tilde{A}^{(i)})^{\mathrm{H}}\tilde{A}^{(i)} - \tilde{S}^{(i)} = -\left((\tilde{A}^{(i)})^{\mathrm{H}}\tilde{B}^{(i)}(\tilde{B}^{(i)})^{\mathrm{H}}\tilde{A}^{(i)} + (\tilde{C}^{(i)})^{\mathrm{H}}\tilde{C}^{(i)}\right), \tag{5.8}$$

for $i = 1, 2, \ldots, p$.

Let $\mathbf{v}$ be an eigenvector of $(\tilde{A}^{(i)})^{\mathrm{H}}\tilde{A}^{(i)}$ and λ be the corresponding eigenvalue for some i, that is, $(\tilde{A}^{(i)})^{\mathrm{H}}\tilde{A}^{(i)}\mathbf{v} = \lambda\mathbf{v}$. Multiplying (5.8) from the right by $\mathbf{v}$ and from the left by $\mathbf{v}^{\mathrm{H}}$ implies

$$(|\lambda|^2 - 1)\mathbf{v}^{\mathrm{H}}\tilde{S}^{(i)}\mathbf{v} = -\left(\mathbf{v}^{\mathrm{H}}(\tilde{A}^{(i)})^{\mathrm{H}}\tilde{B}^{(i)}(\tilde{B}^{(i)})^{\mathrm{H}}\tilde{A}^{(i)}\mathbf{v} + \mathbf{v}^{\mathrm{H}}(\tilde{C}^{(i)})^{\mathrm{H}}\tilde{C}^{(i)}\mathbf{v}\right).$$

Note that the right-hand side is non-positive and $\mathbf{v}^{\mathrm{H}}\tilde{S}^{(i)}\mathbf{v}$ is positive, hence $|\lambda|^2 \le 1$. From the arbitrariness of λ and i we know that

$$\|\mathsf{bcirc}(\mathcal{A})\|_2 = \sqrt{\lambda_{\max}\big(\mathsf{bcirc}(\mathcal{A})^{\mathrm{H}}\mathsf{bcirc}(\mathcal{A})\big)} = \sqrt{\max_i \lambda_{\max}\big((\tilde{A}^{(i)})^{\mathrm{H}}\tilde{A}^{(i)}\big)} \le 1.$$

Suppose that $\|\mathsf{bcirc}(\mathcal{A})\|_2 = 1$, then $\lambda = 1$ is an eigenvalue for some $(\tilde{A}^{(i)})^{\mathrm{H}}\tilde{A}^{(i)}$ and $(\tilde{A}^{(i)})^{\mathrm{H}}\tilde{A}^{(i)} - I$ is singular. Let V be a matrix such that $\big((\tilde{A}^{(i)})^{\mathrm{H}}\tilde{A}^{(i)} - I\big)V = 0$. Multiplying (5.8) from the right by V and left by V^{H} implies

$$\begin{cases} (\tilde{B}^{(i)})^{\mathrm{H}}\tilde{A}^{(i)}V = 0, \\ \tilde{C}^{(i)}V = 0. \end{cases}$$

Then multiplying (5.8) from the right by V implies $\big((\tilde{A}^{(i)})^{\mathrm{H}}\tilde{A}^{(i)} - I\big)\tilde{S}^{(i)}V = 0$. Without loss of generality, we can assume that there exists a diagonal matrix Σ such that $\tilde{S}^{(i)}V = V\Sigma$.

Multiplying $\tilde{A}^{(i)}\tilde{S}^{(i)}(\tilde{A}^{(i)})^{\mathrm{H}} - \tilde{S}^{(i)} + \tilde{B}^{(i)}(\tilde{B}^{(i)})^{\mathrm{H}} = 0$ from the right by $\tilde{A}^{(i)}V$ gives $\tilde{S}^{(i)}\tilde{A}^{(i)}V = \tilde{A}^{(i)}V\Sigma$. Let $\mathbf{v}$ be any column of V and σ be the corresponding diagonal entry of Σ, then

$$\begin{cases} \tilde{S}^{(i)}\mathbf{v} = \sigma\mathbf{v}, \\ \tilde{S}^{(i)}\tilde{A}^{(i)}\mathbf{v} = \sigma\tilde{A}^{(i)}\mathbf{v}. \end{cases}$$

If $\tilde{S}^{(i)}$ has distinct diagonal entries, than $\mathbf{v}$ and $\tilde{A}^{(i)}\mathbf{v}$ must be parallel, that is, there exist α such that $\tilde{A}^{(i)}\mathbf{v} = \alpha\mathbf{v}$. However, from $\tilde{C}^{(i)}V = 0$ we get that $\tilde{C}^{(i)}\mathbf{v} = 0$, which contradicts the observability of the balanced system. Therefore, $\|\mathsf{bcirc}(\mathcal{A})\|_2 < 1$ when $\tilde{S}^{(i)}$ has distinct diagonal entries, for $i = 1, 2, \ldots, p$. ☐

Using Lemma 5.3, we can prove the condition for reachability and observability of the truncated system $(\mathcal{A}_{\mathbf{r}}, \mathcal{B}_{\mathbf{r}}, \mathcal{C}_{\mathbf{r}})$.

THEOREM 5.4. *If the partition satisfies $\sigma_{\min}(\tilde{S}_r^{(i)}) > \sigma_{\max}(\tilde{S}_{n-r}^{(i)})$, for any $i = 1, 2, \ldots, p$, then the subsystem $(\mathcal{A}_{\mathbf{r}}, \mathcal{B}_{\mathbf{r}}, \mathcal{C}_{\mathbf{r}})$ is reachable and observable.*

PROOF. From the upper left corner of (5.1) we can deduce that

$$(\tilde{A}_{11}^{(i)})^{\mathrm{H}}\tilde{S}_r^{(i)}\tilde{A}_{11}^{(i)} + (\tilde{A}_{21}^{(i)})^{\mathrm{H}}\tilde{S}_{n-r}^{(i)}\tilde{A}_{21}^{(i)} - \tilde{S}_r^{(i)} = -(\tilde{C}_1^{(i)})^{\mathrm{H}}\tilde{C}_1^{(i)}, \tag{5.9}$$

for $i = 1, 2, \ldots, p$.

Suppose that the subsystem $(\mathcal{A}_{\mathbf{r}}, \mathcal{B}_{\mathbf{r}}, \mathcal{C}_{\mathbf{r}})$ is not observable, then there exists some j such that $\operatorname{rank}(\tilde{C}_1^{(j)}) < r$. Hence we can find an unit eigenvector $\mathbf{v}$ with corresponding eigenvalue λ of $\tilde{A}_{11}^{(j)}$ such that

$$\begin{cases} \tilde{A}_{11}^{(j)}\mathbf{v} = \lambda\mathbf{v}, \\ \tilde{C}_1^{(j)}\mathbf{v} = 0. \end{cases}$$

For $i = j$, multiplying (5.9) from the right by $\mathbf{v}$ and left by $\mathbf{v}^{\mathrm{H}}$, we have

$$(1 - |\lambda|^2)\mathbf{v}^{\mathrm{H}}\tilde{S}_r^{(j)}\mathbf{v} = \mathbf{v}^{\mathrm{H}}(\tilde{A}_{21}^{(j)})^{\mathrm{H}}\tilde{S}_{n-r}^{(j)}\tilde{A}_{21}^{(j)}\mathbf{v}.$$

Since $\mathbf{v}^{\mathrm{H}}\tilde{S}_r^{(i)}\mathbf{v} \ge \sigma_{\min}(\tilde{S}_r^{(i)})$ and $\mathbf{v}^{\mathrm{H}}(\tilde{A}_{21}^{(i)})^{\mathrm{H}}\tilde{S}_{n-r}^{(i)}\tilde{A}_{21}^{(i)}\mathbf{v} \le \|\tilde{A}_{21}^{(i)}\mathbf{v}\|_2^2\sigma_{\max}(\tilde{S}_{n-r}^{(i)})$, we have $(1 - |\lambda|^2)\sigma_{\min}(\tilde{S}_r^{(i)}) \le \|\tilde{A}_{21}^{(i)}\mathbf{v}\|_2^2\sigma_{\max}(\tilde{S}_{n-r}^{(i)})$. According to Lemma 5.3,

$$\left\|\tilde{A}^{(i)}\begin{bmatrix}\mathbf{v}\\0\end{bmatrix}\right\|_2 \le \|\tilde{A}^{(i)}\|_2\left\|\begin{bmatrix}\mathbf{v}\\0\end{bmatrix}\right\|_2 \le \|\mathsf{bcirc}(\mathcal{A})\|_2\|\mathbf{v}\|_2 \le 1,$$

which is equivalent to $\|\tilde{A}_{21}^{(i)}\|_2^2 \leq 1 - \|\tilde{A}_{11}^{(i)}\mathbf{v}\|_2^2 = 1 - |\lambda|^2$. Hence

$$(1 - |\lambda|^2)\sigma_{\min}(\tilde{S}_r^{(i)}) \leq (1 - |\lambda|_2)\sigma_{\max}(\tilde{S}_{n-r}^{(i)}).$$

Theorem 5.2 implies that the subsystem $(\mathcal{A}_{\mathbf{r}}, \mathcal{B}_{\mathbf{r}}, \mathcal{C}_{\mathbf{r}})$ is asymptotically stable, thus $1-|\lambda|^2 > 0$ and $\sigma_{\min}(\tilde{S}_r^{(i)}) \leq \sigma_{\max}(\tilde{S}_{n-r}^{(i)})$, which leads to a contradiction. Therefore, the subsystem is observable. The reachability of the subsystem can be proved similarly. □

5.3. Error bound. In this part, we discuss the L^∞-norm bound for the balanced truncation. First, we introduce the definition of L^∞-norm of a tDMLTI system (3.1).

DEFINITION 5.5. Let $G(z)$ be the transfer function of a tDMLTI system (3.1), then its L^∞-norm is defined as

$$\|G(z)\|_{L^\infty} = \|\mathsf{bcirc}(G(z))\|_{L^\infty} = \max_{\theta\in[0,2\pi]} \sigma_{\max}\Big(\mathsf{bcirc}\big(G(e^{\iota\theta})\big)\Big), \quad \iota = \sqrt{-1}.$$

THEOREM 5.6. *Suppose that $(\mathcal{A}, \mathcal{B}, \mathcal{C})$ is a balanced realization of the transfer function $G(z)$ and $(\mathcal{A}_{\mathbf{r}}, \mathcal{B}_{\mathbf{r}}, \mathcal{C}_{\mathbf{r}})$ is the multirank-$\mathbf{r}$ truncation of $(\mathcal{A}, \mathcal{B}, \mathcal{C})$. Then an L^∞-norm bound on the absolute error in the frequency domain is given by*

$$E_\infty = \|G(z) - G_{\mathbf{r}}(z)\|_{L^\infty} = \|\mathsf{bcirc}(G(z) - G_{\mathbf{r}}(z))\|_{L^\infty} \leq 2 \max_i \sum_{j=r+1}^{n} s_j^{(i)},$$

where $s_j^{(i)}$ is the jth diagonal element of the matrix $\tilde{S}^{(i)}$.

PROOF. The reachability and observability Gramian $\mathcal{S}$ of the balanced realization $(\mathcal{A}, \mathcal{B}, \mathcal{C})$ satisfies

$$\begin{cases} \mathcal{A} * \mathcal{S} * \mathcal{A}^{\mathrm{H}} - \mathcal{S} + \mathcal{B} * \mathcal{B}^{\mathrm{H}} = 0, \\ \mathcal{A}^{\mathrm{H}} * \mathcal{S} * \mathcal{A} - \mathcal{S} + \mathcal{C}^{\mathrm{H}} * \mathcal{C} = 0. \end{cases}$$

Applying operator bcirc on both sides of these Lyapunov equations and using Lemma 2.4, we get

$$\begin{cases} \tilde{A}^{(i)}\tilde{S}^{(i)}(\tilde{A}^{(i)})^{\mathrm{H}} - \tilde{S}^{(i)} + \tilde{B}^{(i)}(\tilde{B}^{(i)})^{\mathrm{H}} = 0, \\ (\tilde{A}^{(i)})^{\mathrm{H}}\tilde{S}^{(i)}\tilde{A}^{(i)} - \tilde{S}^{(i)} + (\tilde{C}^{(i)})^{\mathrm{H}}\tilde{C}^{(i)} = 0, \end{cases}$$

for $i = 1, 2, \ldots, p$. According to [**1**, Theorem 2], for the LTI system discribed by $(\tilde{A}^{(i)}, \tilde{B}^{(i)}, \tilde{C}^{(i)})$, we have $\|\tilde{C}^{(i)}(zI - \tilde{A}^{(i)})^{-1}\tilde{B}^{(i)} - \tilde{C}_1^{(i)}(zI - \tilde{A}_{11}^{(i)})^{-1}\tilde{B}_1^{(i)}\|_2 \leq 2\sum_{j=r+1}^{n} s_j^{(i)}$, where $s_j^{(i)}$ is the jth diagonal entry of $\tilde{S}^{(i)}$. Therefore,

$$\begin{aligned} \|G(z) - G_{\mathbf{r}}(z)\|_{L^\infty} &= \|\mathsf{bcirc}(G(z) - G_{\mathbf{r}}(z))\|_{L^\infty} \\ &= \max_i \sigma_{\max}\big(\tilde{C}^{(i)}(zI - \tilde{A}^{(i)})^{-1}\tilde{B}^{(i)} - \tilde{C}_1^{(i)}(zI - \tilde{A}_{11}^{(i)})^{-1}\tilde{B}_1^{(i)}\big) \\ &\leq 2 \max_i \sum_{j=r+1}^{n} s_j^{(i)}. \end{aligned}$$

□

5.4. Hankel norm. If the state-space coordinates of a tDMLTI system are changed to $\mathcal{Z}_t \triangleq \mathcal{T} * \mathcal{X}_t$ for some nonsingular $\mathcal{T}$, then the system can be rewritten as

$$\begin{cases} \mathcal{Z}_{t+1} = \mathcal{T} * \mathcal{A} * \mathcal{T}^{-1} * \mathcal{Z}_t + \mathcal{T} * \mathcal{B} * \mathcal{U}_t, \\ \mathcal{Y}_t = \mathcal{C} * \mathcal{T}^{-1} * \mathcal{Z}_t. \end{cases}$$

Therefore, the reachability and observability Gramians become $\mathcal{T} * \mathcal{W}_r * \mathcal{T}^{\mathrm{H}}$ and $(\mathcal{T}^{-1})^{\mathrm{H}} * \mathcal{W}_o * \mathcal{T}^{-1}$, respectively. The forms of the Gramians depend on the state-space coordinates but their product transforms from $\mathcal{W}_r * \mathcal{W}_o$ to $\mathcal{T} * \mathcal{W}_r * \mathcal{W}_o * \mathcal{T}^{-1}$, which means the T-eigenvalues of the product $\mathcal{W}_r * \mathcal{W}_o$ are invariant under the state-space transformations and are hence input/output invariants. Based on this observation, we define the Hankel singular values of $G(z)$ as follows.

DEFINITION 5.7. Let $G(z)$ be the transfer function tensor of a tDMLTI system (3.1), then the *Hankel singualr values* of $G(z)$ are defined as

$$\sigma_i(G(z)) \triangleq \sqrt{\lambda_i(\mathcal{W}_r * \mathcal{W}_o)},$$

where by convention $\sigma_i(G(z)) \geq \sigma_{i+1}(G(z))$ and $\mathcal{W}_r, \mathcal{W}_o$ are the infinite horizon reachability and observability Gramians, respectively.

Define the Hankel tensor $\mathcal{H}_\infty$ as

$$\mathcal{H}_\infty = \mathcal{O}_\infty * \mathcal{R}_\infty, \tag{5.10}$$

where $\mathcal{R}_\infty = \begin{bmatrix} \mathcal{B} & \mathcal{A} * \mathcal{B} & \cdots & \mathcal{A}^k * \mathcal{B} & \cdots \end{bmatrix}$, $\mathcal{O}_\infty = \begin{bmatrix} \mathcal{C} \\ \mathcal{C} * \mathcal{A} \\ \vdots \\ \mathcal{C} * \mathcal{A}^k \\ \vdots \end{bmatrix}$.

Then the singular values of $\mathcal{H}_\infty$ can be evaluated from

$$\begin{aligned} \sigma_i(\mathcal{H}_\infty)^2 &= \lambda_i(\mathcal{H}_\infty^{\mathrm{H}} * \mathcal{H}_\infty) = \lambda_i(\mathcal{R}_\infty^{\mathrm{H}} * \mathcal{O}_\infty^{\mathrm{H}} * \mathcal{O}_\infty * \mathcal{R}_\infty) \\ &= \lambda_i(\mathcal{R}_\infty * \mathcal{R}_\infty^{\mathrm{H}} * \mathcal{O}_\infty^{\mathrm{H}} * \mathcal{O}_\infty) = \lambda_i(\mathcal{W}_r * \mathcal{W}_o). \end{aligned} \tag{5.11}$$

Therefore, the Hankel singular values of $G(z)$ are the singular values of the Hankel tensor $\mathcal{H}_\infty$.

In addition, the Hankel tensor $\mathcal{H}_\infty$ can be seen as the mapping from the past inputs to the future outputs via the state $\mathcal{X}_0$ at time $t = 0$. That is, if $\mathcal{U}_t = 0$ for $t \geq 0$, then

$$\mathcal{X}_0 = \mathcal{R}_\infty * \begin{bmatrix} \mathcal{U}_{-1} \\ \mathcal{U}_{-2} \\ \vdots \end{bmatrix}, \quad \begin{bmatrix} \mathcal{Y}_0 \\ \mathcal{Y}_1 \\ \vdots \end{bmatrix} = \mathcal{O}_\infty * \mathcal{X}_0. \tag{5.12}$$

The induced norm of the mapping $\mathcal{H}_\infty$ is defined by the largest singular value of $\mathcal{H}_\infty$ and hence we can define the Hankel norm of $G(z)$ as follows.

DEFINITION 5.8. Let $G(z)$ be the transfer function of tDMLTI system (3.1), then the Hankel norm of $G(z)$ is defined as

$$\|G(z)\|_H \triangleq \sigma_{\max}(\mathcal{H}_\infty) = \sqrt{\lambda_{\max}(\mathcal{W}_r * \mathcal{W}_o)},$$

where $\mathcal{W}_r, \mathcal{W}_o$ are the infinite horizon reachability Gramian and observability Gramian, respectively.

6. Pole Assignment

We discuss the pole assignment problem of the tDMLTI system (3.1) in this section.

DEFINITION 6.1 (poles). For tDMLTI system (3.1), we regard the T-eigenvalues, T-eigenvectors, T-Jordan canonical form, T-characteristic equation, and T-characteristic polynomial as the eigenvalues, eigenvectors, Jordan canonical form, characteristic equation, and characteristic polynomial of the system, respectively. The eigenvalues of the system are also called the *poles* of the system.

6.1. State feedback control law. In modern control theory, the design of a control system is actually the design for the control law. We define the state feedback control law for the tDMLTI system as follows.

DEFINITION 6.2. The state feedback control law of the tDMLTI system (3.1) has the following form:

$$\mathcal{U}_t = \mathcal{K} * \mathcal{X}_t + \mathcal{G} * \mathcal{V}_t, \tag{6.1}$$

where $\mathcal{K}$ is the state gain tensor, $\mathcal{V}_t$ is external input signal, and $\mathcal{G}$ is the external input tensor.

Under the state feedback control law (6.1), the tDMLTI system (3.1) becomes a closed-loop system

$$\begin{cases} \mathcal{X}_{t+1} = \mathcal{A}_c * \mathcal{X}_t + \mathcal{B}_c * \mathcal{V}_t, \\ \mathcal{Y}_t = \mathcal{C} * \mathcal{X}_t, \end{cases} \tag{6.2}$$

where $\mathcal{A}_c = \mathcal{A} + \mathcal{B} * \mathcal{K}$ and $\mathcal{B}_c = \mathcal{B} * \mathcal{G}$.

6.2. Existence of solutions to pole assignment under state feedback control. First, we provide the algebraic statement for the state feedback pole assignment problem as follows.

PROBLEM 6.1. *Given $\mathcal{A}, \mathcal{B} \in \mathbb{C}^{n\times n\times p}$ and a set of np complex numbers $S = \{s_1, s_2, \ldots, s_{np}\}$ such that if $s_i \in S$ and $\mathrm{Im}(s_i) \neq 0$ (not necessarily distinct), then $\bar{s}_i \in S$, find a tensor $\mathcal{K} \in \mathbb{R}^{n\times n\times p}$ such that*

$$\lambda_i(\mathcal{A} + \mathcal{B} * \mathcal{K}) = s_i, \quad i = 1, 2, \ldots, np. \tag{6.3}$$

THEOREM 6.3. *The tensor pair $(\mathcal{A}, \mathcal{B})$ is reachable if and only if, for any set S, there exists a tensor $\mathcal{K}$ such that $\mathcal{A} + \mathcal{B} * \mathcal{K}$ has S for its set of T-eigenvalues.*

PROOF. "$\Leftarrow$": Let $s_1, s_2, \ldots, s_{np}$ be any distinct real numbers such that $\det(\tilde{A}^{(i)} - s_j I) \neq 0$ for $i = 1, 2, \ldots, p$, $j = 1, 2, \ldots, np$. Suppose $\mathcal{K}$ is the tensor such that

$$(\mathcal{A} + \mathcal{B} * \mathcal{K}) * \mathcal{V}_j = s_j \mathcal{V}_j, \tag{6.4}$$

for $j = 1, 2, \ldots, np$, where $\mathcal{V}_j \neq 0 \in \mathbb{C}^{n\times 1\times p}$ is the corresponding T-eigenvector for s_j. Since $\det(\tilde{A}^{(i)} - s_j I) \neq 0$, the matrix $\tilde{A}^{(i)} - s_j I$ is invertible, for $i = 1, 2, \ldots, p$, $j = 1, 2, \ldots, np$. Hence $\mathcal{A} - s_j \mathcal{I}$ is invertible and (6.4) is equivalent to

$$-(\mathcal{A} - s_j \mathcal{I})^{-1} * \mathcal{B} * \mathcal{K} * \mathcal{V}_j = \mathcal{V}_j, \tag{6.5}$$

for $j = 1, 2, \dots, np$. On the other hand, according to Cayley-Hamilton theorem, there exist suitable functions $\alpha_k^{(i)}(s_j)$ such that

$$-(\tilde{A}^{(i)} - s_j I)^{-1} = \sum_{k=1}^{np-1} \alpha_k^{(i)}(s_j)(\tilde{A}^{(i)})^{k-1}, \tag{6.6}$$

for $i = 1, 2, \dots, p$. Denote

(6.7)

$$\mathcal{K}_k = \mathsf{bcirc}^{-1}\big((F_p \otimes I_n)\,\mathrm{diag}(\alpha_k^{(1)}(s_j)\tilde{K}^{(1)}, \alpha_k^{(1)}(s_j)\tilde{K}^{(2)}, \cdots, \alpha_k^{(p)}(s_j)\tilde{K}^{(p)})(F_p^{\mathrm{H}} \otimes I_n)\big),$$

then combining (6.5), (6.6), and (6.7), we have

$$\mathcal{R} * \begin{bmatrix} \mathcal{K}_1 * \mathcal{V}_j \\ \vdots \\ \mathcal{K}_{np-1} * \mathcal{V}_j \end{bmatrix} = \mathcal{V}_j \quad \text{or} \quad \mathsf{bcirc}(\mathcal{R})\mathsf{unfold}\begin{bmatrix} \mathcal{K}_1 * \mathcal{V}_j \\ \vdots \\ \mathcal{K}_{np-1} * \mathcal{V}_j \end{bmatrix} = \mathsf{unfold}(\mathcal{V}_j),$$

where $\mathcal{R} = \begin{bmatrix} \mathcal{B} & \mathcal{A} * \mathcal{B} & \cdots & \mathcal{A}^{np-1} * \mathcal{B} \end{bmatrix}$. According to [**14**, Page 360], the T-eigenvalue problem of the tensor $\mathcal{A} + \mathcal{B} * \mathcal{K}$ is equivalent to the eigenvalue problem of the matrix $\mathsf{bcirc}(\mathcal{A} + \mathcal{B} * \mathcal{K})$. Thus $\mathsf{unfold}(\mathcal{V}_j)$ is the corresponding eigenvector of the eigenvalue s_j. Since s_j ($j = 1, 2, \dots, np$) are distinct, $\mathsf{unfold}(\mathcal{V}_j)$ ($j = 1, 2, \dots, np$) are linearly independent, hence $\mathrm{rank}(\mathsf{bcirc}(\mathcal{R})) = np$. By the definition of multi-rank, we have $\mathrm{rank}(\begin{bmatrix} \mathcal{B} & \mathcal{A} * \mathcal{B} & \cdots & \mathcal{A}^{np-1} * \mathcal{B} \end{bmatrix}) = [n, n, \dots, n]$. Therefore the tensor pair $(\mathcal{A}, \mathcal{B})$ is reachable.

"$\Rightarrow$": Suppose that $(\mathcal{A}, \mathcal{B})$ is reachable, then

$$\mathrm{rank}(\mathcal{R}) = \mathrm{rank}(\begin{bmatrix} \mathcal{B} & \mathcal{A} * \mathcal{B} & \cdots & \mathcal{A}^{np-1} * \mathcal{B} \end{bmatrix}) = [n, n, \dots, n].$$

Multiplying $\mathsf{bcirc}(\mathcal{R})$ from the left by $F_p^{\mathrm{H}} \otimes I_n$ and from the right by $F_p \otimes I_{n^2p}$, we get $(F_p^{\mathrm{H}} \otimes I_n)\mathsf{bcirc}(\mathcal{R})(F_p \otimes I_{n^2p}) = \tilde{R}$, where $\tilde{R}$ is a block diagonal matrix of size $np \times n^2p^2$ with

$$\tilde{R}((i-1)n+1 : in, (i-1)n^2p+1 : in^2p) = \begin{bmatrix} \tilde{B}^{(i)} & \widetilde{(\mathcal{A} * \mathcal{B})}^{(i)} & \cdots & \widetilde{(\mathcal{A}^{np-1} * \mathcal{B})}^{(i)} \end{bmatrix}$$

for $i = 1, 2, \dots, p$. Thus,

$$\begin{aligned} np &= \mathrm{rank}(\mathsf{bcirc}(\mathcal{R})) \\ &\le \min\left\{\mathrm{rank}((F_p \otimes I_n)), \mathrm{rank}(\tilde{R}), \mathrm{rank}((F_p^{\mathrm{H}} \otimes I_{n^2p}))\right\} \\ &= \min\{np, \mathrm{rank}(\tilde{R}), n^2p^2\}, \end{aligned}$$

which means $\mathrm{rank}(\tilde{R}) \ge np$ and hence

$$\mathrm{rank}\left(\begin{bmatrix} \tilde{B}^{(i)} & \widetilde{(\mathcal{A} * \mathcal{B})}^{(i)} & \cdots & \widetilde{(\mathcal{A}^{np-1} * \mathcal{B})}^{(i)} \end{bmatrix}\right) = n$$

for $i = 1, 2, \dots, p$. Note that

$$\mathsf{bcirc}(\mathcal{A}^k * \mathcal{B}) = (F_p \otimes I_n)\,\mathrm{diag}\left((\tilde{A}^{(1)})^k \tilde{B}^{(1)}, (\tilde{A}^{(2)})^k \tilde{B}^{(2)}, \cdots, (\tilde{A}^{(p)})^k \tilde{B}^{(p)}\right)(F_p^{\mathrm{H}} \otimes I_n)$$

for any k, thus $\widetilde{(\mathcal{A}^k * \mathcal{B})}^{(i)} = (\tilde{A}^{(i)})^k \tilde{B}^{(i)}$ for $i = 1, 2, \dots, p$. Therefore, if we consider the LTI system discribed by $\mathbf{x}_{t+1} = \tilde{A}^{(i)}\mathbf{x}_t + \tilde{B}^{(i)}\mathbf{u}_t$, then the matrix pair $(\tilde{A}^{(i)}, \tilde{B}^{(i)})$ is reachable. According to [**20**, Theorem 3.1], for any set $S^{(i)} = \{s_1^{(i)}, s_2^{(i)}, \dots, s_n^{(i)}\}$, there exists matrix $\tilde{K}^{(i)}$ such that $\tilde{A}^{(i)} + \tilde{B}^{(i)}\tilde{K}^{(i)}$ has $S^{(i)}$ for its set of eigenvalues. This means, for any set $S = \{s_1, s_2, \dots, s_{np}\}$, we can evenly divide it into p subsets

$S^{(i)} = \{s_1^{(i)}, s_2^{(i)}, \dots, s_n^{(i)}\}$ and find matrices $\tilde{K}^{(i)}$ for $i = 1, 2, \dots, p$ such that $\tilde{A}^{(i)} + \tilde{B}^{(i)}\tilde{K}^{(i)}$ has $S^{(i)}$ for its set of eigenvalues. Then we can construct $\mathcal{K}$ by

$$\mathcal{K} = \mathsf{bcirc}^{-1}\left((F_p \otimes I_n)\,\mathrm{diag}(\tilde{K}^{(1)}, \tilde{K}^{(2)}, \cdots, \tilde{K}^{(p)})(F_p^{\mathrm{H}} \otimes I_n)\right),$$

and $\mathcal{A} + \mathcal{B} * \mathcal{K}$ has S for its set of T-eigenvalues. □

7. Numerical Experiments

In this section, we give some examples to illustrate the tDMLTI system theories. All numerical tests are done using MATLAB version R2021a on a laptop computer with an Intel Quad-Core i5-5287U CPU (2.90GHz) and 8 GB of RAM.

7.1. A stable tDMLTI system example. In this part, we construct a tensor $\mathcal{A}$ which is stable under the t-product multilinear system but unstable under the Einstein product multilinear system given by Chen *et al.* in [**3**, Definition 3.1].

Since the Einstein product is defined on even-order tensors, we need to introduce the higher order t-product before the construction of our example.

Suppose $\mathcal{A} \in \mathbb{C}^{n_1 \times n_2 \times \cdots \times n_p}$. Denote $\mathcal{A}_i \in \mathbb{R}^{n_1 \times n_2 \times \cdots \times n_{p-1}}$ as the order-$(p-1)$ tensor created by holding the pth index of $\mathcal{A}$ fixed at i for $i = 1, 2, \dots, n_p$. We can create a tensor in a block circulant pattern, where each block is a tensor of order-$(p-1)$,

$$\mathsf{circ}(\mathcal{A}) = \begin{bmatrix} \mathcal{A}_1 & \mathcal{A}_{n_p} & \mathcal{A}_{n_p-1} & \cdots & \mathcal{A}_2 \\ \mathcal{A}_2 & \mathcal{A}_1 & \mathcal{A}_{n_p} & \cdots & \mathcal{A}_3 \\ \vdots & \ddots & \ddots & \ddots & \vdots \\ \mathcal{A}_{n_p} & \mathcal{A}_{n_p-1} & \ddots & \mathcal{A}_2 & \mathcal{A}_1 \end{bmatrix},$$

where $\mathsf{circ}(\mathcal{A})$ is of the size $(n_1 n_p \times n_2 n_p \times \cdots \times n_{p-2} n_p \times n_{p-1})$. If $\mathcal{A}$ is an order-3 tensor, then $\mathsf{circ}(\mathcal{A})$ is a block circulant matrix. Similarly, define $\mathsf{unfold}(\cdot)$ to take an $n_1 \times n_2 \times \cdots \times n_p$ tensor and return an $n_1 n_p \times n_2 \times \cdots \times n_{p-1}$ block tensor as $\mathsf{unfold}(\mathcal{A}) = [\mathcal{A}_1^\top\ \mathcal{A}_2^\top \cdots \mathcal{A}_{n_p}^\top]^\top$. Reversely, $\mathsf{fold}(\cdot, n_p)$ takes an $n_1 n_p \times n_2 \times \cdots \times n_{p-1}$ block tensor and returns an $n_1 \times n_2 \times \cdots \times n_p$ tensor, i.e., $\mathsf{fold}(\mathsf{unfold}(\mathcal{A}), n_p) = \mathcal{A}$. Now we can extend the t-product to order-p tensors.

Definition 7.1. Let $\mathcal{A}$ be an $n_1 \times n_2 \times \cdots \times n_p$ tensor and $\mathcal{B}$ be an $n_2 \times l \times n_3 \times \cdots \times n_p$ tensor, then the t-product $\mathcal{A} * \mathcal{B}$ is the order-p $(p \geq 4)$ tensor defined recursively as

$$\mathcal{A} * \mathcal{B} = \mathsf{fold}(\mathsf{circ}(\mathcal{A}) * \mathsf{unfold}(\mathcal{B})) \tag{7.1}$$

of size $n_1 \times l \times n_3 \times \cdots \times n_p$.

Under this definition, we can extend the definiton of tDMLTI system (3.1) to order-p $(p \geq 4)$ tensors. Now consider a order-4 tensor $\mathcal{A} \in \mathbb{R}^{2\times2\times2\times2}$, the T-eigenproblem of $\mathcal{A}$ is $\mathcal{A} * \mathcal{X} = \lambda \mathcal{X}$, where λ is the T-eigenvalue of $\mathcal{A}$ and $\mathcal{X}$ is the corresponding T-eigenvector.

By the definition of order-p $(p \geq 4)$ t-product, the above eigenproblem is equivalent to $\mathsf{fold}\big(\mathsf{circ}(\mathcal{A}) * \mathsf{unfold}(\mathcal{X})\big) = \lambda \mathcal{X}$. Applying the $\mathsf{unfold}(\cdot)$ operator on both sides, we have $\begin{bmatrix} \mathcal{A}_1 & \mathcal{A}_2 \\ \mathcal{A}_2 & \mathcal{A}_1 \end{bmatrix} * \begin{bmatrix} \mathcal{X}_1 \\ \mathcal{X}_2 \end{bmatrix} = \lambda \begin{bmatrix} \mathcal{X}_1 \\ \mathcal{X}_2 \end{bmatrix}$, i.e., $\begin{cases} \mathcal{A}_1 * \mathcal{X}_1 + \mathcal{A}_2 * \mathcal{X}_2 = \lambda \mathcal{X}_1 \\ \mathcal{A}_2 * \mathcal{X}_1 + \mathcal{A}_1 * \mathcal{X}_2 = \lambda \mathcal{X}_2 \end{cases}$.

Applying $\mathsf{bcirc}(\cdot)$ on both sides, we get

$$\begin{cases} \mathsf{bcirc}(\mathcal{A}_1)\mathsf{bcirc}(\mathcal{X}_1) + \mathsf{bcirc}(\mathcal{A}_2)\mathsf{bcirc}(\mathcal{X}_2) = \lambda\mathsf{bcirc}(\mathcal{X}_1), \\ \mathsf{bcirc}(\mathcal{A}_2)\mathsf{bcirc}(\mathcal{X}_1) + \mathsf{bcirc}(\mathcal{A}_1)\mathsf{bcirc}(\mathcal{X}_2) = \lambda\mathsf{bcirc}(\mathcal{X}_2). \end{cases}$$

Hence

$$\begin{bmatrix} \mathsf{bcirc}(\mathcal{A}_1) & \mathsf{bcirc}(\mathcal{A}_2) \\ \mathsf{bcirc}(\mathcal{A}_2) & \mathsf{bcirc}(\mathcal{A}_1) \end{bmatrix} \begin{bmatrix} \mathsf{bcirc}(\mathcal{X}_1) \\ \mathsf{bcirc}(\mathcal{X}_2) \end{bmatrix} = \lambda \begin{bmatrix} \mathsf{bcirc}(\mathcal{X}_1) \\ \mathsf{bcirc}(\mathcal{X}_2) \end{bmatrix}.$$

Recall that $\mathsf{vec}(AXB) = (B^\top \otimes A)\mathsf{vec}(X)$, we have

$$\left(I \otimes \begin{bmatrix} \mathsf{bcirc}(\mathcal{A}_1) & \mathsf{bcirc}(\mathcal{A}_2) \\ \mathsf{bcirc}(\mathcal{A}_2) & \mathsf{bcirc}(\mathcal{A}_1) \end{bmatrix}\right) \mathsf{vec}\left(\begin{bmatrix} \mathsf{bcirc}(\mathcal{X}_1) \\ \mathsf{bcirc}(\mathcal{X}_2) \end{bmatrix}\right) = \lambda\mathsf{vec}\left(\begin{bmatrix} \mathsf{bcirc}(\mathcal{X}_1) \\ \mathsf{bcirc}(\mathcal{X}_2) \end{bmatrix}\right),$$

which means the T-spectral radius of $\mathcal{A}$ is the same as the matrix

$$\begin{bmatrix} \mathsf{bcirc}(\mathcal{A}_1) & \mathsf{bcirc}(\mathcal{A}_2) \\ \mathsf{bcirc}(\mathcal{A}_2) & \mathsf{bcirc}(\mathcal{A}_1) \end{bmatrix}.$$

Chen *et al.* [**3**] introduced the multiliear control system under the Einstein product, which has the following form,

$$\begin{cases} \mathcal{X}_{t+1} = \mathcal{A} *_N \mathcal{X}_t + \mathcal{B} *_N \mathcal{U}_t, \\ \mathcal{Y}_t = \mathcal{C} *_N \mathcal{X}_t. \end{cases} \tag{7.2}$$

Here $\mathcal{A} \in \mathbb{R}^{J_1 \times J_1 \times \cdots \times J_N \times J_N}$, $\mathcal{B} \in \mathbb{R}^{J_1 \times K_1 \times \cdots \times J_N \times K_N}$, $\mathcal{C} \in \mathbb{R}^{I_1 \times J_1 \times \cdots \times I_N \times J_N}$ are even-order tensor pairs, and "$*_N$" denotes the Einstein product,

$$(\mathcal{A} *_N \mathcal{B})_{j_1 i_1 \ldots j_N i_N} = \sum_{k_1=1}^{K_1} \cdots \sum_{k_N=1}^{K_N} \mathcal{A}_{j_1 k_1 \ldots j_N k_N} \mathcal{B}_{k_1 i_1 \ldots k_N i_N}$$

for any two even-order tensors $\mathcal{A} \in \mathbb{R}^{J_1 \times K_1 \times \cdots \times J_N \times K_N}$ and $\mathcal{B} \in \mathbb{R}^{K_1 \times I_1 \times \cdots \times K_N \times I_N}$.

DEFINITION 7.2 ([**7**, Definition 1.2]). Let $\mathcal{A} \in \mathbb{R}^{J_1 \times J_1 \times \cdots \times J_N \times J_N}$ be a $2N$th-order tensor. If a nonzero Nth-order tensor $\mathcal{X} \in \mathbb{R}^{J_1 \times J_2 \times \cdots \times J_N}$ and $\lambda \in \mathbb{R}$ satisfy $\mathcal{A} *_N \mathcal{X} = \lambda \mathcal{X}$, then λ and $\mathcal{X}$ are called the U-eigenvalue and U-eigenvector of $\mathcal{A}$.

According to [**4**, Proposition 5.1], the system (7.2) is stable if and only if all the U-eigenvalues of $\mathcal{A}$ have magnitude no larger than 1 and for those equal to 1, the algebraic and geometry multiplicities must be equal. Based on this conclusion, we construct a tensor $\mathcal{A} \in \mathbb{R}^{2 \times 2 \times 2 \times 2}$ which is unstable in the MLTI system (7.2) and stable in the tDMLTI system (3.1) as follows.

$$\begin{cases} \mathcal{A}_1^{(1)} = \begin{bmatrix} -0.3194 & 0.5609 \\ -0.4174 & -0.4322 \end{bmatrix}, & \mathcal{A}_1^{(2)} = \begin{bmatrix} 0.3575 & -0.1492 \\ -0.2751 & -0.3452 \end{bmatrix}, \\ \mathcal{A}_2^{(1)} = \begin{bmatrix} -0.5408 & 0.2778 \\ 0.1442 & 0.4562 \end{bmatrix}, & \mathcal{A}_2^{(2)} = \begin{bmatrix} -0.3441 & -0.6700 \\ 0.7599 & -0.4580 \end{bmatrix}. \end{cases}$$

Then U-eigenvalue problem of $\mathcal{A}$ is equivalent to the matrix eigenvalue problem [**7**, Proposition 2.4] $\begin{bmatrix} \mathcal{A}_1^{(1)} & \mathcal{A}_2^{(1)} \\ \mathcal{A}_1^{(2)} & \mathcal{A}_2^{(2)} \end{bmatrix} \mathbf{x} = \lambda \mathbf{x}$.

The U-spectral radius and T-spectral radius of $\mathcal{A}$ are 1.1417 and 0.8856. Therfore $\mathcal{A}$ is stable in the tDMLTI system but unstable in the MLTI system.

7.2. Reachable and observable example. Consider a tDMLTI system with realization $\mathcal{A}, \mathcal{B}, \mathcal{C} \in \mathbb{C}^{3\times3\times3}$ constructed as follows.

$$A(:,:,1) = \begin{bmatrix} -0.2593 & -0.4397 & -0.1272 \\ -1.1363 & -0.1844 & -0.4165 \\ 1.4420 & 0.8927 & 0.4104 \end{bmatrix}, \quad B(:,:,1) = C(:,:,1) = \begin{bmatrix} 1 & 0 & 0 \\ 0 & 0 & 0 \\ 0 & 0 & 0 \end{bmatrix},$$

$$A(:,:,2) = \begin{bmatrix} 0.4489 + 0.3880\imath & 0.3127 + 0.4147\imath & 0.0959 + 0.1820\imath \\ -0.3829 - 0.6072\imath & -0.6017 - 0.2107\imath & 0.0683 - 0.3800\imath \\ 0.4284 + 0.0680\imath & 0.3717 - 0.2427\imath & -0.0805 + 0.1691\imath \end{bmatrix},$$

$$A(:,:,3) = \begin{bmatrix} 0.4489 - 0.3880\imath & 0.3127 - 0.4147\imath & 0.0959 - 0.1820\imath \\ -0.3829 + 0.6072\imath & -0.6017 + 0.2107\imath & 0.0683 + 0.3800\imath \\ 0.4284 - 0.0680\imath & 0.3717 + 0.2427\imath & -0.0805 - 0.1691\imath \end{bmatrix},$$

$$B(:,:,2) = C(:,:,2) = B(:,:,3) = C(:,:,3) = 0.$$

We compute the T-eigenvalues of $\mathcal{A}$ based on Lemma 2.7:

$$\sigma(\mathcal{A}) = \{-0.9001, 0.5003, -0.1001, -0.8003, \\ 0.6002, -0.1999, 0.8000, -0.2000, 0.1999\}.$$

Note that all these T-eigenvalues have magnitude less than 1. Therefore, according to Lemma 4.1, this tDMLTI system is asymptotically stable. In addition, computing the reachability gramian $\mathcal{R}$ using (4.2) and then transform $\mathsf{bcirc}(\mathcal{R})$ into a block diagonal matrix by Fourier transform, we can easily see that $\mathcal{R}$ is Hermite positive definite. Thus $(\mathcal{A}, \mathcal{B})$ is reachable. Similarly, by computing the observability tensor $\mathcal{O}$ defined as (4.4), we can see that $(\mathcal{A}, \mathcal{C})$ is observable. Therefore, the realization $(\mathcal{A}, \mathcal{B}, \mathcal{C})$ is asymptotically stable, reachable, and observable and hence we can compute the infinite horizon reachability Gramian and observability Gramian by the Lyapunov equations $\mathcal{A} * \mathcal{W}_r * \mathcal{A}^{\mathrm{H}} - \mathcal{W}_r = -\mathcal{B} * \mathcal{B}^{\mathrm{H}}, \quad \mathcal{A}^{\mathrm{H}} * \mathcal{W}_o * \mathcal{A} - \mathcal{W}_o = -\mathcal{C}^{\mathrm{H}} * \mathcal{C}$.

7.3. Illustration for balanced realization and reduction. In this part, we consider an arbitrary tDMLTI system with $\mathcal{A}, \mathcal{B}, \mathcal{C} \in \mathbb{C}^{50\times50\times3}$. We construct the balanced realization $(\bar{\mathcal{A}}, \bar{\mathcal{B}}, \bar{\mathcal{C}})$ and then truncate the balanced realization into $(\mathcal{A}_{\mathbf{r}}, \mathcal{B}_{\mathbf{r}}, \mathcal{C}_{\mathbf{r}})$ using the method introduced in Section V. The relation between the L^∞-norm $\|G(z) - G_{\mathbf{r}}(z)\|_{L^\infty}$ and the multirank $\mathbf{r}$ is illustrated in Figure 1.

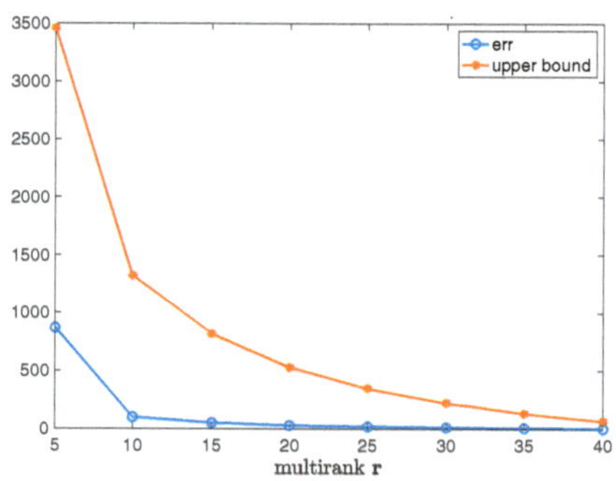

FIGURE 1. Illustration of the L^∞-norm $\|G(z) - G_{\mathbf{r}}(z)\|_{L^\infty}$ and upper bound under different multirank $\mathbf{r}$.

In Figure 1, "err" means $\|G(z) - G_{\mathbf{r}}(z)\|_{L^\infty}$ and "upper bound" equals to $2 \max_i \sum_{j=r+1}^{n} s_j^{(i)}$. This result is consistent with Theorem 5.6.

8. Conclusions

In this paper, we describe the discrete multilinear time-invariant systems under the tensor t-product and develop the basic theories including the system stability, the reachability, the controllability, the observability, the balanced truncation, and the pole assignment under the state feedback control.

References

[1] U. Al-Saggaf and G. Franklin, *An error bound for a discrete reduced order model of a linear multivariable system*, IEEE transactions on Automatic Control, 32(9):815–819, 1987.

[2] Maolin Che and Yimin Wei, *An efficient algorithm for computing the approximate t-URV and its applications*, J. Sci. Comput. **92** (2022), no. 3, Paper No. 93, 27, DOI 10.1007/s10915-022-01956-y. MR4458649

[3] C. Chen, A. Surana, A. Bloch, and I. Rajapakse, *Multilinear time invariant system theory*, in 2019 Proceedings of the Conference on Control and its Applications, pages 118–125. SIAM, 2019.

[4] Can Chen, Amit Surana, Anthony M. Bloch, and Indika Rajapakse, *Multilinear control systems theory*, SIAM J. Control Optim. **59** (2021), no. 1, 749–776, DOI 10.1137/19M1262589. MR4220654

[5] Juefei Chen, Longxiu Huang, and Yimin Wei, *Coseparable nonnegative tensor factorization with t-CUR decomposition*, SIAM J. Matrix Anal. Appl. **45** (2024), no. 4, 1805–1826, DOI 10.1137/23M1625998. MR4805863

[6] J. Chen, W. Ma, Y. Miao, and Y. Wei, *Perturbations of tensor-Schur decomposition and its applications to multilinear control systems and facial recognitions*, Neurocomputing, 547, 2023. Id/No 126359.

[7] Lu-Bin Cui, Chuan Chen, Wen Li, and Michael K. Ng, *An eigenvalue problem for even order tensors with its applications*, Linear Multilinear Algebra **64** (2016), no. 4, 602–621, DOI 10.1080/03081087.2015.1071311. MR3479021

[8] W. Ding, K. Liu, E. Belyaev, and F. Cheng, *Tensor-based linear dynamical systems for action recognition from 3d skeletons*, Pattern Recognition, 77:75–86, 2018.

[9] Feiyang Han, Yimin Wei, and Pengpeng Xie, *Regularized and structured tensor total least squares methods with applications*, J. Optim. Theory Appl. **202** (2024), no. 3, 1101–1136, DOI 10.1007/s10957-024-02507-1. MR4794216

[10] R. C. Hoover, K. Caudle, and K. Braman, *A new approach to multilinear dynamical systems and control*, `arXiv:2108.13583`, 2021.

[11] Misha E. Kilmer, Karen Braman, Ning Hao, and Randy C. Hoover, *Third-order tensors as operators on matrices: a theoretical and computational framework with applications in imaging*, SIAM J. Matrix Anal. Appl. **34** (2013), no. 1, 148–172, DOI 10.1137/110837711. MR3032996

[12] Misha E. Kilmer and Carla D. Martin, *Factorization strategies for third-order tensors*, Linear Algebra Appl. **435** (2011), no. 3, 641–658, DOI 10.1016/j.laa.2010.09.020. MR2794595

[13] M. E. Kilmer, C. D. Martin, and L. Perrone, *A third-order generalization of the matrix SVD as a product of third-order tensors*, Tufts University, Department of Computer Science, Tech. Rep. TR-2008-4, 2008.

[14] Wei-hui Liu and Xiao-qing Jin, *A study on T-eigenvalues of third-order tensors*, Linear Algebra Appl. **612** (2021), 357–374, DOI 10.1016/j.laa.2020.11.004. MR4192241

[15] C. Lu, J. Feng, Y. Chen, W. Liu, Z. Lin, and S. Yan, *Tensor robust principal component analysis with a new tensor nuclear norm*, IEEE transactions on pattern analysis and machine intelligence, 42(4):925–938, 2019.

[16] Kathryn Lund, *The tensor t-function: a definition for functions of third-order tensors*, Numer. Linear Algebra Appl. **27** (2020), no. 3, e2288, 17, DOI 10.1002/nla.2288. MR4157199

[17] Carla D. Martin, Richard Shafer, and Betsy Larue, *An order-p tensor factorization with applications in imaging*, SIAM J. Sci. Comput. **35** (2013), no. 1, A474–A490, DOI 10.1137/110841229. MR3033057

[18] Yun Miao, Liqun Qi, and Yimin Wei, *Generalized tensor function via the tensor singular value decomposition based on the T-product*, Linear Algebra Appl. **590** (2020), 258–303, DOI 10.1016/j.laa.2019.12.035. MR4051879

[19] Yun Miao, Liqun Qi, and Yimin Wei, *T-Jordan canonical form and T-Drazin Inverse based on the T-Product*, Commun. Appl. Math. Comput. **3** (2021), no. 2, 201–220, DOI 10.1007/s42967-019-00055-4. MR4249881

[20] S. K. Mitter and R. Foulkes, *Controllability and pole assignment for discrete time linear systems defined over arbitrary fields*, SIAM J. Control **9** (1971), 1–7. MR290826

[21] Changxin Mo, Weiyang Ding, and Yimin Wei, *Perturbation analysis on T-eigenvalues of third-order tensors*, J. Optim. Theory Appl. **202** (2024), no. 2, 668–702, DOI 10.1007/s10957-024-02444-z. MR4787516

[22] Bruce C. Moore, *Principal component analysis in linear systems: controllability, observability, and model reduction*, IEEE Trans. Automat. Control **26** (1981), no. 1, 17–32, DOI 10.1109/TAC.1981.1102568. MR609248

[23] C. Ozdemir, R. C. Hoover, and K. Caudle, *2dtpca: A new framework for multilinear principal component analysis*, in 2021 IEEE International Conference on Image Processing (ICIP), pages 344–348. IEEE, 2021.

[24] Lars Pernebo and Leonard M. Silverman, *Model reduction via balanced state space representations*, IEEE Trans. Automat. Control **27** (1982), no. 2, 382–387, DOI 10.1109/TAC.1982.1102945. MR680103

[25] M. Rogers, L. Li, and S. J. Russell, *Multilinear dynamical systems for tensor time series*, Advances in Neural Information Processing Systems, 26:2634–2642, 2013.

[26] A. Surana, G. Patterson, and I. Rajapakse, *Dynamic tensor time series modeling and analysis*, in 2016 IEEE 55th Conference on Decision and Control (CDC), pages 1637–1642. IEEE, 2016.

[27] L. Wang, K. Xie, T. Semong, and H. Zhou, *Missing data recovery based on tensor-CUR decomposition*, IEEE Access, 6:532–544, 2017.

[28] Pan Zhou, Canyi Lu, Zhouchen Lin, and Chao Zhang, *Tensor factorization for low-rank tensor completion*, IEEE Trans. Image Process. **27** (2018), no. 3, 1152–1163, DOI 10.1109/TIP.2017.2762595. MR3747081

Shanghai Center for Mathematical Sciences, Fudan University, Shanghai, 200433, P.R. of China.

Email address: `18110840010@fudan.edu.cn`

School of Mathematical Sciences and Key Laboratory of Mathematics for Nonlinear Sciences, Fudan University, Shanghai, 200433, P.R. of China.

Email address: `ymwei@fudan.edu.cn`

Contemporary Mathematics
Volume **833**, 2026
https://doi.org/10.1090/conm/833/16669

Studying optimal designs for multivariate crossover trials

Shubham Niphadkar and Siuli Mukhopadhyay

ABSTRACT. This article discusses A-, D- and E-optimality results for multivariate crossover designs, where more than one response is measured from every period for each subject. The motivation for these multivariate designs comes from a 3×3 crossover trial that investigates how an oral drug affects biomarkers of mucosal inflammation, by analyzing the various gene profiles from each participant. A multivariate response crossover model with fixed effects including direct and carryover effects, and with heteroscedastic error terms is considered to fit the multiple responses measured. It is assumed all throughout the article that there is no correlation between responses but there is presence of correlation within responses. Corresponding to the direct effects, we obtain the information matrix in a multiple response setup. Various results regarding this information matrix are studied. For p periods and t treatments, orthogonal array design of type I and strength 2 is proved as A-, D- and E-optimal, when $p = t \geq 3$.

1. Introduction

In many clinical studies, crossover designs with measurements on multiple responses from each subject in each period are utilized. [**7**] and [**22**] used a multivariate crossover design to measure systolic and diastolic blood pressure, and to measure multiple gene expression profiles, respectively. However in these studies no investigation of optimality of such multivariate crossover designs were done. The researchers instead considered various structures for between and within correlation in responses, and discussed methods of estimation in such crossover trials. We explore optimality of exact multiple response crossover designs, in this study. For a given design, by denoting the number of treatments and periods by t and p, respectively, we obtain A-, D- and E-optimal designs for the case when $p = t \geq 3$.

Crossover designs find application in various areas like pharmaceutical studies, clinical trials, agricultural sciences and biological studies ([**5, 28**]). Various researchers including [**8**], [**9**], [**4**], [**16**], [**30**], [**18**], [**19**], [**17**], [**10**], [**11**] and [**27**] have studied optimality of univariate crossover designs considering a fixed effect model, while [**20**], [**21**], [**2**], [**3**] and [**12**] considered a random subject effect model. Optimal crossover designs, where responses are non-normal in nature were explored by [**28**], [**13**], [**24**] and [**29**], to name a few. However, all these works are based on

2020 *Mathematics Subject Classification.* Primary 62K05, 62K10; Secondary 15A09, 15A15, 05B15.

Key words and phrases. A-optimal, D-optimal, E-optimal, multivariate crossover design, orthogonal array.

single response crossover trials. Recently, [**25**] investigated universal optimality of multivariate crossover design, where the responses are uncorrelated and have same within response correlation structure, whereas [**26**] studied efficient designs when the responses are correlated, under the proportional and the markovian covariance structure.

In this chapter, we introduce a multivariate response model with fixed effects that incorporates both direct as well as carryover effects of treatments for modelling data from a multivariate crossover trial. We consider the responses to be uncorrelated, while each response is considered to be correlated within themselves with different responses having different correlation structure and heteroscedastic variances. In this case, we prove that corresponding to direct effects, the information matrix is not completely symmetric. Therefore, to find A-, D- and E-optimal designs we resort to eigenvalue analysis of the information matrix.

2. Proposed statistical model

We examine $p \times t$ crossover designs where $t \geq 2$. Here t represent treatments which are allocated over p periods and n experimental subjects. We denote the collection of all such crossover designs by $\Omega_{t,n,p}$. In this work, we consider that corresponding to a design $d \in \Omega_{t,n,p}$, for every subject in each period, the observations are recorded on g response variables, where $g \geq 1$. For the observation measured in the ith period and j^{th} subject on the kth response, the corresponding random variable Y_{dijk} is modelled as:

$$Y_{dijk} = \mu_k + \alpha_{i,k} + \beta_{j,k} + \tau_{d(i,j),k} + \rho_{d(i-1,j),k} + \epsilon_{ijk}, \tag{2.1}$$

where $1 \leq i \leq p$, $1 \leq j \leq n$ and $1 \leq k \leq g$. Here μ_k, $\alpha_{i,k}$, $\beta_{j,k}$ denote the intercept, effect due to the ith period and effect due to the jth subject, respectively, on the kth response. The direct and the carryover effect of first order on the kth response due to treatment s is denoted as $\tau_{s,k}$ and $\rho_{s,k}$, respectively, where $1 \leq s \leq t$. We assume all the effects to be fixed effects. The corresponding error term is denoted by ϵ_{ijk}, where we assume that $\mathbb{E}(\epsilon_{ijk}) = 0$, $\text{Cov}(\epsilon_{ijk}, \epsilon_{i'jk'}) = 0$, for $1 \leq i, i' \leq p$ and $1 \leq k \neq k' \leq g$, and $\text{Cov}(\epsilon_{ijk}, \epsilon_{i'j'k'}) = 0$, for $1 \leq i, i' \leq p$, $1 \leq j \neq j' \leq n$ and $1 \leq k, k' \leq g$. These imply that there is no correlation between the observations taken on distinct response variables. Also as in most crossover trials, we assume that the observations from distinct subjects are uncorrelated. In the above model, we have considered that the intercept, effects due to periods, subjects, treatments and first order carryover effects change according to the response variable. We assume that there is no carryover effect associated with the first period, thus $\rho_{d(0,j),k} = 0$.

Next, we discuss steps to write model (2.1) in matrix form. For $d \in \Omega_{t,n,p}$, we let $\boldsymbol{Y}_{dk} = (Y_{d11k}, \cdots, Y_{dp1k}, Y_{d12k}, \cdots, Y_{dp2k}, \cdots, Y_{d1nk}, \cdots, Y_{dpnk})'$, $\boldsymbol{\epsilon}_k = (\epsilon_{11k}, \cdots, \epsilon_{p1k}, \epsilon_{12k}, \cdots, \epsilon_{p2k}, \cdots, \epsilon_{1nk}, \cdots, \epsilon_{pnk})'$, $\boldsymbol{\alpha}_k = (\alpha_{i,k})_{i=1}^{p}$, $\boldsymbol{\beta}_k = (\beta_{j,k})_{j=1}^{n}$, $\boldsymbol{\tau}_k = (\tau_{s,k})_{s=1}^{t}$ and $\boldsymbol{\rho}_k = (\rho_{s,k})_{s=1}^{t}$ which are the column vectors. Here for the kth response, $\boldsymbol{\alpha}_k$, $\boldsymbol{\beta}_k$, $\boldsymbol{\tau}_k$ and $\boldsymbol{\rho}_k$ are the vectors of period, subject, direct and carryover effects, respectively. Note that in the g variable case, we have g number of $\boldsymbol{\tau}_k$'s and $\boldsymbol{\rho}_k$'s. The parameter vector is denoted by $\boldsymbol{\theta}_k = \left(\mu_k, \boldsymbol{\alpha}_k', \boldsymbol{\beta}_k', \boldsymbol{\tau}_k', \boldsymbol{\rho}_k'\right)'$.

The incidence matrix of the period versus direct effects for the jth subject, $\boldsymbol{T}_{dj}$, is a $p \times t$ matrix, while the incidence matrix of the period versus carryover effects, $\boldsymbol{F}_{dj}$, is also of dimension $p \times t$. For each j, $\boldsymbol{F}_{dj}$ can be written as

$\boldsymbol{F}_{dj} = \begin{bmatrix} \boldsymbol{0}'_{p-1\times 1} & 0 \\ \boldsymbol{I}_{p-1} & \boldsymbol{0}_{p-1\times 1} \end{bmatrix} \boldsymbol{T}_{dj}$. Defining $\boldsymbol{X}_{dk} = \begin{bmatrix} \boldsymbol{1}_{np} & \boldsymbol{P} & \boldsymbol{U} & \boldsymbol{T}_d & \boldsymbol{F}_d \end{bmatrix}$, model (2.1) can be represented in matrix form as

$$\boldsymbol{Y}_{dk} = \boldsymbol{X}_{dk}\boldsymbol{\theta}_k + \boldsymbol{\epsilon}_k, \tag{2.2}$$

where $\boldsymbol{T}_d = \begin{bmatrix} \boldsymbol{T}'_{d1} & \cdots & \boldsymbol{T}'_{dn} \end{bmatrix}'$, $\boldsymbol{F}_d = \begin{bmatrix} \boldsymbol{F}'_{d1} & \cdots & \boldsymbol{F}'_{dn} \end{bmatrix}'$, $\boldsymbol{P} = \boldsymbol{1}_n \otimes \boldsymbol{I}_p$ and $\boldsymbol{U} = \boldsymbol{I}_n \otimes \boldsymbol{1}_p$. Using $\boldsymbol{X_1} = \begin{bmatrix} \boldsymbol{P} & \boldsymbol{U} \end{bmatrix}$ and $\boldsymbol{X_2} = \begin{bmatrix} \boldsymbol{T}_d & \boldsymbol{F}_d \end{bmatrix}$, $\boldsymbol{X}_{dk}$ is written as

$$\boldsymbol{X}_{dk} = \begin{bmatrix} \boldsymbol{1}_{np} & \boldsymbol{X_1} & \boldsymbol{X_2} \end{bmatrix}. \tag{2.3}$$

Combining the g models in (2.2), we get

$$\begin{aligned} \begin{bmatrix} \boldsymbol{Y}'_{d1} & \cdots & \boldsymbol{Y}'_{dg} \end{bmatrix}' &= (\oplus_{k=1}^{g} \boldsymbol{X}_{d1}) \begin{bmatrix} \boldsymbol{\theta}'_1 & \cdots & \boldsymbol{\theta}'_g \end{bmatrix}' + \begin{bmatrix} \boldsymbol{\epsilon}'_1 & \cdots & \boldsymbol{\epsilon}'_g \end{bmatrix}' \\ &= (\boldsymbol{I}_g \otimes \boldsymbol{X}_{d1}) \begin{bmatrix} \boldsymbol{\theta}'_1 & \cdots & \boldsymbol{\theta}'_g \end{bmatrix}' + \begin{bmatrix} \boldsymbol{\epsilon}'_1 & \cdots & \boldsymbol{\epsilon}'_g \end{bmatrix}'. \end{aligned} \tag{2.4}$$

Along with the assumptions of zero mean errors and uncorrelated response variables, we consider that the dispersion matrix of $\boldsymbol{\epsilon}_k$ is given as $\mathbb{D}(\boldsymbol{\epsilon}_k) = \sigma_k^2 \boldsymbol{\Sigma}_k = \sigma_k^2 (\boldsymbol{I}_n \otimes \boldsymbol{V}_k)$, where $\boldsymbol{V}_k$ is a $p \times p$ known positive definite and symmetric matrix, and $\sigma_k^2 > 0$ is an unknown constant. We assume that σ_k^2's and $\boldsymbol{V}_k$'s are not the same for different responses.

3. Information matrix

In this section, corresponding to the direct effects we derive the information matrix, when $g > 1$. For $g = 1$, we have the univariate response for which the information matrix is studied in [1] for the model (2.4). For $d \in \Omega_{t,n,p}$, corresponding to the direct effects let the information matrix be denoted by $\boldsymbol{C}_d$.

Note that here we consider within subject correlation in the errors. The dispersion matrix of $\boldsymbol{\epsilon} = \begin{bmatrix} \boldsymbol{\epsilon}'_1 & \cdots & \boldsymbol{\epsilon}'_g \end{bmatrix}'$ is

$$\mathbb{D}(\boldsymbol{\epsilon}) = \boldsymbol{\Sigma} = \sigma_1^2 \boldsymbol{\Sigma}_1 \oplus \cdots \oplus \sigma_g^2 \boldsymbol{\Sigma}_g = \oplus_{k=1}^{g} \sigma_k^2 \boldsymbol{\Sigma}_k, \tag{3.1}$$

where $\boldsymbol{\Sigma}_k = \boldsymbol{I}_n \otimes \boldsymbol{V}_k$, and σ_k^2 and $\boldsymbol{V}_k$ vary with k.

LEMMA 3.1. *The information matrix for direct effects is given by*

$$\boldsymbol{C}_d = \sigma_1^{-2} \boldsymbol{C}_{d(1)} \oplus \cdots \oplus \sigma_g^{-2} \boldsymbol{C}_{d(g)}, \tag{3.2}$$

where

$$\begin{aligned} \boldsymbol{C}_{d(k)} &= \boldsymbol{C}_{d(k)11} - \boldsymbol{C}_{d(k)12} \boldsymbol{C}^{-}_{d(k)22} \boldsymbol{C}_{d(k)21} \\ &= \boldsymbol{T}'_d \boldsymbol{\Sigma}_k^{-1/2} \operatorname{pr}^{\perp}\left(\boldsymbol{\Sigma}_k^{-1/2} \begin{bmatrix} \boldsymbol{X_1} & \boldsymbol{F}_d \end{bmatrix}\right) \boldsymbol{\Sigma}_k^{-1/2} \boldsymbol{T}_d. \end{aligned} \tag{3.3}$$

Here $\boldsymbol{C}_{d(k)11} = \boldsymbol{T}'_d \boldsymbol{A}^*_k \boldsymbol{T}_d$, $\boldsymbol{C}_{d(k)12} = \boldsymbol{C}'_{d(k)21} = \boldsymbol{T}'_d \boldsymbol{A}^*_k \boldsymbol{F}_d$, $\boldsymbol{C}_{d(k)22} = \boldsymbol{F}'_d \boldsymbol{A}^*_k \boldsymbol{F}_d$, $\boldsymbol{A}^*_k = \boldsymbol{H}_n \otimes \boldsymbol{V}^*_k$, $\boldsymbol{V}^*_k = \boldsymbol{V}_k^{-1} - \delta_k \boldsymbol{V}_k^{-1} \boldsymbol{J}_{p\times p} \boldsymbol{V}_k^{-1}$ *and* $\delta_k = \left(\boldsymbol{1}'_p \boldsymbol{V}_k^{-1} \boldsymbol{1}_p\right)^{-1}$.

PROOF. Let $\boldsymbol{\xi}_k = \begin{bmatrix} \mu_k & \boldsymbol{\alpha}'_k & \boldsymbol{\beta}'_k & \boldsymbol{\rho}'_k \end{bmatrix}'$, $\boldsymbol{\eta}^{(1)} = \begin{bmatrix} \boldsymbol{\tau}'_1 & \cdots & \boldsymbol{\tau}'_g \end{bmatrix}'$ and $\boldsymbol{\eta}^{(2)} = \begin{bmatrix} \boldsymbol{\xi}'_1 & \cdots & \boldsymbol{\xi}'_g \end{bmatrix}'$. Then by rearranging the parameters, we represent the model (2.4) as

$$\left[\boldsymbol{Y}_{d1}^{'} \quad \cdots \quad \boldsymbol{Y}_{dg}^{'}\right]^{'} = \left(\boldsymbol{I}_g \otimes \boldsymbol{T}_d\right)\boldsymbol{\eta}^{(1)} + \left(\boldsymbol{I}_g \otimes \left[\boldsymbol{1}_{np} \quad \boldsymbol{X}_1 \quad \boldsymbol{F}_d\right]\right)\boldsymbol{\eta}^{(2)} + \left[\boldsymbol{\epsilon}_1^{'} \quad \cdots \quad \boldsymbol{\epsilon}_g^{'}\right]^{'}. \tag{3.4}$$

Premultiplying (3.4) by $\boldsymbol{\Sigma}^{-1/2}$, the model can be expressed as

$$\left[\boldsymbol{Y}_{d1(new)}^{'} \quad \cdots \quad \boldsymbol{Y}_{dg(new)}^{'}\right]^{'} = \boldsymbol{\Sigma}^{-1/2}\left(\boldsymbol{I}_g \otimes \boldsymbol{T}_d\right)\boldsymbol{\eta}^{(1)} + \boldsymbol{\Sigma}^{-1/2}\left(\boldsymbol{I}_g \otimes \left[\boldsymbol{1}_{np} \quad \boldsymbol{X}_1 \quad \boldsymbol{F}_d\right]\right)\boldsymbol{\eta}^{(2)} + \left[\boldsymbol{\epsilon}_{1(new)}^{'} \quad \cdots \quad \boldsymbol{\epsilon}_{g(new)}^{'}\right]^{'}, \tag{3.5}$$

where $\boldsymbol{Y}_{dk(new)} = \sigma_k^{-1}\boldsymbol{\Sigma}_k^{-1/2}\boldsymbol{Y}_{dk}$ and $\boldsymbol{\epsilon}_{k(new)} = \sigma_k^{-1}\boldsymbol{\Sigma}_k^{-1/2}\boldsymbol{\epsilon}_k$.

Therefore from [**15**], employing the expression for the information matrix, we have

$$\boldsymbol{C}_d = \left(\boldsymbol{I}_g \otimes \boldsymbol{T}_d^{'}\right)\boldsymbol{\Sigma}^{-1/2}\,\mathrm{pr}^{\perp}\left(\boldsymbol{\Sigma}^{-1/2}\left(\boldsymbol{I}_g \otimes \left[\boldsymbol{1}_{np} \quad \boldsymbol{X}_1 \quad \boldsymbol{F}_d\right]\right)\right)\boldsymbol{\Sigma}^{-1/2}\left(\boldsymbol{I}_g \otimes \boldsymbol{T}_d\right).$$

From [**1**], since $\boldsymbol{1}_{np}$ belongs to the column space of $\boldsymbol{X}_1$, we get

$$\boldsymbol{C}_d = \left(\boldsymbol{I}_g \otimes \boldsymbol{T}_d^{'}\right)\boldsymbol{\Sigma}^{-1/2}\,\mathrm{pr}^{\perp}\left(\boldsymbol{\Sigma}^{-1/2}\left(\boldsymbol{I}_g \otimes \left[\boldsymbol{X}_1 \quad \boldsymbol{F}_d\right]\right)\right)\boldsymbol{\Sigma}^{-1/2}\left(\boldsymbol{I}_g \otimes \boldsymbol{T}_d\right). \tag{3.6}$$

Using (3.1) in the above equation, we get

$$\begin{aligned} \boldsymbol{C}_d &= \oplus_{k=1}^{g}\sigma_k^{-2}\boldsymbol{T}_d^{'}\boldsymbol{\Sigma}_k^{-1/2}\,\mathrm{pr}^{\perp}\left(\boldsymbol{\Sigma}_k^{-1/2}\left[\boldsymbol{X}_1 \quad \boldsymbol{F}_d\right]\right)\boldsymbol{\Sigma}_k^{-1/2}\boldsymbol{T}_d \\ &= \sigma_1^{-2}\boldsymbol{C}_{d(1)} \oplus \cdots \oplus \sigma_g^{-2}\boldsymbol{C}_{d(g)}, \end{aligned} \tag{3.7}$$

where

$$\boldsymbol{C}_{d(k)} = \boldsymbol{T}_d^{'}\boldsymbol{\Sigma}_k^{-1/2}\,\mathrm{pr}^{\perp}\left(\boldsymbol{\Sigma}_k^{-1/2}\left[\boldsymbol{X}_1 \quad \boldsymbol{F}_d\right]\right)\boldsymbol{\Sigma}_k^{-1/2}\boldsymbol{T}_d. \tag{3.8}$$

Here $\boldsymbol{\Sigma}_k = \boldsymbol{I}_n \otimes \boldsymbol{V}_k$, where $\boldsymbol{V}_k$ is a known positive definite and symmetric $p \times p$ matrix. So from [**1**], we get

$$\boldsymbol{T}_d^{'}\boldsymbol{\Sigma}_k^{-1/2}\,\mathrm{pr}^{\perp}\left(\boldsymbol{\Sigma}_k^{-1/2}\left[\boldsymbol{X}_1 \quad \boldsymbol{F}_d\right]\right)\boldsymbol{\Sigma}_k^{-1/2}\boldsymbol{T}_d = \boldsymbol{C}_{d(k)11} - \boldsymbol{C}_{d(k)12}\boldsymbol{C}_{d(k)22}^{-}\boldsymbol{C}_{d(k)21}, \tag{3.9}$$

where $\boldsymbol{C}_{d(k)11} = \boldsymbol{T}_d^{'}\boldsymbol{A}_k^{*}\boldsymbol{T}_d$, $\boldsymbol{C}_{d(k)12} = \boldsymbol{C}_{d(k)21}^{'} = \boldsymbol{T}_d^{'}\boldsymbol{A}_k^{*}\boldsymbol{F}_d$, $\boldsymbol{C}_{d(k)22} = \boldsymbol{F}_d^{'}\boldsymbol{A}_k^{*}\boldsymbol{F}_d$, $\boldsymbol{A}_k^{*} = \boldsymbol{H}_n \otimes \boldsymbol{V}_k^{*}$, $\boldsymbol{V}_k^{*} = \boldsymbol{V}_k^{-1} - \delta_k\boldsymbol{V}_k^{-1}\boldsymbol{J}_{p\times p}\boldsymbol{V}_k^{-1}$ and $\delta_k = \left(\boldsymbol{1}_p^{'}\boldsymbol{V}_k^{-1}\boldsymbol{1}_p\right)^{-1}$.

Thus from (3.8) and (3.9), we can prove (3.3). □

REMARK 3.2. The information matrix $\boldsymbol{C}_d$ is symmetric, non-negative definite (n.n.d.) matrix having zero column sums and row sums. Moreover, $\boldsymbol{C}_d$ is invariant with respect to the choice of generalized inverses involved.

PROOF. Evidently, we have by (3.2) and (3.3) that $\boldsymbol{C}_d$ is symmetric and n.n.d. From Lemma 3.1, we get

$$\boldsymbol{C}_d = \sigma_1^{-2}\boldsymbol{C}_{d(1)} \oplus \cdots \oplus \sigma_g^{-2}\boldsymbol{C}_{d(g)}, \tag{3.10}$$

where

$$\boldsymbol{C}_{d(k)} = \boldsymbol{T}_d^{'}\boldsymbol{\Sigma}_k^{-1/2}\,\mathrm{pr}^{\perp}\left(\boldsymbol{\Sigma}_k^{-1/2}\left[\boldsymbol{X}_1 \quad \boldsymbol{F}_d\right]\right)\boldsymbol{\Sigma}_k^{-1/2}\boldsymbol{T}_d.$$

Here $\boldsymbol{\Sigma}_k = \boldsymbol{I}_n \otimes \boldsymbol{V}_k$, where $\boldsymbol{V}_k$ is a $p \times p$ known positive definite and symmetric matrix. Hence from [**1**], we get that $\boldsymbol{C}_{d(k)}\boldsymbol{1}_t = \boldsymbol{0}_{t\times 1}$, $\boldsymbol{1}_t^{'}\boldsymbol{C}_{d(k)} = \boldsymbol{0}_{1\times t}$ and $\boldsymbol{C}_{d(k)}$

is invariant with respect to the choice of generalized inverses involved. Thus from (3.10), we get

$$\begin{aligned} \boldsymbol{C}_d \mathbf{1}_{gt} &= \oplus_{k=1}^{g} \sigma_k^{-2} \boldsymbol{C}_{d(k)} \mathbf{1}_t = \mathbf{0}_{gt\times 1}, \\ \mathbf{1}_{gt}^{'} \boldsymbol{C}_d &= \oplus_{k=1}^{g} \sigma_k^{-2} \mathbf{1}_t^{'} \boldsymbol{C}_{d(k)} = \mathbf{0}_{1\times gt}, \text{ and} \end{aligned}$$

$\boldsymbol{C}_d$ is invariant with respect to the choice of generalized inverses involved. □

4. Optimality results

Various authors including [**14**], [**9**], [**31**], [**6**] and [**1**] have discussed A-, D-, E- and universal optimality criterion for the $g = 1$ setup. For the univariate response case, [**14**] and [**31**] have provided sufficient conditions for universal optimality. In this section, we extend these sufficient conditions for the multivariate response setup. Let for $g > 1$, $\mathcal{B}_{gt}$ be a class of $gt \times gt$ symmetric, non-negative definite (n.n.d.) matrices having zero row sums. Note that for the multivariate response case, the information matrix for the direct effects, $\boldsymbol{C}_d$, is a $gt \times gt$ matrix. From Remark 3.2, it is clear that for $g > 1$, $\boldsymbol{C}_d \in \mathcal{B}_{gt}$. In the next lemmas, we present sufficient conditions for obtaining universally optimal designs in the $g > 1$ setup.

LEMMA 4.1. *Let a design $d^* \in \mathcal{D}$, where $\mathcal{D}$ is a subclass of designs, be such that under model (2.4), the information matrix $\boldsymbol{C}_{d^*}$ is completely symmetric and d^* maximizes $\operatorname{tr}(\boldsymbol{C}_d)$ over $d \in \mathcal{D}$. Then d^* is universally optimal for the direct effects over $\mathcal{D}$.*

PROOF. The proof is on similar lines as that of [**14**]. □

LEMMA 4.2. *Let a design $d^* \in \mathcal{D}$, where $\mathcal{D}$ is a subclass of designs, be such that for $d \in \mathcal{D}$, the corresponding information matrix $\boldsymbol{C}_d \in \mathcal{B}_{gt}$; for any $d \in \mathcal{D}$ and $\boldsymbol{C}_d \neq \mathbf{0}_{gt\times gt}$, there exists scalars $b_{d1}, \cdots, b_{d(gt)!} \geq 0$ satisfying $\boldsymbol{C}_{d^*} = \sum_{\kappa=1}^{(gt)!} b_{d\kappa} \boldsymbol{P}_\kappa \boldsymbol{C}_d \boldsymbol{P}_\kappa^{'}$, where $g > 1$ and $\boldsymbol{P}_1, \cdots, \boldsymbol{P}_{(gt)!}$ are all possible distinct $gt \times gt$ permutation matrices. Then the design d^* is universally optimal for the direct effects over $\mathcal{D}$, if d^* maximizes $\operatorname{tr}(\boldsymbol{C}_d)$ over $d \in \mathcal{D}$.*

PROOF. The proof is similar to that in [**31**]. □

Now, we consider designs with $p = t$. For the univariate response case with correlated error terms, [**17**] proved that a design given as an orthogonal array of type I and strength 2, $OA_I(n = \lambda t(t-1), p, t, 2)$, where $3 \leq p \leq t$ and λ is a positive integer, is universally optimal for the direct effects over the class of binary designs. Next, we check whether the universal optimality of $OA_I(n = \lambda t(t-1), p = t, t, 2)$, where $t \geq 3$ and λ is a positive integer, holds for the multivariate response setup. Let us define the class of competing designs as

$$\mathcal{D} = \{d \in \Omega_{t,n,p=t} : d \text{ is a binary design and } t \geq 3\}.$$

It should be noted that $\mathcal{D}$ is a subclass of designs which are uniform on subjects.

LEMMA 4.3. *Let $d^* \in \Omega_{t,n=\lambda t(t-1),p=t}$ be a design given by $OA_I(n = \lambda t(t-1), p = t, t, 2)$, where λ is a positive integer and $t \geq 3$. Then for $g > 1$, corresponding to design d^*, the information matrix for the direct effects can be expressed as*

$$\boldsymbol{C}_{d^*} = \oplus_{k=1}^{g} \sigma_k^{-2} \left(\det(\boldsymbol{Q}_k)/q_{(k)22}\right) \boldsymbol{H}_t, \tag{4.1}$$

where $\boldsymbol{Q}_k = \frac{n}{t-1}\begin{bmatrix} q_{(k)11} & q_{(k)12} \\ q_{(k)12} & q_{(k)22} \end{bmatrix}$, $\det(\boldsymbol{Q}_k) \neq 0$, $q_{(k)11} = \operatorname{tr}\left(\boldsymbol{T}'_{d^*1}\boldsymbol{V}^*_k\boldsymbol{T}_{d^*1}\right)$, $q_{(k)12} = \operatorname{tr}\left(\boldsymbol{T}'_{d^*1}\boldsymbol{V}^*_k\boldsymbol{\psi}\boldsymbol{T}_{d^*1}\right)$ *and* $q_{(k)22} = \operatorname{tr}\left(\boldsymbol{T}'_{d^*1}\boldsymbol{\psi}'\boldsymbol{V}^*_k\boldsymbol{\psi}\boldsymbol{T}_{d^*1}\right) - \frac{(\boldsymbol{V}^*_k)_{1,1}}{t}$, $\boldsymbol{H}_t = \boldsymbol{I}_t - \frac{1}{t}\boldsymbol{1}_t\boldsymbol{1}'_t$, $\boldsymbol{\psi} = \begin{bmatrix} \boldsymbol{0}'_{p-1\times 1} & 0 \\ \boldsymbol{I}_{p-1} & \boldsymbol{0}_{p-1\times 1} \end{bmatrix}$, $\boldsymbol{V}^*_k = \boldsymbol{V}_k^{-1} - \left(\boldsymbol{1}'_p\boldsymbol{V}_k^{-1}\boldsymbol{1}_p\right)^{-1}\boldsymbol{V}_k^{-1}\boldsymbol{1}_p\boldsymbol{1}'_p\boldsymbol{V}_k^{-1}$, *and* $(\boldsymbol{V}^*_k)_{1,1}$ *is the element corresponding to first row and first column of the matrix* $\boldsymbol{V}^*_k$.

PROOF. Here $d^* \in \Omega_{t,n=\lambda t(t-1),p=t}$ is a design given by $OA_I(n = \lambda t(t-1), p = t, t, 2)$, where λ is a positive integer and $t \geq 3$. From [**23**] and [**1**], we get that for $g = 1$, the information matrix corresponding to d^* for the direct effects is given as

$$\boldsymbol{C}_{d^*} = \boldsymbol{T}'_d\boldsymbol{\Sigma}_1^{-1/2}\operatorname{pr}^{\perp}\left(\boldsymbol{\Sigma}_1^{-1/2}\begin{bmatrix}\boldsymbol{X}_1 & \boldsymbol{F}_d\end{bmatrix}\right)\boldsymbol{\Sigma}_1^{-1/2}\boldsymbol{T}_d = \left(\det(\boldsymbol{Q}_1)/q_{(1)22}\right)\boldsymbol{H}_t,$$

where $\boldsymbol{Q}_1 = \frac{n}{t-1}\begin{bmatrix} q_{(1)11} & q_{(1)12} \\ q_{(1)12} & q_{(1)22} \end{bmatrix}$, $\det(\boldsymbol{Q}_1) \neq 0$, $q_{(1)11} = \operatorname{tr}\left(\boldsymbol{T}'_{d^*1}\boldsymbol{V}^*_1\boldsymbol{T}_{d^*1}\right)$, $q_{(1)12} = \operatorname{tr}\left(\boldsymbol{T}'_{d^*1}\boldsymbol{V}^*_1\boldsymbol{\psi}\boldsymbol{T}_{d^*1}\right)$ and $q_{(1)22} = \operatorname{tr}\left(\boldsymbol{T}'_{d^*1}\boldsymbol{\psi}'\boldsymbol{V}^*_1\boldsymbol{\psi}\boldsymbol{T}_{d^*1}\right) - \frac{(\boldsymbol{V}^*_1)_{1,1}}{t}$, $\boldsymbol{H}_t = \boldsymbol{I}_t - \frac{1}{t}\boldsymbol{1}_t\boldsymbol{1}'_t$, $\boldsymbol{\psi} = \begin{bmatrix} \boldsymbol{0}'_{p-1\times 1} & 0 \\ \boldsymbol{I}_{p-1} & \boldsymbol{0}_{p-1\times 1} \end{bmatrix}$, $\boldsymbol{V}^*_1 = \boldsymbol{V}_1^{-1} - \left(\boldsymbol{1}'_p\boldsymbol{V}_1^{-1}\boldsymbol{1}_p\right)^{-1}\boldsymbol{V}_1^{-1}\boldsymbol{1}_p\boldsymbol{1}'_p\boldsymbol{V}_1^{-1}$, and $(\boldsymbol{V}^*_1)_{1,1}$ is the element corresponding to first row and first column of the matrix $\boldsymbol{V}^*_1$. Here $\boldsymbol{V}_1$ is a known positive definite and symmetric matrix.

For the multivariate response setup, we have $\boldsymbol{V}_k$ to be a known positive definite and symmetric matrix. Using the expression of the information matrix from Lemma 3.1, we get

$$\boldsymbol{C}_{d^*} = \oplus_{k=1}^{g}\sigma_k^{-2}\left(\det(\boldsymbol{Q}_k)/q_{(k)22}\right)\boldsymbol{H}_t,$$

where $\boldsymbol{Q}_k = \frac{n}{t-1}\begin{bmatrix} q_{(k)11} & q_{(k)12} \\ q_{(k)12} & q_{(k)22} \end{bmatrix}$, $\det(\boldsymbol{Q}_k) \neq 0$, $q_{(k)11} = \operatorname{tr}\left(\boldsymbol{T}'_{d^*1}\boldsymbol{V}^*_k\boldsymbol{T}_{d^*1}\right)$, $q_{(k)12} = \operatorname{tr}\left(\boldsymbol{T}'_{d^*1}\boldsymbol{V}^*_k\boldsymbol{\psi}\boldsymbol{T}_{d^*1}\right)$ and $q_{(k)22} = \operatorname{tr}\left(\boldsymbol{T}'_{d^*1}\boldsymbol{\psi}'\boldsymbol{V}^*_k\boldsymbol{\psi}\boldsymbol{T}_{d^*1}\right) - \frac{(\boldsymbol{V}^*_k)_{1,1}}{t}$, $\boldsymbol{V}^*_k = \boldsymbol{V}_k^{-1} - \left(\boldsymbol{1}'_p\boldsymbol{V}_k^{-1}\boldsymbol{1}_p\right)^{-1}\boldsymbol{V}_k^{-1}\boldsymbol{1}_p\boldsymbol{1}'_p\boldsymbol{V}_k^{-1}$, and $(\boldsymbol{V}^*_k)_{1,1}$ is the element corresponding to first row and first column of the matrix $\boldsymbol{V}^*_k$. □

REMARK 4.4. Let $d^* \in \Omega_{t,n=\lambda t(t-1),p=t}$ be a design given by

$$OA_I(n = \lambda t \times (t-1), p = t, t, 2),$$

where λ is a positive integer and $t \geq 3$. Then for $g > 1$, corresponding to the direct effects, the information matrix $\boldsymbol{C}_{d^*}$ is not completely symmetric.

PROOF. From Lemma 4.3, we know that for $g > 1$,

$$\boldsymbol{C}_{d^*} = \oplus_{k=1}^{g}\sigma_k^{-2}\left(\det(\boldsymbol{Q}_k)/q_{(k)22}\right)\boldsymbol{H}_t,$$

where $\boldsymbol{Q}_k = \frac{n}{t-1}\begin{bmatrix} q_{(k)11} & q_{(k)12} \\ q_{(k)12} & q_{(k)22} \end{bmatrix}$, $\det(\boldsymbol{Q}_k) \neq 0$, $q_{(k)11} = \operatorname{tr}\left(\boldsymbol{T}'_{d^*1}\boldsymbol{V}^*_k\boldsymbol{T}_{d^*1}\right)$, $q_{(k)12} = \operatorname{tr}\left(\boldsymbol{T}'_{d^*1}\boldsymbol{V}^*_k\boldsymbol{\psi}\boldsymbol{T}_{d^*1}\right)$ and $q_{(k)22} = \operatorname{tr}\left(\boldsymbol{T}'_{d^*1}\boldsymbol{\psi}'\boldsymbol{V}^*_k\boldsymbol{\psi}\boldsymbol{T}_{d^*1}\right) - \frac{(\boldsymbol{V}^*_k)_{1,1}}{t}$, $\boldsymbol{H}_t = \boldsymbol{I}_t - \frac{1}{t}\boldsymbol{1}_t\boldsymbol{1}'_t$, $\boldsymbol{\psi} = \begin{bmatrix} \boldsymbol{0}'_{p-1\times 1} & 0 \\ \boldsymbol{I}_{p-1} & \boldsymbol{0}_{p-1\times 1} \end{bmatrix}$, $\boldsymbol{V}^*_k = \boldsymbol{V}_k^{-1} - \left(\boldsymbol{1}'_p\boldsymbol{V}_k^{-1}\boldsymbol{1}_p\right)^{-1}\boldsymbol{V}_k^{-1}\boldsymbol{1}_p\boldsymbol{1}'_p\boldsymbol{V}_k^{-1}$, and $(\boldsymbol{V}^*_k)_{1,1}$ is the element corresponding to first row and first column of the matrix $\boldsymbol{V}^*_k$.

Since $\sigma_k^{-2} > 0$, the necessary condition for the matrix $\boldsymbol{C}_{d^*}$ to be completely symmetric is that all the off-diagonal elements of $\left(\det\left(\boldsymbol{Q}_k\right)/q_{(k)22}\right)\boldsymbol{H}_t$ are 0. Since we know that off-diagonal elements of $\boldsymbol{H}_t$ are $-1/t$ and for $t \geq 3$, $\det\left(\boldsymbol{Q}_k\right) \neq 0$, the off-diagonal elements of $\left(\det\left(\boldsymbol{Q}_k\right)/q_{(k)22}\right)\boldsymbol{H}_t$ are non-zero. Thus for $g > 1$ and $t \geq 3$, $\boldsymbol{C}_{d^*}$ is not completely symmetric. □

From Remark 4.4, it is clear that we cannot use Lemma 4.1. Since $(\det(\boldsymbol{Q}_k)/q_{(k)22})\boldsymbol{H}_t$ is a completely symmetric matrix, in order to apply Lemma 4.2, we require that for any $d \in \mathcal{D}$ and $\boldsymbol{C}_{d(k)} \neq \boldsymbol{0}_{t\times t}$, $\sigma_k^{-2}\operatorname{tr}\left(\boldsymbol{C}_{d^*(k)}\right)/\left[t! \times \operatorname{tr}\left(\boldsymbol{C}_{d(k)}\right)\right]$ is same for all k. Since $\boldsymbol{C}_{d(k)}$ depends on k through $\boldsymbol{V}_k$, and σ_k^2 and $\boldsymbol{V}_k$ may vary according to k, this condition does not hold true. So, Lemma 4.2 is not applicable. Thus we resort to identify A-, D- and E-optimal design.

Note that we have gt direct effects. From [**1**], we have that $\boldsymbol{C}_{d(k)}\boldsymbol{1}_t = \boldsymbol{0}_{t\times 1}$. So from Lemma 3.1, it is clear that for $g > 1$, $\operatorname{rank}\left(\boldsymbol{C}_d\right) \leq g(t-1)$. Thus all contrasts are not estimable. Hence, here we only consider designs such that for each variable, $t-1$ independent orthonormal contrasts of the direct effects are estimable.

To combine these $t-1$ contrasts for all variables, we introduce a $(t-1)\times t$ matrix $\boldsymbol{L}_k$, where $\operatorname{rank}\left(\boldsymbol{L}_k\right) = t-1$, $\boldsymbol{L}_k\boldsymbol{1}_t = \boldsymbol{0}_{t\times 1}$ and $\boldsymbol{L}_k\boldsymbol{L}_k^{'} = \boldsymbol{I}_{t-1}$. So, $\boldsymbol{L}_k\boldsymbol{\tau}_k$ is the vector of $t-1$ independent orthonormal contrasts of the direct effects corresponding to the kth response variable. Let us denote $\mathbb{D}\left(\boldsymbol{L}\hat{\boldsymbol{\tau}}\right)$ by $\boldsymbol{G}_d$, where $\boldsymbol{L}\hat{\boldsymbol{\tau}}$ is the best linear unbiased estimator of $\boldsymbol{L}\boldsymbol{\tau}$ and

$$\boldsymbol{L} = \boldsymbol{L}_1 \oplus \cdots \oplus \boldsymbol{L}_g = \oplus_{k=1}^{g}\boldsymbol{L}_k. \tag{4.2}$$

Then a design $d^* \in \mathcal{D}$ is A-, D- and E-optimal for the direct effects over $\mathcal{D}$ under model (2.4), for $g \geq 1$, if d^* minimizes $\operatorname{tr}\left(\boldsymbol{G}_d\right)$, $\det\left(\boldsymbol{G}_d\right)$ and the maximum eigenvalue of $\boldsymbol{G}_d$, respectively, over $d \in \mathcal{D}$.

From Lemma 3.1, we get that for $d \in \Omega_{t,n,p}$,

$$\mathbb{D}\left(\boldsymbol{L}\begin{bmatrix}\hat{\boldsymbol{\tau}}_1^{'} & \hat{\boldsymbol{\tau}}_2^{'} & \cdots & \hat{\boldsymbol{\tau}}_g^{'}\end{bmatrix}^{'}\right) = \boldsymbol{G}_d = \sigma_1^2\boldsymbol{L}_1\boldsymbol{C}_{d(1)}^{+}\boldsymbol{L}_1^{'} \oplus \cdots \oplus \sigma_g^2\boldsymbol{L}_g\boldsymbol{C}_{d(g)}^{+}\boldsymbol{L}_g^{'}, \tag{4.3}$$

where $\boldsymbol{C}_{d(k)}^{+}$ is the Moore-Penrose inverse of the matrix $\boldsymbol{C}_{d(k)}$.

Minimizing $\sum_{k=1}^{g}\operatorname{tr}\left(\sigma_k^2\boldsymbol{L}_k\boldsymbol{C}_{d(k)}^{+}\boldsymbol{L}_k^{'}\right)$, $\sum_{k=1}^{g} log\left(\det\left(\sigma_k^2\boldsymbol{L}_k\boldsymbol{C}_{d(k)}^{+}\boldsymbol{L}_k^{'}\right)\right)$ and the maximum eigenvalue of $\oplus_{k=1}^{g}\sigma_k^2\boldsymbol{L}_k\boldsymbol{C}_{d(k)}^{+}\boldsymbol{L}_k^{'}$ we can find A-, D- and E-optimal design, respectively.

Let $\zeta_{d(k)1} \leq \cdots \leq \zeta_{d(k)(t-1)}$ be the positive eigenvalues of $\boldsymbol{C}_{d(k)}$ and $\eta_{d(k)1} \leq \cdots \leq \eta_{d(k)(t-1)}$ be the eigenvalues of $\sigma_k^2\boldsymbol{L}_k\boldsymbol{C}_{d(k)}^{+}\boldsymbol{L}_k^{'}$. Then following a similar approach as given in [**6**], we get that all the nonzero eigenvalues of $\boldsymbol{C}_{d(k)}^{+}$ are the eigenvalues of $\boldsymbol{L}_k\boldsymbol{C}_{d(k)}^{+}\boldsymbol{L}_k^{'}$. Thus the eigenvalues of $\sigma_k^2\boldsymbol{L}_k\boldsymbol{C}_{d(k)}^{+}\boldsymbol{L}_k^{'}$ are

$$\eta_{d(k)1} = \frac{\sigma_k^2}{\zeta_{d(k)(t-1)}}, \cdots, \eta_{d(k)(t-1)} = \frac{\sigma_k^2}{\zeta_{d(k)1}}. \tag{4.4}$$

Using the above eigenvalue equations from (4.4), we get

$$\sum_{k=1}^{g} \operatorname{tr}\left(\sigma_k^2 \boldsymbol{L}_k \boldsymbol{C}_{d(k)}^{+} \boldsymbol{L}_k^{'}\right) = \sum_{k=1}^{g}\sum_{\omega=1}^{t-1} \eta_{d(k)\omega} = \sum_{k=1}^{g}\sum_{\omega=1}^{t-1} \frac{\sigma_k^2}{\zeta_{d(k)\omega}},$$

$$\sum_{k=1}^{g} log\left(\det\left(\sigma_k^2 \boldsymbol{L}_k \boldsymbol{C}_{d(k)}^{+} \boldsymbol{L}_k^{'}\right)\right) = \sum_{k=1}^{g}\sum_{\omega=1}^{t-1} log\left(\eta_{d(k)\omega}\right) = \sum_{k=1}^{g}\sum_{\omega=1}^{t-1} log\left(\frac{\sigma_k^2}{\zeta_{d(k)\omega}}\right), \text{ and}$$

The maximum eigenvalue of $\oplus_{k=1}^{g} \sigma_k^2 \boldsymbol{L}_k \boldsymbol{C}_{d(k)}^{+} \boldsymbol{L}_k^{'} = \max_{1\le k\le g} \eta_{d(k)(t-1)} = \max_{1\le k\le g} \dfrac{\sigma_k^2}{\zeta_{d(k)1}}$.

Thus a design $d^* \in \mathcal{D}$ is A-, D- and E-optimal for the direct effects over $\mathcal{D}$ if d^* minimizes $\sum_{k=1}^{g}\sum_{\omega=1}^{t-1} \frac{\sigma_k^2}{\zeta_{d(k)\omega}}$, $\sum_{k=1}^{g}\sum_{\omega=1}^{t-1} log\left(\frac{\sigma_k^2}{\zeta_{d(k)\omega}}\right)$ and $\max_{1\le k\le g} \sigma_k^2\zeta_{d(k)1}^{-1}$, respectively, over $d \in \mathcal{D}$.

THEOREM 4.5. *Let $\mathcal{D}$ be a class of binary designs with $p = t$ and $d^* \in \mathcal{D}$ be a design represented as $OA_I\,(n = \lambda t\,(t-1)\,, p = t, t, 2)$, where λ is a positive integer and $t \ge 3$. Then under the $g > 1$ case, d^* is A-, D- and E-optimal for the direct effects over $\mathcal{D}$.*

PROOF. Here $d^* \in \mathcal{D}$ is a design represented as $OA_I\,(n = \lambda t\,(t-1)\,, p = t, t, 2)$, where $\mathcal{D}$ is a class of binary designs with $p = t \ge 3$ and λ is a positive integer. From [**17**], we know that for $g = 1$ setup, d^* is universally optimal for the direct effects over $\mathcal{D}$. Since universal optimality implies A-, D- and E-optimality, we get that for $g = 1$ and $t \ge 3$, d^* minimizes $\sum_{\omega=1}^{t-1} \frac{\sigma_1^2}{\zeta_{d(1)\omega}}$, $\sum_{\omega=1}^{t-1} log\left(\frac{\sigma_1^2}{\zeta_{d(1)\omega}}\right)$ and $\sigma_1^2\zeta_{d(1)1}^{-1}$ over $d \in \mathcal{D}$. Here $\zeta_{d(1)1} \le \cdots \le \zeta_{d(1)(t-1)}$ are the positive eigenvalues of $\boldsymbol{C}_{d(1)}$, where $\boldsymbol{C}_{d(1)}$ is as given in Lemma 3.1, and $\boldsymbol{V}_1$ is a known positive definite and symmetric $p \times p$ matrix.

Here, for $g > 1$, we have $\boldsymbol{V}_k$ as a known positive definite and symmetric $p \times p$ matrix. So for $g > 1$ and $t \ge 3$, for each k, d^* minimizes $\sum_{\omega=1}^{t-1} \frac{\sigma_k^2}{\zeta_{d(k)\omega}}$, $\sum_{\omega=1}^{t-1} log\left(\frac{\sigma_k^2}{\zeta_{d(k)\omega}}\right)$ and $\sigma_k^2\zeta_{d(k)1}^{-1}$ over $d \in \mathcal{D}$. Hence for $g > 1$ and $t \ge 3$, d^* minimizes $\sum_{k=1}^{g}\sum_{\omega=1}^{t-1} \frac{\sigma_k^2}{\zeta_{d(k)\omega}}$, $\sum_{k=1}^{g}\sum_{\omega=1}^{t-1} log\left(\frac{\sigma_k^2}{\zeta_{d(k)\omega}}\right)$ and $\max_{1\le k\le g} \sigma_k^2\zeta_{d(k)1}^{-1}$ over $d \in \mathcal{D}$. Thus for $t \ge 3$, d^* is A-, D- and E-optimal for the direct effects over $\mathcal{D}$ under the $g > 1$ case. □

5. Illustration

In this section, we examine A-, D- and E-efficiency of a 3×3 crossover design using a gene profile study by [**22**]. This design allocates treatment sequences ACB, BAC and CBA each to 6 subjects, where A and B denote 10 mg and 25 mg dose of an oral drug, respectively, and C denotes placebo. From [**25**], it is noted that in the gene profile, there exist pair of genes which are uncorrelated between themselves but may have within-response correlation.

Let d_0 represent the above 3×3 design with treatment sequences ACB, BAC and CBA. To find the A-, D- and E-efficiencies of this d_0, we assume a bivariate response setup, i.e., $g = 2$. Here, we consider $\frac{\sigma_1^2}{\sigma_2^2} = 2$ (any non-negative value other than 2 could have been taken), $\boldsymbol{V}_1$ to have $AR(1)$ structure and $\boldsymbol{V}_2$ to have equi-correlated structure. Thus,

$$\boldsymbol{V}_1 = \begin{bmatrix} 1 & r_{(1)} & r_{(1)}^2 \\ r_{(1)} & 1 & r_{(1)} \\ r_{(1)}^2 & r_{(1)} & 1 \end{bmatrix} \quad \text{and} \quad \boldsymbol{V}_2 = \begin{bmatrix} 1 & r_{(1)} & r_{(1)} \\ r_{(1)} & 1 & r_{(1)} \\ r_{(1)} & r_{(1)} & 1 \end{bmatrix},$$

where $-0.5 < r_{(1)} < 1$. The constraint on the values of $r_{(1)}$ is to ensure positive definiteness of $\boldsymbol{V}_1$ and $\boldsymbol{V}_2$.

From Theorem 4.5, we see that for $g > 1$, a design $d^* \in \mathcal{D}$ represented by $OA_I\left(n = \lambda t\left(t-1\right), p = t, t, 2\right)$ is A-, D- and E-optimal for the direct effects over $\mathcal{D}$. We are interested in studying efficiency of the binary design d_0 for all permissible values of $r_{(1)}$. Suppose for a design d_0, the A-, D- and E-efficiency is given by

$$\text{A-Efficiency}_{d_0} = \frac{\phi_A\left(\boldsymbol{C}_{d^*}\right)}{\phi_A\left(\boldsymbol{C}_{d_0}\right)}, \ \text{D-Efficiency}_{d_0} = \frac{\phi_D\left(\boldsymbol{C}_{d^*}\right)}{\phi_D\left(\boldsymbol{C}_{d_0}\right)}, \ \text{and}$$
$$\text{E-Efficiency}_{d_0} = \frac{\phi_E\left(\boldsymbol{C}_{d^*}\right)}{\phi_E\left(\boldsymbol{C}_{d_0}\right)},$$

where for $d \in \{d_0, d^*\}$, $\phi_A\left(\boldsymbol{C}_d\right) = \sum_{k=1}^{g}\sum_{\omega=1}^{t-1}\frac{\sigma_k^2}{\zeta_{d(k)\omega}}$, $\phi_D\left(\boldsymbol{C}_d\right) = \prod_{k=1}^{g}\prod_{\omega=1}^{t-1}\frac{\sigma_k^2}{\zeta_{d(k)\omega}}$ and $\phi_E\left(\boldsymbol{C}_d\right) = \max_{1\leq k\leq g}\frac{\sigma_k^2}{\zeta_{d(k)1}}$. It should be noted that minimization of $\sum_{k=1}^{g}\sum_{\omega=1}^{t-1} log\left(\frac{\sigma_k^2}{\zeta_{d(k)\omega}}\right)$ is equivalent to minimization of $\prod_{k=1}^{g}\prod_{\omega=1}^{t-1}\frac{\sigma_k^2}{\zeta_{d(k)\omega}}$. Also, note that $\phi_A\left(\boldsymbol{C}_d\right)$, $\phi_D\left(\boldsymbol{C}_d\right)$ and $\phi_E\left(\boldsymbol{C}_d\right)$ depends on $r_{(1)}$ only through the eigenvalues, $\zeta_{d(k)\omega}$. The values of A-, D- and E-Efficiency$_{d_0}$ lie between 0 and 1, and a value closer to 1 indicates that design d is a nearly efficient design. From Figure 1, we see that A-, D- and E-efficiency of d_0 is at most 0.0278, 5.9537×10^{-7} and 0.0278, respectively. Thus, implying that the experimenter would have obtained much higher efficiency if they had used an orthogonal array type design in the gene data example.

6. Conclusion

In this article, we considered a multivariate fixed effect model that accounts for within-response correlation. While the g responses were assumed to be uncorrelated, the within-response correlation structure was allowed to vary between distinct responses. Corresponding to the direct effects, the information matrix deviated from the $g = 1$ case under this error structure, particularly concerning the complete symmetricity. We demonstrated that an orthogonal array design of type I with strength 2 is A-, D-, and E-optimal for the direct effects over a class of binary designs with $p = t \geq 3$ in the $g > 1$ scenario. Under an uncorrelated but heteroscedastic error structure, it can be shown using similar techniques, if the universal optimality holds for a balanced uniform design over a subclass of designs for the direct effects (or carryover effects) in the univariate response case, then over the same subclass of designs it remains A-, D-, and E-optimal for the direct effects (or carryover effects) under the multivariate response setup.

Though in this article, we considered a fixed effect model, a mixed effects model with random subject effects may be considered. In [**12**], exact optimal designs for a univariate mixed effects model with random subject effects having independent and identical normal distribution are studied. This model may be extended to the multivariate mixed effects model by considering independent and identically distributed error terms following normal distribution. In this case, the responses will

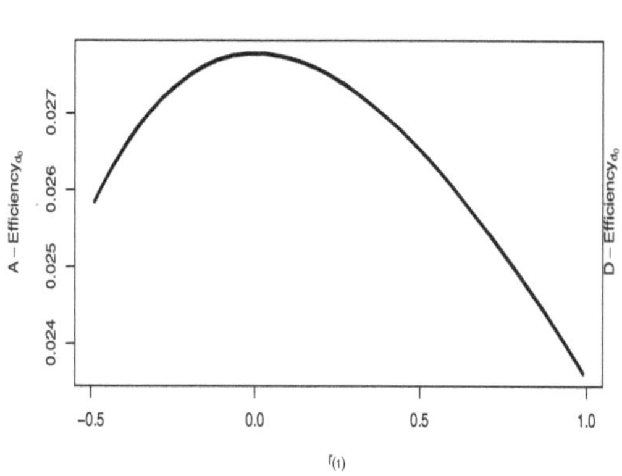

(A) A-efficiency.

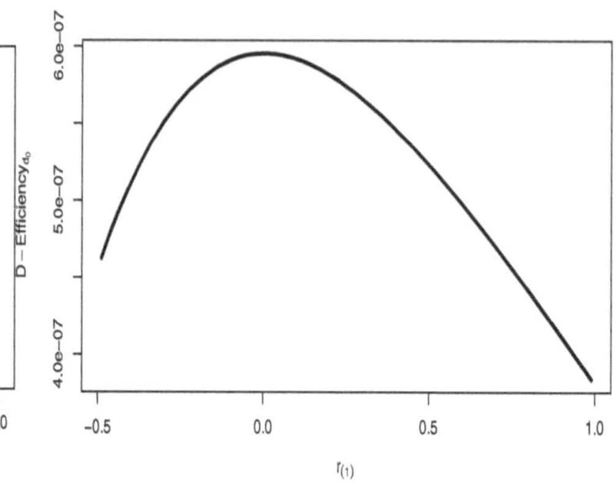

(B) D-efficiency.

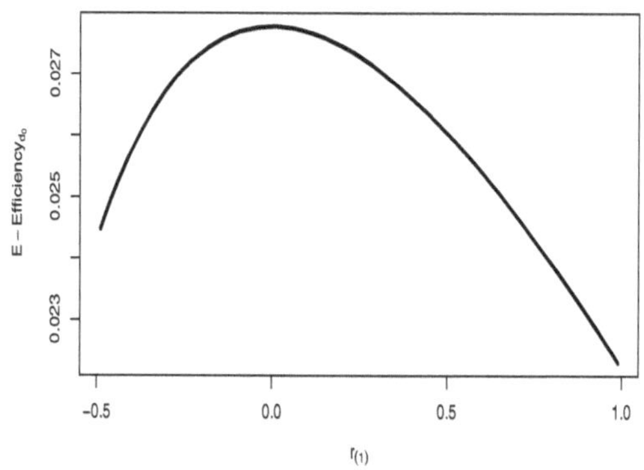

(C) E-efficiency.

FIGURE 1. Efficiency of d_0.

be uncorrelated between themselves, but the random subject effects will introduce correlation within responses, to be specific the within-response correlation structure will be equi-correlated. We can also consider the case where error terms have within-response correlation, but are uncorrelated between responses. In this scenario, the dispersion matrix of the observations on responses will have a complicated structure and hence it may be tedious to identify a theoretical exact optimal design. As a future direction, we intend to explore the A-, D- and E-optimality aspects in these scenarios for the multiple response case.

References

[1] M. Bose and A. Dey, *Optimal crossover designs*, World Scientific Publishing Co. Pte. Ltd., Hackensack, NJ, 2009, DOI 10.1142/9789812818430. MR2524180

[2] K. C. Carrière and G. C. Reinsel, *Optimal two-period repeated measurement designs with two or more treatments*, Biometrika **80** (1993), no. 4, 924–929, DOI 10.1093/biomet/80.4.924. MR1282801

[3] K. C. Carrière and R. Huang, *Crossover designs for two-treatment clinical trials*, J. Statist. Plann. Inference **87** (2000), no. 1, 125–134, DOI 10.1016/S0378-3758(99)00185-8. MR1772044

[4] C. S. Chêng and C.-F. Wu, *Balanced repeated measurements designs*, Ann. Statist. **8** (1980), no. 6, 1272–1283. MR594644

[5] W. G. Cochran, *Long-term agricultural experiments*, Supplement to the Journal of the Royal Statistical Society **6** (1939), no. 2, 104–148, DOI 10.2307/2983686.

[6] A. Das, *An introduction to optimality criteria and some results on optimal block design*, Design Workshop Lecture Notes (Kolkata, West Bengal, India), Theoretical Statistics and Mathematics Unit, 2002, 21 pp.

[7] J. M. Grender and W. D. Johnson, *Analysis of crossover designs with multivariate response*, Statistics in Medicine **12** (1993), no. 1, 69–89, DOI 10.1002/sim.4780120108.

[8] A. Hedayat and K. Afsarinejad, *Repeated measurements designs. I*, A survey of statistical design and linear models (Proc. Internat. Sympos., Colorado State Univ., Ft. Collins, Colo., 1973), North-Holland, Amsterdam-Oxford, 1975, pp. 229–242. MR375678

[9] A. Hedayat and K. Afsarinejad, *Repeated measurements designs. II*, Ann. Statist. **6** (1978), no. 3, 619–628. MR488527

[10] A. S. Hedayat and M. Yang, *Universal optimality of balanced uniform crossover designs*, Ann. Statist. **31** (2003), no. 3, 978–983, DOI 10.1214/aos/1056562469. MR1994737

[11] A. S. Hedayat and M. Yang, *Universal optimality for selected crossover designs*, J. Amer. Statist. Assoc. **99** (2004), no. 466, 461–466, DOI 10.1198/016214504000000331. MR2062831

[12] A. S. Hedayat, J. Stufken, and M. Yang, *Optimal and efficient crossover designs when subject effects are random*, J. Amer. Statist. Assoc. **101** (2006), no. 475, 1031–1038, DOI 10.1198/016214505000001384. MR2324142

[13] J. Jankar, A. Mandal, and J. Yang, *Optimal crossover designs for generalized linear models*, J. Stat. Theory Pract. **14** (2020), no. 2, Paper No. 23, 27, DOI 10.1007/s42519-020-00089-5. MR4087048

[14] J. Kiefer, *Construction and optimality of generalized Youden designs*, A survey of statistical design and linear models (Proc. Internat. Sympos., Colorado State Univ., Ft. Collins, Colo., 1973), North-Holland, Amsterdam-Oxford, 1975, pp. 333–353. MR395079

[15] J. Kunert, *Optimal design and refinement of the linear model with applications to repeated measurements designs*, Ann. Statist. **11** (1983), no. 1, 247–257. MR684882

[16] J. Kunert, *Optimality of balanced uniform repeated measurements designs*, Ann. Statist. **12** (1984), no. 3, 1006–1017, DOI 10.1214/aos/1176346717. MR751288

[17] J. Kunert and R. J. Martin, *Optimality of type I orthogonal arrays for cross-over models with correlated errors*, J. Statist. Plann. Inference **87** (2000), no. 1, 119–124, DOI 10.1016/S0378-3758(99)00182-2. MR1772043

[18] H. B. Kushner, *Optimality and efficiency of two-treatment repeated measurements designs*, Biometrika **84** (1997), no. 2, 455–468, DOI 10.1093/biomet/84.2.455. MR1467060

[19] H. B. Kushner, *Optimal and efficient repeated-measurements designs for uncorrelated observations*, J. Amer. Statist. Assoc. **93** (1998), no. 443, 1176–1187, DOI 10.2307/2669860. MR1649211

[20] E. Laska, M. Meisner, and H. B. Kushner, *Optimal crossover designs in the presence of carryover effects*, Biometrics **39** (1983), no. 4, 1087–1091, DOI 10.2307/2531343.

[21] E. M. Laska and M. Meisner, *A variational approach to optimal two-treatment crossover designs: application to carryover-effect models*, Journal of the American Statistical Association **80** (1985), no. 391, 704–710, DOI 10.1080/01621459.1985.10478172.

[22] B. R. Leaker, V. A. Malkov, R. Mogg, M. K. Ruddy, G. C. Nicholson, A. J. Tan, C. Tribouley, G. Chen, I. De Lepeleire, N. A. Calder, H. Chung, P. Lavender, L. N. Carayannopoulos, and T. T. Hansel, *The nasal mucosal late allergic reaction to grass pollen involves type 2 inflammation (IL-5 and IL-13), the inflammasome (IL-1β), and complement*, Mucosal Immunology **10** (2017), no. 2, 408–420, DOI 10.1038/mi.2016.74.

[23] R. J. Martin and J. A. Eccleston, *Variance-balanced change-over designs for dependent observations*, Biometrika **85** (1998), no. 4, 883–892, DOI 10.1093/biomet/85.4.883. MR1666707

[24] S. Mukhopadhyay, S. P. Singh, and A. Singh, *Locally optimal binary crossover designs*, Stat. Appl. **19** (2021), no. 1, 223–246.

[25] S. Niphadkar and S. Mukhopadhyay, *Universally optimal multivariate crossover designs*, Sankhya B **86** (2024), no. 2, 586–603, DOI 10.1007/s13571-024-00341-z. MR4814149

[26] S. Niphadkar and S. Mukhopadhyay, *Efficient designs for multivariate crossover trials*, `arXiv:2401.04498`, 2024.

[27] R. Singh and J. Kunert, *Efficient crossover designs for non-regular settings*, Metrika **84** (2021), no. 4, 497–510, DOI 10.1007/s00184-020-00780-4. MR4250786

[28] S. P. Singh and S. Mukhopadhyay, *Bayesian crossover designs for generalized linear models*, Comput. Statist. Data Anal. **104** (2016), 35–50, DOI 10.1016/j.csda.2016.06.002. MR3540985

[29] S. P. Singh, S. Mukhopadhyay, and H. Raj, *Min-max crossover designs for two treatments binary and Poisson crossover trials*, Stat. Comput. **31** (2021), no. 5, Paper No. 54, 11, DOI 10.1007/s11222-021-10029-3. MR4291707

[30] J. Stufken, *Some families of optimal and efficient repeated measurements designs*, J. Statist. Plann. Inference **27** (1991), no. 1, 75–83, DOI 10.1016/0378-3758(91)90083-Q. MR1089354

[31] C.-M. Yeh, *Conditions for universal optimality of block designs*, Biometrika **73** (1986), no. 3, 701–706, DOI 10.2307/2336535. MR897862

Department of Mathematics, Indian Institute of Technology Bombay, Powai, Mumbai, Maharashtra 400076, India
Email address: niphadkarshubham@gmail.com

Department of Mathematics, Indian Institute of Technology Bombay, Powai, Mumbai, Maharashtra 400076, India
Email address: siuli@math.iitb.ac.in

Contemporary Mathematics
Volume **833**, 2026
https://doi.org/10.1090/conm/833/16670

Inverse Z-matrices with the bi-diagonal south-west structure

S. Pratihar and K. C. Sivakumar

ABSTRACT. Two new matrix classes are introduced; inverse cyclic matrices and bi-diagonal south-west matrices. An interesting relation is established between these classes. Applications to two classes of inverse Z-matrices are provided.

1. Introduction

1.1. Inverse Z-matrices: a brief survey. The objective of this article is twofold. One is to present a brief survey on the class of inverse Z-matrices, with specific reference to its nonnegativity/nonpositivity properties. The second aim is to propose two matrix classes, motivated by a result on inverse M-matrices, and to understand their relationship. We start with a short survey of the literature on inverse Z-matrices.

The set of real square matrices of order n is denoted by $\mathbf{M}_n(\mathbb{R})$. We refer to a matrix each of whose entry is nonzero as a *full matrix*. For any matrix A, we use $A \geq 0$ to denote the fact that all the entries of A are nonnegative; $A > 0$ ($A < 0$) will signify that all the entries of A are positive (negative). The determinant of A is denoted by $\det A$. For $\alpha, \beta \subseteq \{1, 2, \ldots, n\}$, whose elements are in the ascending order, we use $A \begin{bmatrix} \alpha \\ \beta \end{bmatrix}$ to denote the submatrix of A containing the rows and columns indexed by α and β, respectively. $A[\alpha]$ is referred to as a principal submatrix of A. If $\alpha \subsetneq \{1, 2, \ldots, n\}$, then $A[\alpha]$ will be called a *proper* principal submatrix. Finally, $A(\alpha)$ denotes the principal submatrix in the rows and columns defined by the complement α' of α, in $\{1, 2, \ldots, n\}$. We use $\rho(A)$ to denote the spectral radius of A, which by definition is the maximum of the moduli of the eigenvalues of A.

We shall be concerned with the notion of irreducibility of matrices, which we recall, next.

DEFINITION 1.1. For a matrix $A = (a_{ij}) \in \mathbf{M}_n(\mathbb{R})$, the digraph (directed graph) $\mathcal{D}(A)$ of A, has $\{v_1, v_2, \ldots, v_n\}$ as the vertex set, and whose edge set consists of those ordered pairs (v_i, v_j), if the corresponding entry $a_{ij} \neq 0$.

2020 *Mathematics Subject Classification.* Primary 15A09, 15B48, 05C20.

Key words and phrases. Digraph, Z-matrix, M-matrix, inverse M-matrix, N-matrix, inverse N-matrix.

The first author was supported by the Prime Minister's Research Fellowship (PMRF), Ministry of Education, Government of India.

Note that the above definition allows the digraph to have loops. A (directed) path in $\mathcal{D}(A)$ from vertex v_i to vertex v_j, is a set of distinct vertices $\{v_i, v_{i_1}, \ldots, v_{i_r}, v_j\}$ such that $(v_i, v_{i_1}), (v_{i_1}, v_{i_2}), \ldots, (v_{i_r}, v_j)$ are (directed) edges. A matrix A is called *irreducible* if its associated digraph $\mathcal{D}(A)$ is strongly connected, i.e., there is at least one directed path between any two vertices in $\mathcal{D}(A)$.

We shall be interested in the following class of sign pattern matrices. A matrix $A = (a_{ij}) \in \mathbf{M}_n(\mathbb{R})$ is said to be a *Z-matrix* if $a_{ij} \leq 0$, for all $i \neq j$ (so that all the off-diagonal entries of A are nonpositive). Any such matrix A can be represented as

$$A = tI - B, \text{ where } t \in \mathbb{R} \text{ and } B \geq 0.$$

Denote for $r = 1, 2, \ldots, n$,

$$\rho_r(B) := max\{\rho(\hat{B}) : \hat{B} \text{ is a principal submatrix of } B \text{ of order } r\},$$

with the convention that $\rho_0(B) := -\infty$ and $\rho_{n+1}(B) := \infty$. The following classification of Z-matrices was introduced in [**4**]. Let $L_s (s = 0, 1, 2, \ldots, n)$ denote the class of real $n \times n$ matrices having the form

$$A = tI - B, \text{ where } B \geq 0 \text{ and } \rho_s(B) \leq t < \rho_{s+1}(B).$$

A nonsingular matrix A is called an *inverse Z-matrix*, if A^{-1} is a Z-matrix. In order to understand the behaviour of those inverse Z-matrices that will be considered here, first we recall the corresponding classes of Z-matrices.

First, we recall the notion of an M-matrix. Consider a Z-matrix represented as above. If, in addition, one has $t \geq \rho(B)$, then A is called an *M-matrix*. Such a matrix A is invertible, if $t > \rho(B)$. In that case, it is well known that $A^{-1} \geq 0$. There is a variety of results that identify when a Z-matrix is an M-matrix. In the book [**1**] more than fifty characterizations are presented. Two of them are recalled here.

THEOREM 1.2 ([**1**, Theorem 2.3]). *Let A be any Z-matrix. Then the following are equivalent:*

1. *A is a nonsingular M-matrix;*
2. *all the principal minors of A are positive;*
3. *A^{-1} exists and $A^{-1} \geq 0$.*

It follows that the diagonal entries of any nonsingular M-matrix are *positive.* For the case of irreducible matrices, there is something more to say.

THEOREM 1.3 ([**1**, Thorem 2.7]). *Let A be any Z-matrix. Then A is a nonsingular irreducible M-matrix if and only if $A^{-1} > 0$.*

The other classes of Z-matrices are obtained if the number t in the representation $A = tI - B$, for $B \geq 0$, lies to the left of $\rho(B)$. Let us recall one such class, viz., N-matrices. Once again, let A be given as above. If t satisfies the inequalities $\rho_{n-1}(B) < t < \rho(B)$, for $n \geq 2$ then A is referred to as an *N-matrix.* This matrix class was introduced and investigated in [**3**]. The determinant of an N-matrix is negative and each proper principal submatrix is an invertible M-matrix (so that all its diagonal entries are positive). While an invertible M-matrix is inverse nonnegative, an invertible N-matrix is inverse negative. In particular, it follows that an N-matrix is irreducible. We record this result.

THEOREM 1.4 ([**6**, Corollary 2.8]). *Let A be a Z-matrix. Then A is a N-matrix if and only if $A^{-1} < 0$.*

If equality is allowed in the strict inequality in the definition of an N-matrix i.e., in the given representation of A, if t satisfies the inequalities $\rho_{n-1}(B) \leq t < \rho(B)$, then A is referred to as an N_0-*matrix*. This matrix class was investigated in [**6**] and a number of interesting properties were proved. Clearly, every N-matrix is an N_0-matrix and that the latter matrix class is the topological closure of the former. Interestingly, analogous to the property of any matrix belonging to the class of N-matrices, we have the following result: A matrix A is an N_0-matrix precisely if A is a Z matrix with the property that all proper principal submatrices are (not necessarily invertible) M-matrices, and the determinant of A is negative [**6**, Lemma 2.1]. Here is another result: If A is a Z-matrix, then A is an N_0-matrix if and only if $A^{-1} \leq 0$ and is irreducible [**6**, Theorem 2.7]. Note that a nonsingular matrix is irreducible if and only if its inverse is irreducible.

We turn our attention to yet another class which has received considerable attention. A matrix $A \in \mathbf{M}_n(\mathbb{R})$ with $n \geq 3$, is an F_0-*matrix* if $A = tI - B$, with $B \geq 0$ and $\rho_{n-2}(B) \leq t < \rho_{n-1}(B)$. An equivalent manner in which this class is defined is given by the following. Let $A \in \mathbf{M}_n(\mathbb{R})$ be a Z-matrix. Then A is an F_0-matrix if and only if, it satisfies the following conditions:

(1) all principal submatrices of order at most $n-2$ are M-matrices;
(2) at least one principal submatrix of order $n-1$ is an N_0-matrix.

While there is no nonpositivity or nonnegativity result that is true, in general, for F_0-matrices, we have:

THEOREM 1.5 ([**2**, Theorem 2.4]). *Let A be a Z-matrix. Then A is an F_0-matrix if and only if:*

(1) *$det(A) < 0$;*
(2) *all principal minors of A^{-1} of order at least two are nonpositive;*
(3) *at least one diagonal entry of A^{-1} is positive.*

For more details, we refer the reader to [**2**] and [**6**]. Recalling the definition of L_s, it now clear that, for $s = n$, we obtain the M-matrix class, the N_0-matrix class corresponds to $s = n-1$, while for $s = n-2$, it is the F_0-matrix class that one obtains.

Next, we revisit a matrix class, historically important, due to its relevance to inverse Z-matrices. Let real numbers $a_1, a_2, \dots, a_n$ be such that $a_n > a_{n-1} > \dots > a_1$. A matrix $A := (a_{ij})$ is called a *matrix of type D*, if

$$a_{ij} = \begin{cases} a_i, & \text{if } i \leq j, \\ a_j, & \text{otherwise.} \end{cases} \tag{1.1}$$

A matrix of type D of order $n \times n$ will be denoted by D_n. For instance,

$$D_4 = \begin{pmatrix} a_1 & a_1 & a_1 & a_1 \\ a_1 & a_2 & a_2 & a_2 \\ a_1 & a_2 & a_3 & a_3 \\ a_1 & a_2 & a_3 & a_4 \end{pmatrix}.$$

We also have the following recurrence relation:

$$D_{n+1} = \begin{pmatrix} D_n & b \\ b^T & a_{n+1} \end{pmatrix},$$

where $b = (a_1, a_2, \dots, a_n)^T$. Type D matrices were introduced in [**9**], where the first item below was proved.

THEOREM 1.6. *Let $A \in \mathbf{M}_n(\mathbb{R})$ be a type D-matrix.*

(1) *If $a_1 > 0$, then A is an inverse M-matrix and A^{-1} is tridiagonal* [**9**, Theorem 2.3].
(2) *If $a_n < 0$, then A is an inverse N_0-matrix such that A^{-1} is tridiagonal* [**7**, Theorem 2.4].
(3) *If $a_n > 0$ and $a_{n-1} < 0$, then A^{-1} is a tridiagonal F_0-matrix* [**13**, Theorem 4.21].

The results above, obtained for three special classes of inverse Z-matrices, were extended to the full class L_s, which we recall. We set $L_{-1} := L_n$.

THEOREM 1.7 ([**12**, Theorem 3.12]). *Let $A \in \mathbf{M}_n(\mathbb{R})$ be a matrix of type D with $a_1 \neq 0$. Let s denote the number of nonpositive parameters in the sequence $a_n > a_{n-1} > \cdots > a_1$. Then A^{-1} is a tridiagonal Z-matrix and $A^{-1} \in L_{s-1}$.*

In the next result, we obtain specific consequences, presented explicitly, for the three classes: N-matrices, N_0-matrices and F_0-matrices. It appears that these have not been noticed before (see the para after Theorem 3.12, [**12**]). We skip their proofs. Note that the number of nonpositive parameters in the first and the second items is n, while in the third instance, this number is $n-1$. Observe that the second item below, strengthens the first item above.

THEOREM 1.8. *Let $A \in \mathbf{M}_n(\mathbb{R})$ be a type D-matrix.*

(1) *If $a_n < 0$, then A^{-1} is a tridiagonal N-matrix (and hence a tridiagonal N_0-matrix).*
(2) *If $a_n = 0$, then A^{-1} is a tridiagonal N_0-matrix, which is not an N-matrix.*
(3) *If $a_{n-1} = 0$, then A^{-1} is a tridiagonal F_0-matrix.*

1.2. Objective. In this article, we prove a result each for inverse M-matrices and inverse N-matrices. While the inverse M-matrix result is the same as what is stated in Theorem 1.9 to follow, we prove it as a consequence of the main result of this article. The result for inverse N-matrices is new. It is pertinent to note that the inverse M-matrix problem is to characterize those nonsingular nonnegative matrices that are inverses of M-matrices. While a number of matrix classes have been shown to be inverse M-matrices, the problem in its full generality remains open. We refer the reader to the recent monograph [**5**] for a comprehensive treatment.

Consider the following two results.

THEOREM 1.9 ([**11**, Theorem 4.3]). *Let $A = (a_{ij}) \in \mathbf{M}_n(\mathbb{R})$. Then the following statements are equivalent:*

(1) *A is an inverse M-matrix such that $\mathcal{D}(A^{-1})$ is the simple n-cycle $v_1 \to v_2 \to \cdots \to v_n \to v_1$ (with loops).*
(2) *$A > 0$ and the following hold:*
    (a) $a_{11}a_{22}\cdots a_{nn} > a_{12}a_{23}\cdots a_{(n-1)n}a_{n1}$,
    (b) $a_{ij} = \dfrac{a_{ik}a_{kj}}{a_{kk}}$, *for all distinct vertices v_i, v_j, v_k such that v_k lies on the path from v_i to v_j.*

Set $Z := (e^n, e^1, e^2, \ldots, e^{n-1})$, where e^k denotes the kth standard basis vector of $\mathbb{R}^n$.

THEOREM 1.10 ([**11**, Corollary 4.5]). *Let Z be the matrix defined as above. Let $\alpha_1, \alpha_2, \ldots, \alpha_n$ be nonnegative numbers and*

$$A := p(Z) = \alpha_1 I + \alpha_2 Z + \cdots + \alpha_n Z^{n-1}.$$

Then the following are equivalent:

(1) *A^{-1} is an M-matrix with $\mathcal{D}(A^{-1})$ being the simple n-cycle $v_1 \to v_2 \to \cdots \to v_n \to v_1$, with loops.*

(2) *$A > 0$ satisfies*
   (a) *$\alpha_1 > \alpha_2 > 0$,*
   (b) *$\alpha_r = (\alpha_2)^{r-1}/(\alpha_1)^{r-2}$, for $r = 3, \ldots, n$.*

The motivation for the article comes from the question as to what lies in the background of Theorem 1.9, which when applied to matrices satisfying the second condition, gives rise to the special class of inverse M-matrices, described in the first item. Specifically, inspired by the first item above, we introduce a matrix class, without requiring it to be an M-matrix. We call such a matrix a *bi-diagonal south-west matrix.* Simultaneously, another new class is identified, taking cue from the second item, giving rise to what we refer to as an *inverse cyclic matrix.*

A relationship is established between these two matrix classes in Theorem 2.11. This result, while enabling us to recover Theorem 1.9 as a particular case, further allows us to obtain a new result on inverse N-matrices, which is presented in Theorem 3.1. An analogue of Theorem 1.10 is obtained in Theorem 3.6, for inverse N-matrices.

2. Inverse cyclic matrices and bi-diagonal south-west matrices

We propose two matrix classes and study their relationship. The first notion is that of the *inverse cyclic property* of a matrix.

DEFINITION 2.1. Let $A = (a_{ij}) \in \mathbf{M}_n(\mathbb{R})$ be such that $a_{ii} \neq 0, 1 \leq i \leq n$. Then A is said to have the inverse cyclic property (or is called an inverse cyclic matrix), if the entries of A satisfy the following conditions:

$$a_{ij} = \begin{cases} \frac{a_{ik}a_{kj}}{a_{kk}}, & \text{for } i < k < j, \\ \frac{a_{in}a_{nj}}{a_{nn}}, & \text{for } j < i \neq n, \\ \frac{a_{n1}a_{1j}}{a_{11}}, & \text{for } j < i = n. \end{cases} \tag{2.1}$$

The requirement on the entries above can also be written equivalently, as

$$a_{ij} = \begin{cases} \frac{a_{i(i+1)}a_{(i+1)(i+2)}\cdots a_{(j-1)j}}{a_{(i+1)(i+1)}a_{(i+2)(i+2)}\cdots a_{(j-1)(j-1)}}, & \text{for } i+1 < j, \\ \frac{a_{i(i+1)}a_{(i+1)(i+2)}\cdots a_{(n-1)n}a_{n1}a_{12}\cdots a_{(j-1)j}}{a_{(i+1)(i+1)}\cdots a_{nn}a_{11}\cdots a_{(j-1)(j-1)}}, & \text{for } j < i \neq n, \\ \frac{a_{n1}a_{12}a_{23}\cdots a_{(j-2)(j-1)}a_{(j-1)j}}{a_{11}a_{22}\cdots a_{(j-1)(j-1)}}, & \text{for } j < i = n. \end{cases} \tag{2.2}$$

For an inverse cyclic matrix, it follows that all the elements in the upper triangular part are determined by the elements on the main diagonal and the super diagonal, while the elements in the lower triangular part are determined by those on the main diagonal, the super diagonal and the entry lying in the intersection of the first column and the last row of the matrix.

EXAMPLE 2.2. It may be verified that the matrix $A = \begin{pmatrix} 2 & -2 & -4 & 0 \\ 0 & 1 & 2 & 0 \\ 0 & 0 & -2 & 0 \\ 2 & -2 & -4 & 1 \end{pmatrix}$ has the inverse cyclic property.

For $A = (a_{ij}) \in \mathbf{M}_n(\mathbb{R})$, set

$$d := a_{11}a_{22}\cdots a_{(n-1)(n-1)}a_{nn},$$

i.e. the product of the diagonal entries of A,

$$c := a_{12}a_{23}\cdots a_{(n-1)n}a_{n1},$$

which we may refer to as the *cyclic* product. Let us denote by x^k, the kth column of the matrix $X = (x_{ij})$.

THEOREM 2.3. *Let $A \in \mathbf{M}_n(\mathbb{R})$ be an inverse cyclic matrix. Then*

$$\det A = \frac{(d-c)^{n-1}}{d^{n-2}}.$$

PROOF. First, note that an inverse cyclic matrix is one where all the diagonal entries are nonzero. Thus, $d \neq 0$. Let C be the matrix obtained from A, after employing the following column operations:

$$c^k = a^k - \frac{a_{(k-1)k}}{a_{(k-1)(k-1)}}a^{k-1}, \; 2 \leq k \leq n. \tag{2.3}$$

It is easy to see that C is a lower triangular matrix with diagonal entries

$$c_{ii} = \begin{cases} a_{11}, & \text{for } i = 1, \\ a_{ii} - \frac{a_{12}a_{23}\cdots a_{(n-1)n}a_{n1}}{a_{11}a_{22}\cdots a_{(i-1)(i-1)}a_{(i+1)(i+1)}\cdots a_{nn}}, & \text{for } i \neq 1. \end{cases}$$

Thus, we have

$$c_{ii} = \begin{cases} a_{11}, & \text{for } i = 1, \\ \frac{d-c}{d}a_{ii}, & \text{for } i \neq 1. \end{cases}$$

Then,

$$\begin{aligned} \det C = \prod_{i=1}^{n} c_{ii} &= \left(\frac{d-c}{d}\right)^{n-1} a_{11}a_{22}a_{33}\cdots a_{nn} \\ &= \frac{(d-c)^{n-1}}{d^{n-2}}. \end{aligned}$$

□

Next, we introduce the second new class of matrices called *bi-diagonal south-west* matrices.

DEFINITION 2.4. Let $A \in \mathbf{M}_n(\mathbb{R})$. Suppose that all the entries of A that lie on the main diagonal, the super diagonal and the entry in the south-west corner, are nonzero, while all the other entries are zero. Then A is called a bi-diagonal south-west matrix, or bdsw, for short.

As an illustration, a pattern of the bdsw matrix (for $n = 4$), is given by

$$\begin{pmatrix} * & * & 0 & 0 \\ 0 & * & * & 0 \\ 0 & 0 & * & * \\ * & 0 & 0 & * \end{pmatrix},$$

where the symbol $*$ stands for a nonzero entry.

REMARK 2.5. Let A be a bdsw matrix. Then every proper principal submatrix of A is either an upper triangular matrix or a block lower triangular matrix. Moreover, every proper principal submatrix of a bdsw matrix is nonsingular.

The motivation for considering the class of bdsw matrices arises from the structure of the digraph associated with it. The digraph corresponding to a bdsw matrix (of order $n \times n$) is the simple directed n-cycle $v_1 \to v_2 \to \cdots \to v_n \to v_1$, with loops. The digraph for a bdsw matrix of order 4 is given below.

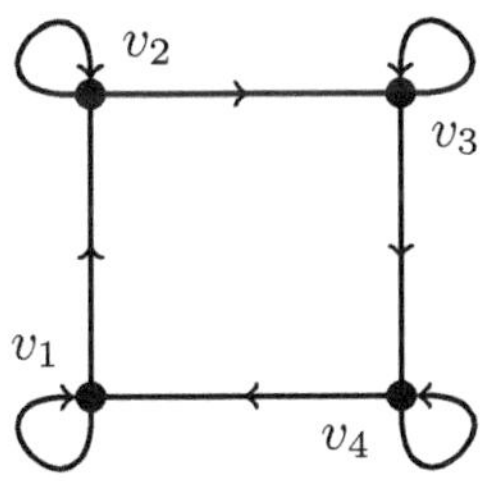

FIGURE 1. Digraph of a 4×4 bdsw matrix.

Recall that *unipathic* digraphs are those digraphs in which there is at most one (directed) path between any two vertices. However, matrices A possessing the bdsw structure are examples of irreducible unipathic matrices. This means that, in $\mathcal{D}(A)$, there is exactly *one* path between any two vertices.

We will make use of an auxiliary result, which presents a graph theoretic formula for the inverse of a nonsingular matrix. Let us recall some terminology, in this context. Let $p(v_i \to v_j)$ denote any (directed) path from v_i to v_j and $l(p)$, the length of the path p. The set of vertices of $\mathcal{D}(A)$ not belonging to the path p will be denoted by $V(p)$, and by $A[p(v_i \to v_j)]$ we signify the product of those elements of A, that lie along the given path p of $\mathcal{D}(A)$, from vertex v_i to vertex v_j. We set, $A[p(v_i \to v_j)] = 0$ if there is no path from vertex v_i to vertex v_j.

THEOREM 2.6 ([**10**, Corollary 9.1]). *Let $A = (a_{ij}) \in \mathbf{M}_n(\mathbb{R})$ be nonsingular. Set $A^{-1} = (\tilde{a}_{ij})$. Then, we have*

$$\tilde{a}_{ii} = \det A(i) / \det A$$

and

$$\tilde{a}_{ij} = \frac{1}{\det A} \sum_{p(v_i \to v_j)} (-1)^{l(p)} A[p(v_i \to v_j)] \det A[V(p)], \quad \text{for } i \neq j. \tag{2.4}$$

We will make use of the following particular case.

COROLLARY 2.7. *Let A be a nonsingular unipathic matrix. Then, the off-diagonal entries of A^{-1} are given by*

$$\tilde{a}_{ij} = (-1)^{l(p)} A[p(v_i \to v_j)] \det A[V(p)]/\det A, \text{ for } i \neq j, \tag{2.5}$$

where p is the unique path from vertex v_i to vertex v_j, if it exists.

LEMMA 2.8. *The inverse of a nonsingular bdsw matrix is a full matrix.*

PROOF. Let A be a nonsingular bdsw matrix. Then A is a unipathic matrix, so that the entries of $A^{-1} = (\tilde{a}_{ij})$ satisfy (2.5). Moreover, A is irreducible and so $A[p(v_i \to v_j)] \neq 0$, for all i, j. Also, from Remark 2.5, $\det A[V(p)] \neq 0$, for any path p. In particular, all the diagonal entries of A^{-1} are nonzero and hence $\tilde{a}_{ij} \neq 0$, for all i, j. Thus, A^{-1} is a full matrix. □

REMARK 2.9. The converse of Lemma 2.8 is false. For example, the matrix

$$A = \begin{pmatrix} 1 & 2 & 2 \\ 1 & 1 & 1 \\ 2 & 2 & 1 \end{pmatrix},$$

is a nonsingular full matrix, But its inverse

$$A^{-1} = \begin{pmatrix} -1 & 2 & 0 \\ 1 & -3 & 1 \\ 0 & 2 & -1 \end{pmatrix},$$

is not a bdsw matrix.

Now, we prove the main result of this article. This brings about the relationship between *nonsingular* matrices possessing the inverse cyclic property and bdsw matrices. This result also justifies the nomenclature "inverse cyclic matrix". We shall make use of the following auxiliary result, which presents a formula for the determinant of a submatrix of a nonsingular matrix and a related submatrix of its inverse.

LEMMA 2.10 ([**8**, pp. 5]). *Let $A \in \mathbf{M}_n(\mathbb{R})$ be nonsingular and $B := A^{-1}$. Let $\alpha = \{i_1, i_2, \ldots, i_p\}$ and $\beta = \{j_1, j_2, \ldots, j_p\}$ with $1 \leq i_1 < i_2 < \cdots < i_p \leq n$ and $1 \leq j_1 < j_2 < \cdots < j_p \leq n$. Set $\gamma_p := (-1)^{\sum_{m=1}^{p}(i_m + j_m)}$. We then have:*

$$\det B \begin{bmatrix} \alpha \\ \beta \end{bmatrix} = \frac{\gamma_p}{\det A} \det A \begin{bmatrix} \beta' \\ \alpha' \end{bmatrix}. \tag{2.6}$$

THEOREM 2.11. *Suppose that $A \in \mathbf{M}_n(\mathbb{R})$ is nonsingular. Then A is a full matrix and has the inverse cyclic property, if and only if A^{-1} is a bdsw matrix.*

PROOF. Let us assume that A is a full matrix and has the inverse cyclic property. Since A is nonsingular, by Theorem 2.3, $d - c \neq 0$. Define $B = (b_{ij})$, where

$$b_{ij} = \frac{1}{d-c} \begin{cases} -a_{ij}\Big(\prod\limits_{\substack{k=1 \\ k \neq i,j}}^{n} a_{kk}\Big), & \text{if } j = i+1, \\ -a_{ij}\Big(\prod\limits_{k=2}^{n-1} a_{kk}\Big), & \text{if } i = n, j = 1, \\ \prod\limits_{\substack{k=1 \\ k \neq i}}^{n} a_{kk}, & \text{if } i = j, \\ 0, & \text{otherwise.} \end{cases} \tag{2.7}$$

We claim that $B = A^{-1}$. Since A is a full matrix, it is clear that B is a bdsw matrix. Now, for all $i \in \{1, 2, \ldots, n\}$,

$$(AB)_{ij} = \begin{cases} a_{i1}b_{11} + a_{in}b_{n1}, & \text{if } j = 1, \\ a_{i(j-1)}b_{(j-1)j} + a_{ij}b_{jj}, & \text{if } j \neq 1. \end{cases} \tag{2.8}$$

We have, for $2 \leq j \leq n$,

$$\frac{b_{(j-1)j}}{b_{jj}} = -\frac{a_{(j-1)j}}{a_{(j-1)(j-1)}}$$

and

$$\frac{b_{n1}}{b_{11}} = -\frac{a_{n1}}{a_{nn}}.$$

Incorporating these in (2.8), we obtain

$$(AB)_{ij} = \begin{cases} \left(a_{i1} - \frac{a_{in}a_{n1}}{a_{nn}}\right)b_{11}, & \text{for } j = 1, \\ \left(a_{ij} - \frac{a_{i(j-1)}a_{(j-1)j}}{a_{(j-1)(j-1)}}\right)b_{jj}, & \text{for } j \neq 1. \end{cases} \tag{2.9}$$

CASE 1 ($i = j$). First, let $i = j = 1$. Then, using (2.2), we have

$$\begin{aligned} (AB)_{11} &= \left(a_{11} - \frac{a_{1n}a_{n1}}{a_{nn}}\right)b_{11} \\ &= \left(a_{11} - \frac{a_{12}a_{23}\cdots a_{(n-1)n}a_{n1}}{a_{22}a_{33}\cdots a_{(n-1)(n-1)}a_{nn}}\right)b_{11} \\ &= \frac{a_{11}a_{22}\cdots a_{nn} - a_{12}a_{23}\cdots a_{(n-1)n}a_{n1}}{a_{22}a_{33}\cdots a_{nn}}b_{11} = 1. \end{aligned}$$

Next, let $i = j \neq 1$. Then, using (2.2) again, we have

$$\begin{aligned} (AB)_{ii} &= \left(a_{ii} - \frac{a_{i(i-1)}a_{(i-1)i}}{a_{(i-1)(i-1)}}\right)b_{ii} \\ &= \left(a_{ii} - \frac{a_{i(i+1)}a_{(i+1)(i+2)}\cdots a_{(n-1)n}a_{n1}a_{12}\cdots a_{(i-2)(i-1)}a_{(i-1)i}}{a_{(i+1)(i+1)}\cdots a_{nn}a_{11}\cdots a_{(i-2)(i-2)}a_{(i-1)(i-1)}}\right)b_{ii} \\ &= \frac{a_{11}a_{22}\cdots a_{nn} - a_{12}a_{23}\cdots a_{(n-1)n}a_{n1}}{a_{11}a_{22}\cdots a_{(i-1)(i-1)}a_{(i+1)(i+1)}\cdots a_{nn}}b_{ii} = 1. \end{aligned}$$

CASE 2 ($i \neq j$). First, let $j = 1$. We have $a_{i1} = \dfrac{a_{in}a_{n1}}{a_{nn}}$, so that $(AB)_{i1} = 0$. Next, we consider $j \neq 1$. We have two cases. First, let $i < j$. Then

$$\frac{a_{i(j-1)}a_{(j-1)j}}{a_{(j-1)(j-1)}} = a_{ij}.$$

Also for $j < i < n$,

$$\begin{aligned} \frac{a_{i(j-1)}a_{(j-1)j}}{a_{(j-1)(j-1)}} &= \frac{a_{i(i+1)}a_{(i+1)(i+2)}\cdots a_{n1}a_{12}\cdots a_{(j-2)(j-1)}a_{(j-1)j}}{a_{(i+1)(i+1)}a_{(i+2)(i+2)}\cdots a_{nn}a_{11}\cdots a_{(j-2)(j-2)}a_{(j-1)(j-1)}} \\ &= a_{ij}, \end{aligned}$$

and for $j < i = n$,

$$\begin{aligned} \frac{a_{n(j-1)}a_{(j-1)j}}{a_{(j-1)(j-1)}} &= \frac{a_{n1}a_{12}\cdots a_{(j-2)(j-1)}a_{(j-1)j}}{a_{11}a_{22}\cdots a_{(j-2)(j-2)}a_{(j-1)(j-1)}} \\ &= a_{nj}. \end{aligned}$$

Then from (2.9), it follows that $(AB)_{ij} = 0$, in this case, too. This completes the proof that $AB = I$.

Conversely, suppose that $B = A^{-1}$ is a bdsw matrix. Then from Lemma 2.8, A is a full matrix and so, in particular all the diagonal entries of A are nonzero. Now, using Lemma 2.10, we will show that A has the inverse cyclic property. This is equivalent to proving that the following are true:

$$\det A \begin{pmatrix} i & k \\ k & j \end{pmatrix} = 0, \text{ for } i < k < j, \tag{2.10}$$

$$\det A \begin{pmatrix} i & n \\ j & n \end{pmatrix} = 0, \text{ for } j < i \neq n \tag{2.11}$$

and

$$\det A \begin{pmatrix} 1 & n \\ 1 & j \end{pmatrix} = 0, \text{ for } j < i = n. \tag{2.12}$$

In view of Lemma 2.10, it suffices to show that the following hold, for the matrix B:

$$\det B \begin{pmatrix} 1,2,\dots,k-1,k+1,\dots,j-1,j+1,\dots,n \\ 1,2,\dots,i-1,i+1,\dots,k-1,k+1\dots,n \end{pmatrix} = 0, \text{ for } i < k < j, \tag{2.13}$$

$$\det B \begin{pmatrix} 1,2,\dots,j-1,j+1,\dots,n-1 \\ 1,2,\dots,i-1,i+1,\dots,n-1 \end{pmatrix} = 0, \text{ for } j < i \neq n, \tag{2.14}$$

$$\det B \begin{pmatrix} 2,3,\dots,j-1,j+1,\dots,n \\ 2,3,\dots,\dots,n-1 \end{pmatrix} = 0, \text{ for } j < i = n. \tag{2.15}$$

Consider the case $i < k < j$. Let E_1 be the submatrix of B of order $n-2$, corresponding to the left hand side of (2.13). Let C_1 be the submatrix of E_1 obtained by taking the *first* $k-1$ rows and *all* the $n-2$ columns. Consider the submatrix F_1 of C_1, obtained by taking all the $k-1$ rows and the *first* $k-2$ columns of C_1. The rank of F_1 does not exceed $k-2$. Since B is a bdsw matrix, and since the entries along all the other columns of C_1, not being in the submatrix F_1, are zero, it follows that the $k-1$ rows of C_1 are linearly dependent. Thus, $\det E_1 = 0$, showing that (2.13) holds.

Next, let $j < i \neq n$. Suppose that E_2 is the submatrix of B of order $n-2$, corresponding to the left hand side of (2.14). Let C_2 be the submatrix of E_2 obtained by taking the *last* $n-i$ rows and *all* the $n-2$ columns. Consider the submatrix F_2 of C_2, obtained by taking *all* the $n-i$ rows and the *last* $n-i-1$ columns of C_2. The rank of F_2 does not exceed $n-i-1$. Moreover, since the entries along all the other columns of C_2, not being in the submatrix F_2, are zero (since B is a bdsw matrix), it follows that the $n-i$ rows of C_2 are linearly dependent. Thus, $\det E_2 = 0$, showing that (2.14) holds.

Finally, let $j < i = n$. Denote by E_3, the submatrix of B of order $n-2$, corresponding to the left hand side of (2.15). Again, since B is a bdsw matrix, all the entries in its *last* row except the first and the last entries, are zero. Thus, all the entries in the *last row* of E_3 are zero, so that $\det E_3 = 0$. Thus (2.15) holds.

We have shown that A has the inverse cyclic property, completing the proof. □

EXAMPLE 2.12. Let $A = \begin{pmatrix} 1 & -1 & -1 \\ -2 & 1 & 1 \\ 2 & -2 & -1 \end{pmatrix}$. Then A satisfies the inverse cyclic property. Also,

$$A^{-1} = \begin{pmatrix} -1 & -1 & 0 \\ 0 & -1 & -1 \\ -2 & 0 & 1 \end{pmatrix},$$

is a bdsw matrix. Consider the matrix C, obtained by changing A in precisely one entry (a_{32}), given by

$$C = \begin{pmatrix} 1 & -1 & -1 \\ -2 & 1 & 1 \\ 2 & 2 & -1 \end{pmatrix}.$$

Since $c_{32} \neq \dfrac{c_{31}c_{12}}{c_{11}}$, the matrix C does not possess the inverse cyclic property. Here,

$$C^{-1} = \frac{1}{3}\begin{pmatrix} -3 & -3 & 0 \\ 0 & 1 & 1 \\ -6 & -4 & -1 \end{pmatrix}$$

is not a bdsw matrix.

REMARK 2.13. It is easy to verify that any principal submatrix of an inverse cyclic matrix inherits the property. Also, if A is a nonsingular matrix having the inverse cyclic property and $B = A^{-1}$, then for all $1 \leq i \leq n-1$,

$$a_{ii}b_{ii} = a_{(i+1)(i+1)}b_{(i+1)(i+1)} = \frac{d}{d-c}.$$

3. Inverse Z-matrices with the bdsw structure

In this section, we present applications of Theorem 2.11 to the two classes of nonsingular Z-matrices under study here, viz., inverse M-matrices and inverse N-matrices.

THEOREM 3.1. *For $A \in \mathbf{M}_n(\mathbb{R})$, we have the following:*

(1) *A is an inverse M-matrix such that A^{-1} possesses the bdsw structure if and only if $A > 0$, $d - c > 0$ and A has the inverse cyclic property.*
(2) *Let A be an even order matrix. Then A is an inverse N-matrix, A^{-1} has the bdsw pattern if and only if $A < 0$, has the inverse cyclic property with $d - c < 0$.*
(3) *Let A be an odd order matrix. Then A is an inverse N-matrix such that A^{-1} is endowed with the bdsw structure if and only if $A < 0$ has the inverse cyclic property and $d - c > 0$.*

PROOF. (1) Let A be an inverse M-matrix such that A^{-1} has the bdsw structure. Since A^{-1} is an M-matrix, then all the diagonal entries of A are nonzero and $A = (A^{-1})^{-1} \geq 0$. Moreover, the bdsw structure of A^{-1}, implies that A is a full matrix i.e., $A > 0$ and A has the inverse cyclic property (by Theorem 2.11). Also, the entries of $A^{-1} =: B = (b_{ij})$ are given by formula (2.7).

Since B is an M-matrix, i.e., $b_{ij} \leq 0$, for $i \neq j$, and the entries of A are positive, it follows from (2.7) that $d - c > 0$.

Conversely, assume that $A > 0$ has the inverse cyclic property and $d - c > 0$. Then A is nonsingular and by Theorem 2.11, A^{-1} is a bdsw matrix. To show that

A^{-1} also an M-matrix, it is enough to show that A^{-1} is a Z-matrix, since $A > 0$. However, this is easy to see, by the formula for A^{-1} given in (2.7).

Items (2) and (3) follow in an entirely similar manner, with the additional observation that the sign of the products appearing in the inverse formula depend on the odd-even parity of the number of factors. □

The following examples illustrate the above Theorem.

EXAMPLE 3.2. Consider the matrix

$$A = \begin{pmatrix} 4 & 4 & 8 & 4 & 4 \\ 1 & 2 & 4 & 2 & 2 \\ 1 & 1 & 4 & 2 & 2 \\ 2 & 2 & 4 & 4 & 4 \\ 2 & 2 & 4 & 2 & 4 \end{pmatrix}.$$

It may be verified that A has the inverse cyclic property and $d - c > 0$. Also

$$A^{-1} = \frac{1}{4} \begin{pmatrix} 2 & -4 & 0 & 0 & 0 \\ 0 & 4 & -4 & 0 & 0 \\ 0 & 0 & 2 & -1 & 0 \\ 0 & 0 & 0 & 2 & -2 \\ -1 & 0 & 0 & 0 & 2 \end{pmatrix},$$

a bdsw M-matrix.

EXAMPLE 3.3. Consider the (even order) matrix

$$A = \begin{pmatrix} -2 & -2 & -4 & -8 \\ -4 & -1 & -2 & -4 \\ -2 & -2 & -1 & -2 \\ -2 & -2 & -4 & -2 \end{pmatrix}.$$

Then, A has the inverse cyclic property and $d - c < 0$. Its inverse is given by

$$A^{-1} = \frac{1}{6} \begin{pmatrix} 1 & -2 & 0 & 0 \\ 0 & 2 & -4 & 0 \\ 0 & 0 & 2 & -2 \\ -1 & 0 & 0 & 1 \end{pmatrix},$$

which is a bdsw N-matrix.

EXAMPLE 3.4. The odd order matrix

$$A = \begin{pmatrix} -2 & -2 & -4 & -8 & -16 \\ -8 & -1 & -2 & -4 & -8 \\ -4 & -4 & -1 & -2 & -4 \\ -2 & -2 & -4 & -1 & -2 \\ -2 & -2 & -4 & -8 & -2 \end{pmatrix},$$

is an inverse cyclic matrix, with $d - c > 0$. Here,

$$A^{-1} = \frac{1}{14} \begin{pmatrix} 1 & -2 & 0 & 0 & 0 \\ 0 & 2 & -4 & 0 & 0 \\ 0 & 0 & 2 & -4 & 0 \\ 0 & 0 & 0 & 2 & -2 \\ -1 & 0 & 0 & 0 & 1 \end{pmatrix},$$

is a bdsw N-matrix.

REMARK 3.5. The following examples illustrate that the odd-even parity in the order of the matrix is indispensable in Theorem 3.1. The matrix

$$A = \begin{pmatrix} -2 & -2 & -2 & -2 \\ -1 & -2 & -2 & -2 \\ -1 & -1 & -2 & -2 \\ -1 & -1 & -1 & -2 \end{pmatrix}$$

of even order has the inverse cyclic property with $d - c > 0$. While

$$A^{-1} = \frac{1}{2} \begin{pmatrix} -2 & 2 & 0 & 0 \\ 0 & -2 & 2 & 0 \\ 0 & 0 & -2 & 2 \\ 1 & 0 & 0 & -2 \end{pmatrix}$$

is a bdsw matrix, it is not an N-matrix. On the other hand, if

$$A = \begin{pmatrix} -2 & -2 & -2 & -2 & -2 \\ -1 & -2 & -2 & -2 & -2 \\ -1 & -1 & -2 & -2 & -2 \\ -1 & -1 & -1 & -2 & -2 \\ -1 & -1 & -1 & -1 & -2 \end{pmatrix},$$

then A is of odd order, $d - c < 0$ and has the inverse cyclic property. Here

$$A^{-1} = \frac{1}{2} \begin{pmatrix} -2 & 2 & 0 & 0 & 0 \\ 0 & -2 & 2 & 0 & 0 \\ 0 & 0 & -2 & 2 & 0 \\ 0 & 0 & 0 & -2 & 2 \\ 1 & 0 & 0 & 0 & -2 \end{pmatrix},$$

is not even a Z-matrix.

We conclude this article with a result for inverse N-matrices, analogous to Theorem 1.10.

THEOREM 3.6. *Let Z be defined as in Theorem 1.10. Let $\alpha_1, \alpha_2, \ldots, \alpha_n$ be nonpositive numbers and*

$$A := p(Z) = \alpha_1 I + \alpha_2 Z + \cdots + \alpha_n Z^{n-1}.$$

Then the following are equivalent:

(1) *A^{-1} is an N-matrix with $\mathcal{D}(A^{-1})$ being the simple n-cycle $v_1 \to v_2 \to \cdots \to v_n \to v_1$, with loops.*

(2) *$A < 0$ satisfies*
   - (a) *$\alpha_2 < \alpha_1 < 0$,*
   - (b) *$\alpha_r = (\alpha_2)^{r-1}/(\alpha_1)^{r-2}$, for $r = 3, \ldots, n$.*

PROOF. It may be verified that

$$A = \begin{pmatrix} \alpha_1 & \alpha_2 & \cdots & \alpha_{n-1} & \alpha_n \\ \alpha_n & \alpha_1 & \alpha_2 & \cdots & \alpha_{n-1} \\ \vdots & \alpha_n & \alpha_1 & \ddots & \vdots \\ \alpha_3 & \vdots & \ddots & \alpha_1 & \alpha_2 \\ \alpha_2 & \alpha_3 & \cdots & \alpha_n & \alpha_1 \end{pmatrix}.$$

Suppose that (1) holds. Then A^{-1} is an N-matrix with the bdsw structure. By Theorem 3.1, we then have $A < 0$ and A possesses the inverse cyclic property. Further, $d - c < 0$, if n even and $d - c > 0$, for n odd. Moreover, the inverse cyclic property of A yields (b). Finally, the sign of $d - c$ for the odd-even parity of the order of the matrix, yields the inequality $\alpha_2 < \alpha_1 < 0$.

Conversely, suppose that $A < 0$ satisfies the conditions (a) and (b). Then, A is a full matrix with the inverse cyclic property. The inequalities $\alpha_2 < \alpha_1 < 0$ imply that $d - c < 0$, if n even and $d - c > 0$, for n odd. By Theorem 3.1 again, A^{-1} is an N-matrix, with the bdsw structure, so that $\mathcal{D}(A^{-1})$ is a simple cycle $v_1 \to v_2 \to \cdots \to v_n \to v_1$, with loops, completing the proof. □

4. Concluding remarks

We propose two new classes of nonsingular matrices and establish a relationship between them. We obtain consequences for two subclasses of inverse Z-matrices. It is natural to ask what versions of these results apply to other inverse Z-matrices, in general and the classes of N_0-matrices and F_0-matrices, in particular. This is a problem for a future study.

Acknowledgments

The authors thank the referee for comments that have improved the readability.

References

[1] A. Berman and R. J. Plemmons, *Nonnegative matrices in the mathematical sciences*, Classics in Applied Mathematics, vol. 9, Society for Industrial and Applied Mathematics (SIAM), Philadelphia, PA, 1994. Revised reprint of the 1979 original, DOI 10.1137/1.9781611971262. MR1298430

[2] Y. Chen, *Notes on F_0-matrices*, Linear Algebra Appl. **142** (1990), 167–172, DOI 10.1016/0024-3795(90)90265-E. MR1077983

[3] K. Fan, *Some matrix inequalities*, Abh. Math. Sem. Univ. Hamburg **29** (1966), 185–196, DOI 10.1007/BF03016047. MR212040

[4] M. Fiedler and T. L. Markham, *A classification of matrices of class Z*, Linear Algebra Appl. **173** (1992), 115–124, DOI 10.1016/0024-3795(92)90425-A. MR1170507

[5] C. R. Johnson, R. L. Smith, and M. J. Tsatsomeros, *Matrix positivity*, Cambridge Tracts in Mathematics, vol. 221, Cambridge University Press, Cambridge, 2020, DOI 10.1017/9781108778619. MR4411340

[6] G. A. Johnson, *A generalization of N-matrices*, Linear Algebra Appl. **48** (1982), 201–217, DOI 10.1016/0024-3795(82)90108-2. MR683219

[7] G. A. Johnson, *Inverse N_0-matrices*, Linear Algebra Appl. **64** (1985), 215–222, DOI 10.1016/0024-3795(85)90278-2. MR776528

[8] S. Karlin, *Total positivity. Vol. I*, Stanford University Press, Stanford, CA, 1968. MR230102

[9] T. L. Markham, *Nonnegative matrices whose inverses are M-matrices*, Proc. Amer. Math. Soc. **36** (1972), 326–330, DOI 10.2307/2039153. MR309970

[10] J. S. Maybee, D. D. Olesky, P. van den Driessche, and G. Wiener, *Matrices, digraphs, and determinants*, SIAM J. Matrix Anal. Appl. **10** (1989), no. 4, 500–519, DOI 10.1137/0610036. MR1016799

[11] J. J. McDonald, M. Neumann, H. Schneider, and M. J. Tsatsomeros, *Inverses of unipathic M-matrices*, SIAM J. Matrix Anal. Appl. **17** (1996), no. 4, 1025–1036, DOI 10.1137/S0895479894277247. MR1410716

[12] R. Nabben, *Z-matrices and inverse Z-matrices*, Linear Algebra Appl. **256** (1997), 31–48, DOI 10.1016/S0024-3795(97)81111-1. MR1438224

[13] S. Mondal and K. C. Sivakumar, *A survey of Z-matrices and new results on the subclass of F_0-matrices*, Inverse problems, regularization methods and related topics—a volume in honour of Thamban Nair, Ind. Appl. Math., Springer, Singapore, [2025] ©2025, pp. 303–327, DOI 10.1007/978-981-97-7989-5_14. MR4890583

Department of Mathematics, Indian Institute of Technology Madras, Chennai, 600036, India

Email address: ma22d019@smail.iitm.ac.in (Samapti Pratihar)

Department of Mathematics, Indian Institute of Technology Madras, Chennai, 600036, India

Email address: kcskumar@iitm.ac.in (K.C. Sivakumar)

Contemporary Mathematics
Volume **833**, 2026
https://doi.org/10.1090/conm/833/16671

Totally positive matrices

T. E. S. Raghavan

ABSTRACT. An m×n matrix $A = (a_{ij})$ is called a nonnegative matrix (≥ 0) when $a_{ij} \geq 0$ for all $i = 1, \ldots, m$; $j = 1, \ldots, n$. The matrix A is called totally nonnegative (TN) if all minors of all order are nonnegative. The matrix A is called totally positive (TP) if all minors of order q, $q = 1, 2, \ldots, \min(m, n)$ are strictly positive. When $m = n$, the matrix A is called oscillatory if A is TN and for some power k, A^k is TP.

While the Russian school headed by M. G. Krein and F. R. Gantmacher was motivated by problems of small oscillations and vibrations of mechanical systems that led to the study of the spectral properties of oscillatory matrices, the American school headed by I. J. Schoenberg and A. M. Whitney concentrated on the sign variation properties of arbitrary vectors $y \in \mathbb{R}^n$ and their image Ay under TN matrices. Influenced by the works of Krein and Schoenberg, Samuel Karlin noticed its applicability to the study of the properties of the transition function of strong Markov processes, and their intimate connections to the count of disjoint paths among paths in a directed graph without cycles. Yet another application appears in statistical decision theory. Virtually many standard one parameter family of probability density functions possess variation diminishing properties. Extensions of the Perron Frobenius theorem to Positive operators by M. G. Krein and M. A. Rutman, are also extendable to the study of totally positive operators.

1. Introduction

1.1. Determinants. Given a matrix $A = (a_{ij})_{n\times n}$ with real entries, we can associate a function $\mathcal{D} : A \to \mathbb{R}$ which satisfies:

(i) $\mathcal{D}(\mathbf{e_1}, \mathbf{e_2}, \ldots \mathbf{e_n}) = 1$ where $\mathbf{e_j}$ is the jth column unit vector (all but the jth co-ordinate are 0 and the jth co-ordinate is 1).

(ii) If $\mathbf{c_1}.\mathbf{c_2}, \cdots, \mathbf{c_n}$ are the columns of A, then $\mathcal{D}$ is affine linear in $\mathbf{c_j}$, when the rest of the coordinates are fixed for $j = 1, 2, \cdots, n$.

(iii) If σ is a *transposition* of $\{1, 2, \cdots, n\}$ resulting in a swap of columns $\mathbf{c_i}$ and $\mathbf{c_j}$ of A, then $\mathcal{D}(\mathbf{c_1}, \cdots, \mathbf{c_j} \cdots, \mathbf{c_i}, \cdots, \mathbf{c_n} = -\mathcal{D}(\mathbf{c_1}, \cdots, \mathbf{c_i} \cdots, \mathbf{c_j}, \cdots, \mathbf{c_n})$.

If the transposition σ is denoted by

$$\sigma = \begin{Bmatrix} 1, & 2, & .. & i, & .., j, & .., & n \\ 1, & 2, & .. & j, & .., i, & .., & n \end{Bmatrix},$$

2020 *Mathematics Subject Classification.* Primary 15B48, 05C50.

Key words and phrases. Nonnegative matrix, totally nonnegative matrix, totally positive matrix, oscillatory matrix.

then let $\varepsilon(\sigma) = -1$. If σ is the permutation given by $\sigma = \begin{pmatrix} 1, 2, \cdots, n \\ \alpha_1, \alpha_2, \cdots, \alpha_n \end{pmatrix}$ then since any permutation is the product of transpositions

$$\begin{aligned} \mathcal{D}(\mathbf{c_1}, \mathbf{c_2}, \cdots, \mathbf{c_n}) &= \mathcal{D}\Big(\sum_{\alpha_1} c_{1\alpha_1} e_{\alpha_1}, \sum_{\alpha_2} c_{2\alpha_2} e_{\alpha_2}, \cdots, \sum_{\alpha_n} c_{n\alpha_n} e_{\alpha_n}\Big) \\ &= \sum \cdots \sum\nolimits_{\alpha_1 \cdots \alpha_n} c_{1\alpha_1} c_{2\alpha_2} \cdots c_{n\alpha_n} \mathcal{D}(e_{\alpha_1}, e_{\alpha_2}, \cdots, e_{\alpha_n}) \\ &= \sum_{\sigma} \varepsilon(\sigma) c_{1\sigma(1)} c_{2\sigma(2)} \cdots c_{n\sigma(n)}, \end{aligned}$$

where $\sigma : \begin{pmatrix} 1, 2, \cdots, n \\ \alpha_1 . \alpha_2, \cdots, \alpha_n \end{pmatrix}$.

We use brackets $\begin{bmatrix} \cdot & \cdot & \cdot \\ \cdot & \cdot & \cdot \end{bmatrix}$ to denote submatrices and use $\begin{pmatrix} \cdot & \cdot & \cdot \\ \cdot & \cdot & \cdot \\ \cdot & \cdot & \cdot \end{pmatrix}$ to denote the determinants of square submatrices. For example if $i_1 < i_2 \cdots < i_k$ are row indices and $j_1 < j_2 \cdots < j_k$ are column indices, we denote by

$$A \begin{bmatrix} i_1 & \cdots & i_k \\ j_1 & \cdots & j_k \end{bmatrix} \text{ the sub matrix } \begin{bmatrix} a_{i_1 j_1} & a_{i_1 j_2} & \cdots & a_{i_1 j_k} \\ a_{i_2 j_1} & a_{i_2 j_2} & \cdots & a_{i_2 j_k} \\ \cdots & \cdots & \cdots & \cdots \\ a_{i_k j_1} & a_{i_k j_2} & \cdots & a_{i_k j_k} \end{bmatrix}$$

and by $A \begin{pmatrix} i_1 & \cdots & i_k \\ j_1 & \cdots & j_k \end{pmatrix}$, the determinant of the above submatrix. We call those determinants as the minors of such submatrices.

The rank of a matrix $A = (a_{ij})_{m \times n}$ is r if for some submatrix of order $r \times r$, the determinant is not zero, but all minors of order $> r$ are zero.

We call the submatrix $A \begin{bmatrix} i_1 & \cdots & i_k \\ j_1 & \cdots & j_k \end{bmatrix}$ a *principal submatrix* when $i_1 = j_1$, $i_2 = j_2$, $\cdots, i_k = j_k$. We call the same submatrix a *leading principal submatrix* when $i_1 = j_1 = 1$, $i_2 = j_2 = 2$, $\cdots, i_k = j_k = k$.

1.2. Compound matrices. Given an $n \times n$ matrix $A = (a_{ij})$ where $i, j = 1, 2, \cdots, n$, we can choose pairs of row indices (i_1, i_2) and column indices (j_1, j_2) and totally order such pairs lexicographically. For example if $n = 4$, the pairs $(1,2), (1,3), (1,4), (2,3), (2,4), (3,4)$ will be totally ordered by $(1,2) \prec (1,3) \prec (1,4) \prec (2,3) \prec (2,4) \prec (3,4)$.

If we have, say $\alpha = (1,4)$ and $\beta = (2,3)$, then the compound matrix $\mathfrak{A}$ has as its αth row β^{th} column entry $\mathfrak{A}_{\alpha\beta}$ given by

$$\mathfrak{A}_{\alpha\beta} = \det \begin{pmatrix} \{1,4\} \\ \{2,3\} \end{pmatrix} = \det \begin{bmatrix} a_{12} & a_{13} \\ a_{42} & a_{43} \end{bmatrix} = a_{12} a_{43} - a_{13} a_{42}.$$

For example if $A = \begin{bmatrix} 2 & -3 & 5 & 1 \\ -6 & 1 & 3 & 2 \\ 5 & 2 & 4 & 1 \\ 1 & 3 & 0 & 2 \end{bmatrix}$ then for $\alpha = (1,4), \beta = (2,3)$ we have

$$\mathfrak{A}_{\alpha\beta} = \det \begin{bmatrix} a_{12} & a_{13} \\ a_{42} & a_{43} \end{bmatrix} = a_{12} a_{43} - a_{13} a_{42} = (-3)(0) - (5)(3) = -15.$$

1.3. Cauchy - Binet Formula. If A is an $m \times n$ matrix and B is an $n \times m$ matrix then the $m \times m$ matrix $C = AB$ has its determinant $\det C$. We can compute $\det C$ as a function of the minors of order m of A and B.

$$c_{ij} = \sum_{\alpha=1}^{m} a_{i\alpha} b_{\alpha j} \quad i, j = 1, 2, \cdots m$$

$$\det C = \det \begin{bmatrix} \sum\limits_{\alpha_1} a_{1\alpha_1} b_{\alpha_1 1} & \cdots & \sum\limits_{\alpha_m} a_{1\alpha_m} b_{\alpha_m m} \\ \sum\limits_{\alpha_1} a_{2\alpha_1} b_{\alpha_1 1} & \cdots & \sum\limits_{\alpha_m} a_{2\alpha_m} b_{\alpha_m m} \\ \cdots & \cdots & \cdots \\ \sum\limits_{\alpha_1} a_{m\alpha_1} b_{\alpha_1 1} & \cdots & \sum\limits_{\alpha_m} a_{m\alpha_m} b_{\alpha_m m} \end{bmatrix}.$$

Which by affine linearity gives

$$\det C = \sum_{\alpha_1, \alpha_2, \cdots, \alpha_m} \det \begin{bmatrix} a_{1\alpha_1} b_{\alpha_1 1} & \cdots & a_{1\alpha_m} b_{\alpha_m m} \\ \cdots & \cdots & \cdots \\ a_{m\alpha_1} b_{\alpha_1 1} & \cdots & a_{m\alpha_m} b_{\alpha_m m} \end{bmatrix}$$

$$(1.1) \qquad = \sum_{\alpha_1, \alpha_2, \cdots, \alpha_m} A \begin{pmatrix} 1, 2, \cdots, m \\ \alpha_1, \alpha_2, \cdots, \alpha_m \end{pmatrix} b_{\alpha_1 1} b_{\alpha_2 2} \cdots b_{\alpha_m m}$$

If $m > n$ some of the α_i's will be same and all those terms vanish. Thus we can as well assume $m \leq n$. Given any subset $\{\alpha_1, \alpha_2, \cdots, \alpha_m\} \subset \{1, 2, \cdots, n\}$ we can rearrange $\{\alpha_1, \alpha_2, \cdots, \alpha_m\}$ as $\{k_1, k_2, \cdots, k_m\}$ where $k_1 < k_2 \cdots < k_m$. Thus

$$A \begin{pmatrix} 1 & 2 & \cdots & m \\ \alpha_1 & \alpha_2 & \cdots & \alpha_m \end{pmatrix} = \varepsilon(\sigma) A \begin{pmatrix} 1 & 2 & \cdots & m \\ k_1 & k_2 & \cdots & k_m \end{pmatrix}.$$

Here $\sigma = \begin{pmatrix} 1 & 2 & \cdots & m \\ k_1 & k_2 & \cdots & k_m \end{pmatrix}$

Thus expression (1.1) is

$$\det C = \sum_{1 \leq k_1 < k_2 \cdots < k_m \leq n} A \begin{pmatrix} 1 & 2 & \cdots & m \\ k_1 & k_2 & \cdots & k_m \end{pmatrix} B \begin{pmatrix} k_1 & k_2 & \cdots & k_m \\ 1 & 2 & \cdots & m \end{pmatrix}.$$

REMARK 1.1. Cauchy - Binet formula extends to compound matrices. If $C = AB$ for $n \times n$ matrices A and B, $\mathcal{C}_p = \mathcal{A}_p \mathcal{B}_p$ where $p =$ all subsets of size p from $\{1, 2, \cdots, n\}$. In fact if $B = A^{-1}$, $\mathcal{B}_p = \mathcal{A}_p^{-1}$, for all $p = 1, 2, \cdots n$.

1.4. Gaussian Elimination and Sylvester's identity. Given an $m \times n$ matrix $A = (a_{ij})$, we have 3 types of row operations called elementary row operations:

Type 1: Interchange any two rows.
Type 2: Multiply any row by a scalar multiple.
Type 3: Add a scalar multiple of any row to another row.

By performing the above operations on an $m \times m$ identity matrix I, we get a matrix, say E, called an *elementary matrix*, The matrix EA corresponds to such operations on A.

An $n \times n$ matrix A is *upper triangular* if all the entries below the main diagonal are zero. By a series of such elementary row operations it is possible to reduce any $n \times n$ matrix to an upper triangular matrix. Suppose leading principal minor of order r is non-zero, we can reduce by elementary row operation the leading $r \times r$ principal submatrix of A to an upper triangular matrix that retains the determinantal value and is reducible to a matrix of the type

$$\begin{bmatrix} a_{11} & a_{12} & \cdots & a_{1r} & \cdots & a_{1k} & \cdots & a_{1n} \\ 0 & a_{22}^{(1)} & \cdots & a_{2r}^{(1)} & \cdots & a_{2k}^{(1)} & \cdots & a_{2n}^{(1)} \\ 0 & 0 & \ddots & \cdots & \cdots & \cdots & \cdots & \cdots \\ 0 & 0 & 0 & a_{rr}^{(r-1)} & \cdots & a_{rk}^{(r-1)} & \cdots & a_{rn}^{(r-1)} \\ 0 & 0 & \cdots & 0 & 0 & & & \\ 0 & 0 & \cdots & 0 & \cdots & a_{ik}^{(r)} & \cdots & a_{in}^{(r)} \\ 0 & 0 & \cdots & 0 & \cdots & \vdots & \cdots & \vdots \end{bmatrix}$$

The matrix $G_r = \begin{bmatrix} a_{11} & a_{12} & \cdots & a_{1r} \\ 0 & a_{22}^{(1)} & \cdots & a_{2r}^{(1)} \\ & \ddots & & \\ 0 & & & a_{rr}^{(r-1)} \end{bmatrix}$ has $\det G_r = a_{11}a_{22}^{(1)} \cdots a_{rr}^{(r-1)}$.

In fact we have

$$A\begin{pmatrix} 1 & 2 & \cdots & r & i \\ 1 & 2 & \cdots & r & k \end{pmatrix} = a_{11}a_{22}^{(1)} \cdots a_{rr}^{(r-1)}a_{ik}^{(r)}$$

and

$$A\begin{pmatrix} 1 & 2 & \cdots & r \\ 1 & 2 & \cdots & r \end{pmatrix} = a_{11}a_{22}^{(1)} \cdots a_{rr}^{(r-1)}.$$

Thus

$$a_{ik}^{(r)} = \frac{A\begin{pmatrix} 1 & 2 & \cdots & r & i \\ 1 & 2 & \cdots & r & k \end{pmatrix}}{A\begin{pmatrix} 1 & 2 & \cdots & r \\ 1 & 2 & \cdots & r \end{pmatrix}}.$$

The determinant of the submatrix $B=(b_{ik})_{i>r,k>r}$ where

$$b_{ik}=A\begin{pmatrix} 1 & 2 & \cdots & r & i \\ 1 & 2 & \cdots & r & k \end{pmatrix}$$

can be computed as follows. Since

$$a_{ik}^{(r)} = \frac{b_{ik}}{A\begin{pmatrix} 1 & 2 & \cdots & r \\ 1 & 2 & \cdots & r \end{pmatrix}}$$

we have $A\begin{pmatrix} 1 & 2 & \cdots & r \\ 1 & 2 & \cdots & r \end{pmatrix} a_{ik}^{(r)} = b_{ik}$ for $i, k > r$.

Since $\det A = A\begin{pmatrix} 1 & 2 & \cdots & n \\ 1 & 2 & \cdots & n \end{pmatrix} = \det G_r . \det \begin{bmatrix} a_{r+1,r+1}^{(r)} & \cdots & a_{r+1,n}^{(r)} \\ \cdots & \cdots & \cdots \\ a_{n,r+1}^{(r)} & \cdots & a_{nn}^{(r)} \end{bmatrix}$.

$$\det(B)_{r+1}^{n} = (\det G_r)^{n-r-1} . \det A \tag{1.2}$$

By continuity arguments it holds good on principal minors with no restrictions. The above identity in (1.2) is called the *Sylvester's identity*.

Another very useful version of Sylvester's identity is the following: Given $A = (a_{ij})_{n \times n}$. Let A_{ij} be the determinant of the $n-1 \times n-1$ submatrix deleting row i

and column j. Similarly let $A_{\{(i_1,i_2);(j_1,j_2)\}}$ be the determinant of the $n-2 \times n-2$ submatrix deleting rows $i_1 < i_2$ and columns $j_1 < j_2$. Then we claim

$$\det A.\ \det A_{\{(i_1,i_2);(j_1,j_2)\}} = \det \begin{bmatrix} A_{i_1 j_1} & A_{i_1 j_2} \\ A_{i_2 j_1} & A_{i_2 j_2} \end{bmatrix}.$$

PROOF. We can assume w.l.g $i_1 = n-1$, $i_2 = n$ and $j_1 = n-1$, $j_2 = n$. In case $A\begin{pmatrix} 1 & 2 & \cdots & n-2 \\ 1 & 2 & \cdots & n-2 \end{pmatrix}$ is non-singular we can identify with B in Sylvester's formula

$$\det B = \det \begin{bmatrix} A_{i_1 j_1} & A_{i_1 j_2} \\ A_{i_2 j_1} & A_{i_2 j_2} \end{bmatrix} = A\begin{pmatrix} 1 & 2 & \cdots & n-2 \\ 1 & 2 & \cdots & n-2 \end{pmatrix} \det A \tag{1.3}$$

□

2. On totally positive and totally non-negative matrices

DEFINITION 2.1. An $n \times n$ matrix A is *totally non-negative* (TN) if all its minors of any order are nonnegative. It is called *totally positive* (TP) if all its minors of any order are totally positive.

EXAMPLE 2.2. $A = \begin{bmatrix} 4 & 3 & 2 & 1 \\ 5 & 4 & 3 & 2 \\ 6 & 5 & 4 & 3 \\ 7 & 6 & 5 & 4 \end{bmatrix}$ has all minors of order 2 positive, while all minors of order 3 are zero and $\det A = 0$. Thus A is TN, but not TP.

We have a class of matrices that are totally positive(TP). The following is one such.

PROPOSITION 2.3. Let $0 < a_1 < a_2 \cdots < a_n$ and let $0 < \lambda_1 < \cdots < \lambda_n$. Then $V = [a_i^{\lambda_j}]_{i,j=1,2\cdots,n}$ is TP.

PROOF. We first claim that $\det V \neq 0$. Suppose not. Then the columns of V are linearly dependent and that $f(x) = \sum_{k=1}^{n} c_k x^{\lambda_k} = 0$ at $x = a_1, a_2. \cdots, a_n$ for some $(c_1, \cdots, c_k) \neq (0 \cdots, 0)$. Let $f_1(x) = x^{-\lambda_1} f(x)$. Then $f_1(a_1) = 0$. Also

$$\begin{aligned} f_1(a_2) &= c_1 + c_2 a_2^{(\lambda_2 - \lambda_1)} + \cdots + c_n a_n^{(\lambda_n - \lambda_1)} \\ &= \frac{c_1 a_2^{\lambda_1} + c_2 a_2^{\lambda_2} + \cdots + c_n a_2^{\lambda_n}}{a_2^{\lambda_1}} = 0. \end{aligned}$$

Thus by Rolle's theorem $\frac{d}{dx} f_1(x)$ has $n-1$ positive roots. We can assume by induction it can't be. Thus $\det V \neq 0$. Continuing with our proof we need the following. □

LEMMA 2.4. *If a principal minor of a TN matrix A vanishes, then all the bordering principal minors vanish. In particular* $\det A = 0$.

PROOF. Let $A\begin{pmatrix} 1 & 2 & \cdots & q \\ 1 & 2 & \cdots & q \end{pmatrix} = 0$, $q < n$.

case (i) $a_{11} = 0$.

Since A is TN, $\det \begin{bmatrix} a_{11} & a_{1k} \\ a_{i1} & a_{ik} \end{bmatrix} \leq 0$ (Note $a_{i1} a_{1k} \geq 0\ \forall\ i, k$)

This shows that $a_{i1}a_{1k} = 0$. In case $a_{1k} \neq 0$ for some k, we have $a_{i1} = 0$ $i = 1, 2, \cdots, n$. $\implies \det A = 0$.

case (ii) Let $A\begin{pmatrix}1 & 2 & \cdots & p-1\\ 1 & 2 & \cdots & p-1\end{pmatrix} \neq 0$, but $A\begin{pmatrix}1 & 2 & \cdots & p\\ 1 & 2 & \cdots & p\end{pmatrix} = 0$.

Let $b_{ik} = A\begin{pmatrix}1 & 2 & \cdots & p-1 & i\\ 1 & 2 & \cdots & p-1 & k\end{pmatrix}$ $i, k \geq p$.

By Sylvester's identity for $p \leq i_1 < i_2 \cdots < i_q \leq n; p \leq k_1 \cdots < k_q \leq n$

$$B\begin{pmatrix}i_1 & \cdots & i_q\\ k_1 & \cdots & k_q\end{pmatrix} = A\begin{pmatrix}1 & 2 & \cdots & p-1\\ 1 & 2 & \cdots & p-1\end{pmatrix}^{q-1} A\begin{pmatrix}1 & 2 & \cdots & p-1, i_1, \cdots i_q\\ 1 & 2 & \cdots & p-1, k_1, \cdots k_q\end{pmatrix}.$$

Thus $B\begin{bmatrix}i_1 & \cdots & i_q\\ k_1 & \cdots & k_q\end{bmatrix}$ is TN. Since by assumption

$b_{pp} = A\begin{pmatrix}1 & 2 & \cdots & p\\ 1 & 2 & \cdots & p\end{pmatrix} = 0$, B falls under case (i) and

$$B\begin{pmatrix}p & p+1 & \cdots & n\\ p & p+1 & \cdots & n\end{pmatrix} = A\begin{pmatrix}1 & 2 & \cdots & p-1\\ 1 & 2 & \cdots & p-1\end{pmatrix}^{n-p} A\begin{pmatrix}1 & 2 & \cdots & n\\ 1 & 2 & \cdots & n\end{pmatrix} = 0.$$

Thus $\det A = 0$.

□

THEOREM 2.5. *Let A be an $n \times n$ TN matrix. Then for any $1 < p < n$,* $A\begin{pmatrix}1 & 2 & \cdots & n\\ 1 & 2 & \cdots & n\end{pmatrix} \leq A\begin{pmatrix}1 & 2 & \cdots & p\\ 1 & 2 & \cdots & p\end{pmatrix} A\begin{pmatrix}p+1 & \cdots & n\\ p+1 & \cdots & n\end{pmatrix}$.

PROOF. Without loss of generality, all principal minors are positive, for otherwise by the previous lemma (Lemma 2.4), $\det A = 0$ and the above inequality is trivial. For $n = 2$, we have $A\begin{pmatrix}1 & 2\\ 1 & 2\end{pmatrix} = a_{11}a_{22} - a_{12}a_{21} \leq a_{11}a_{22}$ $(a_{12} \geq 0, a_{21} \geq 0)$. We will assume by induction that the inequality is valid for all TN matrices of order $<$ n. From Sylvester's identity we have

$$A\begin{pmatrix}1 & 2 & \cdots & n\\ 1 & 2 & \cdots & n\end{pmatrix} A\begin{pmatrix}1 & 2 & \cdots & p-1\\ 1 & 2 & \cdots & p-1\end{pmatrix}^{(n-p)} = B\begin{pmatrix}p & p+1 & \cdots & n\\ p & p+1 & \cdots & n\end{pmatrix}$$

Here $B = (b_{ik})$ $i, k \geq p$ and $b_{ik} = A\begin{pmatrix}1 & 2 & \cdots & p-1 & i\\ 1 & 2 & \cdots & p-1 & k\end{pmatrix}$. By induction

$$\begin{aligned} B\begin{pmatrix}p & p+1 & \cdots & n\\ p & p+1 & \cdots & n\end{pmatrix} &\leq b_{pp} B\begin{pmatrix}p+1 & \cdots & n\\ p+1 & \cdots & n\end{pmatrix}\\ &\leq b_{pp} A\begin{pmatrix}1 & 2 & \cdots & p-1, p+1 & \cdots & n\\ 1 & 2 & \cdots & p-1, p+1 & \cdots & n\end{pmatrix}\\ &\quad \times A\begin{pmatrix}1 & 2 & \cdots & p-1\\ 1 & 2 & \cdots & p-1\end{pmatrix}^{n-p-1}. \end{aligned}$$

Using another round of induction, we get

$$\begin{aligned} B\begin{pmatrix}p & p+1 & \cdots & n\\ p & p+1 & \cdots & n\end{pmatrix} &\leq b_{pp} A\begin{pmatrix}1 & 2 & \cdots & p-1\\ 1 & 2 & \cdots & p-1\end{pmatrix}\\ &\quad \times A\begin{pmatrix}p+1 & \cdots & n\\ p+1 & \cdots & n\end{pmatrix} A\begin{pmatrix}1 & 2 & \cdots & p-1\\ 1 & 2 & \cdots & p-1\end{pmatrix}^{n-p-1} \end{aligned}$$

Thus cancelling $A\begin{pmatrix}1 & 2 & \cdots & p-1\\ 1 & 2 & \cdots & p-1\end{pmatrix}^{(n-p)}$ on both sides we get

$$A\begin{pmatrix}1 & 2 & \cdots & n\\ 1 & 2 & \cdots & n\end{pmatrix} \le A\begin{pmatrix}1 & 2 & \cdots & p\\ 1 & 2 & \cdots & p\end{pmatrix} A\begin{pmatrix}p+1 & \cdots & n\\ p+1 & \cdots & n\end{pmatrix}.$$

□

While the spectral theorem is undoubtedly the most central theorem in linear algebra, the next major theorem in linear algebra is the celebrated Perron - Frobenius theorem for matrices with nonnegative entries that are irreducible. While the applications of the spectral theorem for real symmetric matrices dominate many facets of multivariate statistical analysis, quantum mechanics and many other physical sciences, applications of the Perron - Frobenius theorem are to be found in the areas of Markov Chains, Numerical analysis, Operations Research, Categorical Data analysis, problems of international trade, cooperative game theory and so on.

It offers great insights into the geometry of linear and non-linear transformations in ordered vector spaces. Such transformations are very closely related to war duels and general minimax theorems (M. Shiffman, Ky Fan). We would refer to Bapat and Raghavan (1996) (*Nonnegative Matrices and Applications, Cambridge University Press*) and to the classic papers of Krein and Rutman (*Functional Analysis and Measure Theory, American Mathematical Society, Translations, series 1, vol 10, 1950*).

We recall the Perron-Frobenius theorem for ready reference.

DEFINITION 2.6. An $n \times n$ matrix A with non-negative entries is reducible if we can find a permutation matrix P such that PAP^T is of the form

$$PAP^T = \begin{bmatrix} B & 0 \\ C & D \end{bmatrix}$$

where B and D are square submatrices. Alternatively, we call A irreducible if for any row i, column j, for some power $n = n(i,j)$, $a_{ij}^{(n)} > 0$.

THEOREM 2.7 (Perron - Frobenius Theorem:). *Let A be an irreducible $n \times n$ matrix with non-negative entries. Then*

(i) *$Ay = \lambda_0 y$ for some $\lambda_0 > 0$ and y (a vector with all co-ordinates positive).*
(ii) *The eigenvalue λ_0 is geometrically simple.*
(iii) *Among all eigenvalues of A, λ_0 is maximal in modulus.*
(iv) *The eigenvalue λ_0 is algebraically simple.*
(v) *If $\lambda_0, \lambda_1, \cdots, \lambda_{k-1}$ are all the eigenvalues of A with $|\lambda_i| = \lambda_0$; $i = 0, 1, 2, \cdots, k-1$, then they are precisely the solutions of the equation $\lambda^k - \lambda_0^k = 0$.*
(vi) *In case all entries of A are positive, λ_0 is the only eigenvalue of A, of absolute value λ_0. namely $k = 1$ in (v).*

2.1. On Oscillatory Matrices. A totally nonnegative (TN) matrix $A = (a_{ij})_{n\times n}$ is called *Oscillatory* if and only if there exists some power k such that A^k is totally positive (TP).

Given a TN matrix A how do we know that A is oscillatory? The following seminal theorem characterizes them.

THEOREM 2.8 (Gantmacher and Krein:). *A TN matrix is oscillatory if and only if*

(i) $\det A \neq 0$
(ii) $a_{ii+1} > 0,\ a_{i+1,i} > 0$ *for all* $i = 1, 2, \cdots, n-1$.

Towards its proof we need the following.

LEMMA 2.9. *If A is oscillatory so are its truncated submatrices consisting of adjacent rows and adjacent columns.*

PROOF. Since truncation can occur at either the front or the back, and since A^T is also oscillatory when A is oscillatory, it suffices to prove that $A\begin{pmatrix} 2 & 3 & \cdots & n \\ 2 & 3 & \cdots & n \end{pmatrix}$ is oscillatory.

We know from Sylvester's inequality for TN matrices

$$0 \leq A\begin{pmatrix} 1 & 2 & \cdots & n \\ 1 & 2 & \cdots & n \end{pmatrix} \leq a_{11} A\begin{pmatrix} 2 & 3 & \cdots & n \\ 2 & 3 & \cdots & n \end{pmatrix}.$$

Since A is oscillatory $\det A > 0$ and we have $A\begin{pmatrix} 2 & 3 & \cdots & n \\ 2 & 3 & \cdots & n \end{pmatrix} > 0$.

Continuing with our proof:

(Necessity): From the above lemma we know $C = \begin{bmatrix} a_{ii} & a_{ii+1} \\ a_{i+1i} & a_{i+1,i+1} \end{bmatrix}$ is oscillatory. If $a_{ii+1} = 0$, then C is reducible and so C cannot be oscillatory. Since C^T is also oscillatory $a_{i+1i} \neq 0$ and thus we have the necessary conditions.
(Sufficiency): Since C is TN $a_{ii}.a_{i+1i+1} - a_{i+1i}a_{ii+1} \geq 0$. But $a_{i+1,i}a_{i,i+1} > 0$. Thus $a_{ii} > 0$ $a_{i+1i+1} > 0$, and all diagonal entries of A are positive.

We need the following lemma in the sequel.

LEMMA 2.10. *Let* $A = (a_{ik})_{m\times n}$ *be a TN matrix. Let any* p *rows that contain either the first row* $(a_{11},\ a_{12}, \cdots, a_{1n})$ *or the last row* $(a_{m1},\ a_{m2}, \cdots, a_{mn})$ *be linearly dependent. Let any* $p-1$ *rows of* A *be linearly independent. Then* $\operatorname{rank} A = p - 1$.

PROOF. Rows $i_1 = 1 < i_2 \cdots < i_p = m$ are linearly dependent while rows $i_1 = 1 < i_2 \cdots < i_{p-1}$ are linearly independent. Thus $a_{i_pk} = \sum_{r=1}^{p-1} \lambda_r a_{i_rk}, k = 1, 2, \cdots, n$.

Here $\lambda_1 \neq 0$ for otherwise the $p-1$ rows $i_2 < i_3 \cdots < i_p = m$ will be linearly dependent which contradicts our assumption. Choose any row j that lies in between say rows i_h and i_{h+1} (that is $i_h < j < i_{h+1}$).

For any $k_1 < k_2 \cdots < k_p$ we can consider the determinant of the matrix

$$P = \begin{bmatrix} a_{i_2k_1} & \cdots & a_{i_2k_p} \\ a_{i_3k_1} & \cdots & a_{i_3k_p} \\ a_{jk_1} & \cdots & a_{jk_p} \\ a_{i_pk_1} & \cdots & a_{i_pk_p} \end{bmatrix}$$

Replacing a_{i_pk} with $\sum_{r=1}^{p-1} \lambda_r a_{i_rk}$, we get

$$\det P = \det \begin{bmatrix} a_{i_2k_1} & \cdots & a_{i_2k_p} \\ a_{i_3k_1} & \cdots & a_{i_3k_p} \\ a_jk_1 & \cdots & a_{jk_p} \\ a_{i_{p-1}k_1} & \cdots & a_{i_{p-1}k_p} \\ \lambda_1 a_{i_1k_1} & \cdots & \lambda_1 a_{i_1k_p} \end{bmatrix}$$

$$= \lambda_1(-1)^{p-1} A \begin{pmatrix} i_1 & i_2 & j & i_{p-1} \\ k_1 & k_2 & \cdots & k_p \end{pmatrix} \tag{2.1}$$

Since by assumption rows $i_1 = 1 < i_2 \cdots < i_{p-1}$ are linearly independent, there should be some columns $k_1^0 < k_2^0 \cdots < k_{p-1}^0$ such that when we skip row j from P we should get a non-singular $p-1 \times p-1$ sub matrix with

$$\begin{aligned} &\det A \begin{bmatrix} i_1 & i_2 & i_h & \cdots & i_{p-1} \\ k_1^0 & k_2^0 & k_h^0 & \cdots & k_{p-1}^0 \end{bmatrix} \\ &= \lambda_1(-1)^{p-2} A \begin{pmatrix} i_1 & \cdots & i_h & i_{h+1} & \cdots & i_{p-1} \\ k_1^0 & k_2^0 & \cdots & \cdots & \cdots & k_{p-1}^0 \end{pmatrix} \neq 0 \end{aligned} \tag{2.2}$$

However since A is TN L.H.S in step (2.2) above is ≥ 0. Thus since R.H.S is $\neq 0$ in (2.2), we have $\lambda_1(-1)^{p-2} > 0$ and so is $\lambda_1(-1)^{p-1} < 0$ in (2.1). Since P is TN we have $\det P = 0$. Thus the jth row is a linear combination of rows $i_1, i_2, \cdots, i_{p-1}$. It follows that the rank of A is $p-1$. □

DEFINITION 2.11. Consider any minor $A\begin{pmatrix} i_1 & i_2 & \cdots & i_p \\ k_1 & k_2 & \cdots & k_p \end{pmatrix}$ of $A = \|a_{ik}\|_1^n$. The minor is called an *almost principal minor* if and only if $i_\nu - k_\nu = 0$ except for one ν in $1 \leq \nu \leq p$.

DEFINITION 2.12. A minor $A\begin{pmatrix} i_1 & \cdots & i_p \\ k_1 & \cdots & k_p \end{pmatrix}$ is called a *quasi-principal minor* if $1 \leq i_1, k_1 < i_2, k_2, \cdots < i_p, k_p \leq n$ and $|i_1 - k_1| \leq 1, |i_2 - k_2| \leq 1, \cdots, |i_p - k_p| \leq 1$.

Continuing with our proof for the sufficiency part, we can assume by induction that all quasi-principal minors of order $< p$ are positive. Suppose an almost principal minor of order p is zero. Say it corresponds to the matrix

$$P = A \begin{bmatrix} i_1 & i_2 & \cdots & i_p \\ k_1 & k_2 & \cdots & k_p \end{bmatrix} \quad \text{with } \det P = 0.$$

By our induction assumption

$$A\begin{pmatrix} i_1 & \cdots & i_{p-1} \\ k_1 & \cdots & k_{p-1} \end{pmatrix} A\begin{pmatrix} i_2 & \cdots & i_p \\ k_2 & \cdots & k_p \end{pmatrix} \neq 0.$$

This implies by our above lemma rank $P = p-1$. Let $h = \max(i_1, k_1)$. Then $h+p-1 \leq i_p, k_p$. Thus any pth order minor of P, say $A\begin{pmatrix} h, & h+1 & \cdots & h+p-1 \\ h, & h+1 & \cdots & h+p-1 \end{pmatrix}$ is zero. Since A is TN all of its bordering principal minors vanish and hence $\det A = 0$, which contradicts our assumption. Thus all quasi-principal minors are positive.

LEMMA 2.13. *Any kth order minor of a matrix $A = (a_{ij})_{n\times n}$ is simply a sum of products of quasi-principal minors.*

PROOF. With each kth order minor $A\begin{pmatrix} i_1 & \cdots & i_k \\ j_1 & \cdots & j_k \end{pmatrix}$ we associate a k-tuple of symbols, $+, -, 0$ as follows: The αth symbol is "$+$" if $i_\alpha < j_\alpha$. It is "$-$" if $i_\alpha > j_\alpha$. It is 0 if $i_\alpha = j_\alpha$. Here $\alpha = 1, 2 \cdots, k$. The k-tuple of symbols are grouped into runs of $+$, runs of $-$ and runs of 0. To fix the idea set $k = 8$ and set $n = 40$ and

$$A\begin{pmatrix} i_1 & i_2 & \cdots & i_k \\ j_1 & j_2 & \cdots & j_k \end{pmatrix} = A\begin{pmatrix} (+ & + & +) & (0) & (- & - & -) & (+) \\ 3, & 5, & 8, & 11, & 14, & 17, & 19, & 23 \\ 4, & 7, & 9, & 11, & 12, & 15, & 18, & 25 \end{pmatrix}$$

In each run of "$+$" increase the last i_α to $i_\alpha + 1 = i'_\alpha$. In each run of "$-$" decrease the first i_α to $i'_\alpha = j_\alpha - 1$. All the rest are unchanged. In our example $(i'_1, i'_2 \cdots i'_8) = (3, 5, 9, 11, 13, 17, 19, 24)$. Observe that the minor $A\begin{pmatrix} i_1 & \cdots & i_8 \\ i'_1 & \cdots & i'_8 \end{pmatrix}$ is quasi-principal. Now we can start with the new minor $A\begin{pmatrix} i'_1 & \cdots & i'_k \\ j_1 & \cdots & j_k \end{pmatrix}$. After some finite number s of iterations we will end up with $i_\alpha^{(s)} = j_\alpha \quad \alpha = 1, 2, \cdots, k$. (As many more "0" symbols will appear in each iteration).

We can also observe that since $1 \le i_1 < i_2 < \cdots < i_k \le n; 1 \le j_1 < j_2 \cdots < j_k \le n \le i_\alpha$ and $j_\alpha \le n - k + \alpha$. Thus $|i_\alpha - j_\alpha| \le n - k$ for $\alpha = 1, 2, \cdots, k$. Thus in fact $i_\alpha^{(t)} = j_\alpha$ for some iterate t for $\alpha = 1, 2, \cdots, k$ and $i_\alpha^{(n-1)} = j_\alpha, \alpha = 1, 2, \cdots, k$. Thus

$$A\begin{pmatrix} i_1 & \cdots & i_k \\ j_1 & \cdots & j_k \end{pmatrix} = \sum A\begin{pmatrix} i_1 & \cdots & i_k \\ i'_1 & \cdots & i'_k \end{pmatrix} \times A\begin{pmatrix} i'_1 & \cdots & i'_k \\ i''_1 & \cdots & i''_k \end{pmatrix} \cdots A\begin{pmatrix} i_1^{(n-2)} & \cdots & i_k^{(n-2)} \\ j_1 & \cdots & j_k \end{pmatrix}$$

where each term on the right are quasi-principal minors. □

If A is TN and at least one term on r.h.s is > 0, $A\begin{pmatrix} i_1 & \cdots & i_k \\ j_1 & \cdots & j_k \end{pmatrix} > 0$. When A is assumed oscillatory, all quasi-principal minors are positive and we are done with sufficiency. □

The following theorem of Whitney shows that one can find TP matrices quite close to TN matrices.

THEOREM 2.14. *Let $A = (a_{ij})_{m \times n}$ be a TN matrix. For any given $\epsilon > 0$, we can find a TP matrix $B = (b_{ij})_{m \times n}$ such that $0 < b_{ij} - a_{ij} < \epsilon$ for all i, j.*

PROOF. The proof is based on the following.

LEMMA 2.15. *For any $x_1, x_2, \cdots, x_n$ real,*

$$\det V(x_1, x_2, \cdots, x_n) = \det \begin{bmatrix} 1 & x_1 & x_1^2 & \cdots & x_1^{n-1} \\ 1 & x_2 & x_2^2 & \cdots & x_2^{n-1} \\ .. & .. & .. & \cdots & .. \\ 1 & x_n & x_n^2 & \cdots & x_n^{n-1} \end{bmatrix} = \prod_{i>j} (x_i - x_j)$$

(The above matrix is called a Vandermonde matrix).

PROOF. In $V(x_1, x_2, \cdots, x_n)$, subtracting the first row from the rest of the rows and then evaluating $\det V$ we get

$$\det V(x_1 \cdots x_n) = (x_2 - x_1) \cdots (x_n - x_1) \det \begin{bmatrix} 1 & x_2 + x_1 & x_2^2 + x_1 x_2 + x_1^2 & \cdots \\ \vdots & \vdots & \vdots & \vdots \\ 1 & x_n + x_1 & x_n^2 + x_1 x_n + x_1^2 & \cdots \end{bmatrix}$$

We notice that the right side determinant is a sum of several determinants that have column terms that are multiples of the first column of V and they all vanish except for $\det V$, Thus

$$\begin{aligned} \det V &= (x_2 - x_1) \cdots (x_n - x_1) \det V(x_2 \cdots x_n) \\ &= \prod_{i>j} (x_i - x_j) \end{aligned}$$

by an induction. □

Continuing with the proof of the theorem, we can add zero rows or columns to A and assume A is an $n \times n$ TN matrix. □

Let r be the rank of A. Observe that for any $0 < q < 1$, the matrix $H = H(q) = [q^{(i-j)^2}] = DQD$, where D is the diagonal matrix $= \operatorname{diag}(q^2, q^4, \cdots, q^{2n})$ which is TN and non-singular, while $Q = [q^{-2ij}]$ is a Vandermonde matrix which is TP. Thus $U = HAH$ is TN and has rank r. We have for any minor of order $k \le r$

$$U \begin{pmatrix} i_1 & \cdots & i_k \\ j_1 & \cdots & j_k \end{pmatrix} = \sum_{(\alpha)(\beta)} H \begin{pmatrix} i_1 & \cdots & i_k \\ \alpha_1 & \cdots & \alpha_k \end{pmatrix} A \begin{pmatrix} \alpha_1 & \cdots & \alpha_k \\ \beta_1 & \cdots & \beta_k \end{pmatrix} H \begin{pmatrix} \beta_1 & \cdots & \beta_k \\ j_1 & \cdots & j_k \end{pmatrix}.$$

Here all terms are nonnegative and at least one term is positive. For example for $k = 1$

$$u_{ij} = \sum_{\alpha,\beta} H \begin{pmatrix} i \\ \alpha \end{pmatrix} A \begin{pmatrix} \alpha \\ \beta \end{pmatrix} H \begin{pmatrix} \beta \\ j \end{pmatrix} = a_{ij} + q. \text{ (a sum of nonnegative terms).}$$

Thus $u_{ij} \ge a_{ij}$ and $u_{ij} \to a_{ij}$ as $q \to 0$. If $r = n$, the theorem is proved as U is TP. Suppose $r < n$, we can derive from U a new matrix $V = V_q(U)$ by adding q to the leading element u_{11} of U. Clearly V is TN and all minors of order r or less are TP_r. Further the rank of V is $r+1$ because all minors of order $r+1$ containing v_{11} are strictly positive. The matrix $U' = HVH$ is TP_{r+1} and if $r+1 < n$, $V' = V_q(U')$ has rank $r+2$. After a finite number of steps we reach the required matrix B which is TP provided q is sufficiently small.

2.2. The semigroup of nonsingular TN matrices. The set s_n of non-singular $n \times n$ TN matrices form a semigroup under multiplication with I as its identity element.

Let $U(t)$ be a differentiable matrix function in $0 \le t \le t_0$, where $U(t)$ is a TN matrix in s_n, for each t and $U(0) = I$. we call $\left(\frac{dU(t)}{dt}\right)_{t=0}$ an infinitesimal element of s_n.

Let σ_n be the totality of all such infinitesimal elements of s_n.

Let $\Omega(t)$ be a piecewise continuous function on $[0\ t_0]$ which for each t gives an element of σ_n. The differential equation

$$\frac{dU(t)}{dt} = \Omega(t) U(t) \text{ with } U(0) = I$$

has a unique continuous solution. We say $U(t_0) \in s_n$ is generated by $\Omega(t)$. Even though we cannot in general recover a whole semi-group via its infinitesimal generators, the key assertion we want to prove is that it is possible for our s_n. The following two critical observations on σ_n show us the way.

LEMMA 2.16. *A matrix $\Omega = (w_{ij})$ is an element of σ_n if and only if $w_{ij} = 0 \ \forall \ |i-j| > 1$ and $w_{ij} \geq 0 \ \forall \ |i-j| = 1$.*

PROOF. (Necessary part): By assumption $\Omega = \frac{dU(t)}{dt}|_{t=0}$ where $U(t) = (u_{ij}(t))$ is a TN matrix and $U(0) = I$. Here, for $i \neq j$, $u_{ij}(0) = 0$ and $u_{ij}(t) \geq 0$. Thus $w_{ij} \geq 0$ if $i \neq j$. Consider the 2×2 minor

$$H(t) = \det \begin{bmatrix} u_{kj}(t) & u_{kk}(t) \\ u_{ij}(t) & u_{ik}(t) \end{bmatrix} \quad \text{for } i > k > j.$$

Since U is TN, $u_{kj}(t)u_{ik}(t) - u_{ij}(t)u_{kk}(t) \geq 0, \ \forall \ 0 \leq t \leq t_0$

$$\begin{aligned} \frac{dH}{dt}|_{t=0} &= [u'_{kj}(t)u_{ik}(t) + u_{kj}(t)u'_{ik}(t) - u'_{ij}(t)u_{kk}(t) - u_{ij}(t)u'_{kk}(t)]_{t=0} \\ &= -u'_{ij}(t)|_{t=0} = -w_{ij} \geq 0 \text{ for } i - j > 1. \end{aligned}$$

Notice $u_{ik}(0) = 0, \ u_{kk}(0) = 1$ when $i \neq k$.
Similarly, $\quad -w_{ij} \geq 0$ for $j - i > 1$.
Thus $\quad w_{ij} = 0$ if $|i - j| > 1$.

(Sufficiency part): We claim that σ_n is a convex cone in the matrix space. To see this, let $\lambda_1, \lambda_2 > 0$ and let $\Omega_1, \Omega_2 \in \sigma_n$. We have by assumption $\Omega_1 = \left(\frac{dU_1}{dt}\right)_{t=0}$, $\Omega_2 = \left(\frac{dU_2}{dt}\right)_{t=0}$ where $U_1, U_2 \in s_n$ and $U_1(0) = U_2(0) = I$. Consider $U(t) = U_1(\lambda_1 t)U_2(\lambda_2 t)$. We get

$$\begin{aligned} \frac{dU}{dt}|_{t=0} &= \lambda_1 \frac{dU_1(\lambda_1 t)}{dt}|_{t=0} U_2(\lambda_2 t)|_{t=0} + \lambda_2 \frac{dU_2(\lambda_2 t)}{dt}|_{t=0} U_1(\lambda_1 t) \\ &= \lambda_1 \Omega_1 + \lambda_2 \Omega_2 \quad (\because U_1(\lambda_1 t)|_{t=0} = I, \ U_2(\lambda_2 t)|_{t=0} = I) \end{aligned}$$

We can consider only those Ω's where just one entry alone is positive and the rest zero. In case it is just diagonal entry say $w_{ii} > 0$ and $w_{kl} = 0 \ \forall \ kl$ other than it. We get the unique solution $e^{\Omega t}$ for

$$\frac{dU}{dt} = \Omega(t)U(t), \quad U(0) = I.$$

They are simply diagonal matrices with positive diagonal entries. Thus they are in s_n. In case it is an off-diagonal element $w_{ij} > 0$, we have to have $|i - j| = 1$. Let $E_{ij} = 1$ for the ijth entry and 0 for all other entries. Then $e^{\Omega t} = I + w_{ij} t E_{ij}$ (Observe $E_{ij}^n = 0 \ \forall \ n > 1$.) Clearly $U(t) = e^{\Omega t}$ is TN when $w_{ij} > 0$. □

In fact a theorem of Whitney asserts the following

THEOREM 2.17. *Let $F_{ij}(w) = I + wE_{ij}$. Then every non-singular TN matrix U is of the form*

$$U = U_1 U_2 \cdots U_{n-1} D V_1 V_2 \cdots V_{n-1}$$

where each U_i is of the form

$$U_i = F_{n,n-1}\begin{pmatrix} i \\ w_{n,n-1} \end{pmatrix} F_{n-1,n-2}\begin{pmatrix} i \\ w_{n-1,n-2} \end{pmatrix} \cdots F_{i+1,i}\begin{pmatrix} i \\ w_{i+1,i} \end{pmatrix}$$

$$V_i = F_{n-i,n-i+1}\begin{pmatrix} i \\ w_{n-i,n-i+1} \end{pmatrix} \cdots F_{n-1,n}\begin{pmatrix} i \\ w_{n-1,n} \end{pmatrix}$$

$D =$ *a diagonal matrix with positive diagonal entries*

The theorem we proved shows that U_i's, V_i's and D can be generated from an element of σ_n and thus we have the following.

THEOREM 2.18. *The semigroup s_n is generated by its infinitesimal elements of σ_n.*

We will need in the sequel the following ingeneous lemma by Fekete on TP matrices.

LEMMA 2.19 (Fekete). *Let $A = (a_{ij})_{n\times m}$ TN matrix where $n \geq m \geq 2$. Let all minors of order $m-1$ using the first $m-1$ columns of A be positive. Also let all minors of order m involving only consecutive rows of A be positive. Then all minors of order m are positive.*

PROOF. For any arbitrary rows $1 \leq i_1 < i_2 \cdots < i_m \leq n$, we have to prove $A\begin{pmatrix} i_1 & i_2 & \cdots & i_m \\ 1 & 2 & \cdots & m \end{pmatrix} > 0$.

We define the dispersion of $\{i_1, i_2, \cdots, i_m\}$ as $d(i_1 \cdots i_m) = (i_m - i_1) - (m-1) \geq 0$. In case $d = 0$ it means $i_1, i_2, \cdots, i_m$ are consecutive rows of A. For $d = 0$ the assumption guarantees $A\begin{pmatrix} i_1 & & \cdots & i_m \\ 1 & 2 & \cdots & m \end{pmatrix} > 0$. We will peg our argument by an induction on d. Assume the lemma for $d < r$ and let $d(i_1 \cdots i_m) = r \geq 1$. This means we have an integer j_k between i_1 and i_m and $j_k \neq i_1$ or $i_2 \cdots$ or i_m. We renumber $\{i_1, \cdots, i_m\} \cup \{j_k\}$ as $\{j_1, j_2, \cdots, j_k, \cdots, j_{m+1}\}$. Now consider the $(m+1)\times(m+1)$ matrix

$$B = \begin{bmatrix} a_{j_1 1} & a_{j_1 2} & \cdots & a_{j_1 m} & 0 \\ a_{j_2 1} & a_{j_2 2} & \cdots & a_{j_2 m} & 0 \\ \cdots & \cdots & \cdots & \cdots & 0 \\ a_{j_k 1} & a_{j_k 2} & \cdots & a_{j_k m} & 0 \\ \cdots & \cdots & \cdots & \cdots & 0 \\ a_{j_{m+1} 1} & a_{j_{m+1} 2} & \cdots & a_{j_{m+1} m} & 1 \end{bmatrix}$$

Let $B_{\alpha,\beta}$ be the determinant of the $m \times m$ matrix got by deleting row α and column β of B. We denote by $\hat{j_k}$, skipping row j_k in expression.

$$B_{j_1 m} = B\begin{pmatrix} j_2 & \cdots & j_k & \cdots & j_m \\ 1 & 2 & \cdots & \cdots & m-1 \end{pmatrix}, \quad B_{j_1 (m+1)} = B\begin{pmatrix} j_2 & \cdots & \cdots & j_{m+1} \\ 1 & 2 & \cdots & m \end{pmatrix}$$

$$B_{j_k m} = B\begin{pmatrix} j_1 & \cdots & \hat{j_k} & \cdots & j_{m+1} \\ 1 & 2 & \cdots & \cdots & m-1 \end{pmatrix}, \quad B_{j_k m+1} = B\begin{pmatrix} j_1 & \cdots & \hat{j_k} & \cdots & j_{m+1} \\ 1 & 2 & \cdots & \cdots & m \end{pmatrix}$$

By Sylvester's identity, applied to rows j_1 and j_k and columns m, $m+1$ in B gives

$$\det B.\ B\begin{pmatrix} j_2 & \cdots & \hat{j_k} & \cdots & j_{m+1} \\ 1 & 2 & \cdots & \cdots & m-1 \end{pmatrix} = \det\begin{bmatrix} B_{j_1 m} & B_{j_1 (m+1)} \\ B_{j_k m} & B_{j_k (m+1)} \end{bmatrix}$$

Also

$$\det B = B\begin{pmatrix} j_1 & \cdots & j_k & \cdots & j_m \\ 1 & 2 & \cdots & \cdots & m \end{pmatrix}$$

Thus

$$B\begin{pmatrix} j_1 & \cdots & j_k & \cdots & j_m \\ 1 & 2 & \cdots & \cdots & m \end{pmatrix} B\begin{pmatrix} j_2 & \cdots & \hat{j_k} & \cdots & j_{m+1} \\ 1 & 2 & \cdots & \cdots & m-1 \end{pmatrix}$$

$$= B_{j_1 m}.B_{j_k (m+1)} - B_{j_k m}.B_{j_1 (m+1)}$$

Bringing $B_{j_k m}, B_{j_1(m+1)}$ to the left hand side we get

$$B\begin{pmatrix} j_1 & \cdots & j_k & \cdots & j_m \\ 1 & 2 & \cdots & \cdots & m \end{pmatrix} B\begin{pmatrix} j_2 & \cdots & \hat{j_k} & \cdots & j_{m+1} \\ 1 & 2 & \cdots & \cdots & m-1 \end{pmatrix}$$
$$+ B_{j_k m}.B_{j_1(m+1)} = B_{j_1 m}.B_{j_k(m+1)}$$

i.e

$$\begin{aligned} & B\begin{pmatrix} j_1 & \cdots & j_k & \cdots & j_m \\ 1 & 2 & \cdots & \cdots & m \end{pmatrix} B\begin{pmatrix} j_2 & \cdots & \hat{j_k} & \cdots & j_{m+1} \\ 1 & 2 & \cdots & \cdots & m-1 \end{pmatrix} \\ & + B\begin{pmatrix} j_1 & \cdots & \hat{j_k} & \cdots & j_m \\ 1 & 2 & \cdots & \cdots & m-1 \end{pmatrix} B\begin{pmatrix} j_2 & \cdots & \cdots & j_{m+1} \\ 1 & 2 & \cdots & m \end{pmatrix} \\ & = B\begin{pmatrix} j_2 & \cdots & \hat{j_k} & \cdots & j_m \\ 1 & 2 & \cdots & \cdots & m-1 \end{pmatrix} B\begin{pmatrix} j_1 & \cdots & \hat{j_k} & \cdots & j_{m+1} \\ 1 & 2 & \cdots & \cdots & m \end{pmatrix}. \end{aligned} \tag{2.3}$$

We make the following key observations. All the minors of order $m-1$ involve only the first $m-1$ columns and they are all positive. Further, $d(j_1 \cdots j_k \cdots j_m) < r$. Similarly $d(j_2 \cdots j_{m+1}) < r$. Thus the L.H.S above is positive in (2.3). It shows

$$A\begin{pmatrix} j_1 & \cdots & \hat{j_l} & \cdots & j_{m+1} \\ 1 & 2 & \cdots & \cdots & m \end{pmatrix}$$

is positive. But it is precisely

$$A\begin{pmatrix} i_1 & \cdots & \cdots & i_m \\ 1 & 2 & \cdots & m \end{pmatrix},$$

Hence the lemma. □

2.3. On matrices with simple structure. Let $A = [a_{ij}]_1^n$. Let $\det(A - \lambda I) = 0$ have as its roots, $\lambda_1, \lambda_2, \cdots, \lambda_n$. When $\det(A - \lambda I) = 0$ we have a vector $x \neq 0$ such that $(A - \lambda I)x = 0$. We call λ an eigenvalue and x its eigenvector.

LEMMA 2.20. *If $\lambda_1 \neq \lambda_2 \neq \cdots \neq \lambda_m$ are eigen values with eigenvectors $x_1, x_2, \cdots, x_m$ respectively then $x_1, x_2, \cdots, x_m$ are linearly independent.*

PROOF. Let $\sum_{j=1}^{m} c_j x_j = 0$ for scalars $c_1, \cdots, c_m$.

$$0 = A0 = A\Big(\sum_{j=1}^{m} c_j x_j\Big) = \sum_{j=1}^{m} c_j A x_j = \sum_{j=1}^{m} c_j \lambda_j x_j = 0$$

Also

$$\lambda_1 \sum_{j=1}^{m} c_j x_j = 0$$

Thus

$$\sum_{j=1}^{m} (\lambda_j - \lambda_1) c_j x_j = 0 \implies \sum_{j=2}^{m} c_j (\lambda_j - \lambda_1) x_j = 0$$

Now

$$A\Big(\sum_{2}^{m} c_j(\lambda_j - \lambda_1)x_j\Big) = 0 \text{ and } \lambda_2 \sum_{j=2}^{m} c_j(\lambda_j - \lambda_1)x_j = 0$$

Thus

$$\sum_{j=2}^{n} c_j(\lambda_j - \lambda_1)\lambda_j x_j = 0 \ \& \ \sum_{j=2}^{n} c_j(\lambda_j - \lambda_1)\lambda_2 x_j = 0$$

Subtracting R.H.S expression from L.H.S we get

$$\sum_{j=2}^{n} (\lambda_j - \lambda_1)(\lambda_j - \lambda_2)c_j x_j = 0 \text{ i.e } \sum_{j=3}^{n} (\lambda_j - \lambda_1)(\lambda_j - \lambda_2)c_j x_j = 0$$

Inductively $c_m(\lambda_m - \lambda_{m-1})(\lambda_m - \lambda_{m-2}) \cdots (\lambda_m - \lambda_1)x_m = 0$. Thus $c_m = 0$. Thus by keeping any λ_k as the last we have $c_k = 0,\ k = 1, 2, \cdots, m$. Hence the lemma. □

When $A = [a_{ij}]_1^n$ admits distinct eigenvalues $\lambda_1 \neq \lambda_2 \neq \cdots \neq \lambda_n$ with respective eigenvectors $x_1, x_2, \cdots, x_n$, then by the previous lemma, the matrix $X = [x_1, x_2, \cdots, x_n]$ is non-singular and $AX = X\Lambda$ where Λ is the diagonal matrix with diagonal entries $\lambda_1, \lambda_2, \cdots, \lambda_n$. We say that matrix A has *simple structure.*

Extending to compound matrices we have $\mathcal{A}_p\mathcal{X}_p = \mathcal{X}_p\mathcal{L}_p$ $(1 \leq p \leq n)$ where $\mathcal{L}_p$ is a diagonal matrix of order $\binom{n}{p}$ whose diagonal entries are products of $\lambda_1, \cdots, \lambda_n$ taken p at a time. The eigenvalue $\lambda_{i_1} \cdots \lambda_{i_p}$ of $\mathcal{A}_p$ has as its eigenvector $X\begin{pmatrix} i_1 & \cdots & i_p \\ j_1 & \cdots & j_p \end{pmatrix}, 1 \leq i_1 < i_2 \cdots < i_p \leq n$. Since any matrix $A = [a_{ij}]_1^n$ is the limit of matrices $\{A_m\}$ of a simple structure for $m = 1, 2 \cdots$ we quote without proof the following:

THEOREM 2.21 (Kronecker). *Let $A = [a_{ij}]_1^n$ have $\lambda_1, \lambda_2, \cdots, \lambda_k$ as distinct eigenvalues with respective multiplicity $m_1, m_2, \cdots, m_k$. Then a complete system of eigenvalues for the compound matrix $\mathcal{A}_p$ consists of products of λ_i's chosen with $v_i \leq m_i$ $i = 1, 2 \cdots k$ such that $\sum v_i = p$.*

2.4. On sign changes of vectors. Let $u = (0, 3, 0, 0, -5, 7, 0, -9) \in \mathbb{R}^8$. With each non-zero co-ordinate we associate its sign as $+$ if it is > 0 and $-$ if it is negative. Discarding zero entries we create $\underline{u} = (3, -5, 7, -9)$. We have 3 sign changes and we denote by $S_u^- =$ sign changes in $\underline{u} = 3$.

By giving arbitrary signs to each 0 co-ordinates we can look for the maximum sign changes- Let $u^+ = (-, 3, -, +, -5, 7, -, -9)$ with $S(u^+) = 6$. Thus in general $S^+(u) \geq S^-(u)$. We say the vector u has *exact sign change* if and only if $S^+(u) = S^-(u)$.

LEMMA 2.22. *For any $u \in \mathbb{R}^n$, $S^+(u) = S^-(u)$ if and only if*

(i) $u_1 u_n \neq 0$

(ii) $u_i = 0$ *for any* $1 < i < n$ *implies* $u_{i-1}.u_{i+1} < 0$.

PROOF. (Sufficency): However, suppose $u_1 u_n \neq 0$ and $u_i = 0 \implies u_{i-1}u_{i+1} < 0$. Consider (u_{i-1}, u_i, u_{i+1}). By assumption it is $(+, 0, -)$ or $(-, 0, +)$. We get exactly one sign change in either case. Thus 0 co-ordinates have no additional effect on sign changes and $S_u^+ = S_u^-$.

(Necessity): Suppose $u_1 = 0$. Then let $u_k \neq 0$ for the first time. Thus $S^-(u_1, \cdots, u_n) = S^-(u_k, \cdots u_n)$. However $(u_1, u_2, \cdots u_{k-1})$ gets at least one more sign change beyond $S^-(u_k, \cdots u_n)$. Thus $S^+(u_1 \cdots u_n) \neq S^-(u_1 \cdots u_n)$. Similar argument applies when $u_n = 0$. Suppose $u_{i-1}u_{i+1} > 0$ for any $u_i = 0$. If $u_{i-1} > 0$, then $u_{i+1} > 0$ and we have $+\ 0\ +$ has 2 sign changes when 0 is given either $+$ or $-$ sign etc. Thus $S^+(u) > S^-(u)$. A similar argument applies when $u_{i-1} < 0$ and $u_{i+1} < 0$. □

2.5. On the eigenvalues of Oscillatory matrices.

THEOREM 2.23 (Gantmacher & Krein). *An oscillatory matrix* $A = [a_{ij}]_1^n$ *has only simple eigenvalues and they are all positive. If* $\overset{k}{u} = (u_{1k}, u_{2k}, \cdots, u_{nk})$ *is an eigenvector for the eigenvalue* λ_k *and if* $c_p, c_{p+1}, \cdots, c_q$ *are real scalars with* $\sum_p^q c_j^2 > 0$, *the number of sign changes of* $u = \sum_p^q c_k \overset{k}{u}$ *lies between* $p-1$ *and* $q-1$. *That is*

$$p - 1 \le S_u^- \le S_u^+ \le q - 1$$

In particular $\overset{k}{u}$ *has exactly* $k-1$ *sign changes for* $k = 1, 2, \cdots, n$. *(Here* $\lambda_1 > \lambda_2 \cdots > \lambda_n > 0$*). Further the nodes of 2 successive eigenvectors* $\overset{k}{u}$ *and* $\overset{k+1}{u}$ *for* $k = 1, 2, \cdots, n-1$ *alternate.*

PROOF. We will first deal with A, itself as a TP matrix. Clearly the elements of $\mathcal{A}_q$ are positive. By Kronecker's theorem, the eigenvalues of $\mathcal{A}_q$ are all possible products of eigenvalues of A taken q at a time. If $\lambda_1, \lambda_2, \cdots, \lambda_n$ are the eigenvalues of A, and when A is TP, w.l.g we can assume $|\lambda_1| \ge |\lambda_2| \ge \cdots \ge |\lambda_n|$. But by the Perron-Frobenius theorem of $\mathcal{A}_q$ the largest eigenvalue is $|\lambda_1\lambda_2\cdots\lambda_q| = \lambda_1\lambda_2\cdots\lambda_q$ ($\because \mathcal{A}_q > 0\ \forall q$). For $q = 2$, $\lambda_1\lambda_2 > 0$. But $\lambda_1 > 0 \implies \lambda_2 > 0$. Since for A, λ_1 is simple by the Perron-Frobenius theorem, we have $\lambda_1 > \lambda_2 > 0$. Inductively $\lambda_1 > \lambda_2 \cdots > \lambda_n > 0$.

Since by the Perron-Frobenius theorem all the roots of A are simple, the eigenvectors $\overset{k}{u}$ for λ_k are unique to within scalar multiples. Thus for the eigenvalue $\lambda_1\lambda_2\cdots\lambda_q$ of $\mathcal{A}_q$, the corresponding eigenvector has co-ordinates

$$U\begin{pmatrix} i_1 & i_2 & \cdots i_q \\ 1 & 2 & \cdots q \end{pmatrix} (1 \le i_1 < i_2 \cdots < i_q \le n),\ q = 1, 2 \cdots n.$$

They are all non-zero and are of the same sign. We can as well choose suitable scalar multiples and assume

$$U\begin{pmatrix} i_1 & i_2 & \cdots i_q \\ 1 & 2 & \cdots q \end{pmatrix} > 0\ (1 \le i_1 < i_2 \cdots < i_q \le n;\ q = 1, 2 \cdots n).$$

We claim that for any $(c_p \cdots, c_q) \ne 0$, the vector $u^0 = \sum_{k=p}^{q} c_k \overset{k}{u}$ has $S^+(u^0) \le q-1$. Suppose on the contrary $S^+(u^0) \ge q$. This means u^0 has $q+1$ co-ordinates $i_1, i_2, \cdots, i_{q+1}$ such that $u^0_{i_\alpha} u^0_{i_{\alpha+1}} \le 0\ \alpha = 1, 2 \cdots, q$. Since $(c_p \cdots, c_q) \ne 0$ and $U\begin{pmatrix} i_1 & i_2 & \cdots & i_q \\ 1 & 2 & \cdots & q \end{pmatrix} > 0$, $u^0_{i_\alpha} \ne 0$ for some α, $\alpha \in \{1, 2, \cdots, q\}$. Choosing $c_1 = \cdots = c_{p-1} = 0$ we have $\sum_{k=1}^{q} c_k \overset{k}{u}_{i_\alpha} = u^0_{i_\alpha}$ $\alpha = 1, 2, \cdots q$. Thus

$$\det \begin{bmatrix} u^0_{i_1} & u_{i_1,1} & \cdots & u_{i_1,q} \\ u^0_{i_2} & u_{i_2,1} & \cdots & u_{i_2,q} \\ \vdots & & & \\ u^0_{i_{q+1}} & u_{i_{q+1},1} & \cdots & u_{i_{q+1},q} \end{bmatrix} = 0.$$

When we expand this along the first column we get

$$\sum (-1)^{\alpha+1} u^0_{i_\alpha}\, U\begin{pmatrix} i_1 & i_2 & i_{\alpha-1} & i_{\alpha+1} \cdots & i_{q+1} \\ 1 & 2 & \cdots & \cdots & q \end{pmatrix} = 0.$$

However $(-1)^{\alpha+1}u^0_{i_\alpha}$ have the same sign for all $\alpha = 1, 2 \cdots q$ (This is so because $u^0_{i_\alpha}u^0_{i_{\alpha+1}} \leq 0, \alpha = 1, 2, \cdots q$). Thus all terms on the left side are of the same sign and at least one term is non-zero. We have a contradiction as R.H.S= 0. Thus $S^+(u^0) \leq q - 1$.

We need to show $S^-(u^0) \geq p - 1$. We proceed as follows:

Let $B = A^{-1} = [b_{ik}]^n_1$. Let $C = B^* = [(-1)^{i+k}b_{ik}]^n_1$. We know C is TP. Also we have $A\overset{k}{u} = \lambda_k\overset{k}{u} \implies B\overset{k}{u} = \frac{1}{\lambda_k}\overset{k}{u}$. If we define $\overset{k^*}{u} = (-u^k_1, u^k_2, -u^k_3, \cdots \pm u^k_n)$. Then $C\overset{k^*}{u} = \frac{1}{\lambda_k}\overset{k^*}{u}$ and $(\overset{n^*}{u}, \overset{n-1^*}{u}, \cdots, \overset{1^*}{u})$ is a complete system of eigenvectors of the totally nonnegative matrix C, for the eigenvalues $\frac{1}{\lambda_n} > \frac{1}{\lambda_{n-1}} \cdots > \frac{1}{\lambda_1}$. The role of q is now $n - (p - 1)$ and so $S^+_{u^*} \leq n - p$. When $u_i.u_{i+1} < 0$, $u^*_i.u^*_{i+1} > 0$ and conversely. Thus $S^-_u + S^+_{u^*} = n - 1$ and so $S^-_u \geq (n - 1) - (n - p) = p - 1$.

In particular $S^+_{u^k} \leq k - 1$ and $S^-_{u^k} \geq k - 1 \implies S^+_{u^k} = S^-_{u^k} = k - 1$. When A is not TP but only oscillatory we know A^h is totally positive (TP) for some power h and its pth compound matrix $\mathcal{A}^h_p$ is irreducible, nonnegative and primitive. Thus $\mathcal{A}_p$ has to be irreducible and primitive. We know A and A^h have the same eigenvectors. Thus the theorem goes through as long as A is oscillatory. □

2.6. Testing when a given matrix $A = [a_{ij}]^n_1$ is TN. Given any matrix $[B]_{m\times n}$, we can by adding 0' rows or 0' columns assume $[B]$ as a square matrix and if the original matrix is TN $\iff$ the padded matrix is TN and so we will assume $A = [a_{ij}]^n_1$. Consider the first column a_{i1} $i = 1, 2, \cdots, n$. If A is TN and if for some $k < l$, $a_{k1} = 0$ $a_{l1} > 0$, then we have

$$\det\begin{bmatrix} a_{k1} & a_{kj} \\ a_{l1} & a_{lj} \end{bmatrix} = -a_{kj}a_{l1} \geq 0 \text{ and } a_{kj} \geq 0 \implies a_{kj} = 0$$

for $j = 2, 3, \cdots n$. Thus it is a 0 row and such rows can be discarded. We can as well assume that for a TN matrix with non-zero rows the first column is of the type $a_{11} > 0,\ a_{21} > 0, .., a_{p1} > 0,\ a_{p+1,1} = 0, ..a_{n1} = 0.(0 < p \leq n)$. Dividing the ith row by a_{i1}, we may as well assume

$$A = \begin{bmatrix} 1 & \cdots & a_{1j} & \cdots \\ 1 & \cdots & a_{2j} & \cdots \\ 1 & \cdots & a_{pj} & \cdots \\ 0 & \cdots & a_{p+1j} & \cdots \\ \vdots & & \vdots & \\ 0 & \cdots & a_{nj} & \cdots \end{bmatrix}$$

subtracting from each row the previous row when the first entry is 1, we get a new matrix and the companion C^*

$$C = \begin{bmatrix} 1 & \cdots & a_{1j} & \cdots \\ 0 & \cdots & a_{2j} - a_{1j} & \cdots \\ 0 & \cdots & a_{3j} - a_{1j} & \cdots \\ 0 & \cdots & a_{p+1j} & \cdots \\ 0 & \cdots & a_{nj} & \cdots \end{bmatrix} \quad C^* = \begin{bmatrix} a_{1j} & \cdots & \cdots \\ a_{2j} - a_{1j} & \cdots & \cdots \\ a_{pj} - a_{p-1j} & \cdots & \cdots \\ \vdots & & \\ a_{nj} & \cdots & \cdots \end{bmatrix}$$

THEOREM 2.24. *The matrix A is TN if and only if C^* is TN*

PROOF. (Sufficiency): When C^* is TN, $a_{2j} - a_{1j} \geq 0$, $a_{pj} - a_{p-1,j} \geq 0$, $a_{p+1,j} \geq 0 \cdots a_{nj} \geq 0$ and that C is TN and we can recover A by adding *successively adjacent* *rows, and this operation always preserves TN.* Conversely we will have to show that C^* is TN when A is TN. This we show by splitting it into 3 cases.

Case (i) We will assume A is TP (Totally Positive) and show C^* is TP$_k$, $k \leq n-1$, namely all $k \times k$ submatrices of C^* are TP for $1 \leq k \leq n-1$. In this case the first column of A has all entries positive and so $p = n$. When we say C^* is TP$_k$ this means so for $k = 1, 2 \cdots, n-1$. We prove this by pegging on an induction on k. Since any second order minors of A are positive, the case $k = 1$ shows C^* is TP$_1$. Suppose $C^* \in \mathrm{TP}_{k-1}, 2 \leq k \leq n-1$ we will prove $C^* \in \mathrm{TP}_k$. For $2 \leq j_1 < j_2 \cdots < j_k \leq n$. Let $C^*[j_1 \cdots j_k]$ be the matrix composed of columns $j_1 \cdots j_k$ of C^*. Clearly $C^* \in \mathrm{TP}_k$ if all kth order minors of $C^*[j_1 \cdots j_k]$ are positive. Since by induction $C^* \in \mathrm{TP}_{k-1}$, all minors of order $k-1$ of $C^*[j_1 \cdots j_k]$ are positive. Moreover $C^*\begin{pmatrix} 1 & 2 & \cdots & k \\ j_1 & \cdots & \cdots & j_k \end{pmatrix} = A\begin{pmatrix} 1 & 2 & \cdots & k \\ j_1 & \cdots & \cdots & j_k \end{pmatrix} > 0$. Also $C^*\begin{pmatrix} i+1 & i+2 & \cdots & i+k \\ j_1 & j_2 & \cdots & j_k \end{pmatrix} = A\begin{pmatrix} i & i+2 & \cdots & i+k \\ 1 & j_1 & \cdots & j_k \end{pmatrix} > 0$ $i = 1, 2, \cdots n-k$. We can apply Fekete's lemma and conclude that all k-th order minors are positive.

Case (ii) Let $A \in \mathrm{TN}$ and $p = n$. We can approximate A by B, TP and $b_{ij} - a_{ij} < \epsilon$. A simple limiting argument establishes the assertion, after applying case(i) to the TP matrix B.

Case (iii) Let $A \in \mathrm{TN}$ and $0 < p < n$. We can by the addition of successive rows consider

$$A_1 = \begin{bmatrix} 1 & \\ 1 & \\ \vdots & \\ 1 & a_{pj} \\ 1 & a_{pj} + a_{p+1,j} \\ 1 & a_{pj} + a_{p+1,j} + a_{p+2,j} \\ \cdots & \\ 1 & a_{pj} + a_{p+1,j} + \cdots a_{nj} \end{bmatrix}$$

with companion C_1^*. We have $A_1 \in \mathrm{TN}$ and $C_1^* = C^*$. Hence $C^* \in \mathrm{TN}$.

□

2.7. Influence function of a string: Suppose a string of length L with tension T is fastened at the end points 0 and L. Suppose unit force is placed at a point s of the string. Let us identify the string with the X-axis and $\mathbb{S}$ as the point with coordinate $(s, 0)$. The application of the unit force tends to move all the movable points of the string in the same direction as the line of deflection created by the unit force at s, as in figure 1.

Let us denote by $k(s, s)$, the deflection f at $\mathbb{S}$. If we use the balancing of the unit force at the two ends OL of the string with tension T, we have by its projection on the Y-axis yield the balancing equation.

$$T(\sin\, \alpha + \sin\, \beta) = 1$$

For small values of α and β we can take

$$\sin\, \alpha = \frac{f}{s} \text{ and } \sin\, \beta = \frac{f}{L-s}.$$

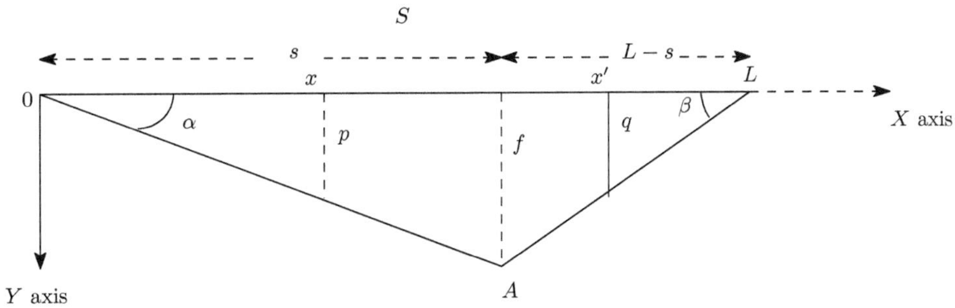

FIGURE 1

$$\text{Thus we find } f = \frac{s(L-s)}{TL}.$$

Using proportionality of sides of similar triangles we get the deflection at x due to unit force at s, given by

$$K(x,s) = p = \frac{p}{f} f = \frac{x}{s} . \frac{s(L-s)}{Tl} = \frac{x(L-s)}{TL} \text{ when } x < s.$$

We therefore have

$$\begin{aligned} K(x,s) &= \frac{x(L-s)}{TL} \text{ if } x \leq s \\ &= \frac{s(L-x)}{TL} \text{ if } x \geq s \end{aligned}$$

Suppose $0 < x_1 < x_2 \cdots < x_n < L$. Then

$$\begin{aligned} K(x_i,x_j) &= \frac{x_i}{TL}, (L-x_j) \text{ if } i \leq j \\ &= \frac{x_j}{TL}, (L-x_i) \text{ if } i \geq j. \end{aligned}$$

We call such matrices $A = (a_{ij})_1^n = (K(x_i,x_j))_{i,j=1,2,\cdots,n}$ *single pair matrices.*

The minors of single pair matrices can be computed by the following formula:

THEOREM 2.25. *Let $u_1, u_2 \cdots, u_n$ and $v_1, v_2, \cdots v_n$ be arbitrary reals. Then the single pair matrix*

$$\begin{aligned} A = [a_{ij}]_1^n &= u_i v_j \ \textit{for } i \leq j \\ &= u_j v_i \ \textit{for } i \geq j \end{aligned}$$

is a symmetric matrix. Further for

$$1 \leq i_1, j_1 < i_2, j_2 < \cdots < i_n, j_n \tag{2.4}$$

$$A\begin{pmatrix} i_1 & \cdots & i_p \\ j_1 & \cdots & j_p \end{pmatrix} = u_{\alpha_1} \det \begin{bmatrix} v_{\beta_1} & v_{\alpha_2} \\ u_{\beta_1} & u_{\alpha_2} \end{bmatrix} . \det \begin{bmatrix} v_{\beta_2} & v_{\alpha_3} \\ u_{\beta_2} & u_{\alpha_3} \end{bmatrix} \cdots \det \begin{bmatrix} v_{\beta_{p-1}} & v_{\alpha_p} \\ u_{\beta_{p-1}} & u_{\alpha_p} \end{bmatrix} .v_{\beta_p}$$

where $\alpha_t = \min(i_t, j_t)$ and $\beta_t = \max(i_t, j_t), t = 1, 2 \cdots \beta$.

In case $1 \leq i_1 < i_2 \cdots < i_p \leq n$ and $1 \leq j_1 < j_2 \cdots j_p \leq n$ alone is satisfied but not condition (2.4) *above then* $\begin{pmatrix} \delta_1 & \cdots & i_p \\ j_1 & \cdots & j_p \end{pmatrix} = 0.$

PROOF. Since A is a symmetric, matrix w.l.g we can assume $i_2 = \alpha_2 = \min(i_2, j_2)$ and $j_2 = \beta_2 = \max(i_2, j_2)$. The first 2 rows of the minor matrix $A\begin{bmatrix} i_1 \cdots i_p \\ j_1 \cdots j_p \end{bmatrix}$ can be written as

$$u_{\alpha_1} v_{\beta_1} \ u_{i_1} v_{j_2} \ u_{i_1} v_{j_3} \ \cdots \ u_{i_1} v_{j_p}$$

$$u_{j_1} v_{i_2} \ u_{i_2} v_{j_2} \ u_{i_2} v_{j_3} \ \cdots \ u_{i_2} v_{j_p}$$

We can evaluate $A\begin{pmatrix} i_1 \cdots i_p \\ j_1 \cdots j_p \end{pmatrix}$ after reducing the first row to $(*, 0, \cdots, 0)$ by subtracting a suitable multiple of row 2 from row 1. The scalar multiple to be chosen is $\frac{u_{i_1}}{u_{i_2}}$. We get

$$A\begin{pmatrix} i_1 & \cdots & i_p \\ j_1 & \cdots & j_p \end{pmatrix} = \left(u_{\alpha_1} v_{\beta_1} - \frac{u_{i_1}}{u_{i_2}} . u_{j_1} v_{i_2}\right) . A \begin{pmatrix} i_2 & \cdots & i_p \\ j_2 & \cdots & j_p \end{pmatrix}.$$

Observe $u_{i_1} u_{j_1} = u_{\alpha_1} u_{\beta_1}$ and so we get

$$\begin{aligned} u_{\alpha_1} v_{\beta_1} - \frac{u_{i_1}}{u_{i_2}} u_{j_1} v_{i_2} &= u_{\alpha_1} v_{\beta_1} - u_{\alpha_1} u_{\beta_1} . \frac{v_{i_2}}{u_{i_2}} \\ &= \frac{u_{\alpha_1}}{u_{\alpha_2}} \left[u_{\alpha_2} v_{\beta_1} - u_{\beta_1} u_{\alpha_2} . \frac{v_{\alpha_2}}{u_{\alpha_2}} \right] \\ &= \frac{u_{\alpha_1}}{u_{\alpha_2}} \det \begin{bmatrix} v_{\beta_1} & v_{\alpha_2} \\ u_{\beta_1} & u_{\alpha_2} \end{bmatrix} \end{aligned}$$

Inductively applying this to $A\begin{pmatrix} i_2 & \cdots & i_p \\ j_2 & \cdots & j_p \end{pmatrix}$ we get the required expression.

In case the condition $1 \le i_1, j_1 < i_2, j_2 \cdots < i_p, j_p \le n$ fails to hold while $1 \le i_1 < i_2 \cdots < i_p \le n$ and $1 \le j_1 < j_2 \cdots < j_p \le n$ holds, we have some i_t, j_t and $i_{t+1} j_{t+1}$ with $j_t > i_{t+1} > i_t$. We can write down the first 2 rows of $A\begin{bmatrix} i_t & i_{t+1} & \cdots & i_p \\ j_t & j_{t+1} & \cdots & j_p \end{bmatrix}$. They are $(u_{i_t} v_{j_t} \ u_{i_t} v_{j_{t+1}} \ \cdots \ u_{i_t} v_p)$ and $(u_{i_{t+1}} v_{j_t} \ u_{i_{t+1}} v_{j_{t+1}} \ \cdots \ u_{i_{t+1}} v_p)$. Thus the two rows are scalar multiple of each other and thus $A\begin{pmatrix} i_t & \cdots & i_p \\ j_t & \cdots & j_p \end{pmatrix} = 0$. □

COROLLARY 2.26. Suppose $u_1, u_2, \cdots, u_n, v_1, v_2, \cdots, v_n$ are non-zero. Then the single pair matrix A is TN if and only if, they all are either positive or negative and

$$\frac{u_1}{v_1} \le \frac{u_2}{v_2} \cdots \le \frac{u_n}{v_n}$$

The rank of the matrix A is $r + 1$ in case r among the above inequalities are strict.

2.8. Graph Theory and totally nonnegative matrices via Lindstrom-Gessel-Viennot Theorem. Consider the following digraph in Figure 2 with finitely many vertices with no loops, no cycles and no multiple a la carte edges. The graph G has vertices $V = \{v_1, v_2, \cdots, v_6\}$, edges $E = \{e_1, e_2 \cdots, e_7\}$. Let edges carry weights $w(e_1), \cdots, w(e_7)$ given in red. There are 3 paths from v_1 to v_6.

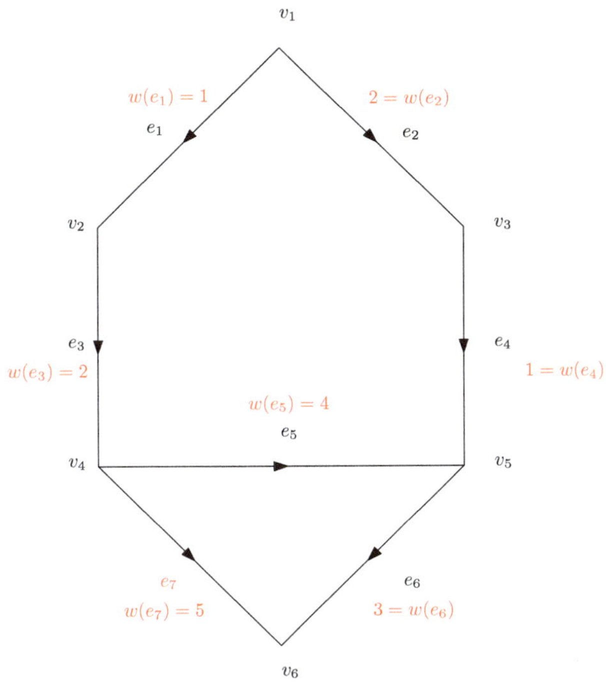

FIGURE 2

Path $P_1 :\ v_1\ e_2\ v_3\ e_4\ v_5\ e_6\ v_6$
$w(P_1) = w(e_2)w(e_4)w(e_6) = 2.1.3 = 6$
Path $P_2 :\ v_1\ e_1\ v_2\ e_3\ v_4\ e_7\ v_6$
$w(P_2) = 1.2.5 = 10$
Path $P_3 :\ v_1\ e_1\ v_2\ e_3\ v_4\ e_5\ v_5\ e_6\ v_6$
$w(P_3) = 1.2.4.3 = 24$

Here the weight of any path is the product of the edge weights along the path. Summing over all path's weights we get $w(v_1v_6) = 6 + 10 + 24 = 40$.

In general given any pair of vertices, say a_i and b_j of G, we define the matrix entry $M(a_i, b_j)$ as the weight sum of all paths from a_i to b_j. Given 2 subsets $\mathcal{A} = \{a_1, a_2 \cdots, a_n\} \subset V$ and $\mathcal{B} = \{b_1, \cdots, b_n\} \subset V$ (not necessarily disjoint), let $\mathcal{M}$ be the path matrix with row set as $\mathcal{A}$ and column set as $\mathcal{B}$.

A path system from $\mathcal{A}$ to $\mathcal{B}$ consists of a permutation σ of $\{1, 2, \cdots, n\}$ and n paths $P_1 : a_1 \to b_{\sigma(1)},\ P_2 : a_2 \to b_{\sigma(2)}, \cdots, P_n : a_n \to b_{\sigma(n)}$. The weight of the path system $\mathcal{P}$ is $W(\mathcal{P}) = \prod_{i=1}^{n} w(P_i)$. Here

$$w(P_i) = w(a_i, b_{\sigma(i)}) = \prod_{e \in P_i} w(e).$$

We define the weight for the trivial path $v \to v$ as $W(P) = 1$.

We are ready to state and prove the following.

THEOREM 2.27 (Lindström–Gessel–Viennot). *Let G be a directed acyclic graph with weight function $w : E \to \mathbb{R}$ and given $\mathcal{A} = \{a_1, a_2, \cdots, a_n\} \subset V$, $\mathcal{B} = \{b_1, \cdots, b_n\} \subset V$, let VD be the set of all vertex disjoint path systems (i.e no* *pair of paths of the system share a common vertex). Then*

$$\det M = \sum_{\mathcal{P} \in VD} \epsilon(\mathcal{P}) w(\mathcal{P}).$$

Here $\epsilon(\mathcal{P})$ corresponds to $\epsilon(\sigma)$, the sign of the permutation σ of the path $P_i : a_i \to b_{\sigma(i)}$ for each path $P_1, \cdots, P_n$ of the path system $\mathcal{P}$.

PROOF.

$$\det M = \sum_{\sigma} \epsilon(\sigma) \prod_{i=1}^{n} m_{i\sigma(i)}.$$

Here

$$\epsilon(\sigma) \prod_{i=1}^{n} m_{i\sigma(i)} = \epsilon(\sigma) \Big(\sum_{P_1 : a_1 \to b_{\sigma(1)}} w(P_1) \Big) \cdots \Big(\sum_{P_n : a_n \to b_{\sigma(n)}} w(P_n) \Big)$$

$$\text{and} \quad \det M = \sum_{\mathcal{P}} \epsilon(\mathcal{P}) w(\mathcal{P})$$

The above sum can include also paths that share a common vector. Let ND be those path systems which have at least 2 common vertices. If VD consists of vertex disjoint path then we can write

$$\sum_{\mathcal{P}} \epsilon(\mathcal{P}) w(\mathcal{P}) = \sum_{\mathcal{P} \in VD} \epsilon(\mathcal{P}) w(\mathcal{P}) + \sum_{\mathcal{P} \in ND} \epsilon(\mathcal{P}) w(\mathcal{P})$$

We will show that $\sum_{\mathcal{P} \in ND} \epsilon(\mathcal{P}) w(\mathcal{P}) = 0$.

Let $\mathcal{T} = (T_1, T_2, \cdots, T_n) \in ND$. Choose the least index i and the next least index $j > i$ such that T_i and T_j have a vertex in common. Let x be the earliest vertex of $T_i \cap T_j$. We can now split the path P_i with vertices

$T_i : \ a_i = u_0 \to u_1 \to x .. \to .. \to b_{\sigma(i)}$

$T_j : \ a_j = v_0 \to v_1 \to x .. \to .. \to b_{\sigma(j)}$

Let $l_i x$ be the sub path in T_i from $a_i = u_0$ to x.

Let $\gamma_i x$ be the sub path in T_i from x to $b_{\sigma(i)}$

Let $l_j x$ be the sub path in T_j from v_0 to x

Let $\gamma_j x$ be the sub path in T_j from x to $b_{\sigma(j)}$.

Observe $w(T_i) = w(l_i x) w(\gamma_i x)$ and $w(T_j) = w(l_j x) w(\gamma_j x)$. We replace the T_i, T_j by new paths H'_i, H'_j where $H'_i = l_i x$ followed by $\gamma_j x$ and $H'_j = l_j x$ followed by $\gamma_i x$

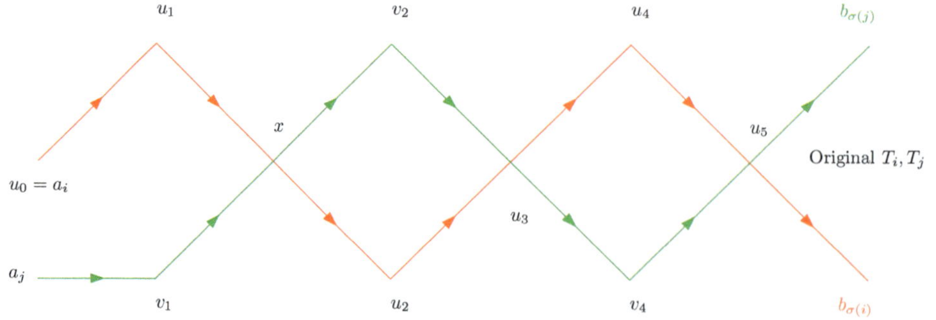

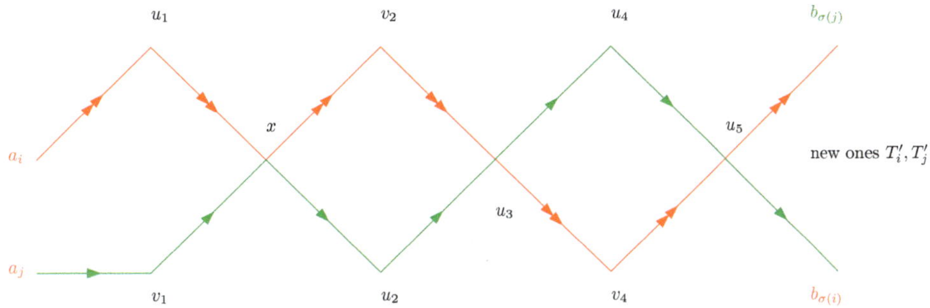

Let $\mathcal{H} = (H_1', H_2' \cdots H_i' \cdots H_j' \cdots H_n')$ be set $H_k' = T_k$ if $k \neq i, j$ $H_i' = T_i'$, $H_j' = T_j'$. If $\sigma' = \sigma \circ (i,j)$ we have $\epsilon(\sigma') = -\epsilon(\sigma)$. Notice $\prod_{\alpha=1}^{n} w(H_\alpha')$ is given by

$$\prod_{i=1}^{n} w(H_i') = \prod_{k \neq i,j} w(H_k').w(H_i')w(H_j').$$

Observe

$$\begin{aligned} w(H_i')w(H_j') &= w(l_i x)w(r_j x).w(l_j x).w(r_j x) \\ &= w(T_i)w(T_j). \end{aligned}$$

while $w(\mathcal{T}) = w(\mathcal{H}')$ $\sigma(\mathcal{T}) = -\sigma(\mathcal{H}')$. Thus $\sum_{\mathcal{P} \in ND} \epsilon(\mathcal{P})w(\mathcal{P}) = 0$ and $\det M = \sum_{\mathcal{P} \in VD} \epsilon(\mathcal{P})w(\mathcal{P})$.

We say a path system is non permutable whenever the only path disjoint system $P_i : \ a_i \to b_{\sigma(i)}$ $i = 1, 2, \cdots n$ correspond to σ as the identity permutate i.e P_i : $a_i \to b_i$ $i = 1, 2, \cdots n$. In that case

$$\det M = \sum_{\mathcal{P}} \epsilon(\mathcal{P})w(\mathcal{P}) = \sum_{\mathcal{P} \in VD} \epsilon(\mathcal{P})w(\mathcal{P})$$

When $w(e) \geq 0$ for all e, $\epsilon(\mathcal{P}) = +1$ for σ =identity permutation and thus $\det M \geq 0$. Since γ is an arbitrary choice this works for $n = 2, 3, \ldots$. i.e. M is totally non-negative. □

We will work out 2 examples, one with all weights of edges $\equiv 1$ and another example with different weights on edges.

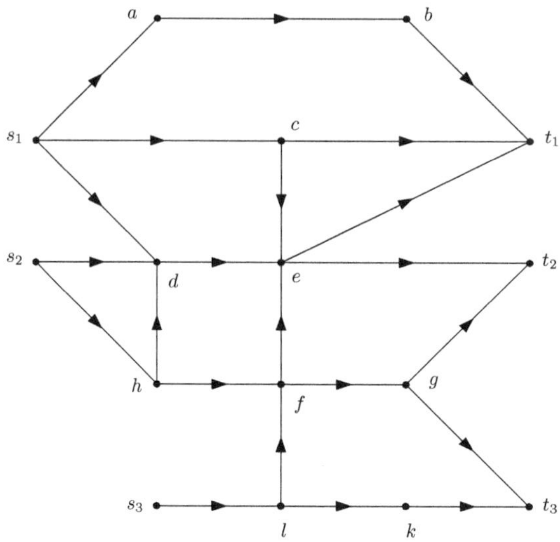

FIGURE 3

Paths from s_1 to t_1	Paths from s_1 to t_2	Paths from s_1 to t_3
$s_1\ a\ b\ t_1$	$s_1\ d\ e\ t_2$	none
$s_1\ c\ t_1$	$s_1\ c\ e\ t_2$	
$s_1\ c\ e\ t_1$		
$s_1\ d\ e\ t_1$		

Paths from s_2 to t_1	Paths from s_2 to t_2	Paths from s_2 to t_3
$s_2\ d\ e\ t_1$	$s_2\ d\ e\ t_2$	$s_2\ h\ f\ g\ t_3$
$s_2\ h\ d\ e\ t_1$	$s_2\ h\ d\ e\ t_2$	
$s_2\ h\ f\ e\ t_1$	$s_2\ h\ f\ e\ t_2$	
	$s_2\ h\ f\ g\ t_2$	

Paths from s_3 to t_1	Paths from s_3 to t_2	Paths from s_3 to t_3
$s_3\ l\ f\ e\ t_1$	$s_3\ l\ f\ e\ t_2$	$s_3\ l\ f\ g\ t_3$
	$s_3\ l\ f\ g\ t_2$	$s_3\ l\ k\ t_3$

The matrix M summarizes the number of paths from s_i to t_j as the ith row jth column entry given by

$$M = \begin{array}{c} \\ s_1 \\ s_2 \\ s_3 \end{array} \begin{array}{c} \begin{array}{ccc} t_1 & t_2 & t_3 \end{array} \\ \begin{bmatrix} 4 & 2 & 0 \\ 3 & 4 & 1 \\ 1 & 2 & 2 \end{bmatrix} \end{array}.$$

We have $\det M = 14$ and M is TN.

For our example there are 4 paths from s_1 to t_1 and four paths from s_2 to t_2 and two paths from s_3 to t_3. These paths are

$\xi_1 = s_1\ a\ b\ t_1$ $\qquad \eta_1 = s_2\ d\ e\ t_2$ $\qquad \zeta_1 = s_3\ l\ f\ g\ t_3$
$\xi_2 = s_1\ c\ t_1$ $\qquad \eta_2 = s_2\ h\ d\ e\ t_2$ $\qquad \zeta_2 = s_3\ l\ k\ t_3$
$\xi_3 = s_1\ c\ e\ t_1$ $\qquad \eta_3 = s_2\ h\ f\ e\ t_2$
$\xi_4 = s_1\ d\ e\ t_1$ $\qquad \eta_4 = s_2\ h\ f\ g\ t_2$

The following are the only paths from s_1 to t_1; s_2 to t_2 and s_3 to t_3 that have no common vertex. They are

$\xi_1,\ \eta_1,\ \zeta_1$ $\qquad \xi_2,\ \eta_1,\ \zeta_1$ $\qquad \xi_3,\ \eta_4,\ \zeta_2$
$\xi_1,\ \eta_1,\ \zeta_2$ $\qquad \xi_2,\ \eta_1,\ \zeta_2$ $\qquad \xi_2,\ \eta_4,\ \zeta_2$
$\xi_1,\ \eta_2,\ \zeta_1$ $\qquad \xi_2,\ \eta_2,\ \zeta_1$
$\xi_1,\ \eta_2,\ \zeta_2$ $\qquad \xi_2,\ \eta_2,\ \zeta_2$
$\xi_1,\ \eta_3,\ \zeta_2$ $\qquad \xi_2,\ \eta_3,\ \zeta_2$
$\xi_1,\ \eta_4,\ \zeta_2$ $\qquad \xi_2,\ \eta_4,\ \zeta_2$

Thus we have 14 disjoint path systems and it coincides with $\det M$.

EXAMPLE 2.28. A weighted graphs with non-negative weights.

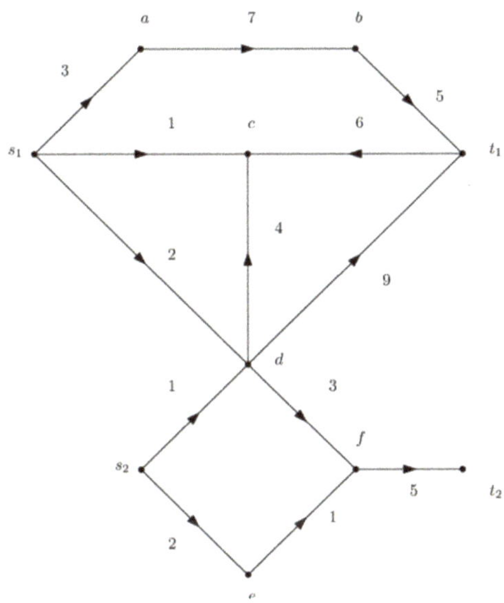

FIGURE 4

Paths and their weights

$s_1\ a\ b\ t_1$ wt$= (3)(7)(5) = 105$ $\qquad s_1\ d\ f\ t_2$ wt$= 30$
$s_1\ c\ t_1$ wt$= 6$ $\qquad s_2\ d\ t_1$ wt$= 9$
$s_1\ d\ c\ t_1$ wt$= 48$ $\qquad s_2\ d\ c\ t_1$ wt$= 24$
$s_1\ a\ t_1$ wt$= 18$

$$M = \begin{bmatrix} 177 & 30 \\ 33 & 25 \end{bmatrix} \quad \det M = 3435.$$

The path disjoint system with weights

(105)$s_1\ a\ b\ t_1$	(105) $s_1\ a\ b\ t_1$	(6) $s_1\ c\ t_1$	(8)$s_1\ c\ t_1$	(48)$s_1\ d\ c\ t_1$	(18)$s_1\ d\ t_1$
(15)$s_2\ d\ f\ t_2$	(10)$s_2\ e\ f\ t_2$	(15)$s_2\ d\ f\ t_2$	(10) $s_2\ e\ f\ t_2$	(10) $s_2\ e\ f\ t_2$	(10) $s_2\ e\ f\ t_2$

$$(105)(15) + (105)(10) + (6)(15) + (6)(10) + (48)(10) + (18)(10) = 3435.$$

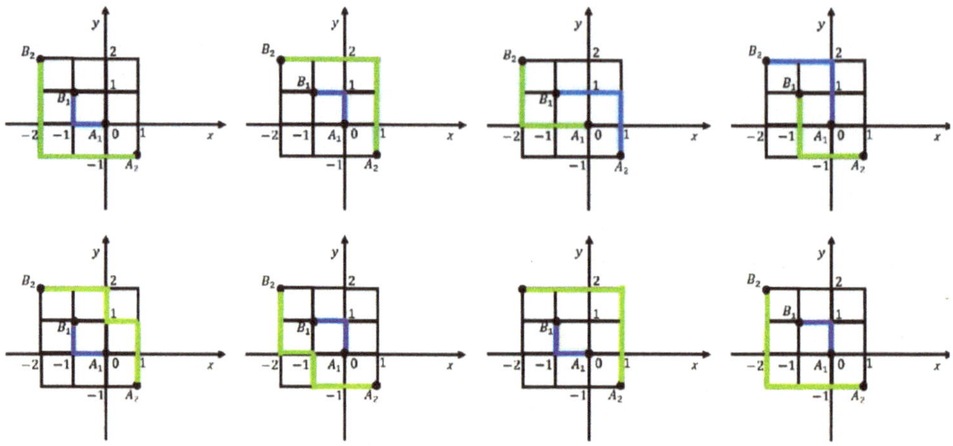

FIGURE 5

Unlike the previous example where we had only identity permutation as the only possible kind of disjoint path system. The example below has 6 identity permutation disjoint path systems and 2 permutations that are transpositions with disjoint path system. While $E(\text{identity}) = +1$, $\epsilon(\text{transposition}) = -1$ and we get $\det M = 6 - 2 = 4$. In fact for this graph you always move either from south to north or from east to west.

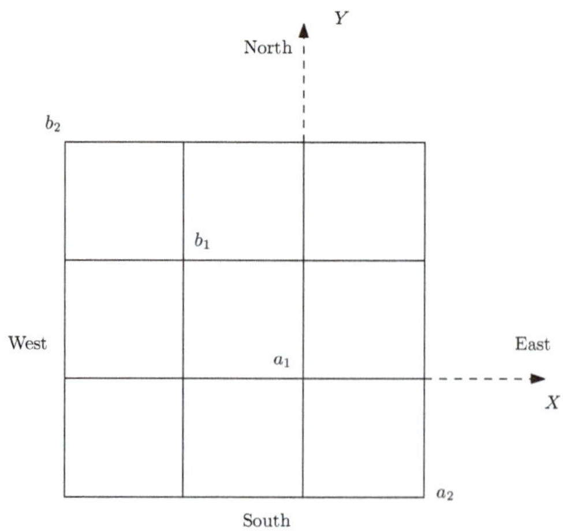

FIGURE 6

The weight matrix is

$$M = \begin{matrix} & b_1 & b_2 \\ a_1 & w(a_1\, b_1) & w(a_1\, b_2) \\ a_2 & w(a_2\, b_1) & w(a_2\, b_2) \end{matrix} = \begin{bmatrix} 2 & 6 \\ 6 & 20 \end{bmatrix}.$$

3. Epilogue

Ever since my graduate student days in I.S.I., Calcutta, (1962-66), the attraction to connect minimax theorem and the Perron-Frobenius theorem and the combinatorial aspects of nonnegative matrices all finally lead me to write a monograph jointly with R.B. Bapat on the spectral as well as on the combinatorial aspects of nonnegative matrices. Even then, my interest was to connect and extend further the theorems of Gantmacher, and Krein on totally positive matrices and look for game theoretic applications.

The theorems of Perron and Frobenius and the identities of Sylvester on determinants, play a fundamental role, once we appreciate the notion of compound matrices and the application of Sylvester's determinantal identities to minors of the original matrix viewed as elements of compound matrices. The seminal book of Gantmacher and Krein on Small oscillations of matrices and vibrations in mechanical systems was available only in Russian and German translations. Ignorance of languages other than just English becomes the biggest handicap in knowledge transfer. Only an analogy can explain such problems.

The philosophical depth and originality of the Upanishads were passed on to scholars in the Middle East and later to Western scholars only by the translation of the Sanskrit works into Persian language. Western scholars were not familiar with Sanskrit language , but were familiar with the Persian language. In deed it was all initiated by Prince Dhara Sukha, the eldest son of the famous Mogul emperor, Shahjahan.

While the spectral aspects of totally positive matrices were further extended to totally positive operators by Karlin and McGregor, the sign variation aspects of the range of totally nonnegative matrices were exploited by Schoenberg and his coresearchers in USA. The first combinatorial connections via Matroids by Lindstrom were further expounded by Gessel and Viennot and this is also related to the earlier work of Karlin and McGregor in Coincidence probability.

With minimal matrix prerequisites, I have closely followed only the approaches of Gantmacher and Krein in many sections. I could not get into the deeper approaches of Schoenburg based on powerful complex analysis arguments to totally positive kernels. Schoenburg's remarkable observation that a function transformed by a totally positive kernel has fewer changes of sign than the original function has apparently far-reaching implications in differential equations. I have included only the discrete version of the same. I have included some references that can hopefully trigger some new directions in the study of totally positive operators.

References

[1] T. Ando, *Totally positive matrices*, Linear Algebra Appl. **90** (1987), 165–219, DOI 10.1016/0024-3795(87)90313-2. MR884118

[2] Bapat, R.B. and T.E.S. Raghavan, *Nonnegative Matrices and Applications*, Cambridge University Press, Cambridge (1996).

[3] Brenti F., *Combinatorics and Total Positivity*, J. Combinatorial Theory, Series A, **71**, pp175-218 (1995).

[4] L. D. Brown, I. M. Johnstone, and K. B. MacGibbon, *Variation diminishing transformations: a direct approach to total positivity and its statistical applications*, J. Amer. Statist. Assoc. **76** (1981), no. 376, 824–832. MR650893

[5] R. A. Brualdi and H. Schneider, *Determinantal identities: Gauss, Schur, Cauchy, Sylvester, Kronecker, Jacobi, Binet, Laplace, Muir, and Cayley*, Linear Algebra Appl. **52/53** (1983), 769–791, DOI 10.1016/0024-3795(83)80049-4. MR1500275

[6] P. N. Choudhury, M. R. Kannan, and A. Khare, *Sign non-reversal property for totally non-negative and totally positive matrices, and testing total positivity of their interval hull*, Bull. Lond. Math. Soc. **53** (2021), no. 4, 981–990, DOI 10.1112/blms.12475. MR4311813

[7] C. W. Cryer, *Some properties of totally positive matrices*, Linear Algebra Appl. **15** (1976), no. 1, 1–25, DOI 10.1016/0024-3795(76)90076-8. MR450303

[8] S. M. Fallat, M. I. Gekhtman, and C. R. Johnson, *Multiplicative principal-minor inequalities for totally nonnegative matrices*, Adv. in Appl. Math. **30** (2003), no. 3, 442–470, DOI 10.1016/S0196-8858(02)00506-7. MR1973953

[9] Ky. Fan, *Fixed-point and minimax theorems in locally convex topological linear spaces*, Proc. Nat. Acad. Sci. U.S.A. **38** (1952), 121–126, DOI 10.1073/pnas.38.2.121. MR47317

[10] Fekete M., and G. Polya, *Uberein Problem von Laguerre*, Rendi. C.M. Palermo 34, pp 89-120 (1912).

[11] S. Fomin and A. Zelevinsky, *Total positivity: tests and parametrizations*, Math. Intelligencer **22** (2000), no. 1, 23–33, DOI 10.1007/BF03024444. MR1745560

[12] Gantmacher F. R., *The Theory of Matrices*, Chelsea, New York, 2 volumes (1959).

[13] Gantmacher F.R., and M.G. Krein, *Oscillation Matrices and Kernels and Small Vibrations of Mechanical Systems*, AMS Chelsea Publ. (2002).

[14] Gessel, I.M and X.G. Viennot, Determinants, paths and plane partitions (1989) preprint.

[15] S. Karlin and J. McGregor, *Coincidence probabilities*, Pacific J. Math. **9** (1959), 1141–1164. MR114248

[16] Karlin, S., *Total Positivity*, vol 1. Stanford University Press, CA (1968).

[17] M. G. Kreĭn and M. A. Rutman, *Linear operators leaving invariant a cone in a Banach space*, Amer. Math. Soc. Translation **1950** (1950), no. 26, 128. MR38008

[18] Koteljanskii, D.M., *Some sufficient conditions for reality and simplicity of the spectrum of a matrix*, Amer. Math Soc. Translations Series 2, 27, pp35-42 (1955).

[19] Loewner, K., *On totally positive matrices*, Math. Z., 63, pp338-340. (1955).

[20] B. Lindström, *On the vector representations of induced matroids*, Bull. London Math. Soc. **5** (1973), 85–90, DOI 10.1112/blms/5.1.85. MR335313

[21] Pinkus, A., *Totally Positive Matrices*, Cambridge University Press, (2010).

[22] I. J. Schoenberg, *On Pólya frequency functions. I. The totally positive functions and their Laplace transforms* (English, with Hebrew summary), J. Analyse Math. **1** (1951), 331–374, DOI 10.1007/BF02790092. MR47732

[23] Raghavan, K.N., *Strums Method for the number of real roots of a real polynomial*, Institute of Mathematical Sciences (India) workshop for school Mathematics Teachers, Sept (2016).

[24] Raghavan, T.E.S., *Extensions of the Theory of Positive Operators and their Relationship to Minimax Games*, PhD Thesis, Indian Statistical Institute (1968).

[25] T. E. S. Raghavan, *On linear operators leaving a convex set invariant in Banach spaces*, Sankhyā Ser. A **27** (1965), 293–302. MR203483

[26] Raghavan, T.E.S., *On Linear Operators leaving invariant a convex set in normed linear spaces*, Mathematika, pp 57-62, (1970).

[27] I. J. Schoenberg and A. Whitney, *On Pólya frequence functions. III. The positivity of translation determinants with an application to the interpolation problem by spline curves*, Trans. Amer. Math. Soc. **74** (1953), 246–259, DOI 10.2307/1990881. MR53177

[28] Shiffman, M., *War duels*, Rand Corporation, Santa Monica, CA (1949).

[29] A. M. Whitney, *A reduction theorem for totally positive matrices* (English, with Hebrew summary), J. Analyse Math. **2** (1952), 88–92, DOI 10.1007/BF02786969. MR53173

Deaprtment of Mathematics, Statistics, and Computer Science, University of Illinois at Chicago, Chicago, Illinois

Email address: terctu@gmail.com

Contemporary Mathematics
Volume **833**, 2026
https://doi.org/10.1090/conm/833/16672

Vertical block matrices and vertical linear complementarity problem

Punit Kumar Yadav, Sajal Ghosh, and S. K. Neogy

ABSTRACT. The vertical block Z and hidden-Z matrices naturally arise in the study of the Vertical Linear Complementarity Problem (VLCP). In particular, hidden-Z matrices provide a foundation for defining hidden-K matrices. Similarly, vertical block N matrices and copositive matrices extend the concepts of N-matrices and Cottle-copositive matrices in a block framework. In this paper, we collect result related to the vertical block matrices. We also explore several classes of vertical block matrices, including vertical block Z, vertical hidden-Z, and Cottle-copositive matrices, and provide a thorough characterization of these matrix classes within the context of VLCP. Furthermore, we investigate the applicability of the Cottle-Dantzig Algorithm for solving VLCP.

1. Introduction

The LCP plays a crucial part in the study of mathematical optimization and computational mathematics. In the 1960s, LCP was developed for use in fields like game theory, computer science, mechanical engineering, and economics, showcasing its utility and wide-ranging influence. The formal definition of LCP is as follows: For a order n real matrix M and a vector $q \in \mathbb{R}^n$, LCP(M, q), is to find vector $u \in \mathbb{R}^n$ such that

$$u \geq 0, \;\; v := Mu + q \geq 0 \text{ and } u * v = 0, \tag{1.1}$$

where $*$ is a pointwise product. Cottle and Pang's monograph [**1**] is widely regarded as the standard reference for the linear complementarity problem (LCP). More on LCP can be found in [**5**, **25**, **29**]. To investigate the structure of the LCP's solution set, various classes of matrices have been extensively studied (see [**2**, **8**, **10**, **15**, **24**, **28**, **32**]). Several important generalizations of the LCP have also been developed, including the horizontal linear complementarity problem (HLCP), the extended HLCP (EHLCP) [**3**], and the vertical linear complementarity problem (VLCP) [**4**, **18**, **22**, **23**]. The concept of a vertical block matrix was introduced to study of the VLCP. Vertical block matrices frequently arise in the study of stochastic games, where the columns represent states and the rows within each block correspond to the actions available in a particular state (see [**17**–**20**]). Similar to the LCP, a variety of block matrix classes have been defined to analyze the VLCP. We now

2020 *Mathematics Subject Classification.* Primary 54C40, 14E20; Secondary 46E25, 20C20.
Key words and phrases. Vertcial Z-matrix, vertical block hidden-Z matrix, vertical linear complementarity problem, Cottle-Dantzig algorithm.

recall the notion of a vertical block matrix, originally introduced by Cottle and Dantzig [**4**], in the context of this generalization. Consider a rectangular matrix A of order $m \times k$ with $m \geq k$. Suppose A is partitioned row-wise into k blocks in the form

$$A = \begin{bmatrix} A^1 \\ A^2 \\ \vdots \\ A^k \end{bmatrix},$$

where each $A^j = (a^j_{rs}) \in R^{m_j \times k}$ with $\sum_{j=1}^k m_j = m$. Then A is called a *vertical block matrix of the type* $(m_1, \ldots, m_k)$. If $m_j = 1, \ j = 1, \ldots, k$, then A is a square matrix. Thus, a vertical block matrix can be viewed as a natural extension of a square matrix.

We now present a generalization of the LCP introduced by Cottle and Dantzig [**4**], which incorporates vertical block matrices. This generalized formulation is known as the VLCP, and it is stated as follows:

Given a vertical block matrix A of type $(m_1, \ldots, m_k)$ and a vector $q \in R^m$, the *vertical linear complementarity problem* (VLCP) is to find $w \in R^m$ and $z \in R^k$ such that

$$w - Az = q, \ w \geq 0, \ z \geq 0$$

$$z_j \prod_{i=1}^{m_j} w_i^j = 0, \text{ for } j = 1, 2, \ldots, k.$$

This problem is denoted as VLCP (q, A). For details on VLCP see [**22, 23, 30**] and references therein.

A submatrix of size k of A is called a *representative submatrix* if its ith row is drawn from the ith block A^i of A. Clearly, a vertical block matrix of type $(m_1, m_2, \ldots, m_k)$ has at most $\prod_{j=1}^k m_j$ distinct representative submatrices.
The various classes of vertical block matrices have been generalized analyze the solution set of VLCP. In this article, we consolidate several results related to vertical block Z, hidden-Z and hidden-K matrices, as well as vertical block N and copositive matrices. Lastly, we present the Cottle-Dantzig algorithm as a solution approach for VLCP and a discussion on bilinear games.

2. Notations and Preliminaries

For a matrix $A \in \mathbb{R}^{m \times k}$, $A_{\cdot j}$ denotes the jth column of A, and $A_{i\cdot}$ denotes the ith row of A. Pos(A) represents the nonnegative cone generated by the columns of A. Any vector $x \in \mathbb{R}^n$ is assumed to be a column vector, and $I_{\cdot j}$ denotes the vector whose jth coordinate is 1 and all other coordinates are 0. For any positive integer n, if $\alpha \subseteq \{1, 2, \ldots, n\}$, then $\bar{\alpha}$ denotes the complement of α in $\{1, 2, \ldots, n\}$. If M is a square matrix of order n, and α, β are two nonempty subsets of $\{1, 2, \ldots, n\}$, then $M_{\alpha\beta}$ denotes the submatrix of M consisting of the rows and columns indexed by α and β, respectively. Similarly, $M_{\cdot\alpha}$ denotes the submatrix consisting of the columns of M indexed by α, and $M_{\alpha\cdot}$ denotes the submatrix consisting of the rows of M indexed by α.

DEFINITION. A matrix $M \in \mathbb{R}^{n\times n}$ is said to be in the class

$\mathcal{S}$	if there exist a vector $x > 0$ such that $Mx > 0$.
P	if all its principal minors are positive.
P_0	if all its principal minors are nonnegative.
N	if all its principal minors are negative.
N_0	if all its principal minors are nonpositive.
R_0	if LCP$(0, M)$ has only the trivial solution.
Z	if all diagonal entries of matrix M are non-positive.
Copositive (C_0)	if $\forall x \geq 0 \Rightarrow x^T Mx \geq 0$.
Strictly Copositive (C)	if $0 \neq \forall x \geq 0 \Rightarrow x^T Mx > 0$.
Copositive Plus (C_0^+)	if $M \in C_0$ and $[x^T Mx = 0,\ x \geq 0]$ $\Rightarrow [(M + M^T)x = 0]$
Copositive Star (C_0^*)	if $M \in C_0$ and $[x^T Mx = 0,\ Mx \geq 0,\ x \geq 0]$ $\Rightarrow [M^T x \leq 0]$

A square submatrix of $A \in \mathbb{R}^{m\times k}$ of order k is called a *representative submatrix*, if its jth row is drawn from the jth block A^j of A, for $j = 1, \ldots, k$. Clearly, a vertical block matrix of type $(m_1, \ldots, m_k)$ has at most $\prod_{j=1}^{k} m_j$ distinct representative submatrices.

A vertical block matrix A of type $(m_1, \ldots, m_k)$ is called a *vertical block* $P\,(P_0)$-*matrix*, if all its representative submatrices are $P\,(P_0)$-matrices. A vertical block matrix A of type $(m_1, \ldots, m_k)$ is called a *vertical block* Q-*matrix* if VLCP(q, A) has a solution for each $q \in R^m$. We conclude this section with following index sets for later use:

$$J_1 = \{1, 2, \ldots, m_1\}$$

and for $2 \leq i \leq k$,

$$J_i = \left\{ \sum_{j=1}^{i-1} m_j + 1, \sum_{j=1}^{i-1} m_j + 2, \ldots, \sum_{j=1}^{i} m_j \right\}.$$

2.1. An Equivalent Formulation. The problem VLCP(q, A) can be formulated as LCP(q, M) as follows:

Let us consider a vertical block matrix A of type $(m_1, m_2, \ldots, m_k)$, where m_j denotes the size of the jth block. We construct a new matrix M by replicating the jth column block $A_{\cdot j}$ exactly m_j times, for each $j = 1, 2, \ldots, k$. That is, $A_{\cdot 1}$ is repeated m_1 times, $A_{\cdot 2}$ is repeated m_2 times, and so on.

Formally, for all $p \in J_s$, we define:

$$M_{\cdot p} = A_{\cdot s}.$$

Through this process, we obtain a square matrix M of order $m \times m$, where $m = \sum_{j=1}^{k} m_j$. This matrix M is referred to as the equivalent square matrix of A.

LEMMA 2.1. *Given the VLCP(q, A), let M be the equivalent square matrix of A. VLCP(q, A) has a solution if and only if LCP(q, M) has a solution.*

PROOF. Suppose (w, z) is a solution to VLCP(q, A). We construct a solution (u, v) to LCP(q, M) as follows:

We take $u = w$. Note that $z_j > 0$ implies $\exists\ p(j) \in J_j$ such that $w_{p(j)} = 0$. Define

$$v_r = \begin{cases} 0, & \text{if } r \neq p(j) \text{ for any } 1 \leq j \leq k \\ z_j, & \text{if } \exists\ a\ j,\ 1 \leq j \leq k, \text{ such that } r = p(j) \end{cases}$$

Note that v_r is well defined. Now it easy to see that (u, v) solves LCP(q, M).

Conversely, suppose (u, v) is a solution to the LCP(q, M). Define the vector $z \in R^k$ by taking

$$z_j = \sum_{i \in J_j} v_i.$$

Note that if $z_j > 0$, $\exists\ i \in J_j$ such that $v_i > 0$ and hence $u_i = 0$. Hence with $w = u$, (w, z) solves VLCP(q, A). □

We shall henceforth refer to the problem LCP(q, M), where M is the equivalent square matrix corresponding to the vertical block matrix A of type $(m_1, \ldots, m_k)$, as the equivalent LCP.

3. The class of the vertical block Z- and P-matrices

The study of square Z-matrices has been motivated by their wide-ranging applications across various fields, as discussed in [**1**]. The Z-matrix serves as a bridge between several important matrix classes, including P, P_0, SSM, and Q-matrices. The Z-matrix property for vertical block matrices was introduced in [**6**]. In this section, we examine vertical block Z- and P-matrices.

DEFINITION 3.1 ([**6**]). We say a vertical block matrix A of type $(m_1, \ldots, m_k)$ is a vertical block Z-matrix if its all representative submatrices are Z-matrices.

To proceed further, we recall a following definition from [**6**].

DEFINITION 3.2. Let A be a vertical block matrix of type $(m_1, \ldots, m_k)$. For each $j = 1, \ldots, k$, let u^j be a nonzero nonnegative row vector of dimension $1 \times m_j$. Let U be a matrix defined by

$$U = \begin{bmatrix} u^1 & \cdots & 0 \\ \vdots & \ddots & \vdots \\ 0 & \cdots & u^k \end{bmatrix}. \tag{3.1}$$

Thus, U is a nonzero nonnegative matrix of order $k \times m$, and it is the direct sum of the vectors u^j.

The following result connects vertical block Z and square Z matrices.

THEOREM 3.3 ([**6**]). *Let A be a vertical block Z-matrix of type $(m_1, \ldots, m_k)$. If U is define as above, then $B = UA$ is a square Z-matrix.*

THEOREM 3.4 ([**6**]). *Let A be a vertical block Z-matrix of type $(m_1, \ldots, m_k)$. Suppose there exists nonzero vectors $y \in \mathbb{R}^m_+$ and $x \in \mathbb{R}^k_+$, such that $Ax \geq 0$ and $y^t Ax = 0$. Then $Ax = 0$.*

PROOF. Assume U is a nonnegative matrix of order $k \times m$ as defined in Definition 3.2. Then, from Theorem 3.3, $B = UA$ is a square Z-matrix of order k. Suppose $Ax \neq 0$. As $Ax \geq 0$, $Bx = UAx \geq 0, \neq 0$. This implies $x^t Bx \geq 0, \neq 0$.

But $x^tBx = x^tUAx = y^t(Ax) = 0$, where $y^t = (x^tU)^t \in \mathbb{R}^m$ and $y \geq 0$, $y \neq 0$. This gives us a contradiction. Hence $Ax = 0$. $\square$

The followings results characterize the vertical block P and P_0 matrices.

THEOREM 3.5 ([**6**]). *Let A be a vertical block Z-matrix of type $(m_1, \ldots, m_k)$. Then the following are equivalent.*

(i) *A be a vertical block P-matrix of type $(m_1, \ldots, m_k)$*
(ii) *There exists a vector $x \geq 0$ such that $Ax \geq 0$.*

THEOREM 3.6 ([**6**]). *Let A be a vertical block Z-matrix of type $(m_1, \ldots, m_k)$. Then the following are equivalent.*

(i) *The eigenvalues of all principal submatrices of A are nonnegative.*
(ii) *All principal minors of A are nonnegative.*
(iii) *A is a P_0-matrix.*
(iv) *$A + \epsilon E$ is a vertical block P-matrix for all $\epsilon > 0$.*

The next corollary is a direct consequences of above theorem.

COROLLARY 3.7 ([**7**]). *Let A be a vertical block Z-matrix of type $(m_1, \ldots, m_k)$. If there exists $x > 0$ such that $Ax > 0$, then A is a P_0-matrix.*

PROOF. Let C be a representative matrix of A. Then $Cx \geq 0$, $x > 0$. Therefore for each $\epsilon > 0$, $(C + \epsilon I)x > 0$. By Theorem 3.6, C is a P_0-matrix, This completes our proof. $\square$

3.1. The class of the vertical block P-matrices. Now we state some more results for the vertical block P-matrix for studying the hidden K-matrices in next section. A vertical block matrix $A \in \mathbb{R}^{m\times k}$ of type $(m_1, \ldots, m_k)$ is said to be strictly vertical if $m > k$. In this section and subsequent sections, a vertical block matrix will imply a strictly vertical block matrix. To proceed further first we look into some vertical block P-matrx properties. All the result of the section are from [**22**].

LEMMA 3.8. *Let* VLCP(q, A) *has a finite number of solutions for all $q \in \mathbb{R}^m$. Then, each of the principal submatrix of any representative submatrix of A is nonsingular.*

Next, we provide a relation between vertical block matrix and the uniqueness of the solution fo VLCP.

THEOREM 3.9. *A is a vertical block P-matrix of type $(m_1, \ldots, m_k)$ if and only if* VLCP(q, A) *has a unique solution for each $q \in \mathbb{R}^m$.*

THEOREM 3.10. *Let $A \in \mathbb{R}^{m\times k}$ be a vertical block matrix of type $(m_1, \ldots, m_k)$. The following are equivalent:*

(i) A is a vertical block P-matrix.

(ii) The equivalent matrix M of A has $(m - k)$ zero eigenvalues and among the remaining k eigenvalues all real eigenvalues are positive. Also eigenvalues of all nontrivial principal submatrices are positive.

THEOREM 3.11. *If A be a vertical block P-matrix of type $(m_1, \ldots, m_k)$, then the equivalent matrix M is a P_0-matrix and for each index set with* $\det(M_{\alpha\alpha}) = 0$, *the columns of $M_{\alpha\alpha}$ are linearly dependent.*

PROOF. As M is a P_0-matrix with all its nontrivial principal minors positive, $\det(M_{\alpha\alpha}) = 0$ for all trivial principal submatrix of $M_{\alpha\alpha}$. Thus, the columns of such principal submatrix are linearly dependent. This completes our proof. □

For the sign nonreversal property related to a given vertical block P-matrix A, we have the following theorem.

THEOREM 3.12. *Let A be an $m \times k$ vertical block matrix of type $(m_1, \ldots, m_k)$. Let M be the equivalent matrix corresponding to A. Then, A is a vertical block P-matrix if and only if M does not reverse the sign of any vector $u \in \mathbb{R}^m$ other than those belonging to the set*

$$S = \left\{ v | v \in \mathbb{R}^m \mid \sum_{r \in J_i} v_r = 0, 1 \leq i \leq k \right\}.$$

4. The class of vertical block hidden Z-matrices

A real matrix M of order n is called a hidden Z-matrix if there exist matrices $X, Y \in \mathbb{R}^{n \times n}$ such that: (i) $MX = Y$, and (ii) there exist vectors $r, s \in \mathbb{R}^n_+$ such that $r^T X + s^T Y > 0$. The class of hidden Z-matrices also plays an important role in solving complementarity problems. For related results, see [**1**]. Mohan and Neogy [**21**] introduced the concept of vertical block hidden Z-matrices, and examined their minimality and complementarity properties. They identified the largest class of vertical block matrices for which the associated VLCP possesses the least element property, thereby allowing it to be solved as a single linear programming problem.

In this section, we present the class of vertical block Z-matrices and state some key results related to the VLCP. All results discussed in this section can be found in [**21**].

DEFINITION 4.1 ([**21**]). Let $A \in \mathbb{R}^{m \times k}$ be a vertical block matrix of type $(m_1, m_2, \ldots, m_k)$. A is called a vertical block hidden Z-matrix if there exists a Z-matrix $X = ((x_{ij})) \in \mathbb{R}^{k \times k}$ and a vertical block Z-matrix $Y = ((y_{ij})) \in \mathbb{R}^{m \times k}$ of the same type as A and nonnegative vectors $r \in \mathbb{R}^k$, $s \in \mathbb{R}^m$ such that

(i) $AX = Y$,
(ii) $r^t X + s^t Y > 0$.

The following results gives us some properties of matrices X and Y as given in Definition 4.1.

LEMMA 4.2. *Let A be a vertical block hidden Z-matrix. Let $X \in \mathbb{R}^{k \times k}$ be a Z-matrix and $Y \in \mathbb{R}^{m \times k}$ be a vertical block Z-matrix of the same type as A satisfying the conditions of Definition 4.1. Then X is nonsingular and there exists an index set $\alpha \subseteq \{1, 2, \ldots, k\}$ such that the matrix*

$$W = \begin{bmatrix} X_{\alpha\alpha} & X_{\alpha\bar{\alpha}} \\ V_{\bar{\alpha}\alpha} & V_{\bar{\alpha}\bar{\alpha}} \end{bmatrix}$$

is in K, where V is a representative submatrix of Y corresponding to a representative submatrix G of A.

PROPOSITION 4.3. *Let A be a vertical block hidden Z-matrix with X and Y as any matrices satisfying the conditions of Definition 4.1. Then there is at least one representative submatrix of A which is hidden Z with respect to X and the corresponding representative submatrix of Y.*

PROOF. As similar to 4.2, we have an index set $\alpha \subseteq \{1, 2, \ldots, k\}$ and a representative submatrix V of Y such that

$$W = \begin{bmatrix} X_{\alpha\alpha} & X_{\alpha\bar{\alpha}} \\ V_{\bar{\alpha}\alpha} & V_{\bar{\alpha}\bar{\alpha}} \end{bmatrix}$$

is a K-matrix. Let G be the corresponding representative submatrix of A. As $W \in K$, $W^t \in K$. Also, there exits some $v \in \mathbb{R}^k$, $v \geq 0$ such that $v^t W > 0$. Let $v = (v_\alpha, v_{\bar{\alpha}})$. Now, we take $r(G)^t = (v_\alpha^t, 0)$ and $s(G)^t = (0, v_{\bar{\alpha}}^t)$. By an easy observation, $GX = V$ and $r(G)^t X + s(G)^t V = v^t W > 0$. Thus, from Definition 4.1, G is a hidden Z-matrix. □

REMARK 4.4. From the proposition 4.3, we conclude that if A is a vertical block hidden Z-matrix with X and Y as any matrices satisfying the conditions of Definition 4.1 then there exists a nonnegative matrix $U \in \mathbb{R}^{k \times m}$ such that UA is a square hidden Z-matrix. Also, if X is a K-matrix then UA is a hidden Z-matrix.

Next we look at some vertical block hidden Z and P-matrix properties.

Let A be a vertical block hidden Z-matrix of type $(m_1, \ldots, m_k)$ and let X and Y be as in Definition 4.1. Let $\alpha \subseteq \{1, 2, \ldots, m\}$ be a row of G. Also note that since $AX = Y$ and $G = A_{\alpha\alpha}X_{\alpha\alpha} + A_{\alpha\overline{\alpha}}X_{\overline{\alpha}\alpha}$, the corresponding Y_α is a representative submatrix of G where $\alpha = \{r \mid \text{the } r\text{th row of } A \text{ is a row of } G\}$.

DEFINITION 4.5. We say that a vertical block matrix A of type $(m_l, m_2, \ldots, m_k)$ has the hidden Z-property, if every representative submatrix of A is a hidden Z-matrix.

One can ask whether a vertical block hidden Z-matrix of type $(m_1, \ldots, m_k)$ has the hidden Z-property. The following example answers in negative.

EXAMPLE 4.6. Consider the vertical block matrix A of type $(2, 3, 2)$ where

$$A = \begin{bmatrix} 2 & -1 & 2 \\ 0 & 0 & 0 \\ -3 & 2 & -2 \\ -5 & 5 & -7 \\ -5 & 3 & -2 \\ -4 & 3 & 5 \\ -1 & -1 & 0 \end{bmatrix}$$

A is a vertical block hidden Z-matrix with respect to X, Y where

$$X = \begin{bmatrix} 1 & 0 & -1 \\ 0 & 1 & -2 \\ 0 & -1 & 0 \end{bmatrix}$$

and

$$Y = \begin{bmatrix} 2 & -3 & 0 \\ 0 & 0 & 0 \\ -3 & 0 & -1 \\ -5 & 12 & -5 \\ -5 & 5 & -1 \\ -4 & -2 & -2 \\ -1 & -1 & 3 \end{bmatrix}$$

We take $\rho' = \begin{bmatrix} 12 & 0 & 0 \end{bmatrix}$ and $\sigma' = \begin{bmatrix} 0 & 0 & 0 & 1 & 0 & 0 & 6 \end{bmatrix}$.

But the representative submatrix

$$G = \begin{bmatrix} 2 & -1 & 2 \\ -3 & 2 & -2 \\ -4 & 3 & 5 \end{bmatrix}$$

is not a hidden Z-matrix ([[**1**], p. 212]).

Although, we have the following relationship.

THEOREM 4.7. *Let A be a vertical block hidden Z-matrix of type $(m_1, \ldots, m_k)$. If X and Y are given as in Definition 4.1 and X is a K-matrix then A has the hidden Z-property. Conversely, if every representative submatrix of A is hidden Z with a common X-matrix, then A is a vertical block hidden Z-matrix of type $(m_1, \ldots, m_k)$.*

LEMMA 4.8. *Let A be a vertical block hidden Z-matrix of type $(m_1, m_2, \ldots, m_k)$ and X and Y be given as in Definition 4.1, A has the hidden Z-property with respect to X which is common for all representative matrices if and only if the following holds.*

For any index set $\alpha \subseteq \{1, 2, \ldots, m\}$ such that $\alpha \cap J_r$ is a singleton set for all $1 \le r \le k$, there is a subset α_1 of α such that

$$\begin{bmatrix} X_{\alpha_2\alpha_2} & X_{\alpha_2\overline{\alpha}_2} \\ Y_{\alpha_1\alpha_2} & Y_{\alpha_1\overline{\alpha}_2} \end{bmatrix}$$

is a K-matrix where $\alpha_2 = \{i \mid (\alpha \setminus \alpha_1) \cap J_i \neq \emptyset\}$.

The follow up question is obvious: If every representative submatrix of a vertical block matrix A is hidden Z, can we say that A is a vertical block hidden Z-matrix? The following example answers in negative.

EXAMPLE 4.9. Let A be a vertical block matrix of type $(1, 2, 1)$ given as:

$$A = \begin{bmatrix} 4 & -8 & -9 \\ 8 & 2 & -2 \\ -6 & -8 & -5 \\ -10 & -2 & 0 \end{bmatrix}$$

A has two representative submatrices G_1, G_2 where

$$G_1 = \begin{bmatrix} 4 & -8 & -9 \\ 8 & 2 & -2 \\ -10 & -2 & 0 \end{bmatrix}$$

and

$$G_2 = \begin{bmatrix} 4 & -8 & -9 \\ -6 & -8 & -5 \\ -10 & -2 & 0 \end{bmatrix}$$

G_1 is a hidden Z-matrix with respect to X_1, Y_1 where

$$X_1 = \begin{bmatrix} 1 & 0 & -1 \\ -6 & 1 & 0 \\ -1 & 0 & 2 \end{bmatrix}$$

and

$$Y_1 = \begin{bmatrix} 61 & -8 & -22 \\ -2 & 2 & -12 \\ 2 & -2 & 10 \end{bmatrix}$$

We take $\rho_1^t = \begin{bmatrix}90 & 0 & 90\end{bmatrix}$ and $\sigma_1^t = \begin{bmatrix}1 & 6 & 1\end{bmatrix}$.
G_2 is a hidden Z-matrix with respect to X_2, Y_2 where

$$X_2 = \begin{bmatrix} 3 & 0 & -1 \\ 0 & 1 & -1 \\ -1 & 0 & 3 \end{bmatrix}$$

and

$$Y_2 = \begin{bmatrix} 21 & -8 & -23 \\ -13 & -8 & -1 \\ -30 & -2 & 12 \end{bmatrix}$$

We take $\rho_2^t = \begin{bmatrix}10 & 20 & 25\end{bmatrix}$ and $\sigma_2^t = \begin{bmatrix}2 & 0 & 1\end{bmatrix}$.
Let us consider the following inequalities:

$$8x_{11} + 2x_{21} - 2x_{31} \leq 0$$

$$-6x_{11} - 8x_{21} - 5x_{31} \leq 0$$

$$x_{21}, x_{31} \leq 0.$$

The system of inequalities admits only the trivial solution: $x_{11} = x_{21} = x_{31} = 0$. However, choosing the zero vector as the first column of X implies that no vectors ρ and σ exist such that $\rho^t X + \sigma^t Y > 0$. Therefore, A is not a vertical block hidden Z-matrix.

Next, we define the vertical block hidden K-matrix. To define, the vertical block hidden K-matrix, first we look at hidden K-matrix.

DEFINITION 4.10. We say that a matrix $M \in \mathbb{R}^{n\times n}$ is a hidden K-matrix if it is a hidden Z-matrix and a P-matrix.

DEFINITION 4.11. Let A be a vertical block hidden Z-matrix. We say that A is a vertical block hidden K-matrix if every representative submatrix of A is a P-matrix.

In next, we provide an example of the vertical block hidden K-matrix.

EXAMPLE 4.12. Let A be the following vertical block matrix of type $(3, 2, 2)$:

$$\begin{bmatrix} 1.8 & 0.40 & 0.10 \\ .90 & 0 & 0 \\ 0.9 & -0.19 & -0.19 \\ \\ 0.33 & 1.53 & 0.119 \\ 0.43 & 1.839 & 0.039 \\ \\ -1.57 & -1.17 & 0.05 \\ -0.70 & -0.70 & 0.50 \end{bmatrix}$$

To show that A is a vertical block hidden K-matrix $X, Y \in \mathbb{R}^{m\times n}$ such that

$$X = \begin{bmatrix} 2.1 & -0.90 & -0.90 \\ -1 & 2.9 & -1 \\ -1 & -1.9 & 7 \end{bmatrix} \text{ and } Y = \begin{bmatrix} 3.28 & -0.65 & -1.32 \\ 1.89 & -0.81 & -0.81 \\ 2.27 & -1 & -1.95 \\ \\ -0.956 & 3.9139 & -0.994 \\ -0.975 & 4.872 & -1.953 \\ \\ -2.177 & -2.075 & 2.933 \\ -1.27 & -2.35 & 4.83 \end{bmatrix}$$

For $r^t = \begin{bmatrix} 4 & 2 & 1 \end{bmatrix}$ and $s^t = \begin{bmatrix} 1 & 1 & 1 & 1 & 1 & 1 & 1 \end{bmatrix}$, we conclude that A is a vertical block hidden K-matrix with respect to X, Y.

The following theorem gives us a characterization of vertical block hidden K-matrix.

THEOREM 4.13 ([**21**]). *Let A be a vertical block hidden Z-matrix of type $(m_1, m_2, \ldots, m_k)$. If X and Y be as in Definition 4.1, then the following are equivalent:*

(i) *A is a vertical block hidden K-matrix.*
(ii) *There exists an $u \in \mathbb{R}^k$, $u > 0$ such that $Au > 0$.*
(iii) *There exists a vector $y \in \mathbb{R}^k$, $y > 0$ such that for any given index set $\alpha \subseteq \{1, 2, \ldots, k\}$, $Wy > 0$, where*

$$W = \begin{bmatrix} X_{\alpha\alpha} & X_{\alpha\bar{\alpha}} \\ V_{\bar{\alpha}\alpha} & V_{\bar{\alpha}\bar{\alpha}} \end{bmatrix}$$

and V is the representative submatrix of Y corresponding to any given representative submatrix G of A. Furthermore, $W \in K$.

(iv) *Every representative submatrix G of A is completely hidden K; i.e., for every index set $\beta \subseteq \{1, 2, \ldots, k\}$, $G_{\beta\beta}$ is hidden K.*

5. The class of vertical block N-matrix

In this section, we look into vertical block N-matrix and copositive result. All the results of this section can be found in [**16**].

DEFINITION 5.1. Suppose A is a vertical block matrix of type $(m_1, \ldots, m_k)$ with $\sum_{i=1}^{k} m_i = m$. We say that A is a vertical block N-matrix if every representative matrix is an N-matrix. We say that A is a vertical block N-matrix of the first category if A is a vertical block N-matrix and A has at least one positive entry. If A is a vertical block N-matrix all of whose entries are negative, then we say that A is a vertical block N-matrix of the second category.

LEMMA 5.2. *A is a vertical block N-matrix of the first category if and only if every representative matrix is an N-matrix of the first category.*

LEMMA 5.3. *Suppose that A is a vertical block category 1 N-matrix of type $(m_1, \ldots, m_k)$. Then A is a Q-matrix.*

LEMMA 5.4. *Suppose A is a vertical block N-matrix of the first category, then there exists a subset $K \subseteq \{1, 2, \ldots, k\}$ such that, with $J = \bigcup_{i\in K} J_i$, A has the* *following partitioned form after a rearrangement of columns and blocks using the same permutation*

$$A = \begin{bmatrix} A_{JK} & A_{J\bar{K}} \\ A_{JK} & A_{J\bar{K}} \end{bmatrix}$$

where $A_{JK} < 0$, $A_{J\bar{K}} < 0$, $A_{J\bar{K}} > 0$, $A_{JK} > 0$ *and* $K, \bar{K}$ *are nonempty sets.*

LEMMA 5.5. *Let A be a vertical block N-matrix of type $(m_1, m_2, \ldots, m_k)$. Then the equivalent matrix $M \in R_0$ and $A \in E^{\cdot}(0)$.*

For copositive matrices we have the following result.

THEOREM 5.6. *Let A be an $m \times k$ vertical block matrix of type $(m_1, \ldots, m_k)$ and let M be the equivalent matrix of A of order m. The following statements are equivalent:*

(i) *Every representative submatrix of A is copositive;*
(ii) *M is copositive.*

LEMMA 5.7. *Suppose A is an $m \times k$ vertical block matrix of type $(m_1, \ldots, m_k)$ and M is its equivalent matrix. If M is copositive-plus, then every representative submatrix of A is copositive-plus.*

PROOF. This follows from the inheritance property of copositive-plus matrices. See [**4**]. □

The converse of the above lemma is not true, is illustrated in the following.

EXAMPLE 5.8. Let $A = \begin{bmatrix} 2 & -2 \\ 0 & -2 \\ 2 & 0 \end{bmatrix}$ be a vertical block matrix of type $(2, 1)$. Both the representative submatrices of A are clearly copositive-plus. However, the equivalent matrix M is not copositive-plus.

COROLLARY 5.9. *If a vertical block matrix A of order $m \times k$ and type $(m_1, \ldots, m_k)$ has copositive-star property. Then the equivalent matrix M is a copositive-star matrix.*

However, if all the representative submatrices of a vertical block matrix A are copositive-plus, then A is a vertical block Q_0 matrix, even though its equivalent matrix M need not be a copositive-plus matrix. A constructive proof of this is given by [**4**].

THEOREM 5.10. *If A is a vertical block N-matrix of type $(m_1, \ldots, m_k)$ and U is given, then UA is an N-matrix.*

LEMMA 5.11. *A is a vertical block N-matrix of the first category if and only if every representative submatrix is an N-matrix of the first category.*

DEFINITION 5.12. A matrix $A \in \mathbb{R}^{n \times n}$ is said to be an $\mathbb{N}$-matrix if there exists a sequence $\{A^{(k)}\}$ where $A^{(k)} = [a_{ij}^{(k)}]$ are N-matrices such that $a_{ij}^{(k)} \to a_{ij}$ for all $i, j \in \{1, 2, \ldots, n\}$.

DEFINITION 5.13. We say A is a vertical block N-matrix of type $(m_1, \ldots, m_k)$ if every representative submatrix is an N-matrix.

THEOREM 5.14. *Let $A \in \mathbb{R}^{m\times k}$ be a vertical block N-matrix of type $(m_1, \ldots, m_k)$. Then the equivalent matrix M of A has exactly one real negative eigenvalue and has at least $(m-k)$ zero eigenvalues. Also all nontrivial principal submatrices have exactly one real negative eigenvalue.*

LEMMA 5.15. *Suppose A is a vertical block N-matrix of type $(m_1, \ldots, m_k)$. Let M be the equivalent matrix. Then there exists a nonempty subset v of $\{1, 2, \ldots, n\}$ such that M can be written in the partitioned form as (if necessary, after a principal rearrangement of its rows and columns)*

$$M = \begin{bmatrix} M_{vv} & M_{v\overline{v}} \\ M_{\overline{v}v} & M_{\overline{v}\overline{v}} \end{bmatrix}$$

where $M_{vv} \leq 0$, $M_{v\overline{v}} \leq 0$, $M_{\overline{v}v} \geq 0$, $M_{\overline{v}\overline{v}} \geq 0$.

THEOREM 5.16. *Let A be a vertical block N-matrix of type $(m_1, \ldots, m_k)$ with R_0-property and M be the equivalent matrix. Assume $v(M) > 0$. Then A is a vertical block Q-matrix.*

6. Cottle-Dantzig Pivot Algorithm For solving The VLCP

In [**4**], Cottle and Dantzig describe a procedure for the processing a generalized linear complementarity problem, which is an extension of Lemke's algorithm for the ordinary linear complementarity problem. We present below the steps of the algorithm, for the sake of completeness.

Cottle-Dantzig algorithm for the VLCP(q, A) starts with the initial solution to (1.3) and (1.4) as

$$w = q + d\, z_0$$

$$z = 0$$

where z_0 is large enough so that $w > 0$ and $d \in R^m$ is any positive vector.

Step 1: Reduce z_0 to $\bar{z}_0 = \min\{z_0 \mid q + dz_0 \geq 0,\ z_0 \geq 0\}$, ensuring that at least one variable w_i (for some $1 \leq i \leq m$), say w_p, reaches zero.

This adjustment results in a basic feasible solution where z_0 takes the place of w_p, forming the initial almost proper basic feasible solution. Next, let r be the unique index satisfying $1 \leq r \leq k$ such that $p \in J_r$. As a result, there exists exactly one pair of non-basic variables, (z_r, w_p), that belong to the same set of related variables.

Step 2: In each iteration, one non-basic variable is removed from the basis, and its related counterpart is added. For instance, in the second iteration, z_r enters the basis.

Step 3: If the variable chosen in Step 2 can increase indefinitely, the process terminates in an *almost proper ray*, called a *secondary proper ray.* Otherwise, one or more basic variables reach zero, and one is replaced by the entering variable, yielding a new almost proper or proper basic feasible solution.

If this new solution has $z_0 = 0$ or z_0 is non-basic, we have solved given equation, providing a solution for VLCP(q, A). Otherwise, we obtain a new almost proper basic feasible solution and a new pair of non-basic variables, (x_β, y_r), belonging to the same related set, say the sth set. This pair is either (z_s, w_t) with $t \in J_s$, or (w_{t_1}, w_{t_2}) with $t_1, t_2 \in J_s$.

We then repeat Step 2.

The Cottle-Dantzig algorithm (Algorithm CD) repeatedly applies Steps 2 and 3. We assume the *standard nondegeneracy condition*, meaning all almost proper basic feasible solutions are nondegenerate. Under this assumption, the leaving variable in Step 3 is uniquely determined at each iteration, ensuring the procedure either finds a solution to VLCP(q, A) or terminates in a secondary proper ray.

An alternative method of solving VLCP(q, A) is to apply Lemke's algorithm to the LCP(q, M). See Lemke [**13**]. One can find out the difference in the two approaches. The following observation is immediate.

REMARK 6.1. *Suppose B is an almost complementary basis matrix generated by the application of Lemke's algorithm with the artificial vector d to the equivalent problem LCP(q, M) then there is at most one index i, $i \in J_s$ such that $-M_{\cdot i}$ is a column of B, for each $1 \leq s \leq k$. This follows from the non-singularity of B.*

The following lemmas are also immediate.

LEMMA 6.2. *Suppose B is an almost complementary basis matrix generated by the application of Lemke's algorithm with the artificial vector d to the equivalent problem LCP(q, M) and that $-M_{\cdot p} = -A_{\cdot s}$ where $p \in J_s$ is the rth column of B. Then the columns corresponding to the variables z_i, for $i \in J_s$ in the tableau corresponding to this basis are all equal to $I_{\cdot r}$, where r is the column index of $-M_{\cdot p}$ in B.*

LEMMA 6.3. *Suppose B is an almost complementary basis matrix generated by the application of Lemke's algorithm with the artificial vector d to the equivalent problem LCP(q, M) and suppose $-M_{\cdot p} = -A_{\cdot s}$ where $p \in J_s$ is the rth column of B. Suppose now, the complementary pivot rule selects $-M_{\cdot j}$ where $j \in J_s$ to be included in the next iteration. The resulting new almost complementary basis matrix and the new tableau have the same set of columns as B, except that the set of basic variables now includes z_j in the place of z_p.*

We say that Lemke's algorithm applied to the equivalent LCP(q, M) executes a trivial pivot step, when a variable z_j replaces another variable z_p, where j and p are in the same index set J_s corresponding to the sth block, for some s, $1 \leq s \leq k$.

REMARK 6.4. The generalized complementary pivot rule of Cottle and Dantzig may identify a unique non-basic column pair $(I_{\cdot j}, I_{\cdot p})$, where $j, p \in J_s$, and select $I_{\cdot p}$ to enter the basis. In contrast, Lemke's algorithm for the equivalent LCP(q, M) first chooses $-M_{\cdot j}$ to enter, which then replaces $-M_{\cdot p}$ through a trivial pivot. In the next iteration, $I_{\cdot p}$ joins the basis matrix.

Let B_1 and B_2 be two almost complementary basis matrices generated by Lemke's algorithm for LCP(q, M). They are *indistinct* if they contain the same set of columns; otherwise, they are *distinct*.

THEOREM 6.5. *The sequence of almost proper basis matrices generated by the application of Cottle-Dantzig algorithm with the artificial vector d to the VLCP(q, A) is the same as the sequence of distinct almost complementary basis matrices generated by applying Lemke's algorithm with the artificial vector d to the equivalent LCP(q, M).*

It is already known that Cottle-Dantzig algorithm can process vertical block matrices in the generalized copositive plus class (see Cottle and Dantzig [**4**]).

The equivalent LCP matrix M, for the above vertical block matrix A is of the form

$$M = \begin{bmatrix} 0 & C \\ D & 0 \end{bmatrix}$$

where $C > 0,\ D > 0$.

It is well known that Lemke's algorithm with any positive vector d as the artificial vector cannot process the equivalent LCP$(-e, M)$, since M has the form $\begin{bmatrix} 0 & C \\ D & 0 \end{bmatrix}$ where C and D are positive and consist of sets of identical columns. However we can apply the algorithm of Lemke and Howson [**14**] to this problem and compute an equilibrium point.

We thus have the following theorem.

THEOREM 6.6 ([**16**]). *A generalized Nash equilibrium point of the generalized bimatrix game can be computed by a generalized Lemke-Howson algorithm in which the complementary pivot rule is replaced by the generalized complementary pivot rule given by Cottle and Dantzig.*

Note that the vertical block matrix A associated with the generalized bimatrix game has the form

$$A = \begin{bmatrix} 0 & C^1 \\ 0 & C^2 \\ \vdots & \vdots \\ 0 & C^m \\ D^1 & 0 \\ D^2 & 0 \\ \vdots & \vdots \\ D^n & 0 \end{bmatrix}$$

where each $C^j > 0$ and each $D^j < 0$.

We then have the following:

THEOREM 6.7 ([**16**]). *Suppose the problem of computing a generalized Nash equilibrium point of the generalized bimatrix game is formulated as finding a solution to the VLCP$(-e, A)$ where $A = \begin{bmatrix} 0 & X \\ Y & 0 \end{bmatrix}$ with $X > 0$ and $Y < 0$. Then we can compute an equilibrium point by applying Cottle-Dantzig algorithm to such a VLCP.*

EXAMPLE 6.8. Consider a vertical block matrix A of type $(2, 2, 1)$ and the vector q as follows:

$$A = \begin{bmatrix} -1 & -3 & -5 \\ -3 & 2 & -1 \\ -4 & 0 & -1 \\ -4 & -2 & -7 \\ -1 & -1 & -3 \end{bmatrix} \ ; \ q = \begin{bmatrix} -1 \\ -1 \\ -4 \\ -4 \\ 2 \end{bmatrix}$$

Here is the list of related variables:

$$\begin{matrix} z_1 & w_1 & w_2 \\ z_2 & w_3 & w_4 \\ z_3 & w_5 & \end{matrix}$$

Table 1: Iteration Number: 0

	w_1	w_2	w_3	w_4	w_5	z_1	z_2	z_3	z_0	q
w_1	1	0	0	0	0	−1	−3	−5	−1	−1
w_2	0	1	0	0	0	−3	2	−1	−1	−1
w_3	0	0	1	0	0	−4	0	−1	−1	−4
w_4	0	0	0	1	0	−4	−2	−7	−1	−4
w_5	0	0	0	0	1	−1	−1	−3	−1	2

Table 2: Iteration Number: 1, Entering Variable: z_0, Leaving Variable: w_4

	w_1	w_2	w_3	w_4	w_5	z_1	z_2	z_3	z_0	q	ratio
w_1	1	0	0	−1	0	3	−1	2	0	3	–
w_2	0	1	0	−1	0	1	4	6	0	3	0.75
w_3	0	0	1	−1	0	0	2	6	0	0	0
z_0	0	0	0	−1	0	4	2	7	1	4	2
w_5	0	0	0	−1	1	3	1	4	0	6	6

Table 3: Iteration Number: 2, Entering Variable: z_2, Leaving Variable: w_3

	w_1	w_2	w_3	w_4	w_5	z_1	z_2	z_3	z_0	q	ratio
w_1	1	0	0.5	−1.5	0	3	0	5	0	3	–
w_2	0	1	−2	1	0	1	0	−6	0	3	3
z_2	0	0	0.5	−0.5	0	0	1	3	0	0	–
z_0	0	0	−1	0	0	4	0	1	1	4	–
w_5	0	0	−0.5	−0.5	1	3	0	1	0	6	–

Table 4: Iteration Number: 3, Entering Variable: w_4, Leaving Variable: w_2

	w_1	w_2	w_3	w_4	w_5	z_1	z_2	z_3	z_0	q	ratio
w_1	1	1.5	−2.5	0	0	4.5	0	−4	0	7.5	1.6667
w_4	0	1	−2	1	0	1	0	−6	0	3	3
z_2	0	0.5	−0.5	0	0	0.5	1	0	0	1.5	3
z_0	0	0	−1	0	0	4	0	1	1	4	1
w_5	0	0.5	−1.5	0	1	3.5	0	−2	0	7.5	2.1429

Table 5: Iteration Number: 4, Entering Variable: z_1, Leaving Variable: z_0

	w_1	w_2	w_3	w_4	w_5	z_1	z_2	z_3	z_0	q
w_1	1	1.5	−1.375	0	0	0	0	−5.125	−1.13	3
w_4	0	1	−1.75	1	0	0	0	−6.25	−0.25	2
z_2	0	0.5	−0.375	0	0	0	1	−0.125	−0.13	1
z_1	0	0	−0.25	0	0	1	0	0.25	0.25	1
w_5	0	0.5	−0.625	0	1	0	0	−2.875	−0.88	4

Therefore the final solution is:

$$\begin{array}{lll} z_1 = 1 & w_1 = 3 & w_2 = 0 \\ z_2 = 1 & w_3 = 0 & w_4 = 2 \\ z_3 = 0 & w_5 = 4 & \end{array}$$

7. Nash equilibrium for Generalized Bimatrix Games

Gowda and Sznajder [**9**], Mohan et al. [**16**] consider the question of computing a generalized Nash equilibrium point for the generalized bimatrix game introduced by M.S. Gowda and R. Sznajder[**9**]). More on Nash equilibrium and bimatrix Games can be found in [**11, 13, 26, 27, 31**].

The generalized bimatrix game introduced in [**9**] is described as follows:

Let $\mathcal{A}$ and $\mathcal{B}$ be two finite sets of matrices, where $\mathcal{A}$ contains s matrices of size $m \times n$, and $\mathcal{B}$ contains r matrices of the same order. Player I constructs his payoff matrix by selecting the ith row from some matrix $A \in \mathcal{A}$ as the ith row of his matrix. He then chooses a mixed strategy over the set $\{1, 2, \ldots, m\}$.

Similarly, Player II (the column player) constructs his payoff matrix by selecting the jth column from some matrix $B \in \mathcal{B}$ as the jth column of his matrix. He then plays a mixed strategy over $\{1, 2, \ldots, n\}$. The remainder of the game proceeds as in a standard bimatrix game.

DEFINITION 7.1. A pair of mixed strategies $(x^*, y^*)_{(M,\,N)}$ is said to be a *generalized Nash equilibrium pair* if there exists a there exists a row representative submatrix M and a column representative submatrix N such that

$$(x^*)^T M y^* \leq x^T R y^*$$

and

$$(x^*)^T N y^* \leq (x^*)^T S y,$$

for all mixed strategies x and y, all row representative submatrices R of $\mathcal{A}$ and all column representative submatrices S of $\mathcal{B}$.

Mohan et al. [**16**] formulated the problem of computing the equilibrium point of a generalized bimatrix game as a VLCP. Lemke [**12**] anticipated a wide range of meaningful applications for the VLCP, originally introduced by Cottle and Dantzig. Although only a few applications have been explored in the literature, the framework holds significant potential.

For the sake of completeness, we first present the formulation of computing a Nash equilibrium point as a Vertical Linear Complementarity Problem.

Suppose, $\mathcal{A}= \{A^p \mid p = 1, 2, \ldots, s\}$ and $\mathcal{B} = \{B^p \mid p = 1, 2, \ldots, r\}$. Consider the matrices C^j, $j = 1, 2, \ldots, m$ and D^j, $j = 1, 2, \ldots, n$ defined as follows:

$$\begin{aligned} C^j_{i\cdot} &= A^i_{j\cdot}, 1 \leq i \leq s \\ D^j_{i\cdot} &= (B^i)^t_{j\cdot}, 1 \leq i \leq r \end{aligned}$$

Without loss of generality, we may assume that each A^p, $p = 1, 2, \ldots, s$ and each B^p, $p = 1, 2, \ldots, r$ are positive matrices. Hence each C^j, $j = 1, 2, \ldots, m$ and each D^j, $j = 1, 2, \ldots, n$ are positive matrices.

Let $X = \begin{bmatrix} C^1 \\ C^2 \\ \vdots \\ C^m \end{bmatrix}$ and $Y = \begin{bmatrix} D^1 \\ D^2 \\ \vdots \\ D^n \end{bmatrix}$ where each C^j is of order $s \times n$ and each D^j is of order $r \times m$ and by our assumption $X > 0,\ Y > 0$.

Consider the matrix

$$A = \begin{bmatrix} 0 & C^1 \\ 0 & C^2 \\ \vdots & \vdots \\ 0 & C^m \\ D^1 & 0 \\ D^2 & 0 \\ \vdots & \vdots \\ D^n & 0 \end{bmatrix} = \begin{bmatrix} 0 & X \\ Y & 0 \end{bmatrix}$$

where the '0' blocks are of appropriate orders. This is a vertical block matrix of type $(s, \ldots, s, r, \ldots, r)$ where the number of blocks is $(m+n)$.

Note that a generalized Nash equilibrium point of the game $(\mathcal{A}, \mathcal{B})$ can be computed by obtaining a solution to the VLCP $(-e, A)$ where $-e$ is the column vector of order $(ms+nr)$, each of whose coordinate is -1 and vice versa (see [**9, 23**]).

Also the representative submatrix

$$A_G = \begin{bmatrix} 0 & C_G \\ D_G & 0 \end{bmatrix} \tag{7.1}$$

obtained from the vertical block matrix A of type $(s, \ldots, s, r, \ldots, r)$ will be of order $(m+n) \times (m+n)$.

EXAMPLE 7.2. Consider the generalized bimatrix game (A, B) where the sets A and B are as follows:

$$A = \left\{ \begin{bmatrix} 3 & 1 \\ 5 & 1 \end{bmatrix}, \begin{bmatrix} 1 & 6 \\ 2 & 8 \end{bmatrix} \right\}, \quad B = \left\{ \begin{bmatrix} 3 & 4 \\ 8 & 3 \end{bmatrix}, \begin{bmatrix} 5 & 6 \\ 7 & 8 \end{bmatrix} \right\}.$$

The related matrices C^j and D^j are as follows:

$$C^1 = \begin{bmatrix} 3 & 1 \\ 1 & 6 \end{bmatrix}, \quad C^2 = \begin{bmatrix} 1 & 5 \\ 2 & 8 \end{bmatrix}, \quad D^1 = \begin{bmatrix} 3 & 8 \\ 5 & 7 \end{bmatrix}, \quad D^2 = \begin{bmatrix} 4 & 3 \\ 6 & 8 \end{bmatrix}.$$

Note that

$$A = \begin{bmatrix} 0 & C^1 \\ 0 & C^2 \\ D^1 & 0 \\ D^2 & 0 \end{bmatrix}$$

is a vertical block matrix of type (2, 2, 2, 2).

A solution obtained by solving VLCP$(-e_8, A)$ is:

$$z_1 = 0.2174, \quad z_2 = 0.0435, \quad z_3 = 0.2857, \quad z_4 = 0.1429,$$

and the corresponding generalized Nash equilibrium is:

$$x_1 = 0.83, \quad x_2 = 0.17, \quad y_1 = 0.67, \quad y_2 = 0.33.$$

The corresponding row representative matrix is:

$$\begin{bmatrix} 3 & 1 \\ 1 & 5 \end{bmatrix} \quad \text{and column representative matrix is} \quad \begin{bmatrix} 3 & 8 \\ 4 & 3 \end{bmatrix}.$$

8. Conclusion

Overall, we present a comprehensive study of various vertical block matrices and their connection to the VLCP. We provide a detailed compilation of results related to vertical block Z and vertical block hidden Z-matrices. Finally, we offer a brief overview of the Cottle-Dantzig algorithm and its relevance to Nash equilibrium. We hope this article helps readers establish a deeper understanding of matrix classes associated with vertical block matrices.

References

[1] Richard W. Cottle, Jong-Shi Pang, and Richard E. Stone, *The linear complementarity problem*, Classics in Applied Mathematics, vol. 60, Society for Industrial and Applied Mathematics (SIAM), Philadelphia, PA, 2009. Corrected reprint of the 1992 original [MR1150683], DOI 10.1137/1.9780898719000.ch1. MR3396730

[2] Katta G. Murty, *On the number of solutions to the complementarity problem and spanning properties of complementary cones*, Linear Algebra Appl. **5** (1972), 65–108, DOI 10.1016/0024-3795(72)90019-5. MR291183

[3] Punit Kumar Yadav and Palpandi Karuppaiah, *Generalizations of R_0 and SSM properties for extended horizontal linear complementarity problem*, J. Optim. Theory Appl. **199** (2023), no. 1, 392–414, DOI 10.1007/s10957-023-02262-9. MR4651072

[4] Richard W. Cottle and George B. Dantzig, *A generalization of the linear complementarity problem*, J. Combinatorial Theory **8** (1970), 79–90. MR254064

[5] B. Curtis Eaves, *The linear complementarity problem*, Management Sci. **17** (1971), 612–634, DOI 10.1287/mnsc.17.9.612. MR282663

[6] A. A. Ebiefung and M. M. Kostreva, *Generalized P_0- and Z-matrices*, Linear Algebra Appl. **195** (1993), 165–179, DOI 10.1016/0024-3795(93)90262-M. MR1253275

[7] A. A. Ebiefung and M. M. Kostreva, *Generalized P_0- and Z-matrices*, Linear Algebra Appl. **195** (1993), 165–179, DOI 10.1016/0024-3795(93)90262-M. MR1253275

[8] M. Seetharama Gowda and Roman Sznajder, *The generalized order linear complementarity problem*, SIAM J. Matrix Anal. Appl. **15** (1994), no. 3, 779–795, DOI 10.1137/S0895479892237859. MR1282694

[9] M. Seetharama Gowda and Roman Sznajder, *A generalization of the Nash equilibrium theorem on bimatrix games*, Internat. J. Game Theory **25** (1996), no. 1, 1–12, DOI 10.1007/BF01254380. MR1375785

[10] C. B. Garcia, *Some classes of matrices in linear complementarity theory*, Math. Programming **5** (1973), 299–310, DOI 10.1007/BF01580135. MR337312

[11] Howson JT Jr (1972), "Equilibria of potymatrix games", Manag Sci 18:318

[12] C. E. Lemke, *Recent results on complementarity problems*, Nonlinear Programming (Proc. Sympos., Univ. of Wisconsin, Madison, Wis., 1970), Academic Press, New York-London, 1970, pp. 349–384. MR272400

[13] C. E. Lemke, *Bimatrix equilibrium points and mathematical programming*, Management Sci. **11** (1964/65), 681–689, DOI 10.1287/mnsc.11.7.681. MR189823

[14] C. E. Lemke and J. T. Howson Jr., *Equilibrium points of bimatrix games*, J. Soc. Indust. Appl. Math. **12** (1964), 413–423. MR173556

[15] S. R. Mohan and R. Sridhar, *On characterizing N-matrices using linear complementarity*, Linear Algebra Appl. **160** (1992), 231–245, DOI 10.1016/0024-3795(92)90449-K. MR1137854

[16] S. R. Mohan, S. K. Neogy, and R. Sridhar, *The generalized linear complementarity problem revisited*, Math. Programming **74** (1996), no. 2, Ser. A, 197–218, DOI 10.1016/0025-5610(96)82467-1. MR1406749

[17] S. R. Mohan, S. K. Neogy, and T. Parthasarathy, *Linear complementarity and the irreducible polystochastic game with the average cost criterion when one player controls transitions*, Game theoretical applications to economics and operations research (Bangalore, 1996), Theory Decis. Lib. Ser. C Game Theory Math. Program. Oper. Res., vol. 18, Kluwer Acad. Publ., Boston, MA, 1997, pp. 153–170. MR1798969

[18] S. R. Mohan, S. K. Neogy, T. Parthasarathy, and S. Sinha, *Vertical linear complementarity and discounted zero-sum stochastic games with $ARAT$ structure*, Math. Program. **86** (1999), no. 3, Ser. A, 637–648, DOI 10.1007/s101070050108. MR1733740

[19] S. R. Mohan, S. K. Neogy, and T. Parthasarathy, *Linear complementarity and discounted polystochastic game when one player controls transitions*, Complementarity and variational problems (Baltimore, MD, 1995), SIAM, Philadelphia, PA, 1997, pp. 284–294. MR1445086

[20] S. R. Mohan, S. K. Neogy, and T. Parthasarathy, *Pivoting algorithms for some classes of stochastic games: a survey*, Int. Game Theory Rev. **3** (2001), no. 2-3, 253–281, DOI 10.1142/S0219198901000385. Special issue on operations research and game theory with economic and industrial applications (Chennai, 2000). MR1855135

[21] S. R. Mohan and S. K. Neogy, *Vertical block hidden Z-matrices and the generalized linear complementarity problem*, SIAM J. Matrix Anal. Appl. **18** (1997), no. 1, 181–190, DOI 10.1137/S0895479894271147. MR1428207

[22] S. R. Mohan and S. K. Neogy, *The role of representative submatrices in vertical linear complementarity theory*, Linear and Multilinear Algebra **41** (1996), no. 2, 175–187, DOI 10.1080/03081089608818471. MR1430490

[23] S. R. Mohan, S. K. Neogy, and R. Sridhar, *The generalized linear complementarity problem revisited*, Math. Programming **74** (1996), no. 2, Ser. A, 197–218, DOI 10.1016/0025-5610(96)82467-1. MR1406749

[24] Katta G. Murty, *On the number of solutions to the complementarity problem and spanning properties of complementary cones*, Linear Algebra Appl. **5** (1972), 65–108, DOI 10.1016/0024-3795(72)90019-5. MR291183

[25] K. G. Murty, *Linear complementarity, linear and nonlinear programming*, Sigma Series in Applied Mathematics, vol. 3, Heldermann Verlag, Berlin, 1988. MR949214

[26] John Nash, *Non-cooperative games*, Ann. of Math. (2) **54** (1951), 286–295, DOI 10.2307/1969529. MR43432

[27] Janovskaya EB, "Equilibrium points in polymatrix games", Lithuanian Mathematical Journal, Vol-8 (1968), 381-384.

[28] G. S. R. Murthy and T. Parthasarathy, *Some properties of fully semimonotone, Q_0-matrices*, SIAM J. Matrix Anal. Appl. **16** (1995), no. 4, 1268–1286, DOI 10.1137/S0895479893253604. MR1351470

[29] Min Sun, *Monotonicity of Mangasarian's iterative algorithm for generalized linear complementarity problems*, J. Math. Anal. Appl. **144** (1989), no. 2, 474–485, DOI 10.1016/0022-247X(89)90347-8. MR1027048

[30] Szanajder R (1994) Degree-theoretic Analysis of the Vertical and Horizontal Linear Complementarity Problems, Ph.D. Thesis. University of Maryland, Maryland

[31] S. K. Neogy, A. K. Das, and P. Das, *Complementarity problem involving a vertical block matrix and its solution using neural network model*, Mathematical programming and game theory for decision making, Stat. Sci. Interdiscip. Res., vol. 1, World Sci. Publ., Hackensack, NJ, 2008, pp. 113–130, DOI 10.1142/9789812813220_0008. MR2459155

[32] Arie Tamir, *An application of Z-matrices to a class of resource allocation problems*, Management Sci. **23** (1976/77), no. 3, 317–323, DOI 10.1287/mnsc.23.3.317. MR441332

Indian Statistical Institute, 7, S.J.S. Sansanwal Marg, New Delhi 110016, India
Email address: punitjrf@gmail.com

Indian Statistical Institute, 7, S.J.S. Sansanwal Marg, New Delhi 110016, India
Email address: sajalg@isid.ac.in

Indian Statistical Institute, 7, S.J.S. Sansanwal Marg, New Delhi 110016, India
Email address: skn@isid.ac.in

Contemporary Mathematics
Volume **833**, 2026
https://doi.org/10.1090/conm/833/16673

Solving fuzzy linear systems using weighted Moore-Penrose and Drazin inverse of extended matrix

Xiaozhou Ye, Predrag S. Stanimirović, and Yimin Wei

This paper is in memory of Professor Jiao-xun Kuang.

ABSTRACT. The main goal of the present study is to solve fuzzy linear systems (FLS) represented as $A\widetilde{\mathbf{x}}(\rho) = \widetilde{\mathbf{y}}(\rho)$, where the coefficient matrix A is a rectangular complex matrix and $\widetilde{\mathbf{x}}(\rho)$, $\widetilde{\mathbf{y}}(\rho)$ are appropriate vectors with fuzzy entries. The standard approach to solving FLS with a crisp coefficient matrix involves solving the equivalent crisp linear system (CLS) $S\mathbf{x}(\rho) = \mathbf{y}(\rho)$, where coefficient matrix is given in a suitable block matrix form. Proper block representations of the weighted Moore-Penrose inverse (WMPI) and the W-weighted Drazin inverse (WDI) of the coefficient matrix S are derived.

1. Introduction and motivation

In scientific and technological computing, various equations describing real-world problems, such as natural phenomena or engineering challenges, often lead to linear systems of equations. Accurate numerical data may not always be attainable, but uncertain data can represent different facets of real-world issues. Fuzzy sets and systems provide a natural framework for describing and managing uncertain data. Zadeh, Dubois, and Prade were the pioneers who introduced the concepts of fuzzy numbers and fuzzy arithmetic operations Since their initial work, many papers and books on fuzzy system theory have been published. Fuzzy systems have been utilized to study a wide range of issues, including fuzzy topological spaces to controlling chaotic systems, fuzzy metric spaces, fuzzy differential equations, fuzzy linear systems, and particle physics. As a result, addressing fuzzy linear systems (FLS), in which some or all parameters are represented as fuzzy numbers, has become a matter of considerable interest.

There are well-established theoretical results for solving FLS with a nonsingular coefficient matrix A. The common method for solving FLS has been to replace

2020 *Mathematics Subject Classification.* Primary 15B15, 03B52; Secondary 94D05, 08A72.

Key words and phrases. Fuzzy linear system, weighted Moore-Penrose inverse, W-weighted Drazin inverse.

The first author is supported by the Ministry of Science and Technology of China under grant H20250939. The second author is supported by the Ministry of Education, Science and Technological Development, Republic of Serbia (No. 451-03-137/2025-03/200124). The third author is supported by the Joint Research Project between China and Serbia under the grant 2024-6-7.

the original $m \times n$ FLS into acorresponding $2m \times 2n$ crisp linear system (CLS) [**7**]. Various methods were proposed for the case where matrix A is singular. These include the block representation of the Moore–Penrose inverse [**9**], the block representation of the group inverse [**10**], the block representation of the core inverse [**11**] as well as the block representation of the DMP inverse and the BT inverse [**14**]. In reference [**12**], general strong fuzzy matrix solutions for a consistent general fuzzy matrix equation were derived, and an algorithm for calculating the Core-EP inverse [**15**] was presented. The Moore-Penrose weak group (MPWG) inverse was investigated in [**30**]. Article [**17**] examined some aspects of the WDI of a CLS, but there are still some areas that require further improvement. The results in [**19**] propose generalizations of composite outer inverses. Wei et al. proposed the weighted Moore-Penrose inverse (WMPI) [**16, 25, 28**] with its applications and the WDI [**29**] for traditional linear systems of the form $Ax = y$, where A is a rectangular matrix.

In this paper, we extend the theory of WMPI and the WDI for the rectangular coefficient matrix S in the CLS $S\mathbf{x}(\rho) = \mathbf{y}(\rho)$, which arises from the FLS $A\widetilde{\mathbf{x}}(\rho) = \widetilde{\mathbf{x}}(\rho)$ defined in (2.3).

Section 2 covers the fundamentals of fuzzy numbers and FLS and revisits some essential concepts related to the WMPI and WDI. In Section 3, we present the expressions for calculating the WMPI and WDI of coefficient matrix S of the CLS $S\mathbf{x}(\rho) = \mathbf{y}(\rho)$, which is related to the FLS $A\widetilde{\mathbf{x}}(\rho) = \widetilde{\mathbf{y}}(\rho)$. Numerical examples are provided in Section 4.

2. Preliminaries

In this section, we present some essential preliminaries regarding basic facts and notions.

DEFINITION 2.1 (Fuzzy numbers). An arbitrary fuzzy number is represented by an ordered pair of functions $\widetilde{\mathbf{f}} = \left(\underline{f}(\rho), \overline{f}(\rho)\right), \rho \in [0,1]$ where the following constraints are met:

(a) $\underline{f}(\rho)$ is a bounded and left-continuous non-decreasing function defined over on the domain $[0,1]$,

(b) $\overline{f}(\rho)$ is a bounded and right-continuous non-increasing function over $[0,1]$,

(c) $\underline{f}(\rho) < \overline{f}(\rho), \quad \rho \in [0,1]$.

DEFINITION 2.2 (Operations on fuzzy numbers). For each fuzzy number $\widetilde{\mathbf{x}} = \left(\underline{x}(\rho), \overline{x}(\rho)\right), \widetilde{\mathbf{y}} = \left(\underline{y}(\rho), \overline{y}(\rho)\right)$ and a real number ζ, the scalar multiplication and the addition of fuzzy numbers are defined by the following rules:

(a) $\widetilde{\mathbf{x}}(\rho) + \widetilde{\mathbf{y}}(\rho) = \left(\underline{x}(\rho) + \underline{y}(\rho), \overline{x}(\rho) + \overline{y}(\rho)\right)$;

(b) $\zeta\,\widetilde{\mathbf{x}}(\rho) = \begin{cases} \left(\zeta\,\underline{\mathbf{x}}(\rho), \lambda\overline{x}(\rho)\right), \zeta \geqslant 0; \\ \left(\zeta\,\overline{\mathbf{x}}(\rho), \lambda\underline{x}(\rho)\right), \zeta < 0, \end{cases}$

(c) $\widetilde{\mathbf{x}}(\rho) = \widetilde{\mathbf{y}}(\rho) \Longleftrightarrow \underline{x}(\rho) = \underline{y}(\rho) \wedge \overline{x}(\rho) = \overline{y}(\rho)$.

Following standard notation, the set of $m \times n$ matrices over the set of real numbers $\mathbb{R}$ is denoted by $\mathbb{R}^{m\times n}$.

DEFINITION 2.3 ([**26**]). For any matrix $A \in \mathbb{R}^{m\times n}$ and Hermitian positive-definite (Hpd) matrices M and N of sizes m and n, respectively, there exists a unique matrix $Q \in \mathbb{C}^{n\times m}$ that satisfies the following matrix equations with the unknown Q:

$$(1)\ AQA{=}A,\quad (2)\ QAQ{=}Q,\quad (3^M)\ (MAQ)^*{=}MAQ,\quad (4^N)\ (NQA)^*{=}NQA.$$

The matrix Q is referred to as the WMPI of A and is commonly denoted as $A^{\dagger}_{M,N} \equiv Q$.

In the specific context $M = I_m$ and $N = I_n$, the matrix $A_{M,N}$ becomes the Moore-Penrose inverse (MPI) $A^{\dagger}$, which represents the unique solution to (1),(2) and

$$(3)\ (AQ)^* = AQ, \quad (4)\ (QA)^* = QA.$$

Lemma 2.1 provides a computationally efficient representation of the WMPI.

LEMMA 2.1 ([**26, 28**]). Consider a matrix $A \in \mathbb{C}^{m\times n}$ with rank $\operatorname{rank}(A) = r$ and Hpd matrices M, N of dimensions $m \times m$ and $n \times n$, respectively. Then there exist matrices $U \in \mathbb{C}^{m\times m}$, $V \in \mathbb{C}^{n\times n}$ satisfying $U^*MU = I_m$, $V^*N^{-1}V = I_n$, that determine the weighted singular weighted decomposition (WSVD) of A

$$A = U \begin{pmatrix} \Sigma & 0 \\ 0 & 0 \end{pmatrix} V^*.$$

In such environment, the WMPI of A is represented as

$$A^{\dagger}_{M,N} = N^{-1}V \begin{pmatrix} \Sigma^{-1} & 0 \\ 0 & 0 \end{pmatrix} U^*M,$$

where $\Sigma = \operatorname{diag}(\mu_1, \ldots, \mu_r)$, $\mu_1 \geqslant \cdots \geqslant \mu_r > 0$ and μ_i^2 is the nonzero eigenvalue of $A^{\sharp}A$ for $A^{\sharp} = N^{-1}A^*M$. The quantities $\mu_i, i = 1, \ldots, r$ are known as weighted singular values of A.

DEFINITION 2.4 ([**26**]). Let $A \in \mathbb{R}^{m\times n}$, $W \in \mathbb{R}^{n\times m}$. A matrix X is called the WDI of A, denoted by $X = A^{\mathrm{D},W}$, if and only if it satisfies the following equations:

$$(1^{k,W})\ (AW)^{k+1}XW = (AW)^k, \quad (2^W)\ XWAWX = X, \quad (5^W)\ AWX = XWA,$$

where $k = \max\{\operatorname{Ind}(AW), \operatorname{Ind}(WA)\}$ and $\operatorname{Ind}(Q)$ denotes the index of Q [**4**].

The WDI can be represented in the following dual forms (see [**4, 20, 27, 29**]):

$$A^{\mathrm{D},W} = A\left[(WA)^{\mathrm{D}}\right]^2 = \left[(AW)^{\mathrm{D}}\right]^2 A, \tag{2.1}$$

The choice $W = I$, reduces $A^{\mathrm{D},W}$ to the usual Drazin inverse A^{D} [**3, 23, 26**], known as the unique solution to the matrix equations (2) and

$$(1^k)\ A^{k+1}X = A^k, \quad (5)\ AX = XA.$$

Lemma 2.2 restates a representation of the WDI.

LEMMA 2.2 ([**29**]). Let $A \in \mathbb{C}^{m\times n}$, $W \in \mathbb{C}^{n\times m}$. The Jordan canonical forms of AW and WA are defined as

$$AW = P \begin{pmatrix} R_1 & 0 \\ 0 & N_1 \end{pmatrix} P^{-1}, \quad WA = Q \begin{pmatrix} R_2 & 0 \\ 0 & N_2 \end{pmatrix} Q^{-1}, \tag{2.2}$$

respectively, where R_1 and R_2 are nonsingular matrices of the same order and N_1 and N_2 are nilpotent matrices.

Then A and W can be represented as

$$A = P\begin{pmatrix} A_{11} & 0 \\ 0 & A_{22} \end{pmatrix} Q^{-1}, \quad W = Q\begin{pmatrix} W_{11} & 0 \\ 0 & W_{22} \end{pmatrix} P^{-1},$$

where P, Q, A_{11} and W_{11} are nonsingular matrices.

Based on the definition $A^{\mathrm{D,W}} = \left[(AW)^{\mathrm{D}}\right]^2 A = A\left[(WA)^{\mathrm{D}}\right]^2$, the WDI of A possesses the representation

$$A^{\mathrm{D,W}} = P\begin{pmatrix} (W_{11}A_{11}W_{11})^{-1} & 0 \\ 0 & 0 \end{pmatrix} Q^{-1}.$$

The FLS $A\widetilde{\mathbf{x}}(\rho) = \widetilde{\mathbf{y}}(\rho)$ is elementwise given by

$$\begin{pmatrix} a_{11} & \cdots & a_{1n} \\ \vdots & \ddots & \vdots \\ a_{m1} & \cdots & a_{mn} \end{pmatrix} \begin{pmatrix} \widetilde{x}_1(\rho) \\ \vdots \\ \widetilde{x}_n(\rho) \end{pmatrix} = \begin{pmatrix} \widetilde{y}_1(\rho) \\ \vdots \\ \widetilde{y}_m(\rho) \end{pmatrix}, \tag{2.3}$$

such that $A = (a_{ij})$ is an $m \times n$ crisp real matrix, $\widetilde{x}_i(\rho), i = 1, \ldots, m$ is a column vector with fuzzy variables, and $\widetilde{y}_i(\rho), i = 1, \ldots, n$ are given fuzzy numbers.

The core-EP inverse of A is the unique matrix defined by $A^{\textcircled{\dagger}} = A^D A^k (A^k)^\dagger$ [**8**, **18**]. The core inverse $A^{\textcircled{\#}} = A^\# A A^\dagger$ satisfies (1),(3) and (5) [**1**]. The weak group (WG) inverse $A^{\textcircled{w}}$ was proposed in [**22**], and further the Moore-Penrose WG (MPWG) inverse inverse $A^{\dagger,WG} = A^\dagger A^{\textcircled{w}} A$ was introduced in [**21**].

3. The solution of fuzzy linear systems

The traditional approach to solving FLS involves transformation of the coefficient matrix $A \in \mathbb{R}^{m\times n}$ into its extended matrix $S \in \mathbb{R}^{2m\times 2n}$. If the extended matrix S is nonsingular, FLS can be solved using its inverse; otherwise, we need to use a generalized inverse of S to solve it. In this research, we focus on the WMPI and WDI of the block coefficient matrix S.

3.1. Transition from FLS TO CLS.

DEFINITION 3.1. For fuzzy linear systems $A\widetilde{\mathbf{x}}(\rho) = \widetilde{\mathbf{y}}(\rho)$ defined in (2.3), consider a CLS $S\mathbf{x}(\rho) = \mathbf{y}(\rho)$ given elemenwise by

$$\begin{pmatrix} s_{1,1} & \cdots & s_{1,n} & \cdots & s_{1,2n} \\ \cdots & \cdots & \cdots & \ddots & \cdots \\ s_{m,1} & \cdots & s_{m,n} & \cdots & s_{m,2n} \\ \cdots & \cdots & \cdots & \ddots & \cdots \\ s_{2m,1} & \cdots & s_{2m,n} & \cdots & s_{2m,2n} \end{pmatrix} \begin{pmatrix} \underline{x}_1(\rho) \\ \vdots \\ \underline{x}_n(\rho) \\ -\overline{x}_1(\rho) \\ \vdots \\ -\overline{x}_n(\rho) \end{pmatrix} = \begin{pmatrix} \underline{y}_1(\rho) \\ \vdots \\ \underline{y}_m(\rho) \\ -\overline{y}_1(\rho) \\ \vdots \\ -\overline{y}_m(\rho) \end{pmatrix}, \tag{3.1}$$

such that s_{ij} are determined using the following cases:

$$\begin{aligned} s_{i,j} = s_{i+m,j+n} &= \begin{cases} a_{ij}, & a_{ij} > 0 \\ 0, & a_{ij} \le 0, \end{cases} \\ s_{i,j+n} = s_{i+m,j} &= \begin{cases} -a_{ij}, & a_{ij} < 0 \\ 0, & a_{ij} \ge 0, \end{cases} \end{aligned}, \quad i = 1, \ldots, m, \; j = 1, \ldots, n. \tag{3.2}$$

For $A = (a_{ij}) \in \mathbb{R}^{m\times n}$, it is useful to define the matrix

$$\rceil A \lceil = (|a_{ij}|) \in \mathbb{R}^{m\times n},$$

where $|a_{ij}|$ denotes the absolute value of $a_{ij} \in \mathbb{R}$.

From Definition 3.1, the matrix S can be represented as

$$S = \begin{pmatrix} \overset{+}{A} & \overset{-}{A} \\ \overset{-}{A} & \overset{+}{A} \end{pmatrix}, \tag{3.3}$$

where $\overset{+}{A}$ and $\overset{-}{A}$ $m \times n$ matrices defined as follows

$$\overset{+}{A} = \left(\overset{+}{a_{ij}}\right) = \frac{1}{2}\left(\rceil A\lceil + A\right), \quad \overset{-}{A} = \left(\overset{-}{a_{ij}}\right) = \frac{1}{2}\left(\rceil A\lceil - A\right),$$

where $\overset{+}{a_{ij}} = \max\{a_{ij}, 0\}$, $\overset{-}{a_{ij}} = \max\{-a_{ij}, 0\}$. Clearly,

$$\rceil A\lceil = \overset{+}{A} + \overset{-}{A}, \quad A = \overset{+}{A} - \overset{-}{A}. \tag{3.4}$$

The matrix S is known as the extended matrix of A.

COROLLARY 3.1. The extended matrix S is invertible if and only if both $\rceil A\lceil$ and A are invertible.

COROLLARY 3.2. If the CLS (3.1) does not have a unique solution, the associated FLS (2.3) does not have one either.

3.2. Inverse and generalized inverses of extended matrix. Let us define the transformation $\partial : \mathbb{R}^{m\times n} \to \mathbb{R}^{m\times n}$ by the rule

$$\partial(A) := S = \begin{pmatrix} \frac{1}{2}\left(\rceil A\lceil + A\right) & \frac{1}{2}\left(\rceil A\lceil - A\right) \\ \frac{1}{2}\left(\rceil A\lceil - A\right) & \frac{1}{2}\left(\rceil A\lceil + A\right) \end{pmatrix} = \begin{pmatrix} \overset{+}{A} & \overset{-}{A} \\ \overset{-}{A} & \overset{+}{A} \end{pmatrix}$$

and the transformation $\partial^{(\sigma)} : \mathbb{R}^{m\times n} \to \mathbb{R}^{m\times n}$ by the rule

$$\partial^{(\sigma)}(A) = \begin{pmatrix} \frac{1}{2}\left(\rceil A\lceil^{(\sigma)} + A^{(\sigma)}\right) & \frac{1}{2}\left(\rceil A\lceil^{(\sigma)} - A^{(\sigma)}\right) \\ \frac{1}{2}\left(\rceil A\lceil^{(\sigma)} - A^{(\sigma)}\right) & \frac{1}{2}\left(\rceil A\lceil^{(\sigma)} + A^{(\sigma)}\right) \end{pmatrix}, \tag{3.5}$$

where $(\cdot)^{(\sigma)}$ denotes the matrix inverse in nonsingular case and corresponding generalized inverse in singular case.

Starting point of our investigation is the inverse of the matrix S restated in Theorem 3.1.

THEOREM 3.1 ([**7**]). If $S := \partial(A)$ is nonsingular, S^{-1} possesses the same structure as S, i.e.

$$S^{-1} = \begin{pmatrix} \mathcal{D} & \mathcal{E} \\ \mathcal{E} & \mathcal{D} \end{pmatrix} = \partial^{-1}(A).$$

where

$$\mathcal{D} = \frac{1}{2}\left[(\overset{+}{A} + \overset{-}{A})^{-1} + (\overset{+}{A} - \overset{-}{A})^{-1}\right] = \frac{1}{2}\left(\rceil A\lceil^{-1} + A^{-1}\right),$$

$$\mathcal{E} = \frac{1}{2}\left[(\overset{+}{A} + \overset{-}{A})^{-1} - (\overset{+}{A} - \overset{-}{A})^{-1}\right] = \frac{1}{2}\left(\rceil A\lceil^{-1} - A^{-1}\right).$$

If $\widetilde{\mathbf{x}}$ and $\widetilde{\mathbf{x}}$ are fuzzy vectors of proper sizes and S is nonsingular, the solution vector of the CLS $S\mathbf{x} = \mathbf{y}$ is

$$\mathbf{x} = S^{-1}\mathbf{y}. \tag{3.6}$$

Our goal is to investigate the transformation $\Game^{(\sigma)}$. It is known from [**9**] that $(\Game(A))^{(\sigma)} = \Game^{(\sigma)}(A)$ for different generalized inverses σ. In Table 1 we surveyed so far investigated generalized inverses $(\Game(A))^{(\sigma)}$ available in the literature.

TABLE 1. Survey of relations $(\Game(A))^{(\sigma)} = \Game^{(\sigma)}(A)$

σ	Relation	Reference
S^{-1}	$(\Game(A))^{-1} = \Game^{-1}(A)$	[**7**]
$S^{(1)}$	$(\Game(A))^{(1)} = \Game^{(1)}(A)$	[**11**]
$S^{\textcircled{\#}}$	$(\Game(A))^{\textcircled{\#}} = \Game^{\textcircled{\#}}(A)$	[**11**]
$S^{\textcircled{\dagger}}$	$(\Game(A))^{\textcircled{\dagger}} = \Game^{\textcircled{\dagger}}(A)$	[**12**]
$S^{\dagger}$	$(\Game(A))^{\dagger} = \Game^{\dagger}(A)$	[**9, 24**]
$S^{\#}$	$(\Game(A))^{\#} = \Game^{\#}(A)$	[**10**]
$S^{(1,3)}$	$(\Game(A))^{(1,3)} = \Game^{(1,3)}(A)$	[**24**]
$S^{\dagger,WG} = S^{\dagger}S^{\textcircled{w}}S$	$(\Game(A))^{\textcircled{w}} = \Game^{\textcircled{w}}(A)$	[**30**]

In the present article we will investigate relations $(\Game(A))^{\dagger}_{M,N} = \Game^{\dagger}_{M,N}(A)$ and $(\Game(A))^{\mathrm{D},W} = \Game^{\mathrm{D},W}(A)$.

3.3. Weighted generalized inverses in the case of singular extended matrix. In the rest of this section, we investigate the CLS $S\mathbf{x} = \mathbf{y}$ determined by a singular extended matrix S. Two cases are investigated in details, namely the WMPI and WDI solutions.

3.3.1. *Weighted Moore-Penrose inverse solution.* Theorem 3.2 gives the WMPI for the extended matrix S.

THEOREM 3.2. For a fuzzy linear system defined by (2.3), if $A \in \mathbb{C}^{m\times n}$ and both M and N are Hpd matrices of orders m and n, respectively, then the WMPI of the extended matrix S, as defined in (3.3) is expressed in the following form

$$S^{\dagger}_{M,N} = \begin{pmatrix} \mathcal{D} & \mathcal{E} \\ \mathcal{E} & \mathcal{D} \end{pmatrix} = \Game^{\dagger}_{M,N}(A), \tag{3.7}$$

where

$$\mathcal{D} = \frac{1}{2}\left(\upharpoonleft A\upharpoonright^{\dagger}_{M,N} + A^{\dagger}_{M,N}\right), \quad \mathcal{E} = \frac{1}{2}\left(\upharpoonleft A\upharpoonright^{\dagger}_{M,N} - A^{\dagger}_{M,N}\right). \tag{3.8}$$

PROOF. **Verification of the equations** (1) **and** (2) is presented in the reference [**9**].

The verification of the equation (3^M) **follows.**

After performing some calculations, we derive

$$\begin{pmatrix} M & 0 \\ 0 & M \end{pmatrix} S\,S^{\dagger}_{M,N} = \begin{pmatrix} MA^{+}\mathcal{D} + MA^{-}\mathcal{E} & MA^{-}\mathcal{D} + MA^{+}\mathcal{E} \\ MA^{-}\mathcal{D} + MA^{+}\mathcal{E} & MA^{+}\mathcal{D} + MA^{-}\mathcal{E} \end{pmatrix},$$

where

$$
\begin{aligned}
&MA^{+}\mathcal{D}+MA^{-}\mathcal{E}\\
&=\frac{1}{4}M\left(\upharpoonleft A\upharpoonright+A\right)\left(\upharpoonleft A\upharpoonright_{M,N}^{\dagger}+A_{M,N}^{\dagger}\right)+\frac{1}{4}M\left(\upharpoonleft A\upharpoonright-A\right)\left(\upharpoonleft A\upharpoonright_{M,N}^{\dagger}-A_{M,N}^{\dagger}\right)\\
&=\frac{1}{2}\left(M\upharpoonleft A\upharpoonright\upharpoonleft A\upharpoonright_{M,N}^{\dagger}+MAA_{M,N}^{\dagger}\right).
\end{aligned}
$$

Similarly,

$$MA^{-}\mathcal{D}+MA^{+}\mathcal{E}=\frac{1}{2}\left(M\upharpoonleft A\upharpoonright\upharpoonleft A\upharpoonright_{M,N}^{\dagger}-MAA_{M,N}^{\dagger}\right),$$

which implies

$$
\begin{aligned}
&\left[\begin{pmatrix}M&0\\0&M\end{pmatrix}SS_{M,N}^{\dagger}\right]^{*}\\
&=\frac{1}{2}\begin{pmatrix}\left(M\upharpoonleft A\upharpoonright\upharpoonleft A\upharpoonright_{M,N}^{\dagger}+MAA_{M,N}^{\dagger}\right)^{*}&\left(M\upharpoonleft A\upharpoonright\upharpoonleft A\upharpoonright_{M,N}^{\dagger}-MAA_{M,N}^{\dagger}\right)^{*}\\\left(M\upharpoonleft A\upharpoonright\upharpoonleft A\upharpoonright_{M,N}^{\dagger}-MAA_{M,N}^{\dagger}\right)^{*}&\left(M\upharpoonleft A\upharpoonright\upharpoonleft A\upharpoonright_{M,N}^{\dagger}+MAA_{M,N}^{\dagger}\right)^{*}\end{pmatrix}\\
&=\frac{1}{2}\begin{pmatrix}M\upharpoonleft A\upharpoonright\upharpoonleft A\upharpoonright_{M,N}^{\dagger}+MAA_{M,N}^{\dagger}&M\upharpoonleft A\upharpoonright\upharpoonleft A\upharpoonright_{M,N}^{\dagger}-MAA_{M,N}^{\dagger}\\M\upharpoonleft A\upharpoonright\upharpoonleft A\upharpoonright_{M,N}^{\dagger}-MAA_{M,N}^{\dagger}&M\upharpoonleft A\upharpoonright\upharpoonleft A\upharpoonright_{M,N}^{\dagger}+MAA_{M,N}^{\dagger}\end{pmatrix}\quad\text{(based on }(3^{M})\text{)}\\
&=\begin{pmatrix}M&0\\0&M\end{pmatrix}SS_{M,N}^{\dagger}.
\end{aligned}
$$

Verification of the equation (4^{N}) follows.

The results from the matrix calculation indicate that

$$\begin{pmatrix}N&0\\0&N\end{pmatrix}S_{M,N}^{\dagger}S=\begin{pmatrix}N\mathcal{D}A^{+}+N\mathcal{E}A^{-}&N\mathcal{E}A^{+}+N\mathcal{D}A^{-}\\N\mathcal{E}A^{+}+N\mathcal{D}A^{-}&N\mathcal{D}A^{+}+N\mathcal{E}A^{-}\end{pmatrix},$$

where

$$
\begin{aligned}
&N\mathcal{D}A^{+}+N\mathcal{E}A^{-}\\
&=\frac{1}{4}N\left(\upharpoonleft A\upharpoonright_{M,N}^{\dagger}+A_{M,N}^{\dagger}\right)\left(\upharpoonleft A\upharpoonright+A\right)+\frac{1}{4}N\left(\upharpoonleft A\upharpoonright_{M,N}^{\dagger}-A_{M,N}^{\dagger}\right)\left(\upharpoonleft A\upharpoonright-A\right)\\
&=\frac{1}{2}\left(N\upharpoonleft A\upharpoonright_{M,N}^{\dagger}\upharpoonleft A\upharpoonright+NA_{M,N}^{\dagger}A\right).
\end{aligned}
$$

Similarly, $N\mathcal{E}A^{+}+N\mathcal{D}A^{-}=\frac{1}{2}\left(N\upharpoonleft A\upharpoonright_{M,N}^{\dagger}\upharpoonleft A\upharpoonright-NA_{M,N}^{\dagger}A\right)$,

Then

$$
\begin{aligned}
&\left[\begin{pmatrix}N&0\\0&N\end{pmatrix}S_{M,N}^{\dagger}S\right]^{*}\\
&=\frac{1}{2}\begin{pmatrix}\left(N\upharpoonleft A\upharpoonright_{M,N}^{\dagger}\upharpoonleft A\upharpoonright+NA_{M,N}^{\dagger}A\right)^{*}&\left(N\upharpoonleft A\upharpoonright_{M,N}^{\dagger}\upharpoonleft A\upharpoonright-NA_{M,N}^{\dagger}A\right)^{*}\\\left(N\upharpoonleft A\upharpoonright_{M,N}^{\dagger}\upharpoonleft A\upharpoonright-NA_{M,N}^{\dagger}A\right)^{*}&\left(N\upharpoonleft A\upharpoonright_{M,N}^{\dagger}\upharpoonleft A\upharpoonright+NA_{M,N}^{\dagger}A\right)^{*}\end{pmatrix}\\
&=\frac{1}{2}\begin{pmatrix}N\upharpoonleft A\upharpoonright_{M,N}^{\dagger}\upharpoonleft A\upharpoonright+NA_{M,N}^{\dagger}A&N\upharpoonleft A\upharpoonright_{M,N}^{\dagger}\upharpoonleft A\upharpoonright-NA_{M,N}^{\dagger}A\\N\upharpoonleft A\upharpoonright_{M,N}^{\dagger}\upharpoonleft A\upharpoonright-NA_{M,N}^{\dagger}A&N\upharpoonleft A\upharpoonright_{M,N}^{\dagger}\upharpoonleft A\upharpoonright+NA_{M,N}^{\dagger}A\end{pmatrix}\quad\text{(based on }(4^{N})\text{)}\\
&=\begin{pmatrix}N&0\\0&N\end{pmatrix}S_{M,N}^{\dagger}S.
\end{aligned}
$$

Based on the above proof and the existence and uniqueness of the WMPI [**28**], we conclude (3.7) □

3.3.2. *W-weighted Drazin inverse solution.* For simplicity, we use the notations $F = \rceil A \lceil W$ and $G = AW$

LEMMA 3.1. For a fuzzy linear systems (2.3), let S be defined in (3.3). Then for each $i = 0, 1, 2, \dots$ it follows

$$S\begin{pmatrix} W & 0 \\ 0 & W \end{pmatrix}\begin{pmatrix} F^iP + G^iQ & F^iP - G^iQ \\ F^iP - G^iQ & F^iP + G^iQ \end{pmatrix} = \begin{pmatrix} F^{i+1}P + G^{i+1}Q & F^{i+1}P - G^{i+1}Q \\ F^{i+1}P - G^{i+1}Q & F^{i+1}P + G^{i+1}Q \end{pmatrix},$$

where P and Q are arbitrary matrices of order m.

PROOF. Using

$$S\begin{pmatrix} W & 0 \\ 0 & W \end{pmatrix} = \begin{pmatrix} A^+W & A^-W \\ A^-W & A^+W \end{pmatrix},$$

in conjunction with (3.4) and definitions of F and G, one concludes

$$\begin{aligned} & S\begin{pmatrix} W & 0 \\ 0 & W \end{pmatrix}\begin{pmatrix} F^iP + G^iQ & F^iP - G^iQ \\ F^iP - G^iQ & (\rceil A \lceil W)^i P + (AW)^iQ \end{pmatrix} \\ &= \begin{pmatrix} (A^+ + A^-)WF^iP + (A^+ - A^-)WG^iQ & (A^+ + A^-)WF^iP - (A^+ - A^-)WG^iQ \\ (A^+ + A^-)WF^iP - (A^+ - A^-)WG^iQ & (A^+ + A^-)WF^iP + (A^+ - A^-)WG^iQ \end{pmatrix} \\ &= \begin{pmatrix} F^{i+1}P + G^{i+1}Q & F^{i+1}P - G^{i+1}Q \\ F^{i+1}P - G^{i+1}Q & F^{i+1}P + G^{i+1}Q \end{pmatrix}, \end{aligned}$$

the proof is finished. □

THEOREM 3.3. For the FLS (2.3), let $A \in \mathbb{R}^{m\times n}$ be a rectangular matrix. Then for arbitrary $W \in \mathbb{R}^{n\times m}$, the WDI of the extended matrix S can be represented as

$$S^{\mathrm{D},W} = \begin{pmatrix} \mathcal{D} & \mathcal{E} \\ \mathcal{E} & \mathcal{D} \end{pmatrix} = \eth^{\mathrm{D},W}(A), \tag{3.9}$$

where

$$\mathcal{D} = \frac{1}{2}\left(\rceil A \lceil^{\mathrm{D},W} + A^{\mathrm{D},W}\right),\ \mathcal{E} = \frac{1}{2}\left(\rceil A \lceil^{\mathrm{D},W} - A^{\mathrm{D},W}\right). \tag{3.10}$$

PROOF. **The verification of the equation $(1^{k,W})$ follows.**
Basic matrix calculus gives

$$\begin{aligned} S\begin{pmatrix} W & 0 \\ 0 & W \end{pmatrix} S^{\mathrm{D},W}\begin{pmatrix} W & 0 \\ 0 & W \end{pmatrix} &= \begin{pmatrix} A^+W & A^-W \\ A^-W & A^+W \end{pmatrix}\begin{pmatrix} \mathcal{D}W & \mathcal{E}W \\ \mathcal{E}W & \mathcal{D}W \end{pmatrix} \\ &= \begin{pmatrix} A^+W\mathcal{D}W + A^-W\mathcal{E}W & A^-W\mathcal{D}W + A^+W\mathcal{E}W \\ A^-W\mathcal{D}W + A^+W\mathcal{E}W & A^+W\mathcal{D}W + A^-W\mathcal{E}W \end{pmatrix}. \end{aligned}$$

Further calculus in conjunction with $F = \rceil A \lceil W$ and $G = AW$ lead to

$$\begin{aligned}
&\overset{+}{A} W\mathcal{D}W + \overset{-}{A} W\mathcal{E}W \\
&= \frac{1}{4}\left(\rceil A\lceil + A\right) W \left(\rceil A\lceil^{\mathrm{D},W} + A^{\mathrm{D},W}\right) W + \frac{1}{4}\left(\rceil A\lceil - A\right) W \left(\rceil A\lceil^{\mathrm{D},W} - A^{\mathrm{D},W}\right) W \\
&= \frac{1}{2}\left(F \rceil A\lceil^{\mathrm{D},W} W + GA^{\mathrm{D},W} W\right) \\
&= \frac{1}{2}\left(F\left((\rceil A\lceil W)^{\mathrm{D}}\right)^2 \rceil A\lceil W + G\left((AW)^{\mathrm{D}}\right)^2 AW\right) \quad \text{(from the property (2.1) of WDI)} \\
&= \frac{1}{2}\left(FF^{\mathrm{D}}FF^{\mathrm{D}} + GG^{\mathrm{D}}GG^{\mathrm{D}}\right) \quad \text{(based on } (5^W)\text{)} \\
&= \frac{1}{2}\left(FF^{\mathrm{D}} + GG^{\mathrm{D}}\right) \quad \text{(based on (2))}
\end{aligned}$$

Analogous calculation leads to

$$\overset{-}{A} W\mathcal{D}W + \overset{+}{A} W\mathcal{E}W = \frac{1}{2}\left(FF^{\mathrm{D}} - GG^{\mathrm{D}}\right).$$

Thus,

$$S\begin{pmatrix} W & 0 \\ 0 & W \end{pmatrix} S^{\mathrm{D},W} \begin{pmatrix} W & 0 \\ 0 & W \end{pmatrix} = \frac{1}{2}\begin{pmatrix} FF^{\mathrm{D}} + GG^{\mathrm{D}} & FF^{\mathrm{D}} - GG^{\mathrm{D}} \\ FF^{\mathrm{D}} - GG^{\mathrm{D}} & FF^{\mathrm{D}} + GG^{\mathrm{D}} \end{pmatrix}.$$

The left-hand side of the equation $(1^{k,W})$ satisfies

$$\begin{aligned}
&\left[S\begin{pmatrix} W & 0 \\ 0 & W \end{pmatrix}\right]^{k+1} S^{\mathrm{D},W} \begin{pmatrix} W & 0 \\ 0 & W \end{pmatrix} \\
&= \left[S\begin{pmatrix} W & 0 \\ 0 & W \end{pmatrix}\right]^{k} S\begin{pmatrix} W & 0 \\ 0 & W \end{pmatrix} S^{\mathrm{D},W} \begin{pmatrix} W & 0 \\ 0 & W \end{pmatrix} \\
&= \frac{1}{2}\left[S\begin{pmatrix} W & 0 \\ 0 & W \end{pmatrix}\right]^{k} \begin{pmatrix} FF^{\mathrm{D}} + GG^{\mathrm{D}} & FF^{\mathrm{D}} - GG^{\mathrm{D}} \\ FF^{\mathrm{D}} - GG^{\mathrm{D}} & FF^{\mathrm{D}} + GG^{\mathrm{D}} \end{pmatrix} \\
&= \frac{1}{2}\left[S\begin{pmatrix} W & 0 \\ 0 & W \end{pmatrix}\right]^{k-1} \left[S\begin{pmatrix} W & 0 \\ 0 & W \end{pmatrix} \begin{pmatrix} FF^{\mathrm{D}} + GG^{\mathrm{D}} & FF^{\mathrm{D}} - GG^{\mathrm{D}} \\ FF^{\mathrm{D}} - GG^{\mathrm{D}} & FF^{\mathrm{D}} + GG^{\mathrm{D}} \end{pmatrix}\right] \\
&= \frac{1}{2}\left[S\begin{pmatrix} W & 0 \\ 0 & W \end{pmatrix}\right]^{k-1} \begin{pmatrix} F^2F^{\mathrm{D}} + G^2G^{\mathrm{D}} & F^2F^{\mathrm{D}} - G^2G^{\mathrm{D}} \\ F^2F^{\mathrm{D}} - G^2G^{\mathrm{D}} & F^2F^{\mathrm{D}} + G^2G^{\mathrm{D}} \end{pmatrix} \quad \text{(based on Lemma 3.1)} \\
&= \frac{1}{2}\left[S\begin{pmatrix} W & 0 \\ 0 & W \end{pmatrix}\right]^{k-2} \left[S\begin{pmatrix} W & 0 \\ 0 & W \end{pmatrix} \begin{pmatrix} F^2F^{\mathrm{D}} + G^2G^{\mathrm{D}} & F^2F^{\mathrm{D}} - G^2G^{\mathrm{D}} \\ F^2F^{\mathrm{D}} - G^2G^{\mathrm{D}} & F^2F^{\mathrm{D}} + G^2G^{\mathrm{D}} \end{pmatrix}\right] \\
&= \frac{1}{2}\left[S\begin{pmatrix} W & 0 \\ 0 & W \end{pmatrix}\right]^{k-2} \begin{pmatrix} F^3F^{\mathrm{D}} + G^3G^{\mathrm{D}} & F^3F^{\mathrm{D}} - G^3G^{\mathrm{D}} \\ F^3F^{\mathrm{D}} - G^3G^{\mathrm{D}} & F^3F^{\mathrm{D}} + G^3G^{\mathrm{D}} \end{pmatrix}.
\end{aligned}$$

Calculations based on the same principle, after completing $k-2$ steps lead to the following results

$$\begin{aligned}
&\left[S\begin{pmatrix} W & 0 \\ 0 & W \end{pmatrix}\right]^{k+1} S^{\mathrm{D},W} \begin{pmatrix} W & 0 \\ 0 & W \end{pmatrix} \\
&= \frac{1}{2}\begin{pmatrix} F^{k+1}F^{\mathrm{D}} + G^{k+1}G^{\mathrm{D}} & F^{k+1}F^{\mathrm{D}} - G^{k+1}G^{\mathrm{D}} \\ F^{k+1}F^{\mathrm{D}} - G^{k+1}G^{\mathrm{D}} & F^{k+1}F^{\mathrm{D}} + G^{k+1}G^{\mathrm{D}} \end{pmatrix} \\
&= \frac{1}{2}\begin{pmatrix} F^k + G^k & F^k - G^k \\ F^k - G^k & F^k + G^k \end{pmatrix} \quad \text{(based on } (1^k)\text{)}.
\end{aligned}$$

Similarly, the right-hand side of the equation $(1^{k,W})$ is transformed into

$$
\begin{aligned}
&\left[S\begin{pmatrix} W & 0 \\ 0 & W \end{pmatrix}\right]^k = \left[S\begin{pmatrix} W & 0 \\ 0 & W \end{pmatrix}\right]^{k-1} \begin{pmatrix} A^{+}W & A^{-}W \\ A^{-}W & A^{+}W \end{pmatrix} \\
&= \frac{1}{2}\left[S\begin{pmatrix} W & 0 \\ 0 & W \end{pmatrix}\right]^{k-1} \begin{pmatrix} F+G & F-G \\ F-G & F+G \end{pmatrix} \\
&= \frac{1}{2}\left[S\begin{pmatrix} W & 0 \\ 0 & W \end{pmatrix}\right]^{k-2} \left[S\begin{pmatrix} W & 0 \\ 0 & W \end{pmatrix}\begin{pmatrix} F+G & F-G \\ F-G & F+G \end{pmatrix}\right] \\
&= \frac{1}{2}\left[S\begin{pmatrix} W & 0 \\ 0 & W \end{pmatrix}\right]^{k-2} \begin{pmatrix} F^2+G^2 & F^2-G^2 \\ F^2-(AW)^2 & F^2+G^2 \end{pmatrix} \quad \text{(based on Lemma 3.1).}
\end{aligned}
$$

After $k-2$ steps the final conclusion is

$$
\left[S\begin{pmatrix} W & 0 \\ 0 & W \end{pmatrix}\right]^k = \frac{1}{2}\begin{pmatrix} F^k+G^k & F^k-G^k \\ F^k-G^k & F^k+G^k \end{pmatrix}.
$$

Thus,

$$
\left[S\begin{pmatrix} W & 0 \\ 0 & W \end{pmatrix}\right]^{k+1} S^{\mathrm{D},W}\begin{pmatrix} W & 0 \\ 0 & W \end{pmatrix} = \left[S\begin{pmatrix} W & 0 \\ 0 & W \end{pmatrix}\right]^k.
$$

The verification of the equation (2^W) follows.
Let

$$
S^{\mathrm{D},W}\begin{pmatrix} W & 0 \\ 0 & W \end{pmatrix} S \begin{pmatrix} W & 0 \\ 0 & W \end{pmatrix} S^{\mathrm{D},W} = \begin{pmatrix} \mathcal{L} & \mathcal{R} \\ \mathcal{R} & \mathcal{L} \end{pmatrix}.
$$

Then

$$
\begin{aligned}
\mathcal{L} &= \mathcal{D}WA^{+}W\mathcal{D} + \mathcal{E}WA^{-}W\mathcal{D} + \mathcal{E}WA^{+}W\mathcal{E} + \mathcal{D}WA^{-}W\mathcal{E} \\
&= \mathcal{D}W(A^{+}W\mathcal{D} + A^{-}W\mathcal{E}) + \mathcal{E}W(A^{-}W\mathcal{D} + A^{+}W\mathcal{E}) \\
&= \frac{1}{4}\left(\urcorner A \ulcorner^{\mathrm{D},W} W + A^{\mathrm{D},W}W\right)\left(\urcorner A \ulcorner W \urcorner A \ulcorner^{\mathrm{D},W} + AWA^{\mathrm{D},W}\right) \\
&\quad + \frac{1}{4}\left(\urcorner A \ulcorner^{\mathrm{D},W} W - A^{\mathrm{D},W}W\right)\left(\urcorner A \ulcorner W \urcorner A \ulcorner^{\mathrm{D},W} - AWA^{\mathrm{D},W}\right) \\
&= \frac{1}{2}\left(\urcorner A \ulcorner^{\mathrm{D},W} W \urcorner A \ulcorner W \urcorner A \ulcorner^{\mathrm{D},W} + A^{\mathrm{D},W}WAWA^{\mathrm{D},W}\right) \\
&= \frac{1}{2}\left(\urcorner A \ulcorner^{\mathrm{D},W} + A^{\mathrm{D},W}\right) \\
&= \mathcal{D}.
\end{aligned}
$$

Similarly, it can be concluded

$$
\begin{aligned}
\mathcal{R} &= \mathcal{E}WA^{+}W\mathcal{D} + \mathcal{D}WA^{-}W\mathcal{D} + \mathcal{D}WA^{+}W\mathcal{E} + \mathcal{E}WA^{-}W\mathcal{E} \\
&= \mathcal{E}.
\end{aligned}
$$

Thus,

$$
S^{\mathrm{D},W}\begin{pmatrix} W & 0 \\ 0 & W \end{pmatrix} S \begin{pmatrix} W & 0 \\ 0 & W \end{pmatrix} S^{\mathrm{D},W} = \begin{pmatrix} \mathcal{D} & \mathcal{E} \\ \mathcal{E} & \mathcal{D} \end{pmatrix} = S^{\mathrm{D},W}.
$$

The verification of the equation (5^W) follows.
In order to verify (5^W) we calculate

$$S\begin{pmatrix} W & 0 \\ 0 & W \end{pmatrix} S^{\mathrm{D},W} = \begin{pmatrix} A^{+}W\mathcal{D}+A^{-}W\mathcal{E} & A^{+}W\mathcal{E}+A^{-}W\mathcal{D} \\ A^{+}W\mathcal{E}+A^{-}W\mathcal{D} & A^{+}W\mathcal{D}+A^{-}W\mathcal{E} \end{pmatrix},$$

$$S^{\mathrm{D},W}\begin{pmatrix} W & 0 \\ 0 & W \end{pmatrix} S = \begin{pmatrix} \mathcal{D}WA^{+}+\mathcal{E}WA^{-} & \mathcal{E}WA^{+}+\mathcal{D}WA^{-} \\ \mathcal{E}WA^{+}+\mathcal{D}WA^{-} & \mathcal{D}WA^{+}+\mathcal{E}WA^{-} \end{pmatrix},$$

where

$$\begin{aligned}
&A^{+}W\mathcal{D}+A^{-}W\mathcal{E} \\
&= \frac{1}{4}\big[(\rceil A\lceil\, W \rceil A\lceil^{\mathrm{D},W} + \rceil A\lceil\, W A^{\mathrm{D},W} + AW \rceil A\lceil^{\mathrm{D},W} + AWA^{\mathrm{D},W}) \\
&\quad + (\rceil A\lceil\, W \rceil A\lceil^{\mathrm{D},W} - \rceil A\lceil\, W A^{\mathrm{D},W} - AW \rceil A\lceil^{\mathrm{D},W} + AWA^{\mathrm{D},W})\big] \\
&= \frac{1}{2}(\rceil A\lceil\, W \rceil A\lceil^{\mathrm{D},W} + AWA^{\mathrm{D},W})
\end{aligned}$$

and

$$\begin{aligned}
&\mathcal{D}WA^{+}+\mathcal{E}WA^{-} \\
&= \frac{1}{4}\big[(\rceil A\lceil^{\mathrm{D},W}\, W \rceil A\lceil + \rceil A\lceil^{\mathrm{D},W}\, WA + A^{\mathrm{D},W}W \rceil A\lceil + A^{\mathrm{D},W}WA) \\
&\quad + (\rceil A\lceil^{\mathrm{D},W}\, W \rceil A\lceil - \rceil A\lceil^{\mathrm{D},W}\, WA - A^{\mathrm{D},W}W \rceil A\lceil + A^{\mathrm{D},W}WA)\big] \\
&= \frac{1}{2}(\rceil A\lceil^{\mathrm{D},W}\, W \rceil A\lceil + A^{\mathrm{D},W}WA),
\end{aligned}$$

Based on $\rceil A\lceil\, W \rceil A\lceil^{\mathrm{D},W} = \rceil A\lceil^{\mathrm{D},W}\, W \rceil A\lceil$ and $AWA^{\mathrm{D},W} = A^{\mathrm{D},W}WA$, we can infer

$$A^{+}W\mathcal{D}+A^{-}W\mathcal{E} = \mathcal{D}WA^{+}+\mathcal{E}WA^{-},$$

Similarly, $A^{+}W\mathcal{E}+A^{-}W\mathcal{D} = \mathcal{E}WA^{+}+\mathcal{D}WA^{-}$.
Thus

$$S\begin{pmatrix} W & 0 \\ 0 & W \end{pmatrix} S^{\mathrm{D},W} = S^{\mathrm{D},W}\begin{pmatrix} W & 0 \\ 0 & W \end{pmatrix} S.$$

Based on the above proof and the existence and uniqueness of the WDI [**29**], its representation (3.9)–(3.10) is confirmed. □

4. Numerical Examples

In this section, we provide numerical examples for WMPI and WDI.

4.1. Examples on weighted Moore-Penrose inverse. Consider the following 3×4 FLS:

$$\begin{pmatrix} 3 & -4 & -8 & 9 \\ 8 & 0 & 1 & 7 \\ -9 & 8 & -1 & 1 \end{pmatrix} \begin{pmatrix} \widetilde{x}_1(\rho) \\ \widetilde{x}_2(\rho) \\ \widetilde{x}_3(\rho) \\ \widetilde{x}_4(\rho) \end{pmatrix} = \begin{pmatrix} (2+3\rho, & 7-2\rho) \\ (3+\rho, & 5-\rho) \\ (1+2\rho, & 4-\rho) \end{pmatrix}. \tag{4.1}$$

The extended matrix is equal to

$$S = \begin{pmatrix} A^{+} & A^{-} \\ A^{-} & A^{+} \end{pmatrix},$$

where $A^{+} = \begin{pmatrix} 3 & 0 & 0 & 9 \\ 8 & 0 & 1 & 7 \\ 0 & 8 & 0 & 1 \end{pmatrix}$, $A^{-} = \begin{pmatrix} 0 & 4 & 8 & 0 \\ 0 & 0 & 0 & 0 \\ 9 & 0 & 1 & 0 \end{pmatrix}$ and the vector $\mathbf{y}(\rho)$ is equal to

$$\mathbf{y} = \begin{pmatrix} 2 & +3\rho \\ 3 & +\rho \\ 1 & +2\rho \\ -7 & +2\rho \\ -5 & +\rho \\ -4 & +\rho \end{pmatrix}.$$

Let

$$M = \begin{pmatrix} 6 & -2 & 1 \\ -2 & 9 & 3 \\ 1 & 3 & 2 \end{pmatrix}, \quad N = \begin{pmatrix} 6 & -1 & 2 & 1 \\ -1 & 10 & 7 & 6 \\ 2 & 7 & 13 & 9 \\ 1 & 6 & 9 & 7 \end{pmatrix},$$

It can be obtained that

$$\upharpoonleft A \upharpoonleft^{\dagger}_{M,N} = 10^{-1} \times \begin{pmatrix} -0.3119 & 0.3093 & 0.5556 \\ 0.1955 & -0.3345 & 0.6834 \\ 1.0345 & -1.3800 & 0.1957 \\ 0.2086 & 1.2723 & -0.6629 \end{pmatrix},$$

$$A^{\dagger}_{M,N} = 10^{-1} \times \begin{pmatrix} -0.5727 & 1.3280 & -0.0593 \\ -0.7685 & 1.4640 & 1.1129 \\ -0.2965 & -0.2880 & -0.4328 \\ 0.6969 & -0.0480 & 0.1297 \end{pmatrix}.$$

So,

$$S^{\dagger}_{M,N} = \begin{pmatrix} D & E \\ E & D \end{pmatrix}$$

where

$$D = 10^{-2} \times \begin{pmatrix} -4.4231 & 8.1864 & 2.4811 \\ -2.8653 & 5.6477 & 8.9815 \\ 3.6901 & -8.3402 & -1.1856 \\ 4.5278 & 6.1212 & -2.6662 \end{pmatrix},$$

and

$$E = 10^{-2} \times \begin{pmatrix} 1.3044 & -5.0936 & 3.0746 \\ 4.8199 & -8.9924 & -2.1477 \\ 6.6549 & -5.4601 & 3.1429 \\ -2.4414 & 6.6013 & -3.9628 \end{pmatrix}.$$

Then, the solution of (4.1) can be generated in terms of $S^{\dagger}_{M,N}$ as follows:

$$\mathbf{x} = S^{\dagger}_{M,N}\mathbf{y} = \begin{pmatrix} 0.2223 & +0.0047\rho \\ 0.4001 & +0.1351\rho \\ -0.5068 & +0.1135\rho \\ 0.2469 & +0.1213\rho \\ -0.2949 & +0.0679\rho \\ -0.6359 & +0.1007\rho \\ 0.2068 & +0.1864\rho \\ -0.4068 & +0.0386\rho \end{pmatrix}.$$

So that,

$$\widetilde{\mathbf{x}}(\rho) = \begin{pmatrix} (0.2223 + 0.0047\rho, & 0.2949 - 0.0679\rho) \\ (0.4001 + 0.1351\rho, & 0.6359 - 0.1007\rho) \\ (-0.5068 + 0.1135\rho, & -0.2068 - 0.1864\rho) \\ (0.2469 + 0.1213\rho, & 0.4068 - 0.0386\rho) \end{pmatrix}.$$

The graph of the fuzzy vector $\widetilde{x}(\rho)$ is presented in Figure 1

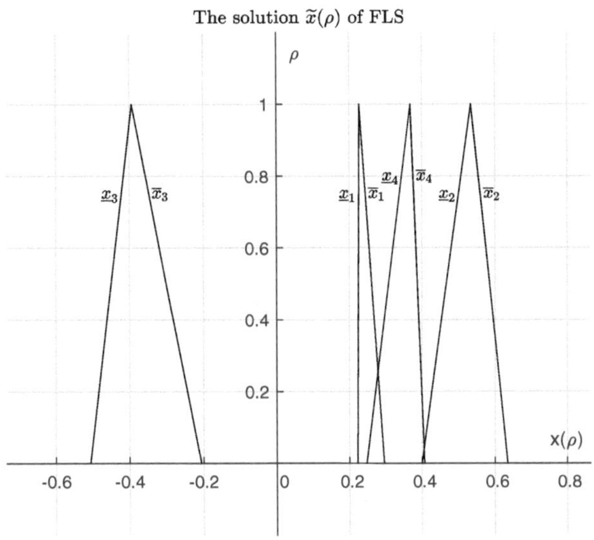

FIGURE 1. The graph of the fuzzy vector $\widetilde{\mathbf{x}}(\rho)$ generated by $S^{\dagger}_{M,N}$.

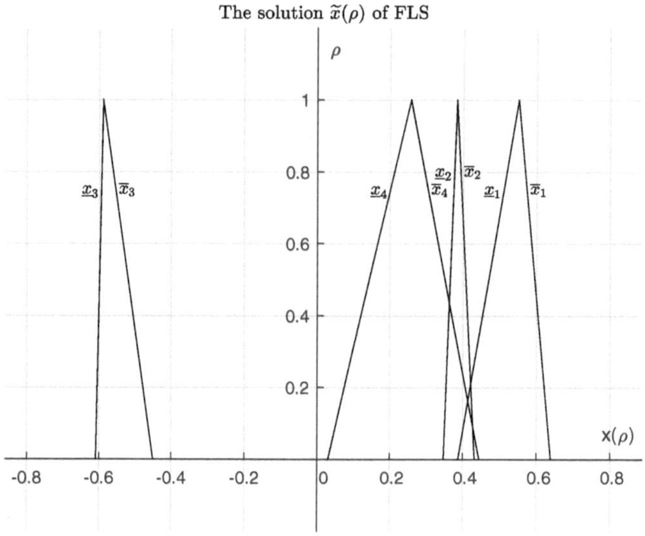

FIGURE 2. The graph of the fuzzy vector $\widetilde{\mathbf{x}}(\rho)$ generated by $S^{\mathrm{D},W}$.

4.2. Examples on W-weighted Drazin inverse. Consider the following 3×4 fuzzy linear system:

$$\begin{pmatrix} 8 & -4 & -1 & 6 \\ -1 & 9 & -1 & 2 \\ 0 & -4 & -9 & -3 \end{pmatrix} \begin{pmatrix} \widetilde{x}_1(\rho) \\ \widetilde{x}_2(\rho) \\ \widetilde{x}_3(\rho) \\ \widetilde{x}_4(\rho) \end{pmatrix} = \begin{pmatrix} (2+3\rho, & 7-2\rho) \\ (3+\rho, & 5-\rho) \\ (1+2\rho, & 4-\rho) \end{pmatrix}. \tag{4.2}$$

Then, extended matrix S from (3.3) is equal to

$$S = \begin{pmatrix} A^+ & A^- \\ A^- & A^+ \end{pmatrix},$$

where

$$A^+ = \begin{pmatrix} 8 & 0 & 0 & 6 \\ 0 & 9 & 0 & 2 \\ 0 & 0 & 0 & 0 \end{pmatrix}, \quad A^- = \begin{pmatrix} 0 & 4 & 1 & 0 \\ 1 & 0 & 1 & 0 \\ 0 & 4 & 9 & 3 \end{pmatrix}.$$

Clearly,

$$\mathbf{y} = \begin{pmatrix} 2 & +3\rho \\ 3 & +\rho \\ 1 & +2\rho \\ -7 & +2\rho \\ -5 & +\rho \\ -4 & +\rho \end{pmatrix}.$$

Let

$$W = \begin{pmatrix} -7 & 6 & -4 \\ 5 & 9 & -8 \\ 5 & -1 & 7 \\ -9 & 4 & 2 \end{pmatrix},$$

It can be obtained

$$\rceil A \lceil^{\mathrm{D},W} = 10^{-3} \begin{pmatrix} 0.7926 & 0.0641 & 0.8233 & 0.7586 \\ -0.0241 & 0.2042 & 1.3620 & 0.3908 \\ -0.3107 & -0.7584 & 0.9727 & -0.0202 \end{pmatrix},$$

$$A^{\mathrm{D},W} = 10^{-3} \begin{pmatrix} 0.2720 & -0.1503 & -0.5103 & 0.1138 \\ -0.3044 & 0.7979 & -0.6552 & -0.1282 \\ -0.6221 & -0.1528 & -2.1185 & -1.0208 \end{pmatrix}.$$

So,

$$S^{\mathrm{D},W} = \begin{pmatrix} D & E \\ E & D \end{pmatrix}$$

where

$$D = 10^{-3} \begin{pmatrix} 0.5323 & -0.0431 & 0.1565 & 0.4362 \\ -0.1642 & 0.5010 & 0.3534 & 0.1313 \\ -0.4664 & -0.4556 & -0.5729 & -0.5205 \end{pmatrix},$$

and

$$E = 10^{-3} \begin{pmatrix} 0.2603 & 0.1072 & 0.6668 & 0.3224 \\ 0.1402 & -0.2969 & 1.0086 & 0.2595 \\ 0.1557 & -0.3028 & 1.5456 & 0.5003 \end{pmatrix}.$$

Then, the solution of (4.2) is obtained as

$$\mathbf{x} = \begin{pmatrix} W & 0 \\ 0 & W \end{pmatrix} S^{\mathrm{D},W} \begin{pmatrix} W & 0 \\ 0 & W \end{pmatrix} \mathbf{y} = \begin{pmatrix} 0.3870 & +0.1632\rho \\ 0.3475 & +0.0355\rho \\ -0.6090 & +0.0198\rho \\ 0.0302 & +0.2267\rho \\ -0.6393 & +0.0891\rho \\ -0.4321 & +0.0491\rho \\ 0.4513 & +0.1379\rho \\ -0.4445 & +0.1875\rho \end{pmatrix},$$

so that,

$$\widetilde{\mathbf{x}} = \begin{pmatrix} (0.3870 + 0.1632\rho, & 0.6393 - 0.0891\rho) \\ (0.3475 + 0.0355\rho, & 0.4321 - 0.0491\rho) \\ (-0.6090 + 0.0198\rho, & -0.4513 - 0.1379\rho) \\ (0.0302 + 0.2267\rho, & 0.4445 - 0.1875\rho) \end{pmatrix}.$$

The graph of the fuzzy vector $\widetilde{\mathbf{x}}(\rho)$ is presented in Figure 2

5. Concluding remarks and future research

In this paper, we derive and prove the expression for WMPI and WDI of the extension matrix of an FLS. These methods can be applied in fields such as physics, engineering, statistics, and social sciences to solve singular fuzzy linear systems, playing an important role in these domains.

Intuitionistic fuzzy linear systems (IFLS) are similar to FLS, and similar methods can be used to solve them. In addition, it is interesting to investigate the conditions for solvability and the general solutions of FLS and IFLS.

References

[1] O. M. Baksalary and G. Trenkler, *Core inverse of matrices*, Linear Multilinear Algebra **58** (2010), no. 5-6, 681–697, DOI 10.1080/03081080902778222. MR2722752

[2] S. Banerjee, S. Biswas, and T.K. Roy, *Intuitionistic Fuzzy Linear System*, Advances in Fuzzy Mathematics, **12(3)** (2017), 475–487.

[3] A. Ben-Israel and T.N.E. Greville, *Generalized inverses: theory and applications*, Canadian Mathematical Society, Springer, New York, Beflin, Heidelberg, Hong Kong, London, Milan, Paris, Tokyo, 2003.

[4] R. E. Cline and T. N. E. Greville, *A Drazin inverse for rectangular matrices*, Linear Algebra Appl. **29** (1980), 53–62, DOI 10.1016/0024-3795(80)90230-X. MR562749

[5] J. Li, Z. Jin, J. Zhou, et al. *On the classic solution of fuzzy linear matrix equations*, Soft Comput., **28** (2024), 9295–9305.

[6] D. S. Djordjević, P. S. Stanimirović, and Y. Wei, *The representation and approximations of outer generalized inverses*, Acta Math. Hungar. **104** (2004), no. 1-2, 1–26, DOI 10.1023/B:AMHU.0000034359.98588.7b. MR2069959

[7] M. Friedman, M. Ming, and A. Kandel, *Fuzzy linear systems*, Fuzzy Sets and Systems **96** (1998), no. 2, 201–209, DOI 10.1016/S0165-0114(96)00270-9. MR1614810

[8] Y. Gao and J. Chen, *Pseudo core inverses in rings with involution*, Comm. Algebra **46** (2018), no. 1, 38–50, DOI 10.1080/00927872.2016.1260729. MR3764841

[9] B. Mihailović, V. Miler Jerković, and B. Malešević, *Solving fuzzy linear systems using a block representation of generalized inverses: the Moore-Penrose inverse*, Fuzzy Sets and Systems **353** (2018), 44–65, DOI 10.1016/j.fss.2017.11.007. MR3853483

[10] B. Mihailović, V. Miler Jerković, and B. Malešević, *Solving fuzzy linear systems using a block representation of generalized inverses: the group inverse*, Fuzzy Sets and Systems **353** (2018), 66–85, DOI 10.1016/j.fss.2018.04.015. MR3853484

[11] H. Jiang, H. Wang, and X. Liu, *Solving fuzzy linear systems by a block representation of generalized inverse: the core inverse*, Comput. Appl. Math. **39** (2020), no. 2, Paper No. 133, 20, DOI 10.1007/s40314-020-01156-0. MR4091993

[12] H. Jiang, X. Liu, and C. Jiang, *On the general strong fuzzy solutions of general fuzzy matrix equation involving the Core-EP inverse*, AIMS Math. **7** (2022), no. 2, 3221–3238, DOI 10.3934/math.2022178. MR4347810

[13] M. He, H. Jiang, and X. Liu, *General strong fuzzy solutions of fuzzy Sylvester matrix equations involving the BT inverse*, Fuzzy Sets and Systems **480** (2024), Paper No. 108862, 20, DOI 10.1016/j.fss.2024.108862. MR4693200

[14] J. Li, *A Block Representation involving the DMP inverse and BT inverse of the associated matrix in fuzzy linear system*, Pure Mathematics, **11** (2021) 2105–2110.

[15] H. Ma and P. S. Stanimirović, *Characterizations, approximation and perturbations of the core-EP inverse*, Appl. Math. Comput. **359** (2019), 404–417, DOI 10.1016/j.amc.2019.04.071. MR3950511

[16] H. Ma, W. Wang, and P. S. Stanimirović, *Weighted Moore-Penrose inverses for dual matrices and its applications*, Appl. Math. Comput. **489** (2025), Paper No. 129145, 14, DOI 10.1016/j.amc.2024.129145. MR4815055

[17] M. Nikuie and M. Z. Ahmad. *Singular fuzzy linear systems and the W-weighted Drazin inverse*, AIP Conference Proceedings, **1660** (2015), 050010; doi: 10.1063/1.4915643.

[18] K. Manjunatha Prasad and K. S. Mohana, *Core-EP inverse*, Linear Multilinear Algebra **62** (2014), no. 6, 792–802, DOI 10.1080/03081087.2013.791690. MR3195967

[19] P. S. Stanimirović, D. Mosić, and Y. Wei, *Generalizations of composite inverses with certain image and/or kernel*, Appl. Math. Comput. **428** (2022), Paper No. 127155, 19, DOI 10.1016/j.amc.2022.127155. MR4414717

[20] P. S. Stanimirović, V. N. Katsikis, and H. Ma, *Representations and properties of the W-weighted Drazin inverse*, Linear Multilinear Algebra **65** (2017), no. 6, 1080–1096, DOI 10.1080/03081087.2016.1228810. MR3615530

[21] C. Wang, X. Liu, and H. Jin, *The MP weak group inverse and its application*, Filomat **36** (2022), no. 18, 6085–6102. MR4560248

[22] H. Wang and J. Chen, *Weak group inverse*, Open Math. **16** (2018), no. 1, 1218–1232, DOI 10.1515/math-2018-0100. MR3871485

[23] H. Wang, T. Jiang, Q. Ling, and Y. Wei, *Dual core-nilpotent decomposition and dual binary relation*, Linear Algebra Appl. **684** (2024), 127–157, DOI 10.1016/j.laa.2023.12.014. MR4683929

[24] K. Wang and B. Zheng, *Inconsistent fuzzy linear systems*, Appl. Math. Comput. **181** (2006), no. 2, 973–981, DOI 10.1016/j.amc.2006.02.019. MR2269976

[25] T. Wei, W. Ding, and Y. Wei, *Singular value decomposition of dual matrices and its application to traveling wave identification in the brain*, SIAM J. Matrix Anal. Appl. **45** (2024), no. 1, 634–660, DOI 10.1137/23M1556642. MR4704634

[26] Y. Wei, P. Stanimirović, and M. Petković, *Numerical and symbolic computations of generalized inverses*, World Scientific Publishing Co. Pte. Ltd., Hackensack, NJ, 2018. MR3837117

[27] Y. Wei and G. Wang, *The perturbation theory for the Drazin inverse and its applications*, Linear Algebra Appl. **258** (1997), 179–186, DOI 10.1016/S0024-3795(96)00159-0. MR1444102 *Expression for the perturbation of the weighted Moore-Penrose inverse*, Comput. Math. Appl., **39** (2000), 13–18.

[28] Y. Wei and H. Wu, *Expression for the perturbation of the weighted Moore-Penrose inverse*, Comput. Math. Appl. **39** (2000), no. 5-6, 13–18, DOI 10.1016/S0898-1221(00)00041-9. MR1742470

[29] Y. Wei, *Integral representation of the W-weighted Drazin inverse*, Appl. Math. Comput. **144** (2003), no. 1, 3–10, DOI 10.1016/S0096-3003(02)00386-7. MR1990039

[30] J. Wu, H. Jiang, M. He, and X. Liu. *General strong fuzzy solutions of complex fuzzy matrix equations involving the Moore-Penrose weak group inverse*, Information Sciences, **654** (2024), 119832.

School of Mathematical Sciences, Fudan University, Shanghai, 200433, P. R. of China
Email address: 22210180114@m.fudan.edu.cn

Department of Computer Science, Faculty of Sciences and Mathematics, University of Niš, Višegradska 33, 18000 Niš, Serbia
Email address: pecko@pmf.ni.ac.rs

School of Mathematical Sciences and Key Laboratory of Mathematics for Nonlinear Sciences, Fudan University, Shanghai, 200433, P. R. of China
Email address: ymwei@fudan.edu.cn

SELECTED PUBLISHED TITLES IN THIS SERIES

833 **Shaun Michael Fallat, Surender K. Jain, Manjunatha Prasad Karantha, Stephen James Kirkland, and T. S. S. R. K. Rao, Editors,** Linear Algebra, Matrices, and Their Applications, 2026

831 **Irena Swanson and Kei-ichi Watanabe, Editors,** Lectures on Tight Closure and Its Applications, 2025

830 **O. Bezushchak, Yu. Drozd, P. Etingof, V. Futorny, R. Grigorchuk, and E. Zelmanov, Editors,** Modern Algebra Volume 2, 2025

829 **O. Bezushchak, Yu. Drozd, P. Etingof, V. Futorny, R. Grigorchuk, and E. Zelmanov, Editors,** Modern Algebra Volume 1, 2025

828 **Hafedh Herichi, Maria Rosaria Lancia, Therese Basa Landry, Anna Rozanova-Pierrat, and Steffen Winter, Editors,** Fractal Geometry in Pure and Applied Mathematics, 2025

827 **Carmen Caprau, J. Scott Carter, Neslihan Gügümcü, and Sam Nelson, Editors,** Algebraic Structures in Knot Theory, 2025

826 **Ahmed Laghribi and André Leroy, Editors,** Algebraic Structures and Applications, 2025

825 **Saurabh Verma, María A. Navascués, and Amit Priyadarshi, Editors,** Advances in Dimension Theory, Fractal Functions and Measures, 2025

824 **Hanna E. Makaruk and Robert M. Owczarek, Editors,** Inverse Problems: In Memory of Professor Zbigniew Oziewicz, 2025

823 **Mikhail Bershtein, Anton Dzhamay, and Andrei Okounkov, Editors,** The Versatility of Integrability, 2025

822 **Ahmad Barhoumi, Roozbeh Gharakhloo, and Andrei Martinez-Finkelshtein, Editors,** Recent Developments in Orthogonal Polynomials, 2025

821 **Bogdan D. Suceavă, Adara M. Blaga, Marian Ioan Munteanu, Cezar Oniciuc, Shoo Seto, Gabriel-Eduard Vîlcu, and Lihan Wang, Editors,** Recent Advances in Differential Geometry and Related Areas, 2025

820 **Sudeb Mitra, Rakesh Kumar Parmar, Saminathan Ponnusamy, and Toshiyuki Sugawa, Editors,** Geometric Function Theory and Related Topics, 2025

819 **Howard S. Cohl, Roberto S. Costas-Santos, and Robert S. Maier, Editors,** Applications and q-Extensions of Hypergeometric Functions, 2025

818 **Howard S. Cohl, Roberto S. Costas-Santos, and Robert S. Maier, Editors,** Classical Hypergeometric Functions and Generalizations, 2025

817 **Mikhail Khovanov, Joshua Sussan, and Anton Zeitlin, Editors,** From Representation Theory to Mathematical Physics and Back, 2025

816 **Luca F. Di Cerbo and Laurenţiu G. Maxim, Editors,** Geometry and Topology of Aspherical Manifolds, 2025

815 **Daniel Orr and Joshua Jeishing Wen, Editors,** Macdonald Theory and Beyond, 2025

814 **Susan Montgomery and Siu-Hung Ng, Editors,** Quantum Groups, Hopf Algebras, and Applications, 2025

813 **Thomas Creutzig, Julia Plavnik, and David Ridout, Editors,** Quantum Symmetries, 2025

812 **Ross Geoghegan, Craig R. Guilbault, and Kim Ruane, Editors,** Topology at Infinity of Discrete Groups, 2025

811 **Krishnendu Gongopadhyay and Shashikant A. Katre, Editors,** Geometry, Groups and Mathematical Philosophy, 2025

810 **Robert J. Berman and Yanir A. Rubinstein, Editors,** Convex and Complex: Perspectives on Positivity in Geometry, 2025

For a complete list of titles in this series, visit the
AMS Bookstore at **www.ams.org/bookstore/conmseries/**.

FSC
www.fsc.org
MIX
Papier | Fördert gute Waldnutzung
FSC® C083411

Kolibri 360 GmbH
Ferdinand-Jühlke-Straße 7
99095 Erfurt, Deutschland
produktsicherheit@kolibri360.de